Studies in Organic Chemistry 32

ORGANIC SOLID STATE CHEMISTRY

Studies in Organic Chemistry

Other titles in this series:

1 **Complex Hydrides** by A. Hajós
2 **Proteoglycans — Biological and Chemical Aspects in Human Life** by J.F. Kennedy
3 **New Trends in Heterocyclic Chemistry** edited by R.B. Mitra, N.R. Ayyangar, V.N. Gogte, R.M. Acheson and N. Cromwell
4 **Inositol Phosphates: Their Chemistry, Biochemistry and Physiology** by D.J. Cosgrove
5 **Comprehensive Carbanion Chemistry. Part A. Structure and Reactivity** edited by E. Buncel and T. Durst
Comprehensive Carbanion Chemistry. Part B. Selectivity in Carbon-Carbon Bond Forming Reactions edited by E. Buncel and T. Durst
6 **New Synthetic Methodology and Biologically Active Substances** edited by Z.-I. Yoshida
7 **Quinonediazides** by V.V. Ershov, G.A. Nikiforov and C.R.H.I. de Jonge
8 **Synthesis of Acetylenes, Allenes and Cumulenes: A Laboratory Manual** by L. Brandsma and H.D. Verkruijsse
9 **Electrophilic Additions to Unsaturated Systems** by P.B.D. de la Mare and R. Bolton
10 **Chemical Approaches to Understanding Enzyme Catalysis: Biomimetic Chemistry and Transition-State Analogs** edited by B.S. Green, Y. Ashani and D. Chipman
11 **Flavonoids and Bioflavonoids 1981** edited by L. Farkas, M. Gábor, F. Kállay and H. Wagner
12 **Crown Compounds: Their Characteristics and Applications** by M. Hiraoka
13 **Biomimetic Chemistry** edited by Z.-I. Yoshida and N. Ise
14 **Electron Deficient Aromatic- and Heteroaromatic-Base Interactions. The Chemistry of Anionic Sigma Complexes** by E. Buncel, M.R. Crampton, M.J. Strauss and F. Terrier
15 **Ozone and its Reactions with Organic Compounds** by S.D. Razumovskii and G.E. Zaikov
16 **Non-benzenoid Conjugated Carbocyclic Compounds** by D. Lloyd
17 **Chemistry and Biotechnology of Biologically Active Natural Products** edited by Cs. Szántay, Á. Gottsegen and G. Kovács
18 **Bio-Organic Heterocycles: Synthetic, Physical Organic and Pharmacological Aspects** edited by H.C. van der Plas, L. Ötvös and M. Simonyi
19 **Organic Sulfur Chemistry: Theoretical and Experimental Advances** edited by F. Bernardi, I.G. Czismadia and A. Mangini
20 **Natural Products Chemistry 1984** edited by R.I. Zalewski and J.J. Skolik
21 **Carbocation Chemistry** by P. Vogel
22 **Biocatalysts in Organic Syntheses** edited by J. Tramper, H.C. van der Plas and P. Linko
23 **Flavonoids and Bioflavonoids 1985** edited by L. Farkas, M. Gábor and F. Kállay
24 **The Organic Chemistry of Nucleic Acids** by Y. Mizuno
25 **New Synthetic Methodology and Functionally Interesting Compounds** edited by Z.-I. Yoshida
26 **New Trends in Natural Products Chemistry 1986** edited by A.-ur-Rahman and P.W. Le Quesne
27 **Bio-Organic Heterocycles 1986. Synthesis, Mechanisms and Bioactivity** edited by H.C. van der Plas, M. Simonyi, F.C. Alderweireldt and J.A. Lepoivre
28 **Perspectives in the Organic Chemistry of Sulfur** edited by B. Zwanenburg and A.J.H. Klunder
29 **Biocatalysis in Organic Media** edited by C. Laane, J. Tramper and M.D. Lilly
30 **Recent Advances in Electroorganic Synthesis** edited by S. Torii
31 **Physical Organic Chemistry 1986** edited by M. Kobayashi

Studies in Organic Chemistry 32

ORGANIC SOLID STATE CHEMISTRY

Edited by

G.R. Desiraju

School of Chemistry, University of Hyderabad, Central University P.O., Hyderabad — 500 134, India

ELSEVIER

Amsterdam — Oxford — New York — Tokyo 1987

ELSEVIER SCIENCE PUBLISHERS B.V.
Sara Burgerhartstraat 25
P.O. Box 211, 1000 AE Amsterdam, The Netherlands

Distributors for the United States and Canada:

ELSEVIER SCIENCE PUBLISHING COMPANY INC.
52, Vanderbilt Avenue
New York, NY 10017, U.S.A.

ISBN 0-444-42844-5 (Vol. 32)
ISBN 0-444-41737-0 (Series)

Printed in The Netherlands

FOREWORD

Over a century has elapsed since it was discovered that, when cinnamic acid is crystallised from organic solvents and exposed to sunlight, the ease and course of the ensuing photodimerization is dependent upon the nature of the solvent used. It was not until the advent of X-ray crystallography, so brilliantly exploited by W.H. and W.L. Bragg in the Davy-Faraday Laboratory of this Institution and elsewhere, and the routine application of X-rays to clarify the atomic structure of organic solids, that it was fully realized how richly polymorphic organic crystals can be. Thanks to this important insight, we can now appreciate why some solids are photoactive, others not. It is, in part, a question of how close to one another and in what disposition of symmetry are the reactive groups (ethylenic linkages in the cinnamic acids) on the adjacent molecules locked within the crystal. This simple realization has led to a massive expansion in the study and industrial utilization of organic solids. Not only is there a growing interest in elucidating the precise course of photochemical transformation; there exists also, wide scope for the design and production of novel photoactive solids, a class of material of growing technical promise. This, in turn, has led to a further dimension of interest in organic solids.

There are, however, myriad other reasons why organic solids are so attractive as objects of study for both physical and biological scientists. This is what has impelled Dr. Desiraju, himself a significant contributor to the field, to edit this timely volume. Contributions from far and wide have covered an extensive range of related topics which, assuredly, will interest the novitiate and experienced researcher as well as the generalist reader concerned with keeping abreast of developments in a fascinating field of enquiry.

The Royal Institution
London

J.M. Thomas

PREFACE

Fulfilling the early predictions of Gerhardt Schmidt, organic solid state chemistry has grown rapidly and vigorously during the last decade towards a synergistic combination of physical sciences which attempts to study in the widest sense, the packing of molecular solids and the chemical consequences of this packing.

Most of us would agree that papers and reviews in a field as interdisciplinary as organic solid state chemistry, tend to be scattered in a variety of journals which cater to distinct reader groups. This dispersal of such a considerable amount of exciting work and the inexplicable lack of definitive books in the area have been the motivating reasons for the publication of this multi-author treatise.

This book appears in a series entitled "Studies in Organic Chemistry". It might well have figured in the parallel series on physical and theoretical chemistry since the contributions range from what would conventionally be termed 'organic chemistry' to what might be referred to as 'chemical physics'. However, such a categorisation is not important in a bridging area and in spite of the absence of clearly-defined labels, organic solid state chemistry today promises to have a significant impact on apparently unrelated areas of research.

It is my privilege to have been associated in this venture with several well-known and distinguished scientists. The selection of topics, in other words, the choice of the contributing authors, was based on an attempt to compile a book which would have as wide a range as possible while also serving as a reference work for specialists and non-specialists alike. It is a pleasure to mention here that almost everyone who was invited to write a chapter, responded affirmatively and enthusiastically. Even so, a comprehensive coverage was scarcely possible. A conscious omission, for instance, is a discussion of organic metals and superconductors. Though the methods developed by organic solid state chemists are clearly appropriate in studies of these low dimensional materials, it was still felt that the voluminous literature in this area is easily accessible through alternative sources. Chapters on non-linear optics and polydiacetylene chemistry were commissioned originally but subsequently withdrawn. Other gaps there certainly are --- on account of the inability of a prospective author to contribute or an unavoidable exclusion by the editor.

I am particularly grateful to Professor D.Y. Curtin who, in a sense, initiated this project and Professor J.M. Thomas whose encouragement and valuable advice steered it through its early days. I would like to thank Professor C.N.R. Rao who, in spite of his many commitments, was able to contribute a chapter. I would also like to record my remembrance of the late Professor M. Simonetta who was the first to accept an invitation to write for this book and who discussed its plan and structure with me at length in Erice, Sicily in June 1985. It is a matter of regret that he did not live to see the completed collection.

Hyderabad
March, 1987

Gautam R. Desiraju

CONTENTS

INTERMOLECULAR INTERACTIONS: FROM CRYSTALLOGRAPHY TO CHEMICAL PHYSICS

CONTRIBUTORS

J. Bernstein, Department of Chemistry, Ben Gurion University of the Negev, Beer Sheva 84105, Israel.

D.Y. Curtin, School of Chemical Sciences, University of Illinois, 1205 W. California Avenue, Urbana, IL 61801, U.S.A.

G. R. Desiraju, School of Chemistry, University of Hyderabad, P.O. Central University, Hyderabad 500 134, India.

A. Gavezzotti, Dipartimento di Chimica Fisica ed Elettrochimica e Centro CNR, Universita di Milano, via Golgi 19, 20133, Milano, Italy.

K.D.M. Harris, Department of Physical Chemistry, University of Cambridge, Lensfield Road, Cambridge, CB2 1EP, U.K.

M. Hasegawa, Department of Synthetic Chemistry, University of Tokyo, Hongo, Bunkyo-ku, Tokyo, Japan.

W. Jones, Department of Physical Chemistry, University of Cambridge, Lensfield Road, Cambridge CB2 1EP, U.K.

S.K. Kearsley, Department of Chemistry, Yale University, P.O.B. 6666, New Haven, CT 06511, U.S.A.

R. Lamartine, Universite Claude Bernard, Laboratoire de Chimie Industrielle, 43 Boulevard du 11 novembre 1918, 69622 Villeur-banne Cedex, France.

I.C. Paul, School of Chemical Sciences, University of Illinois, 501 S. Mathews Avenue, Urbana, IL 61801, U.S.A.

M. Perrin, Universite Claude Bernard, Laboratoire de Chimie Industrielle, 43 Boulevard du 11 novembre 1918, 69622 Villeur-banne Cedex, France.

R. Perrin, Universite Claude Bernard, Laboratoire de Chimie Industrielle, 43 Boulevard du 11 novembre 1918, 69622 Villeurbanne Cedex, France.

P.N. Prasad, Department of Chemistry, State University of New York at Buffalo, Buffalo, N.Y. 14214, U.S.A.

S. Ramdas, British Petroleum plc., BP Research Centre, Chertsey Road, Sunbury-on-Thames, Middlesex TW16 7LN, U.K.

C.N.R. Rao, Solid State and Structural Chemistry Unit, Indian Institute of Science, Bangalore 560 012, India.

J.R. Scheffer, Department of Chemistry, University of British Columbia, Vancouver, B.C. V6T 1Y6, Canada.

M. Simonetta, Dipartimento di Chimica Fisica ed Elettrochimica e Centro CNR, Universita di Milano, via Golgi 19, 20133 Milano, Italy.

C.R. Theocharis, Department of Chemistry, Brunel University Uxbridge, Middlesex UB8 3PH, U.K.

J.M. Thomas, Davy Faraday Research Laboratory, The Royal Institution, 21 Albemarle Street, London W1X 4BS, U.K.

N.W Thomas, Department of Ceramics, The University of Leeds, Leeds, LS2 9JT, U.K.

A. Thozet, Universite Claude Bernard, Laboratoire de Chimie Industrielle, 43 Boulevard du 11 novembre 1918, 69622 Villeurbanne Cedex, France.

G. Tsoucaris, Universite Paris Sud, Centre Pharmaceutique, 92290 Chatenay Malabry, France.

Chapter 1

GEOMETRIC REQUIREMENTS FOR INTRAMOLECULAR PHOTOCHEMICAL HYDROGEN ATOM ABSTRACTION: STUDIES BASED ON A COMBINATION OF SOLID STATE CHEMISTRY AND X-RAY CRYSTALLOGRAPHY

JOHN R. SCHEFFER

1. INTRODUCTION

1.1 Intermolecular vs intramolecular hydrogen atom abstraction

The process whereby a hydrogen atom is transferred from one atom to another is one of the most general reactions in all of chemistry. The transfer can occur intermolecularly between atoms located in different molecules (Fig. 1.1a) or intramolecularly between atoms located in the same molecule (Fig. 1.1b).

Fig. 1.1. The two classes of hydrogen atom abstraction. (a) Intermolecular (b) Intramolecular.

The process is termed hydrogen atom abstraction, and the atom Y to which the hydrogen atom is transferred is referred to as the abstracting atom. The atom X, from which transfer occurs, is most frequently an sp^3-hybridized carbon atom; the abstracting atom Y is most often an oxygen atom and less frequently carbon or nitrogen. This review will be concerned primarily with abstractions for which X = C(sp^3) and Y = O (sp^2) or C (sp^2).

1.2 Oxygen as the abstracting atom

There are three main ways in which an oxygen atom can be activated toward hydrogen atom abstraction. The first is by conversion to an oxygen centered free radical (alkoxy radical R-O·), the second is by generation of a radical cation on oxygen by removal of a non-bonding electron from the oxygen atom of a carbonyl group, and the third is by photochemical conversion of a carbonyl compound to its (n, π^*) excited state. Each of these methods of initiating hydrogen atom abstraction is associated with a name reaction as summarized in Table 1.1.

TABLE 1.1
Classification of name reactions involving hydrogen atom abstraction by oxygen.

Abstracting Species	inter- or intramolecular	Name	Ref.
R-O·	Inter	Radical Initiation	1
R-O·	Intra	Barton Reaction	2
$R_2C=O^{\cdot +}$	Intra	McLafferty Rearrangement	3
$R_2C=O$ (n, π^*)	Inter	Photoreduction	4
$R_2C=O$ (n, π^*)	Intra	Norrish Type II Reaction	5

This review will concern itself mainly with the last reaction listed in Table 1.1, the Norrish type II reaction. Such questions as (1) over what distances can abstraction occur, and (2) what is the preferred angular relationship between the abstracted and abstracting atoms, will be addressed. The review will consist of a brief summary of previous investigations, both experimental and theoretical, into the geometric requirements for each type of hydrogen atom abstraction, and will then turn to a more detailed description of our solid state photochemical and crystallographic results that bear on these questions. Throughout the review, the emphasis will be on the initial, hydrogen atom abstraction processes involved, not on the fate of the radicals and biradicals so produced.

1.3 Carbon as the abstracting atom

Hydrogen atom abstraction by carbon is rare in comparison to hydrogen atom abstraction by oxygen (ref 6). This is due in part to the fact that the transfer of a hydrogen atom between two carbon atoms is more nearly thermoneutral than the transfer from a carbon atom to an oxygen atom. For example, a simple thermochemical calculation reveals that the intermolecular abstraction of a hydrogen atom from ethane by the methoxy radical is approximately 12 kcal/mole exothermic compared to the thermoneutral process of hydrogen atom abstraction from ethane by the ethyl radical. The process of alkene photoexcitation followed by hydrogen atom abstraction is also much rarer than its carbonyl counterpart, one reason being that alkene excited states tend to dissipate their energy by cis,trans isomerization rather than abstraction. Those alkenes which do abstract hydrogen photochemically are generally cyclic (restriction of cis,trans isomerization) and conjugated to carbonyl groups. Conjugation serves two purposes: (1) the absorption maximum is shifted to an experimentally more accessible wavelength, and (2) the radical produced by hydrogen abstraction can be stabilized by resonance with the carbonyl group. As we shall see, the intramolecular photochemical hydrogen atom abstractions involving C=C whose geometric parameters we have elucidated are of this type.

2. BACKGROUND

2.1 General aspects of hydrogen atom abstraction

There is general agreement that, where possible, hydrogen atom abstractions prefer linear transition state geometries (ref. 7). The simplest hydrogen abstraction reaction, the transfer of a hydrogen atom from molecular hydrogen to atomic hydrogen (Fig. 2.1), has in fact been shown to be linear (ref. 8). Detailed

$$\text{H-H} + \cdot\text{H} \longrightarrow [\text{H}\cdots\text{H}\cdots\text{H}] \longrightarrow \text{H}\cdot + \text{H-H}$$

Fig. 2.1. Linear transition state for H_2 + ·H reaction.

molecular orbital calculations indicate an interatomic separation of 1.76 Å at the linear symmetric saddle point (ref. 9). However, most <u>intramolecular</u> hydrogen abstraction reactions involve transition state geometries that cannot possibly be linear because of the geometric constraints of the system. For example, the Norrish type II reaction proceeds through a cyclic, six-membered transition state, and a geometry in which three of the six atoms are in a linear arrangement would require prohibitive bond angle deformations.

Recently Kwart (ref. 10) has postulated that the temperature-dependence of the primary kinetic hydrogen isotope effect may be used as a probe for linear versus non-linear hydrogen transfer processes. Specifically, a temperature-independent KIE coupled with an anomalous A_H/A_D ratio was taken as an indication of a non-linear transition state. This approach, which has been applied to proton and hydride transfer processes as well as hydrogen atom abstraction reactions, has been criticized on theoretical grounds by Anhede and Bergman (ref. 11). Their calculations indicate that the KIE's of both linear and bent transition state hydrogen transfers should be temperature-dependent, although the non-linear process is predicted to have a smaller temperature coefficient.

The magnitude of the primary kinetic deuterium isotope effect in a hydrogen transfer reaction is also an indication of the presence or absence of tunneling. It has long been recognized that in addition to the classical picture of the reactants in a chemical process accumulating thermal energy and surmounting an activation barrier of height E_a, there is a finite probability that, quantum mechanically, the barrier can be penetrated by reactants with thermal energies smaller than E_a (ref. 12). This mechanism depends critically on the wave properties of the reactants and hence on their mass. Tunneling is only important for light atoms such as hydrogen atoms and hydrogen ions, and the rate of tunneling is strongly isotope-dependent, that is, much greater for hydrogen than for deuterium. The observation of a much larger than normal deuterium isotope effect ($k_H/k_D = 10^2$ to 10^4 depending on temperature) indicates that tunneling is occurring. There is no evidence that tunneling is important in photochemical hydrogen atom abstractions, either bimolecular or unimolecular. The k_H/k_D value for the photoreduction of acetophenone by toluene and toluene-d_3 is <u>ca.</u> 6 (ref. 13a), and a value of k_H/k_D = 3 was found by Wagner and Scaiano (ref. 13b) for acetophenone using cyclohexane and cyclohexane-d_{12} as the hydrogen (deuterium) atom donors. In the case of the Norrish type II reaction of phenyl alkyl ketones, k_H/k_D ranges from 1.7 to 5.5 (ref. 14) depending on structure. Interestingly, it has been shown by Grellmann, Weller and Tauer (ref. 15) that reverse type II hydrogen transfer in the case of 2-methylacetophenone (i.e., the ground state conversion of the photo-enol to the keto form) does proceed by tunneling. The tunneling distance in this case was estimated to be 1.8 ± 0.2 Å.

2.2 The 1.8 Å rule - the McLafferty rearrangement

Based on studies of the McLafferty rearrangement of steroidal ketones, Djerassi and co-workers established an upper limit of 1.8 Å for the hydrogen atom transfer distance (ref. 16). This value has also been applied to theoretical treatments of photochemical hydrogen atom abstraction reactions (*vide infra*). The McLafferty rearrangement is a general process undergone by carbonyl compounds in the mass spectrometer (ref. 3). Electron bombardment of the carbonyl compound results in ionization of an n-electron on oxygen to afford the corresponding radical cation. If the carbonyl compound possesses a nearby γ-hydrogen atom, a six-membered transition state hydrogen transfer reaction occurs to afford, after fragmentation, a neutral alkene and an enol radical cation (Fig. 2.2).

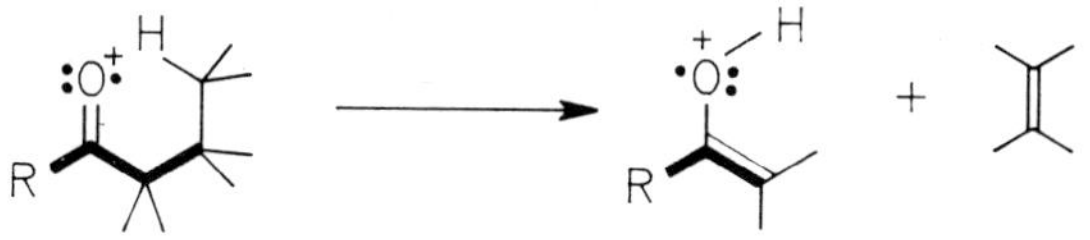

Fig. 2.2. The McLafferty rearrangement.

Djerassi's studies were concerned with ketosteroids of general structure 1, 2, and 3 (Fig. 2.3). In the 16-ketosteroid system 1, the oxygen to γ-hydrogen abstraction distance is variable owing to the conformational flexibility of the alkyl side chain. Using Dreiding models, a minimum abstraction distance of 1.5 Å was established, and these ketones undergo smooth McLafferty rearrangement followed by loss of a methyl radical to afford fragment ions of m/e 259. In the conformationally rigid 11-keto and 15-ketosteroids 2 and 3, the oxygen to γ-hydrogen distances were estimated using molecular models as 1.8 and 2.3 Å, respectively, and neither of these molecules undergo significant McLafferty rearrangement. Based on these results, Djerassi and

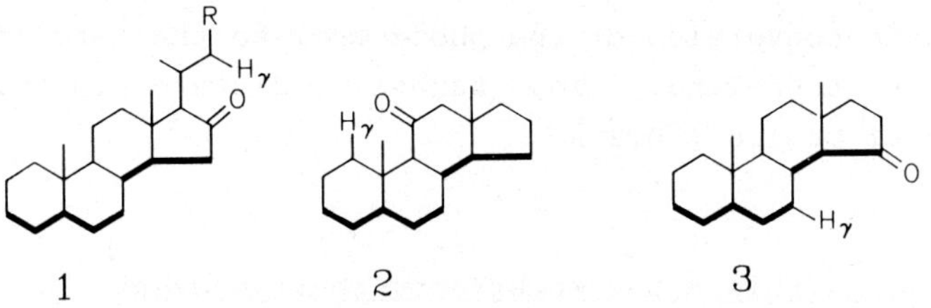

Fig. 2.3. Ketosteroids studied by Djerassi (ref. 16).

co-workers concluded that the upper limit for the abstraction distance in the McLafferty rearrangement is between 1.5 and 1.8 Å.

Similar conclusions were reached by Thomas and Willhalm in a study of the McLafferty rearrangement of a series of _endo_ and _exo_ 2-acetylnorbornane derivatives (ref. 17). These authors found that the _endo_ epimers (e.g., 4, Fig. 2.4) gave fragment ions consistent with a McLafferty rearrangement involving the C(6) _endo_ hydrogen atom (minimum distance _ca_. 1.2 Å), whereas the _exo_ isomers, 5, failed to undergo rearrangement. The minimum abstraction distance in the case of the _exo_ compounds was _ca._ 2.0 Å with the C(7)-_syn_ hydrogen atom.

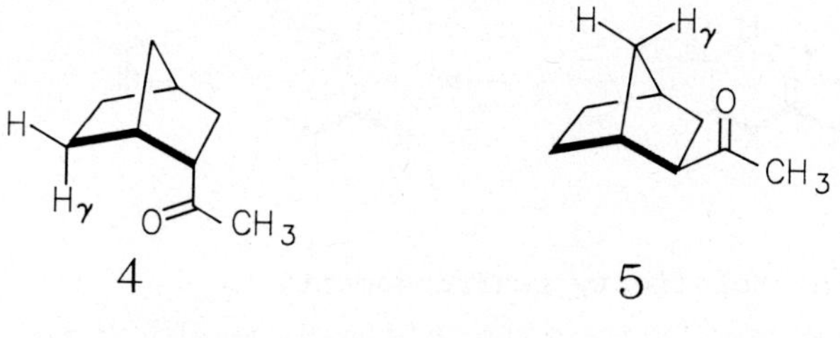

Fig. 2.4. McLafferty rearrangement of _exo_ and _endo_-2-acetylnorbornane.

The mass spectrometric fragmentation patterns of carboxylic acid esters have also been found to depend in several instances on the presence or absence of a sterically accessible γ-hydrogen atom (ref. 18). For example, the epimeric keto-esters 6a and 6b (Fig. 2.5) exhibit fragment ions at $(M-OCH_3)^{\cdot+}$ and $(M-CH_3OH)^{\cdot+}$, respectively. This difference was attributed to the accessibility of the tertiary γ-hydrogen atom at C(9) in the case of 6b (<u>ca</u>. 1.6 Å) but not 6a (ref. 19). In the case of a series of gibberelic acid esters (e.g., 7, Fig. 2.5), fragmentation of the molecular ion was suggested to involve initial abstraction of the benzylic hydrogen atom <u>cis</u> to the ester group (ref. 20). In this case, the distance as estimated using Dreiding molecular models was over 2 Å. When the benzylic hydrogen atom was <u>trans</u> to the ester group, the characteristic fragmentation was not observed. The abstraction distance of > 1.8 Å in the case of 7 was suggested to be consistent with the greater reactivity of benzylic hydrogens toward abstraction compared to unactivated hydrogen atoms.

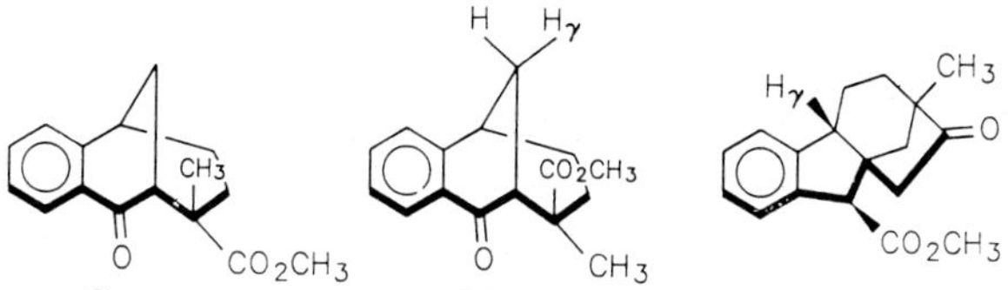

Fig. 2.5. Esters studied by mass spectrometry.

These ester fragmentation pattern differences are sufficiently well established that they have been used to make stereochemical assignments in compounds of unknown configuration (ref. 18); they have also been suggested to occur, not only through six-membered transition states as above, but <u>via</u> five-membered and seven-membered transition states as well (ref. 21).

The stereoelectronic features of the McLafferty rearrangement have been investigated theoretically as well as experimentally. Based on nonempirical molecular orbital calculations, Boer et al. (ref. 22) concluded that for the conformationally flexible molecule 2-pentanone, rearrangement proceeds through the eclipsed conformer 8 (Fig. 2.6), which has a C-H···O distance of 1.6 Å. The transition state for abstraction is reached by a 60° rotation about the β,γ-carbon-carbon bond to give the planar species 9 in which the hydrogen atom is approximately 1.3 Å from carbon and 1.2 Å from oxygen; non-planar transition state geometries were found to have considerably higher energies than that calculated for 9.

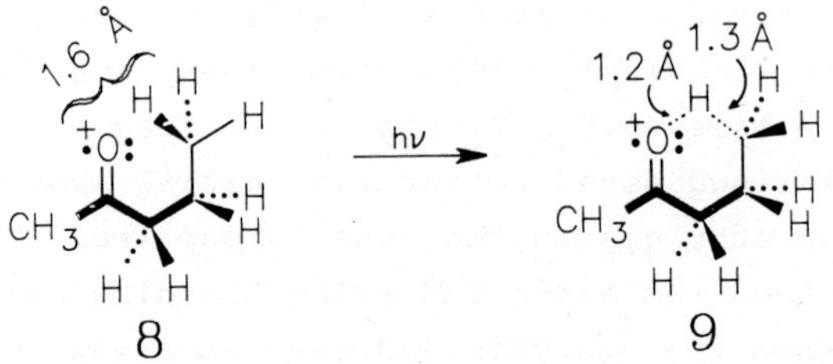

Fig. 2.6. Calculated transition state geometry for McLafferty rearrangement of 2-pentanone.

2.3 The Barton reaction

Named after its discoverer and developer, D.H.R. Barton, the Barton reaction consists of the generation of an alkoxy radical followed by an intramolecular hydrogen atom transfer (ref. 2). The reaction is of direct relevance to the Norrish type II and other photochemical hydrogen atom abstraction reactions because of the remarkable similarity in the chemical behavior of alkoxy radicals and the n,π^* excited states of ketones (ref. 5). The most usual method of generating the alkoxy radical is _via_ photolysis of the corresponding nitrite ester, and under these conditions, the final product of the reaction is the rearranged nitroso compound and/or the tautomeric oxime (Fig. 2.7); hypohalites have also been used frequently as the alkoxy radical precursors. Applied most often in the steroid field, the reaction is important because it allows the functionalization of a specific

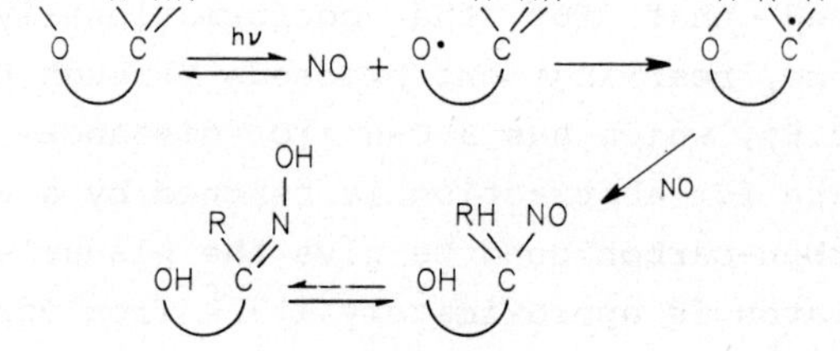

Fig. 2.7. The Barton reaction.

saturated carbon atom in the midst of many others of equal reactivity toward an external reagent.

As with the McLafferty rearrangement, nearly all examples of the Barton reaction involve six-membered transition states. Even in molecules such as 5-phenyl-1-pentylnitrite (10, Fig. 2.8), for which seven-membered transition state abstraction is favored by considerations of radical stability, only γ-hydrogen abstraction was observed (ref. 23). Similarly, no products resulting from five-membered transition state abstraction could be detected from photolysis of 3-phenyl-1-propylnitrite (11).

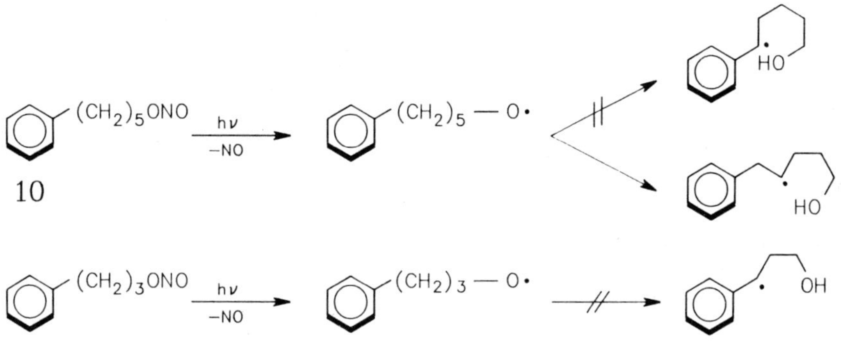

$$\text{PhCH}_2\text{(CH}_2)_5\text{ONO} \; (10) \xrightarrow[-NO]{h\nu} \text{Ph(CH}_2)_5\text{—O}\cdot$$

$$\text{Ph(CH}_2)_3\text{ONO} \; (11) \xrightarrow[-NO]{h\nu} \text{Ph(CH}_2)_3\text{—O}\cdot \nrightarrow$$

Fig. 2.8. The preference for six-membered transition state hydrogen atom abstraction in the Barton reaction.

Many steroidal Barton reactions involve abstraction of a hydrogen atom from an angular methyl group that is in a 1,3-diaxial relationship with the alkoxy radical. This is represented schematically in Fig. 2.9(a) for axial alkoxy groups at positions 2, 4, 6, 11 and 20. Heusler and Kalvoda (ref. 24) measured the methyl carbon to oxygen distances in these and other systems using Dreiding models and found that they were all in the range 2.5 to 2.7 Å. Assuming normal bond angles, this corresponds to an oxygen···hydrogen distance of at least 2.1 Å; this minimum distance is realized when the C-O and C-H bonds lie in the same plane. It should also be noted that this distance is significantly longer than the upper limit of 1.8 to 2.0 Å suggested for the McLafferty rearrangement.

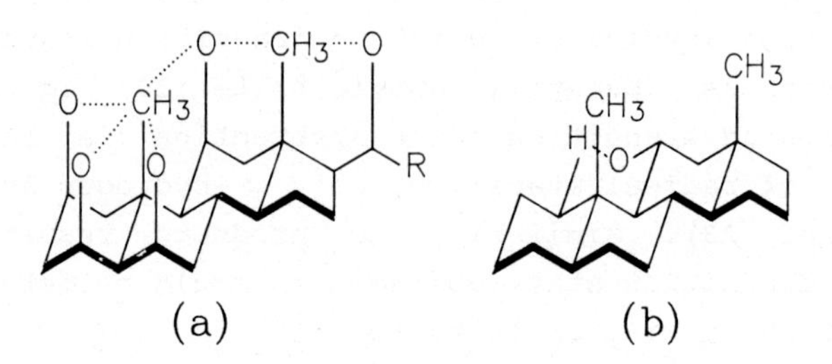

Fig. 2.9. (a) 1,3-Diaxial relationships in the steroidal Barton reaction. (b) Chairlike abstraction geometry in a conformationally rigid molecule.

The Barton reaction of steroids also provides information on the shape of the six-membered transition state involved in hydrogen abstraction. The rotational freedom of the angular methyl groups in systems such as that illustrated in Fig. 2.9(a) does not permit conclusions to be drawn concerning preferred geometry, but in systems where the hydrogen being abstracted is in a fixed conformation, such as that shown in Fig. 2.9(b), it is apparent that a chairlike six atom transition state is involved. Other rigid steroidal ketones that undergo the McLafferty rearrangement also involve chairlike transition state geometries (ref. 24a).

A similar conclusion for a conformationally flexible system was reached by Green et al. (refs. 24b,c) in a study on the Barton reaction of the alkoxy radical derived from 2-hexanol. By deuterium labeling, the preference for abstraction of H_a over H_b was determined to be 1.2 (see Fig. 2.10). This was suggested to result from a chairlike transition state in which abstraction of H_a involves an equatorial R = CH_3 group, whereas H_b abstraction requires that the R group be in the axial position. Similar results (k_a/k_b = 1.3) were obtained for R = cyclohexyl.

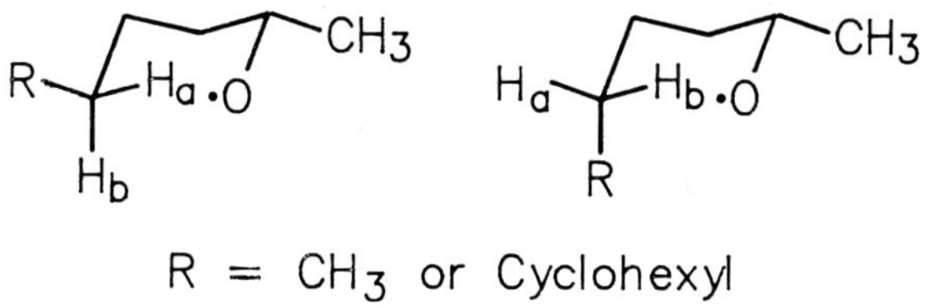

$R = CH_3$ or Cyclohexyl

Fig. 2.10. Stereoselectivity in the Barton reaction.

2.4 Angular Relationships in Hydrogen Atom Abstraction

The distance, d, between the abstracted and abstracting atoms is not the only factor to be considered in defining the geometry of intramolecular hydrogen atom abstraction; the angular relationship must be determined as well. The oxygen atoms of carbonyl groups are not able to abstract hydrogen atoms equally well in all directions. There is general agreement that in both the McLafferty rearrangement and the Norrish type II reaction, it is one of the non-bonding atomic orbitals on oxygen, and not the pi-bond atomic orbital, that is involved in hydrogen abstraction (refs. 4, 22). The carbonyl oxygen n-orbital lies in the plane of the carbonyl group and makes an angle of between 90° and 120° with the C=O axis. Whether the angle is 90° or 120° depends on the nature of the atomic orbital assumed to contain the n-electrons. Kasha (ref. 25) has suggested that the carbonyl lone pairs are non-equivalent, one pair residing in a largely 2s atomic orbital and the other in an AO that is essentially 2p in character (Fig. 2.11a). Of the two, the latter orbital, which forms an angle of 90° with the C=O axis, would certainly be the one involved in abstraction. Alternatively, the carbonyl lone pairs may be viewed as being located in equivalent "rabbit ear" sp^2 hybrid atomic orbitals that are oriented 120° with respect to the C=O axis (Fig. 2.11b). Evidence for this lone pair hybridization comes from X-ray crystallographic studies on the directionality of the hydrogen bonds involving carbonyl groups (ref. 26).

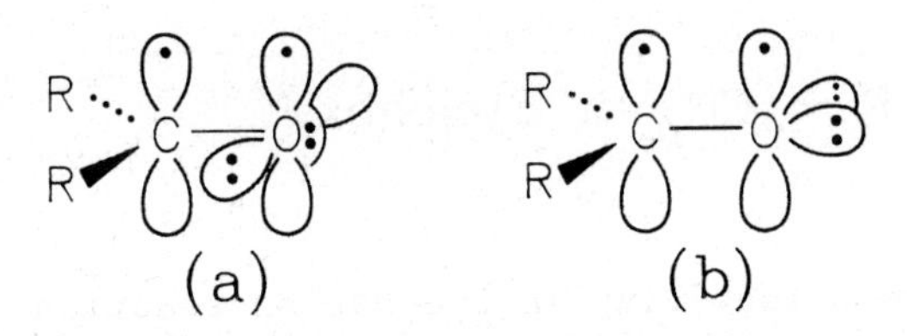

Fig. 2.11. Arrangement of atomic orbitals in carbonyl groups. (a) Kasha model, (b) rabbit ear model.

Abstraction is expected to be most facile when the hydrogen atom approaches the carbonyl group with maximum overlap with the non-bonded atomic orbital on oxygen. Fig. 2.12 depicts the general situation for an intramolecular abstraction process. The angle τ_0 is defined as the degree to which the abstracted hydrogen lies outside the mean plane of the carbonyl group, and Δ_0 is the C=O···H angle. Accordingly, abstraction should be best for $\tau_0 = 0^\circ$ and $\Delta_0 = 90\text{-}120^\circ$, and least favorable when $\tau_0 = 90^\circ$ or $\Delta_0 = 180^\circ$ Similar considerations led Wagner (ref. 5b) to suggest that there may be a $\cos^2\tau_0$ dependence on reactivity in the Norrish type II reaction. Our analysis leads to the additional prediction of a possible $\sin^2\Delta_0$ reactivity correlation.

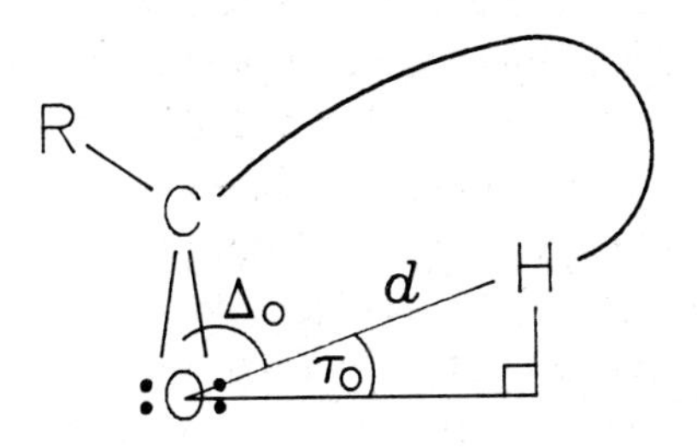

Fig. 2.12. Definition of d, τ_o , and Δ_o .

There is some experimental evidence that large values of τ_o disfavor hydrogen atom abstraction. Henion and Kingston (ref. 27) interpreted the lack of McLafferty rearrangement for ketone 12 (Fig. 2.13) as being due to an unfavorable τ_o angle of 80°. By way of comparison, ketone 13, with a τ_o angle of 50°, did undergo the McLafferty rearrangement. In both cases, d was estimated to be 1.6 Å. The values of d and τ_o were measured at the position of closest approach of H to O using molecular models. A second example is found in the work of Aoyama et al. (ref. 28). These authors observed that keto-alcohol 14 (Fig. 2.13) does not undergo observable chemical change upon irradiation despite a very close γ-hydrogen/oxygen contact. This was interpreted as being due to the fact that the γ-hydrogen atom lies almost precisely in the n-orbital nodal plane (τ_o = 90°). This interpretation should be treated cautiously, however, as γ-hydrogen atom abstraction is known to be reversible, and the lack of observable chemistry does not require the absence of abstraction.

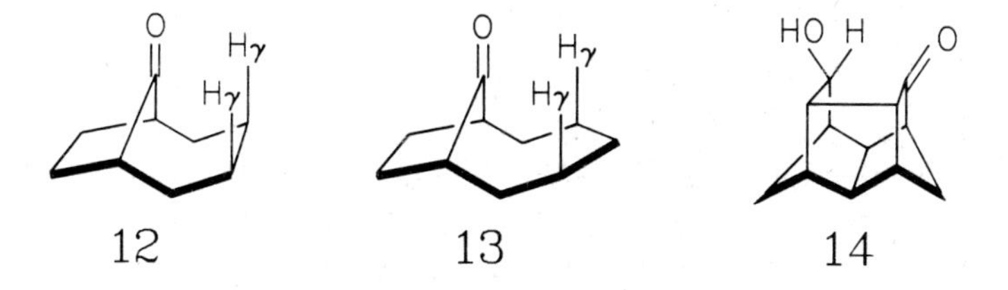

Fig. 2.13. Effect of the angle τ_o on hydrogen atom abstraction.

2.5 Norrish type II transition state geometry

The Norrish type II photochemical reaction (Fig. 2.14), arguably the most ubiquitous and thoroughly-studied of all photochemical reactions, consists of intramolecular γ-hydrogen atom abstraction by an excited carbonyl oxygen atom to produce a 1,4-biradical which has three fates: cyclization (cyclobutanol formation), cleavage to an alkene and an enol (isolated as the corresponding keto compound), and reverse hydrogen transfer to regenerate the ground state ketone. The reaction has been the subject of a number of review articles (ref. 5), and is of great importance for several reasons: used frequently to prepare unusual strained ring compounds, it is also responsible for much of our current knowledge of the properties of 1,4-biradicals and enols. In addition, it has been implicated as an important contributor to polymer photodegradation.

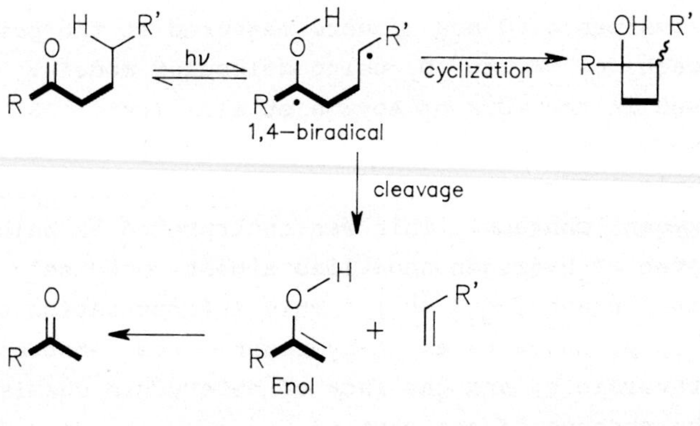

Fig. 2.14. The Norrish type II reaction.

As with both the McLafferty rearrangement and the Barton reaction, six-membered transition states are far more common in the Norrish type II reaction than their five-membered or seven-membered counterparts. As pointed out by Wagner (ref. 29), this reactivity order is inconsistent with transition states possessing linear O···H-C geometries, because a linear geometry should favor the seven-membered over the six-membered process (reduced angle strain).

Boer's calculations (ref. 22) indicate that, like the McLafferty rearrangement, the Norrish type II reaction proceeds through a planar, eclipsed transition state geometry (Fig. 2.6). As both Wagner (ref. 29) and Lewis (ref. 30) point out, however, this conclusion is not consistent with the experimental observation that in acyclic ketones, bulky substituents at the α and β-carbon atoms have little effect on the hydrogen abstraction rate constants, as they should if an eclipsed geometry was involved. Based on this evidence, Wagner (ref. 29) concluded that a staggered, chairlike abstraction geometry (Fig. 2.15a) is most likely. In this view, the preference for six-membered over seven-membered transition state hydrogen abstraction reflects the greater torsional strain of the latter.

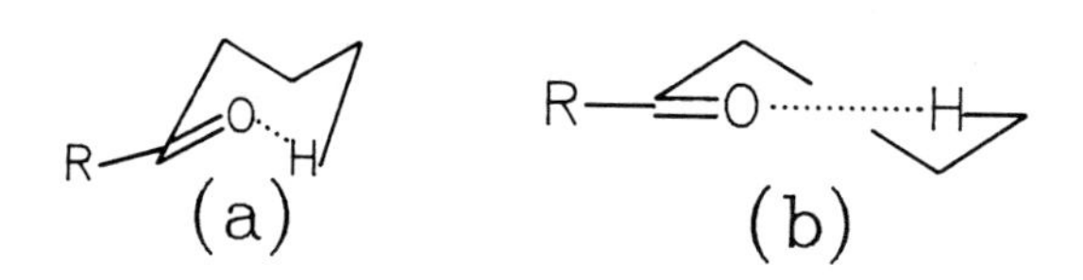

Fig. 2.15. Chairlike (a) and half-chairlike (b) six-membered transition states.

The chairlike geometry, however, is achieved at the expense of the abstraction distance, d. Boer's calculations reveal that even a half-chair geometry (Fig. 2.15b), while relieving eclipsing, increases d by approximately 0.7 Å compared to the fully eclipsed, planar arrangement of Fig. 2.6. Lewis (ref. 30) has suggested that this increase in d for staggered conformations may be a contributing factor in the approximately 50-fold slower rate of hydrogen abstraction for valerophenone compared to the more rigid, bicyclic ketones 15a-b (Fig. 2.16), for which d = 1.7 Å. The

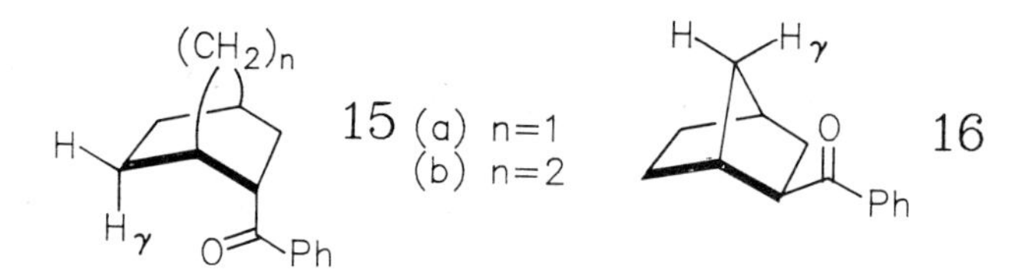

Fig. 2.16. Bicyclic ketones studied by Lewis.

exo ketone 16 (Fig. 2.16) reacts some 10 times more slowly than valerophenone (ref. 30), a fact that was attributed to the relatively long abstraction distance of 2.2 Å.

More recent molecular orbital calculations by Salem (ref. 31) and Dewar (ref. 32) are of interest in connection with the Norrish type II transition state geometry. Salem's calculations were carried out in order to test the symmetry-derived state correlation diagram for photochemical hydrogen atom abstraction and showed, in agreement with theory, that n-orbital abstraction proceeds from n, π^* excited state to biradical along a single, continuous potential energy surface. The abstraction geometry used in these calculations was d = 1.56 Å, $\tau_0 = 0°$, $\Delta_0 = 120°$, and O···H-C angle 180°.

Dewar's MINDO/3 calculations on the Norrish type II reaction of butanal (ref. 32) introduce an aspect of photochemical hydrogen abstraction that was not considered in the earlier calculations, namely, excited state pyramidalization at the carbonyl carbon, an experimentally well-documented occurrence in the case of formaldehyde (ref. 33). Fig. 2.17 depicts the calculated geometries of the ground state, the $(n, \pi^*)^1$ and $(n, \pi^*)^3$ excited states, and the transition states in each case. The O···H transition state distances are 1.6 Å in the case of the singlet reaction and 1.5 Å for the triplet process. The results also indicate that the C=O bond is elongated in its n, π^* excited state from 1.19 Å (ground state) to 1.25 Å (singlet) and 1.23 Å (triplet).

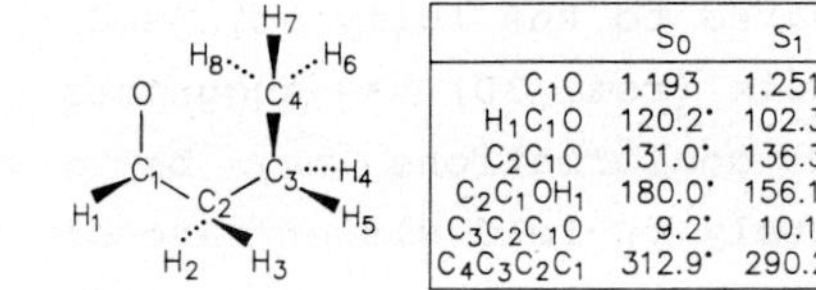

	S_0	S_1	T_1
C_1O	1.193	1.251	1.230
H_1C_1O	120.2°	102.3°	95.5°
C_2C_1O	131.0°	136.3°	134.4°
$C_2C_1OH_1$	180.0°	156.1°	131.0°
$C_3C_2C_1O$	9.2°	10.1°	10.2°
$C_4C_3C_2C_1$	312.9°	290.2°	296.6°

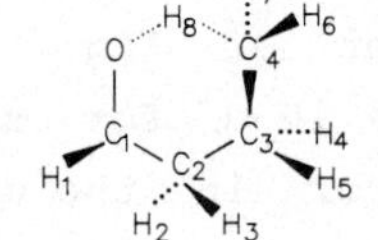

	S_1	T_1
OH_8	1.600	1.500
C_1O	1.273	1.280
H_8OC_1	110.3°	110.0°
$C_2C_1OH_1$	156.6°	162.1°
$C_3C_2C_1O$	27.3°	21.5°
$C_4C_3C_2C_1$	337.3°	339.0°

Fig. 2.17. Calculated ground state, excited state and transition state geometries for Norrish type II reaction of butanal.

While carbonyl carbon pyramidalization is probably a general feature of the n, π^* excited states of aldehydes and aliphatic ketones, such is not the case for aromatic and α, β-unsaturated carbonyl compounds. Hoffmann and Swenson, using extended Huckel and CNDO/2 calculations, showed that the carbonyl group of benzophenone remains planar in its n,π^* excited state (ref. 34). A similar conclusion was reached by Wagner et al. in studies on the viscosity-dependent dual phosphoresence of phenyl alkyl ketones (ref. 35). In this work, the n,π^* triplet was suggested to relax by a slight rotation about the carbonyl carbon-α-carbon bond. Other experimental (ref. 36) and theoretical (ref. 37) studies confirm the essential planarity of the n,π^* excited states of α,β-unsaturated systems such as acrolein and methyl vinyl ketone. The theoretical studies are unanimous in predicting an elongation of the C=O bond in the excited states of conjugated carbonyl compounds.

2.6 Dynamic considerations in intramolecular photochemical hydrogen abstraction reactions

Because bond rotation rates for conformationally flexible molecules in solution are generally competitive with, and often faster than, the rates of intramolecular photochemical hydrogen atom abstraction, it is misleading to think of abstraction as occurring through a single, fixed transition state geometry. The kinetic situation and the various boundary conditions have been analyzed in detail by Wagner (ref. 38). The measured hydrogen abstraction rate constant, k_{obs}, is actually a summation of the intrinsic rate constant, k^o, for each reactive conformation, F, times the equilibrium mole fraction of that conformation, X_F.

$$k_{obs} = \sum_F X_F k_F^o$$

An added complication is that X_F for the ground state may differ from the value of X_F in the excited state. For aralkyl ketones, however, which have very similar ground and excited state geometries (vide supra), the assumption is generally made that $X_F(es) = X_F(gs)$. Only bond rotations that change the abstraction distance, d, or the angles τ_o or Δ_o should affect k^o. The ultimate goal is thus to provide a quantitative relationship between k^o and the abstraction geometry. In practice, this can be accomplished in two ways: (1) design conformationally rigid systems for which $X_F = 1$ and measure k_{obs}, and (2) use conformationally flexible systems and physically restrict their bond rotations such that $X_F = 1$. We have adopted the second approach, and the method of restraint we have selected is to immobilize or "freeze" the reactant in a reactive conformation in its own crystal lattice. The great advantage of this approach is that the geometry of the system is determinable by X-ray diffraction methods.

3. THE SOLID STATE METHOD

The study of organic chemical reactions in the solid state and the interpretation of the results in terms of the shape and packing arrangement of the molecules that make up the crystal lattice have provided unparalleled insights into the mechanistic features of both bimolecular and unimolecular processes (ref. 39). This arises from two properties that are unique to the crystalline phase: (a) determination of the crystal structure by X-ray diffraction methods provides a detailed, three-dimensional view of the reaction ensemble immediately prior to reaction, and (b) because of the relatively strong forces that hold crystals together, atomic and molecular motions are restricted, and chemical reactions in crystals tend to be least motion in character such that the transition states, intermediates, and products closely resemble the reactants in shape and volume. This is the well-known "topochemical principle," of Cohen and Schmidt (ref. 40) and is based on their work on the [2+2] photodimerizations of the cinnamic acids in the solid state. In these studies, the regio- and stereochemistry of the photoproducts were found to be governed by the crystal lattice packing arrangement of the reactants. Compounds that packed with mirror symmetry between neighboring molecules gave mirror-symmetric dimers, and reactants that packed with a center of symmetry between the pre-dimer pair afforded centrosymmetric photoproducts. It was also found that there is a geometric requirement for intermolecular [2+2] photocycloaddition in the solid state. The double bonds undergoing reaction must be parallel with center-to-center distances of less than approximately 4.1 Å. The topochemical principle was restated in somewhat different terms by Cohen who introduced the concept of the solid state reaction cavity (ref. 39f). The reaction cavity was defined as the space occupied by the reacting species and bounded by the surrounding, stationary molecules. Cohen viewed the topochemical principle as resulting from the preference for chemical processes to occur with minimal distortion of the reaction cavity, either in the formation of voids within it or extrusions from it.

Since the pioneering work of Cohen and Schmidt in the early 1960's, the validity of the topochemical principle has been demonstrated time and time again (ref. 39). By using the crystalline state as the reaction medium, therefore, we limit the possible transition state geometries for reaction to those that closely resemble the reactant, and since the reactant shape can be determined accurately by X-ray crystallography, the combined result is an extremely powerful method for the development of structure-reactivity correlations. The application of this solid state method to the question of the geometry of intramolecular photochemical hydrogen atom abstraction is the subject of the remainder of this review article.

4. STUDIES ON ENE-DIONES

We begin by discussing our studies on the photochemistry of a class of compounds having the general tetrahydro-1,4-naphthoquinone (ene-dione) structure 17 (Fig. 4.1). The photochemistry of a large number of these compounds, differing in the location and nature of the various R groups, has been investigated both in solution and the solid state (ref. 41). Two general, hydrogen abstraction-initiated processes were found: (1) abstraction of an allylic (or benzylic) hydrogen atom by carbonyl oxygen (five-membered transition state) to give biradical 18 (Fig. 4.1), and (2) abstraction of an allylic or benzylic hydrogen atom by a C=C carbon atom (six-membered transition state) to afford biradical 23. The final photoproducts are the biradical coupling products 19-22 and 24-25. The photoproduct ratios, their dependence on the nature and location of the R groups, and their variation in proceeding from solution to the solid state, will not be discussed in this review except to point out that compounds 20 and 24 are nearly always the major products in the crystal photolyses. As we shall see, this is due to the fact that these photoproducts have shapes that are very similar to the conformation adopted by ene-dione 17 in the solid state, that is, their formation is topochemically allowed. Formation of the other photoproducts requires conformational changes that are not permitted by the rigid crystal lattice. As a result, these products are observed only in the solution phase irradiations.

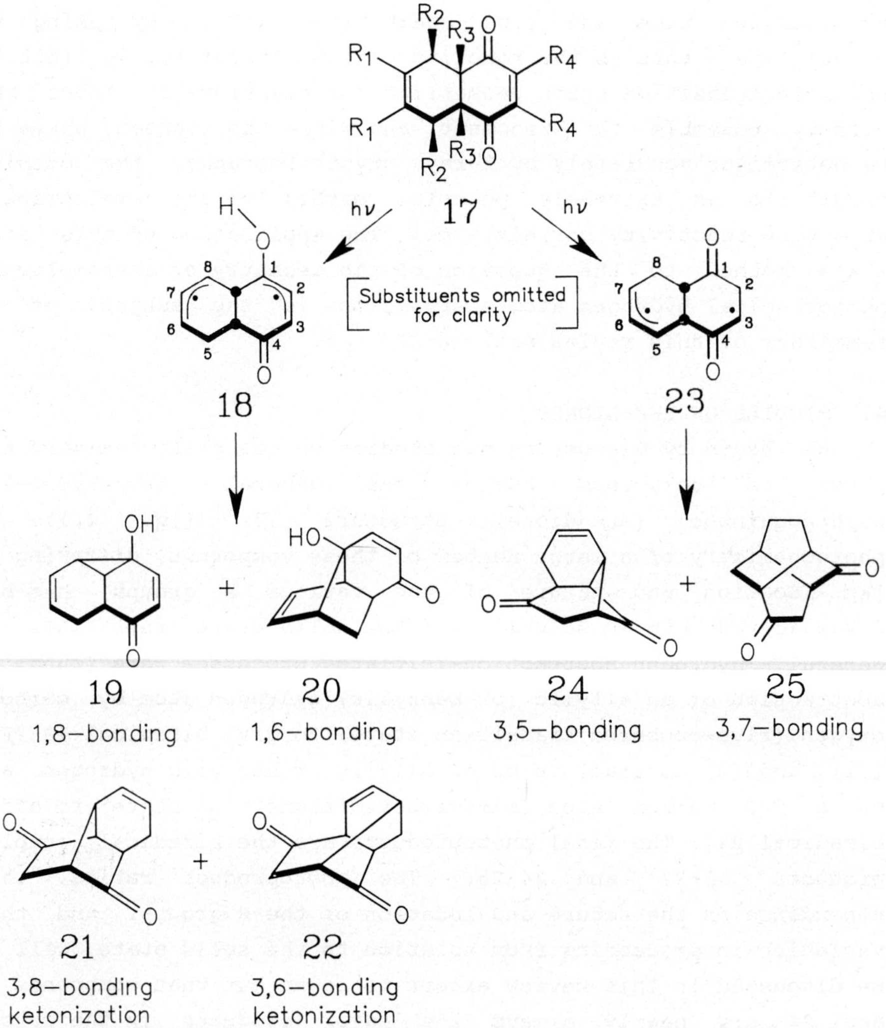

17 (R1, R2, R3, R4)
hν → 18 (Substituents omitted for clarity) → 19 1,8–bonding + 20 1,6–bonding; 21 3,8–bonding ketonization + 22 3,6–bonding ketonization
hν → 23 → 24 3,5–bonding + 25 3,7–bonding

Fig. 4.1. Photochemical reactions of ene-diones.

Concomitant with the photochemical studies, the crystal and molecular structures of the ene-diones were determined by Trotter and co-workers (ref. 39p). These studies showed that ene-diones of general structure 17 crystallize in a common "twist" conformation, regardless of the nature or location of the substituent R groups. The twist conformation is characterized by a staggered arrangement about the central C(4a)-C(8a) carbon-carbon bond, and both six-membered rings have half-chair conformations with approximately planar π-bond systems; the ene-dione ring, with two more sp^2 hybridized carbon atoms, is slightly flatter than the cyclohexene ring. An idealized drawing of the twist conformation is shown in Fig. 4.2.

Knowing the molecular conformation makes the photochemical pathways clear. Biradical 18 is formed <u>via</u> five-membered transition state abstraction by O(1) of H(8a). The alternative process of abstraction of H(5a) by O(4) is sterically impossible. Closure of biradical 18 by C(1)-C(6) bonding can occur from the same basic twist conformation to give the observed solid state product 20, and thus is a topochemically allowed process. Biradical 23 is produced by six-membered transition state abstraction of H(5a) by C(2). The C(3) carbon is also a potential abstractor, as H(5a) lies above, and almost equidistant from, C(2) and C(3). However, no photoproducts corresponding to this mode of reaction have been observed. Biradical 23, like biradical 18, is born in a conformation ideal for closure, and bonding between C(3) and C(5) forms the major photoproduct in the solid state, cyclobutanone 24. These structure-reactivity correlations are summarized in Fig. 4.2.

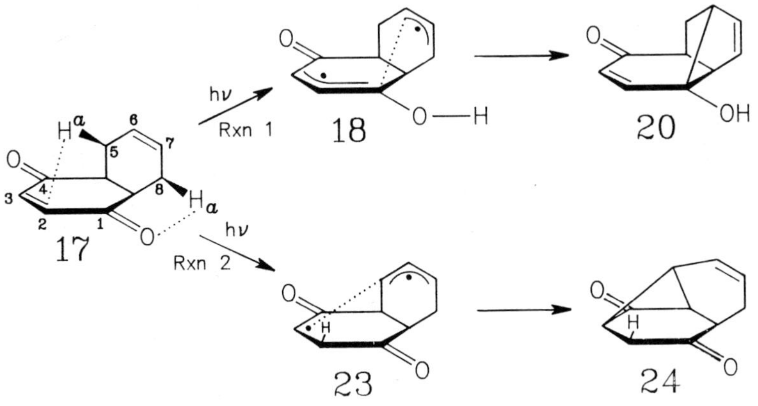

Fig. 4.2. Twist conformation of ene-dione 17 and solid state reaction pathways.

Seven ene-diones (17a-g, Table 4.1) were found to undergo reaction 1 (five-membered transition state hydrogen abstraction by oxygen), and five (17b-c and 17h-j, Table 4.2) took part in reaction 2 (six-membered transition state hydrogen abstraction by carbon). The crystal and molecular structure of each ene-dione was determined, and from this data the values of d, τ , and Δ were calculated. Fig. 4.3 defines d, τ , and Δ for the specific case of ene-dione 17, and Table 4.1 summarizes the results for reaction 1; Table 4.2 does the same for reaction 2.

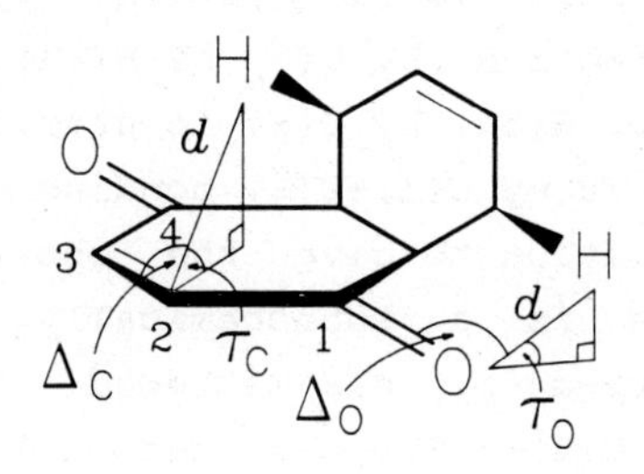

Fig. 4.3. Definition of d, τ , and Δ for ene-diones of general structure 17.

TABLE 4.1

Geometric data for five-membered hydrogen abstraction by carbonyl oxygen (reaction 1).

Ene Dione	R_1	R_2	R_3	R_4	d (Å)	τ_O (°)	Δ_O (°)
17a	Ph	H	H	H	2.5	3	81
17b	Me	H	Me	Me	2.5	0	85
17c	H	Me	Me	Me	2.3	1	86
17d	Me	H	CN	H	2.6	8	84
17e	Me	H	H	benzo	2.6	5	81
17f[x]	Me	H	H	Me/H[y]	2.5	6	85
17g	Me	H	Me/Et	Me/Et[z]	2.4	5	83

[x]Average for two molecules in the asymmetric unit. [y]H at C(2), Me at C(3). [z]Me at C(2), Et at C(3), Me at C(4a), Et at C(8a).

The results are remarkably consistent internally. Five-membered transition state hydrogen abstraction by oxygen occurs over distances between 2.3-2.6 Å, with τ_O varying from 0-8° and Δ_O ranging between 81-86° . Six-membered transition state abstraction by carbon involves somewhat longer distances, 2.7-2.9 Å, with τ_C and Δ_C ranging from 47-52° and 73-75°, respectively. The hydrogen abstraction distances are considerably longer than those involved in the McLafferty rearrangement or the Barton reaction (vide supra). The greater distance for carbon abstraction suggests that the van der Waals radius of the abstracting atom may be important; the van der Waals radius for carbon is 1.7 Å, whereas that for oxygen is 1.5 Å (ref. 42).

TABLE 4.2
Geometric data for six-membered transition state hydrogen abstraction by C=C (reaction 2).

Ene Dione	R_1	R_2	R_3	R_4	d (Å)	τ_C (°)	Δ_C (°)
17b	Me	H	Me	Me	2.9	52	73
17c	H	Me	Me	Me	2.7	50	74
17h	benzo	H	Me	Me	2.8	51	74
17i	benzo	H	Me/H[x]	H/Me[x]	2.9	47	75
17j	benzo	H	Me/H[y]	Me	2.9	50	74

[x]Me at C(2) and C(4a), H at C(3) and C(8a). [y]Me at C(4a), H at C(8a).

In fact, there is a striking correlation between the experimental abstraction distances and the sum of the van der Waals radii of the abstracting and abstracted atoms. Hydrogen has a van der Waals radius of 1.2 Å (ref. 42), giving an O···H sum of 2.7 Å and a C···H sum of 2.9 Å. The data in Tables 4.1 and 4.2 indicate that these sums may represent an upper limit to hydrogen atom abstraction. As we shall see, this attractive and intuitively reasonable idea is, with few exceptions, borne out by our solid state studies on other systems as well.

The angles reported in Tables 4.1 and 4.2 are also

interesting. As discussed earlier, the optimum value of τ_O is 0°. Thus ene-diones 17a-17g, with τ_O ranging from 0-8°, are nearly perfectly arranged for hydrogen abstraction involving the oxygen n-orbital. The angle Δ_O (experimental range 81-86°) is also close to the optimum range of 90-120° for n-orbital abstraction. In the case of hydrogen atom abstraction by the central ene-dione double bond, the abstracting orbital is a p-orbital which, in the ground state, is orthogonal to the plane of the double bond. In this case, therefore, the optimum values of τ_C and Δ_C are 90° and 90°, respectively. The data in Table 4.2 show that ene-diones are less ideally arranged for carbon abstraction than for oxygen abstraction; the average τ_C value is 50°, and the average value of Δ_C is 74°.

The analysis above assumes that the reactive <u>excited state</u> geometry is close to the ground state geometry as determined by X-ray crystallography. Leaving aside the fact that the solid state medium will tend to restrict all but the most minor geometry changes, the evidence is that this assumption is valid in the case of reaction 1, five-membered transition state hydrogen abstraction by oxygen. Photophysical studies (ref. 41b,d) indicate that the excited state responsible for this reaction is the $(n, \pi^*)^1$, and as discussed earlier, the n, π^* excited states of conjugated carbonyl compounds remain planar at the carbonyl carbon. The situation in the case of reaction 2 (carbon abstraction) is less certain. This reaction involves the ene-dione $(\pi, \pi^*)^3$ excited state (ref. 41b,d). It is generally agreed that in solution, the $(\pi, \pi^*)^3$ excited states of both acyclic and cyclic α,β-unsaturated ketones (and presumably cyclohex-2-en-1,4-diones as well) are twisted about the carbon-carbon double bond, the torsion angle and consequent triplet energy varying as the structural constraints of the system (ref. 43 and 44). Recently, however, both Schuster (45) and Pienta (46) have come to the conclusion that the hydrogen abstraction and photorearrangement reactions of cyclohexenones may bypass the transient detected by flash photolysis and assigned to the twisted $(\pi, \pi^*)^3$ species. Another factor to be considered is the suggestion due to Wiesner (ref. 47) that the $(\pi, \pi^*)^3$ excited states of α,β-unsaturated ketones possess pyramidalized β-carbon atoms. This idea was advanced to explain the sterochemistry of the [2+2] photocycloaddition reactions between enones and allenes. Although a number of cases are known that are consistent with this notion, there are some that are not (ref. 48), and a recent ab initio SCF and CI study (ref. 49) concludes that for acrolein, the β-carbon atom of the $(\pi, \pi^*)^3$ excited state remains planar. In light of the above, we prefer to leave the question of the excited state geometry involved in reaction 2 open for the present, and note only that if twisting about the ene-dione double bond is present in the solid state (a process less likely in the solid state than in solution, ref. 50), such twisting may actually facilitate the process of abstraction by carbon by tilting the abstracting p-orbital at C(2) more directly toward the hydrogen atom being abstracted.

The suggestion (above) that hydrogen atom abstraction occurs over distances less than or equal to the sum of the van der Waals radii of the atoms involved has been substantiated by the recent work of Mandelbaum et al. (ref. 51a) on the solid state photochemistry of ene-diones 26a-e (Fig. 4.4). Of these five compounds, only the bis-seven-membered ring derivative 26e was found to be photoreactive in the solid state. The reaction was one of seven-membered transition state hydrogen atom abstraction of an allylic hydrogen atom followed by C(1)···C(6) bonding of the biradical so produced (Fig. 4.4). Unlike ene-diones of general structure 17, ene-diones 26a-e cannot undergo type 1 or type 2 photochemistry because there are no allylic hydrogen atoms on C(5) or C(8) with the proper stereochemistry for abstraction.

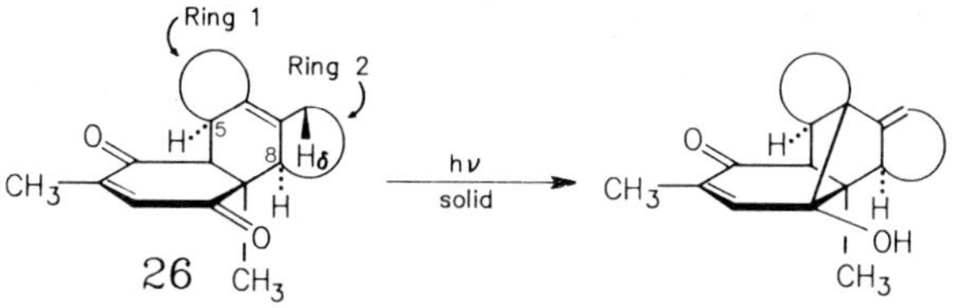

	Ring 1	Ring 2	O····H_δ Distance (Å)	Photorxn
(a)	5	5	3.2	no
(b)	5	6	4.1	no
(c)	6	6	4.8	no
(d)	6	7	4.0	no
(e)	7	7	2.7	yes

Fig. 4.4. Compounds studied by Mandelbaum et al. (ref. 51).

Strikingly, the photoreactive compound 26e was the only one of the five ene-diones studied that had an allylic hydrogen atom within van der Waals distance (d = 2.7 Å, τ_o = 19°, Δ_o = 98°) of an oxygen atom; the others had O···H distances ranging from 3.2-4.8 Å and were unreactive. Two δ-hydrogens are lost during the mass spectrometric retro Diels-Alder fragmentation of ene-diones 26a, 26c, 26e and related compounds (ref. 51b). The fact that compound 26e exhibits the lowest "RDA-2" fragment intensity despite having the most favorable O···H(δ) distance would seem to indicate either (1) H-transfer from a different conformer, (2) H-transfer after carbon-carbon bond cleavage, or (3) a migration terminus other than oxygen. Much less likely is the possibility that the mass spectrometric H-transfer occurs over greater distances than the photochemical process. The distances and angles were determined by X-ray crystallography (ref. 52).

5. STUDIES ON ENONES

A simple chemical modification of the compounds studied above, namely sodium borohydride reduction of one of the two carbonyl groups, transforms the ene-dione chromophore into a 4-hydroxycyclohex-2-en-1-one system of general structure 27 (Fig. 5.1); the use of methyl lithium rather than sodium borohydride affords the 4-methyl-4-hydroxy derivatives. As in the ene-dione series, we have investigated the solid state as well as the solution phase photochemistry of a number of these compounds and have correlated the results with the conformation and packing of the molecules in the crystal as determined by X-ray crystallography (ref. 53). The compounds studied differ in the nature and location of the R groups as well as in the relative configuration at C(4).

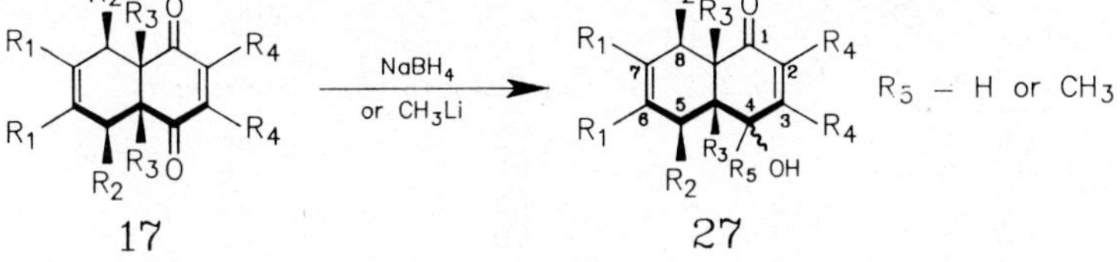

Fig. 5.1. Preparation of enone starting materials.

The transformation from ene-dione to enone brings about striking changes in the photochemistry of these systems, both in solution and the solid state. In solution, only intramolecular [2+2] photocycloaddition leading to cage compounds of general structure 28 is observed (Fig. 5.2). On the other hand, irradiation in the crystalline phase gives rearrangement products 31 and 32. The fact that only traces of cage compounds are formed in the solid state proves that the crystal reactions are true solid state processes. Photoproducts 31 and 32 are formed via intramolecular hydrogen atom abstraction by the β-carbon atom of the enone double bond (Fig. 5.2). Abstraction of an allylic hydrogen atom from C(5) through a five-membered transition state leads to biradical 29 and hence to keto-alcohol 31. Abstraction of a C(8) allylic hydrogen atom (six-membered transition state) gives biradical 30 which then collapses to keto-alcohol 32. This latter process is exactly analogous to the formation of diketone 24 from ene-dione 17 (Fig. 4.1).

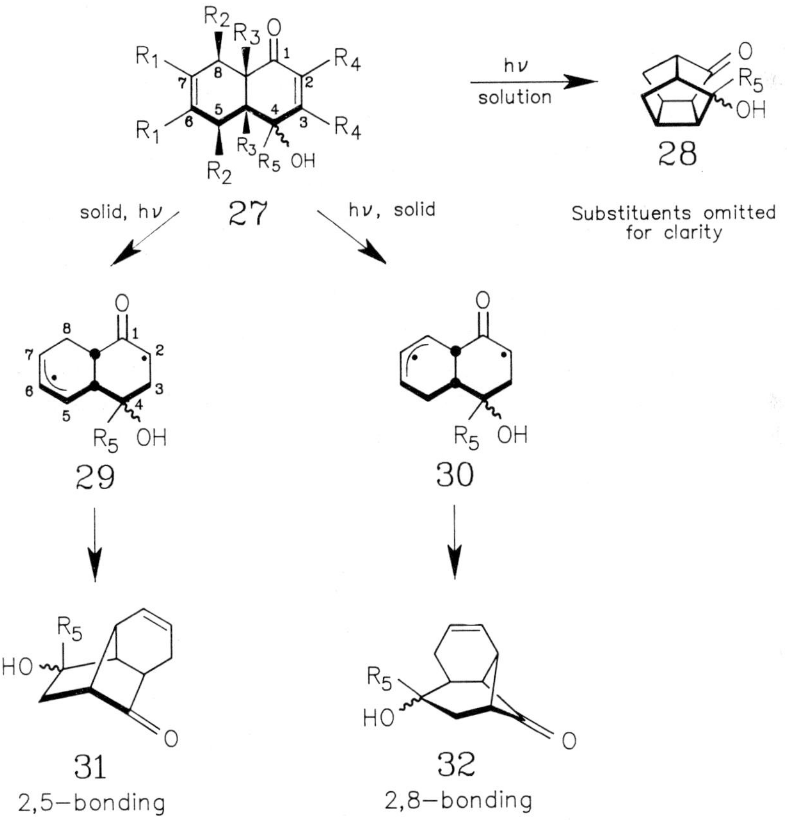

Fig. 5.2. Photochemical reactions of enones.

The solid state photorearrangement of any given enone was found to be completely regiospecific; only one of the two possible products, 31 or 32, was formed. It was not until the crystal and molecular structures of a number of enones had been determined, however, that the correlation between structure and reactivity became clear. It was found that the regioselectivity of hydrogen atom abstraction was determined by the conformation adopted by the starting enone in the solid state and that this conformation was governed in turn by conformational energetics, the minimum energy conformer being the exclusive solid state species in all cases.

Unlike the ene-diones studied earlier, cyclohexenones of structure 27 no longer possess a lateral plane of symmetry. As a result, each compound has _two_ twist conformations, A and B (Fig. 5.3), with different conformational energies (in the ene-diones, conformers A and B have identical conformational energies, and in fact are enantiomers). In solution these conformers are in rapid equilibrium _via_ a double half-chair-to-half-chair ring inversion (Fig. 5.3). Substituents that are pseudo-equatorial in one conformer are pseudo-axial in the other. The factor that determines whether the enones crystallize in conformation A or conformation B is the preference for the bulkier substituent at C(4) to lie in the pseudo-equatorial position. Thus enones that have opposite configurations at this position have different conformations and exhibit different solid state photochemistry. In eight out of the nine compounds studied, the C(4) substituents are hydrogen and hydroxyl, the latter of course being the larger and therefore occupying the pseudoequatorial position. In one instance, the substituents are methyl and hydroxyl, and here the methyl group is dominant.

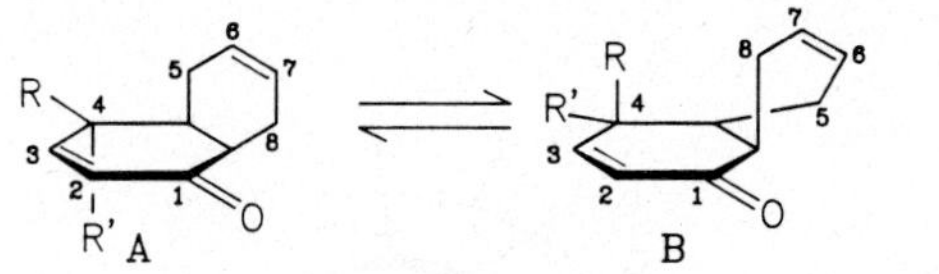

R	R'	Lower Energy (Solid State) Conformer
OH	H	A
H	OH	B
CH_3	OH	A
OH	CH_3	B

Fig. 5.3. Non-equivalent twist conformers of enones of general structure 27.

The solid state photochemical mechanisms involved are very similar to those observed earlier for the $(\pi, \pi^*)^3$ excited states of the ene-diones. In conformation A, the endo allylic hydrogen atom at C(5) lies directly over the enone double bond, and abstraction of this hydrogen by the β-carbon atom followed by C(2)-C(5) bonding of the biradical 29 so produced gives photoproduct 31 in a topochemically allowed process (Fig. 5.4). In conformer B, it is the endo allylic hydrogen atom at C(8) that is situated above the C(2)-C(3) double bond, and abstraction and biradical closure leads to the alternative photoproduct 32.

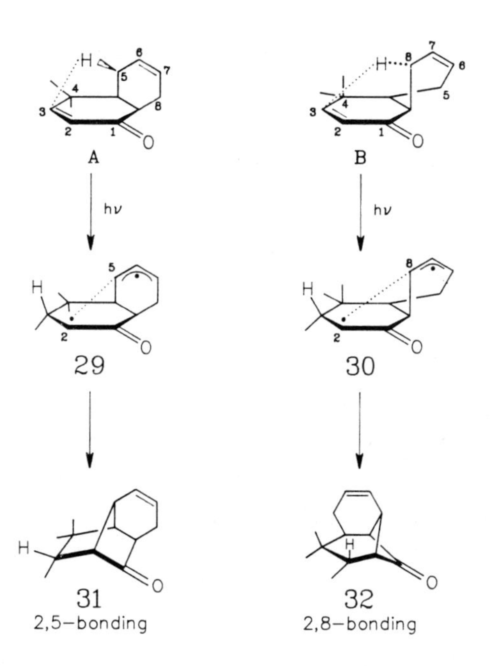

Fig. 5.4. Solid state reaction pathways for enone conformers A and B.

Table 5.1 summarizes the compounds studied, their conformation in the solid state as determined by X-ray crystallography, and the values of d, τ_C, and Δ_C. Also included are the values of θ, the C···H-C angle. The data in Table 5.1 show remarkable internal consistency and fully corroborate the conclusions reached earlier in our studies on ene-diones, namely that intramolecular hydrogen atom abstraction can occur over distances that are less than or equal to the sum of the van der Waals radii of the atoms involved

(2.9 Å in the case of carbon and hydrogen). One entry in Table 5.1 deserves further comment. Cyclohexenone 27b is photochemically inert in the solid state despite having a conformation and abstraction geometry that differ insignificantly from the other (reactive) enones. This lack of reactivity has been attributed tentatively (ref. 54) to a unique crystal lattice steric effect accompanying hydrogen abstraction. The change in hybridization at the β-carbon atom from sp^2 to sp^3, which accompanies hydrogen transfer to this atom, forces the attached methyl group into close contacts with certain hydrogen atoms on

TABLE 5.1

Geometric data for five and six-membered transition state hydrogen abstraction by enone β-carbon atom.[x]

Enone	R_1	R_2	R_3	R_4	R_5	Conf.	d(Å)	τ_C(°)	Δ_C(°)	θ(°)
27a	Me	H	Me	Me	H (cis/R_3)	A	2.7	53	79	107
27b	Me	H	Me	Me	H (trans/R_3)	B	2.9	50	75	108
27c	Me	H	Me	Me	Me(trans/R_3)	A	2.8	50	78	104
27d	H	H	Me	Me	H (cis/R_3)	A	2.8	52	78	105
27e	H	H	Me	Me	H (trans/R_3)	B	2.9	51	72	107
27f	Me	H	H	Me	H (cis/R_3)	A	2.8	54	79	122
27g	H	H	H	Me	H (cis/R_3)	A	2.8	54	80	101
27h	Me	H	H	H	H (cis/R_3)	A	2.8	57	81	99
27i	H	H	H	H	H (cis/R_3)	A	2.8	56	82	102

[x]Refer to Fig. 4.3 for definition of d, τ_C, and Δ_C; θ is the C···H-C angle.

neighboring molecules and thus sterically impedes the reaction (Fig. 5.5). The alternative process of abstraction by the α-carbon atom to give the less stable biradical 33 is also sterically impeded in the case of enone 27b. Similar steric effects are present in the other enones listed in Table 3, but are not as severe as they are in the case of 27b. Fascinatingly, in one instance (R_1 = Me, R_2 = H, R_3 = Me, R_4 = Me, R_5 = H cis to R_3, OH changed to OAc) where path a is impeded and path b is not, reaction <u>is</u> observed at the α-carbon with an abstraction distance of 2.7 Å (ref. 54).

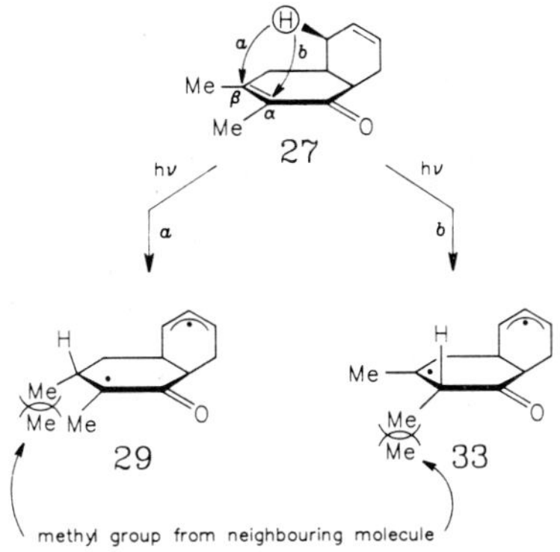

Fig. 5.5. Steric compression resulting from pyramidalization at the α and the β-carbon atoms. The substituents have been omitted for clarity.

6. STUDIES ON α-CYCLOALKYLACETOPHENONES

The final class of compounds we shall discuss in this review is the α-cycloalkylacetophenone system of general structure 34 (Fig. 6.1). A large number of ketones of this structure that differ both in cycloalkane ring size as well as in the nature of the aromatic substituent has been synthesized, photolyzed in the solid state and solution, and the results correlated with their crystal and molecular structures as determined by X-ray diffraction methods (ref. 55).

Without exception, ketones of general structure 34 undergo the Norrish type II reaction, both in the solid state and solution. Six-membered transition state hydrogen atom abstraction affords biradical 35 (Fig. 6.1) which can either return to starting material, cyclize to the corresponding bicyclo[n.2.0]alkanol (36) or cleave to afford cycloalkene plus p-substituted acetophenone (37). As in sections 4 and 5, the emphasis in this review will be on the hydrogen abstraction geometry present in the starting ketones, not on the variation in photoproduct ratio with changes in reaction medium.

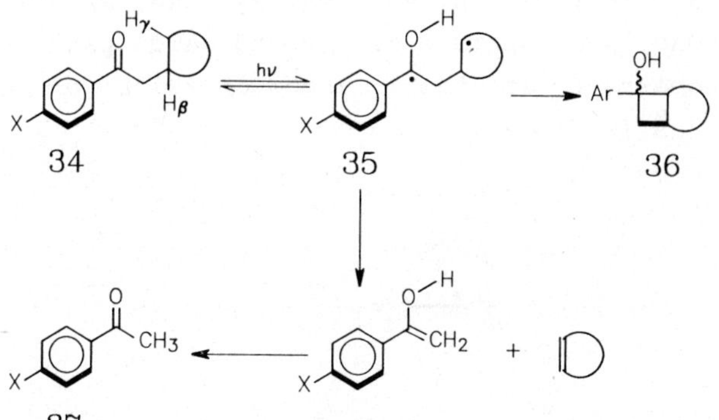

Fig. 6.1. Norrish type II photorearrangement of α-cycloalkylacetophenones.

Table 6.1 lists the compounds studied and the values of d, τ_O, and Δ_O as determined by X-ray crystallography. Also included in Table 6.1 is an entry that categorizes the overall six atom abstraction geometry as being chairlike, boatlike or half-chairlike. The geometric parameters are those for the γ-hydrogen atom that lies closest to the ketone oxygen atom. In most cases, the choice is unequivocal, as there is one unique γ-hydrogen atom that is closest by 0.3-0.4 Å or more. Exceptions are ketones 34a, 34m, and 34n, which have additional γ-hydrogen atoms that lie 0.2 Å, 0.1 Å, and 0.1 Å further away from oxygen than those listed in Table 6.1. In these cases, it is not certain which of the two available γ-hydrogen atoms is actually abstracted. A related aspect is that nearly every compound listed in Table 6.1 has a geometry that is more favorable for five-membered transition state hydrogen abstraction than for the experimentally observed six-membered transition state process. For example, ketone 34b has a tertiary β-hydrogen atom that lies only 2.5 Å from oxygen with $\tau_O = 17°$ and $\Delta_O = 78°$. These values are clearly more favorable than those for the secondary hydrogen atom that is actually abstracted (d = 2.8 Å, $\tau_O = 31°$, $\Delta_O = 96°$).

Only one of the ketones investigated in this study has a type II quantum yield of 1 in solution, even in aqueous acetonitrile, which is known to give a limiting quantum yield of 1 for valerophenone (ref. 5). We suggest that part of this inefficiency in the case of ketones 34a-34q may arise from unproductive β-hydrogen atom abstraction. The 1,3-biradical that would be produced by β-abstraction cannot cleave, and closure would give a highly strained cyclopropanol. As a result, reverse transfer is a logical fate of this hypothetical species. Strikingly, the one ketone that has a quantum yield of 1 in aqueous acetonitrile (34d) is the only one of the series for which the β and the γ-hydrogen atoms are equidistant from the carbonyl oxygen atom.

The data in Table 6.1 indicate that the van der Waals radii sum (2.7 Å in the case of oxygen and hydrogen) is not an absolute upper limit for hydrogen atom abstraction. Five of the seventeen compounds studied in the α-cycloalkylacetophenone series have abstraction distances that exceed the van der Waals radii sum by 0.1 to 0.4 Å. A worry is that these results might be due to sample melting during photolysis. Several of the ketones studied are low melting solids, and this coupled with the fact that most of the cleavage photoproducts are liquids themselves indicates that non-crystalline phases could be a problem. We have adopted three strategems to avoid this possibility: (1) photolysis to very low conversions, usually in the range 1-2%, (2) the use of low photolysis temperatures, generally 100-200 °C below the melting point of the ketone being photolyzed, and (3) the inclusion in the series of high melting ketones (e.g., p-COOH derivatives). Perhaps the best indication that the photolyses listed in Table 6.1 are true solid state processes comes from an independent study of the photochemistry of α-3-methyladamantyl-p-chloroacetophenone (34r, Fig. 6.2). This material, mp 44-45 °C, crystallizes in the chiral space group $P2_12_12_1$, and irradiation of large single crystals affords the cyclobutanol 38 in 70% yield (ref. 56). This cyclobutanol was shown by chiral NMR shift reagent studies to consist of 91% of a single enantiomer (82% enantiomeric excess). This remarkable result indicates that the Norrish type II reaction of ketone 34r (and by extension, those listed in Table 6.1) is a stereospecific, topochemically controlled process in the solid state; no optical activity is generated _via_ photolysis in solution because the ketone is achiral in this medium.

TABLE 6.1
Geometric data for Norrish type II reaction of p-substituted α-cycloalkylacetophenones. Values in parentheses are for a second conformer in the asymmetric unit.

Starting Ketone	Cycloalkane Ring Size	p-Substituent	Abstraction Geometry	d(Å)	τ_o(°)	Δ_o(°)
34a	4	Cl	half-chair	3.1	23	78
34b	5	Cl	boat	2.8	31	96
34c	5	COOH	boat	2.7	29	103
				(2.9)	(48)	(77)
34d	6	Cl	boat	2.6	42	90
34e	6	CN	boat	2.7	42	88
34f	6	COOH	boat	2.6	44	90
34g	6	Me	boat	2.6	50	88
34h	6	OMe	boat	2.6	43	91
34i	7	Cl	boat	2.7	42	82
34j	7	Me	boat	2.7	49	76
34k	8	Cl	boat	2.7	46	77
34l	8	COOH	boat	2.7	49	77
34m	exo-2-norb	Cl	half-chair	3.0	44	75
34n	exo-2-norb	OMe	half-chair	3.0	36	74
34o	1-adamant (plates)	Cl	chair	2.5	43	92
34p	1-adamant (needles)	Cl	chair	2.8	62	77
34q	1-adamant	OMe	chair	2.7	59	80

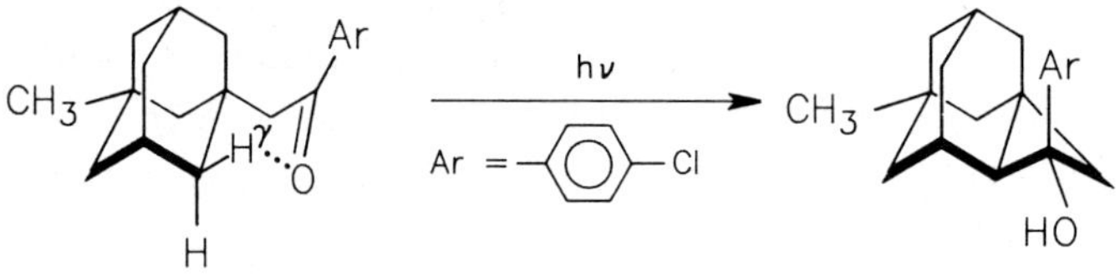

Fig. 6.2. Chirality transfer in the solid state.

The results summarized in Table 6.1 also indicate that a chairlike abstraction geometry is not an essential feature of the Norrish type II reaction and that transition states that deviate considerably from coplanarity ($\tau_o = 0°$) can be tolerated. In one case (34p), a τ_o value of $62°$ was observed. We have also calculated the angle θ (the C-H···O angle) for several of the ketones listed in Table 6.1. An average value of $114°$ was found. This differs from the value of $\theta > 150°$ suggested by Green et al. (ref. 24b) for the Barton reaction of the 2-hexyloxy radical. Fig. 6.3 shows stereodiagrams of typical boatlike and chairlike abstraction geometries observed in this study. In these stereodiagrams, which are taken directly from the crystal structure data, the cyclic, six-atom abstraction geometry is indicated by heavy lines.

With few exceptions, organic molecules crystallize in their lowest energy conformations (ref. 57). Therefore, the conformations adopted by ketones 34a-34q in the crystalline state should represent the predominant conformers present in solution as well, although the occurrence of dimorphs such as 34o and 34p with significantly different abstraction geometries indicates that this conclusion should be treated cautiously. If the Norrish type II reaction in solution occurs predominantly through the minimum energy ketone conformation, we might expect to see a correlation between the solution phase hydrogen abstraction rate constants and the geometric data as determined by X-ray crystallography, i.e., k_H should increase as d decreases and as τ_o and Δ_o more closely approach their optimum values. To test this idea, the hydrogen abstraction rate constants for the p-chloro-substituted ketones

34a, 34b, 34d, 34i, and 34k (cycloalkyl groups C_4-C_8) were determined in benzene solution using standard Stern-Volmer quenching techniques and compared with the geometric data (ref. 58). No correlation was found. For example, ketones 34b (C_5) and 34d (C_6) have identical hydrogen abstraction rate constants in benzene but have structural parameters that differ markedly (d =

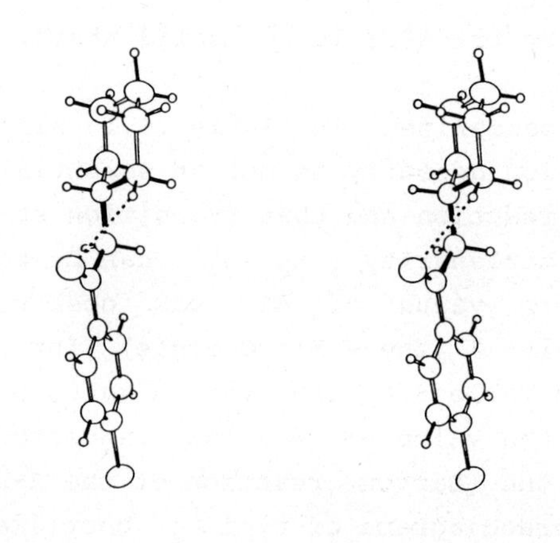

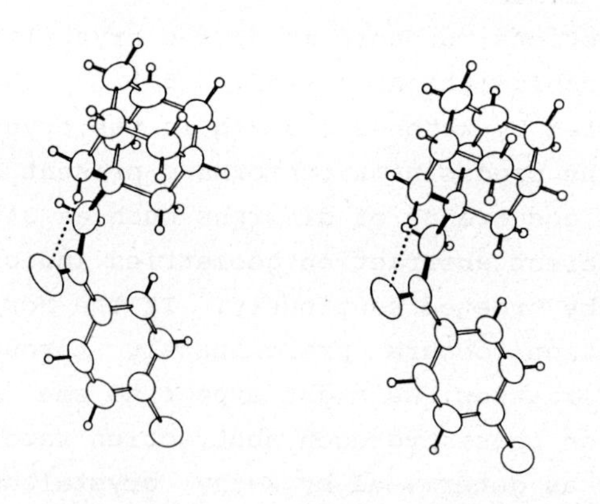

Fig. 6.3. Stereodiagrams depicting boatlike (ketone 34d, top figure) and chairlike (ketone 34o, bottom figure) abstraction geometries.

2.8 Å and 2.6 Å, respectively). Ketone 34i (C_7) reacts nearly five times as rapidly as ketone 34d (C_6) but has less favorable geometric parameters (d = 2.7 Å vs 2.6 Å). In contrast, the solution rate constants correlate beautifully with what is known concerning the relative rates of bimolecular free radical hydrogen atom abstraction from the cycloalkane homologous series cyclobutane, cyclopentane, cyclohexane, cycloheptane, cyclooctane. Several studies have established, using different free radical initiators, that the relative hydrogen abstraction reactivity order is cyclooctane > cycloheptane > cyclohexane ~ cyclopentane > cyclobutane (ref. 59). Our intramolecular hydrogen abstraction rate constants show exactly the same reactivity order: C(4) = 0.3 x 10^8 sec^{-1}; C(5) = C(6) = 1.2 x 10^8 sec^{-1}; C(7) = 5.7 x 10^8 sec^{-1}; C(8) = 6.7 x 10^8 sec^{-1}. Our results also parallel those of Lewis (ref. 60) who studied the solution phase photochemistry of a series of homologous α-cycloalkoxyacetophenones (Fig. 6.4). The hydrogen abstraction rate constants were found to vary with ring size as follows: n = 2, 2 x 10^8 sec^{-1}; n = 3, 3 x 10^9 sec^{-1}; n = 4, 7 x 10^9 sec^{-1}; n = 5, 9 x 10^9 sec^{-1}. This correlation between the relative rate of inter- and intramolecular hydrogen abstraction argues convincingly for a process in solution in which the reacting molecules are able to explore many abstraction geometries during their excited state lifetimes (see discussion in section 2.6).

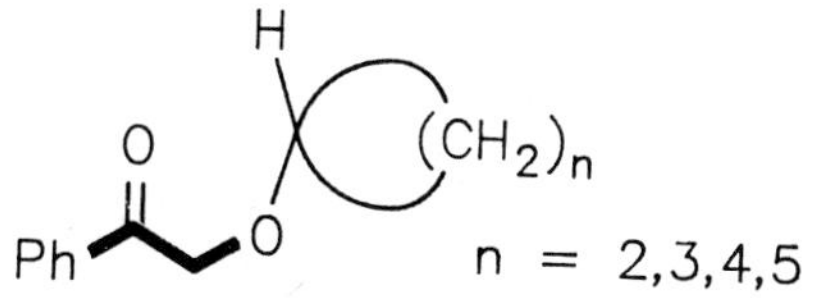

Fig. 6.4. α-Cycloalkoxyacetophenones studied by Lewis.

Using a computer program written by Mr. Stephen Evans at UBC, we are able to simulate the conformational motions of ketones of structure 34 and obtain plots of the variation in d, τ_O and Δ_O as a function of these motions. Studies of this type show that rotation about the α, β -carbon-carbon bond is most effective in reducing the value of d and improving τ_O and Δ_O. Fig. 6.5 shows a typical plot for ketone 34d. A 50° counter-clockwise rotation changes d, τ_O and Δ_O from their crystal conformation values of 2.6 Å, 42° and 90° to 1.8 Å, 4° and 113° respectively. These latter values are essentially perfect for abstraction. However, they are achieved at the expense of increased torsional strain, as the 50° rotated conformation is nearly eclipsed about the α, β -carbon-carbon bond.

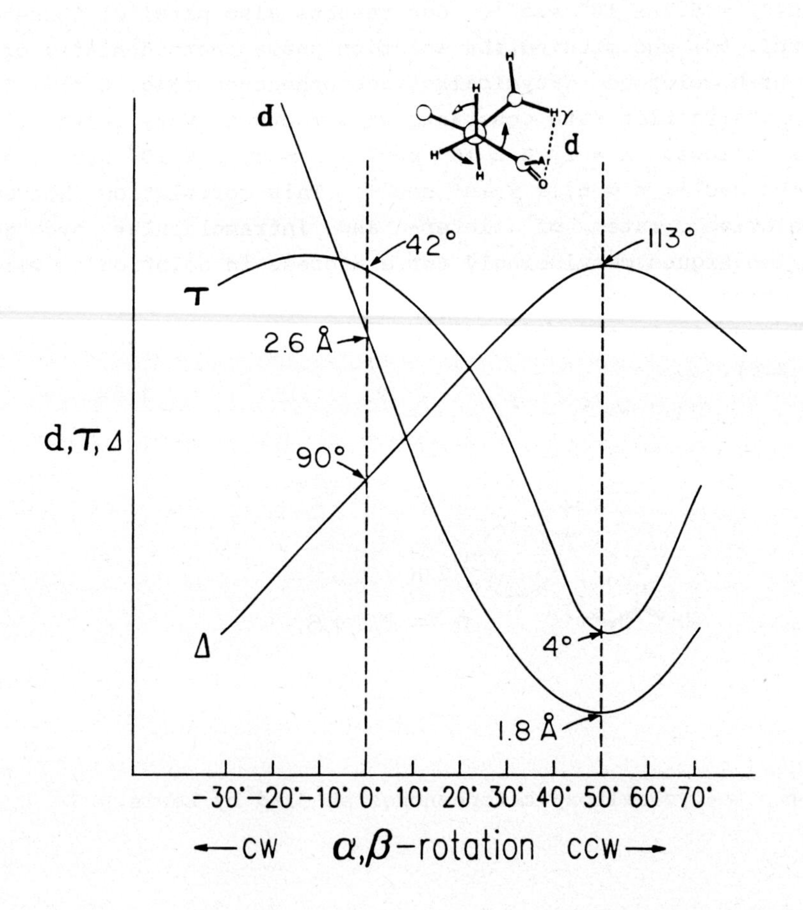

Fig. 6.5. Variation in d, τ_O, and Δ_O as a function of rotation about the α, β -carbon-carbon bond of ketone 34d.

7. CONCLUDING REMARKS

Our results with ene-diones (section 4), cyclohexenones (section 5) and aromatic ketones (section 6) leave no doubt that intramolecular photochemical hydrogen atom abstraction can occur over distances that are much greater than the upper limit of 1.8 Å suggested for the McLafferty rearrangement. What is the source of this large difference? Aside from the crudeness with which the McLafferty distances were measured, one possible factor that has not been considered is the elongation of the C=O bond that accompanies excitation into the n,π^* manifold (ref. 34). Another is that X-ray crystallography consistently underestimates the length of carbon-hydrogen bonds. This is due to the fact that hydrogen atoms, being extraordinarily light, have minimal intrinsic electron density, and the electron density map for these atoms is shifted toward the C-H bonding pair, which is closer to carbon than the hydrogen nucleus. If we assign a generous correction factor of 0.1 Å for each of these effects (ref. 34, 61), the actual excited state abstraction distance could be as much as 0.2 Å shorter than that measured by X-ray crystallography, but only when the C=O and H-C bonds are collinear. This is far from the case for the abstractions discussed in this review. Applying these elongation factors to a typical Norrish type II system, ketone 34d, does not change the abstraction distance, d, at all; the same is true for five-membered transition state abstraction in the case of ene-dione 17b. We feel that the most likely explanation for the closer O···H approach required for the McLafferty rearrangement arises from the fact that the abstracting oxygen atom bears a positive charge in the McLafferty rearrangement, but is neutral in the type II process. A positive charge on oxygen will certainly shrink the orbitals associated with this atom, including the orbital responsible for hydrogen atom abstraction, the singly occupied n-orbital; smaller orbital size necessitates closer hydrogen atom approach.

We conclude by pointing out that our results are supported by those recently obtained by Wagner et al. on the photochemistry of o-tert-butylbenzophenone (39, Fig. 7.1) in the solid state (ref. 62). The reaction is one of seven-membered transition state abstraction by oxygen of a tert-butyl hydrogen atom followed by coupling of the biradical so produced to afford the alcohol 40. X-ray crystallography revealed the existence of two hydrogen atoms located within abstraction distance of oxygen. One had d =

2.2-2.4 Å with τ_o = 30°. The other was situated at a distance of 2.7 Å with τ_o = 90°. Undoubtedly it is the former hydrogen atom that is abstracted. Mohr (ref. 63) has found that the diketone 41 (Fig. 7.1) also undergoes a γ-hydrogen atom abstraction in the solid state. In this case, X-ray crystallography reveals that there are four γ-hydrogen atoms within an abstraction distance of 2.5 to 2.6 Å of a ketone oxygen atom. These four hydrogen atoms are chemically equivalent in solution, and it has not been possible to determine which of the four is abstracted in the crystal photolysis.

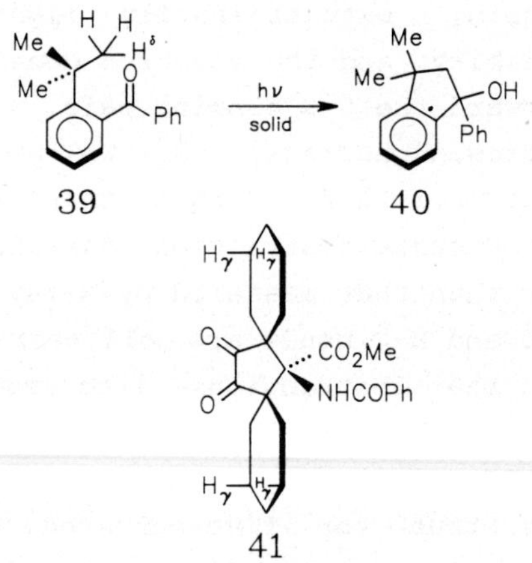

Fig. 7.1. Compounds studied by Wagner et al. (39) and Mohr (41).

8. ACKNOWLEDGEMENTS

Most of the credit for the work reported in sections 4, 5, and 6 should go to the people who actually did it. The photochemical studies were carried out by W.K. Appel, S.H. Askari, A.A. Dzakpasu, B. Harkness, Z.Q. Jiang, N. Omkaram, V. Ramamurthy, and L. Walsh. The crystallographic studies were performed by S. Ariel, S.V. Evans, T.J. Greenhough, S.E.V. Phillips, and A.S. Secco under the supervision of my friend and colleague, J. Trotter. I am grateful to the Natural Sciences and Engineering Research Council of Canada and the Petroleum Research Fund for financial support.

9. REFERENCES

1. (a) C. Walling and M.J. Mintz, J. Am. Chem. Soc., 89, 1515 (1967); (b) C. Walling and B.B. Jacknow, J. Am. Chem. Soc., 82, 6113 (1960); (c) J.K. Kochi in "Free Radicals," J.K. Kochi, Ed., Vol. II, Wiley, New York, 1973, p. 665; (d) R.D. Small, Jr. and J.C. Scaiano, J. Am. Chem. Soc., 100, 296 (1978).
2. (a) R.H. Hesse, Adv. Free Radical Chem., 3, 83 (1967); (b) M. Akhtar, Adv. Photochem., 2, 263 (1964).
3. F.W. McLafferty, "Interpretation of Mass Spectra," W.A. Benjamin, Reading, Mass, 1973, Ch. 4.
4. N.J. Turro, "Modern Molecular Photochemistry," Benjamin/Cummings, Menlo Park, California, 1978, Ch. 10.
5. (a) P.J. Wagner, Acc. Chem. Res., 4, 168 (1971); (b) P.J. Wagner, Top. Curr. Chem., 66, 1 (1976); (c) P.J. Wagner in "Rearrangements in Ground and Excited States," P. de Mayo, Ed., Academic, New York, 1980, Vol. 3, Ch. 20.
6. For a summary of the hydrogen abstraction reactions of the methyl radical, see W.A. Pryor, D.L. Fuller and J.P. Stanley, J. Am. Chem. Soc., 94, 1632 (1972).
7. E.S. Lewis in "Isotopes in Organic Chemistry," E. Buncel and C.C. Lee, Eds., Elsevier, Amsterdam, 1976, Vol. 2, p. 134.
8. B.M. Gimarc, "Molecular Structure and Bonding," Academic Press, New York, 1979, p. 27.
9. B. Liu, J. Chem. Phys., 58, 1925 (1973) and references cited therein.
10. H. Kwart, Acc. Chem. Res., 15, 401 (1982).
11. B. Anhede and N-A. Bergman, J. Am. Chem. Soc., 106, 7634 (1984).
12. R.P. Bell, "The Tunnel Effect in Chemistry," Chapman and Hall, London, 1980.
13. (a) P.J. Wagner and R.A. Leavitt, J. Am. Chem. Soc., 95, 3669 (1973); (b) P.J. Wagner R.J. Truman and J.C. Scaiano, J. Am. Chem. Soc., 107, 7093 (1985).
14. F.D. Lewis, R.W. Johnson and D.R. Kory, J. Am. Chem. Soc., 96, 6100 (1974).
15. K-H. Grellman, H. Weller and E. Tauer, Chem. Phys. Lett., 95, 195 (1983). See also W. Siebrand, T.A. Wildman and M.Z. Zgierski, J. Am. Chem. Soc.,, 106, 4083 and 4089 (1984).
16. (a) C. Djerassi, Pure Appl. Chem., 9, 159 (1964); (b) C.

Djerassi, G. von Mutzenbecher, J. Fajkos, D.H. Williams and H. Budzikiewicz, J. Am. Chem. Soc., 87, 817 (1965).

17. A.F. Thomas and B. Willhalm, Helv. Chim. Acta, 50, 826 (1967).

18. Review: M.M. Green in "Topics in Stereochemistry," N.L. Allinger and E.L. Eliel, Eds., Wiley, N.Y., 1976, Vol. 9, pp 35-110.

19. I. Lengyel and V.R. Ghatak in "Advances in Mass Spectrometry," A.R. West, Ed., Applied Science, Essex, England, 1974, Vol. 6, p 47 ff.

20. R.T. Gray and R.J. Pryce, J. Chem. Soc., Perkin II, 955 (1974).

21. (a) J. Deutsch and A. Mandelbaum, J. Am. Chem. Soc., 92, 4288 (1970); (b) A. Mandelbaum, J. Deutsch, A. Karpati and I Merksammer, Adv. Mass Spectrometry, 5, 672 (1970); (c) S. Weinstein, E. Gil-Av, J.H. Leftin, E.C. Levy and A. Mandelbaum, Org. Mass Spectrometry, 9, 774 (1974).

22. F.P. Boer, T.W. Shannon and F.W. McLafferty, J. Am. Chem. Soc., 90, 7239 (1968).

23. P. Kabasakalian, E.R. Townley and M.D. Yudis, J. Am. Chem. Soc., 84, 2716 (1962).

24. (a) K. Heusler and J. Kalvoda, Angew. Chem., Int. Ed. Engl., 3, 525 (1964); (b) M.M. Green, B.A. Boyle, M. Vairamani, T. Mukhopadhyay, W.H. Saunders, Jr., P. Bowen and N.L. Allinger, J. Am. Chem. Soc., 108, 2381 (1986); (c) M.M. Green, Tetrahedron, 36, 2687 (1980).

25. M. Kasha, Radiation Research, Suppl. 2, 243 (1960). See also J.G. Calvert and J.N. Pitts, Jr., "Photochemistry," Wiley, New York, 1966, pp 249-258.

26. (a) R. Taylor, O. Kennard and W. Versichel, J. Am. Chem. Soc., 105, 5761 (1983); (b) I. Olovsson, Croat. Chem. Acta, 55, 171 (1982); (c) R. Taylor and O. Kennard, Acc. Chem. Res., 17, 320 (1984).

27. J.D. Henion and D.G.I. Kingston, J. Am. Chem. Soc., 96, 2532 (1974).

28. N. Sugiyama, T. Nishio, K. Yamada and H. Aoyama, Bull Chem. Soc. Jpn., 43, 1879 (1970).

29. P.J. Wagner, P.A. Kelso, A.E. Kemppainen and R.G. Zepp, J. Am. Chem. Soc., 94, 7500 (1972).

30. F.D. Lewis, R.W. Johnson and R.A. Ruden, J. Am. Chem. Soc., 94, 4292 (1972).

31. (a) L. Salem, J. Am. Chem. Soc., 96, 3486 (1974); (b) W.G. Dauben, L. Salem and N.J. Turro, Acc. Chem. Res., 8, 41 (1975).
32. M.J.S. Dewar and C. Doubleday, J. Am. Chem. Soc., 100, 4395 (1978).
33. (a) J.C.D. Brand and D.G. Williamson, Adv. Phys. Org. Chem., 1, 365 (1963); (b) D.E. Freeman and W. Klemperer, J. Chem. Phys., 45, 52 (1966).
34. R. Hoffmann and J.R. Swenson, J. Phys. Chem., 74, 415 (1970).
35. P.J. Wagner, M. May and A. Haug, Chem. Phys. Lett., 13, 545 (1972).
36. R.R. Birge, W.C. Pringle and P.A. Leermakers, J. Am. Chem. Soc., 93, 6715 (1971).
37. R.R. Birge and P.A. Leermakers, J. Am. Chem. Soc., 93, 6726 (1971).
38. P.J. Wagner, Acc. Chem. Res., 16, 461 (1983).
39. The following is a partial list of review articles on the subject of chemical studies in organic crystals. (a) M. Hasegawa, Chem. Rev., 83, 507 (1983); (b) J.M. Thomas, S.E. Morsi and J.P. Desvergne, Adv. Phys. Org. Chem., 15, 63 (1977); (c) L. Addadi, S. Ariel, M. Lahav, L. Leiserowitz, R. Popovitz-Biro and C.P. Tang, "Chemical Physics of Solids and their Surfaces," M.W. Roberts and J.M. Thomas, Eds, The Royal Society of Chemistry, London, 1980, Specialist Periodical Reports, Vol. 8, Ch. 7; (d) A. Gavezzotti and M. Simonetta, Chem. Rev., 82, 1 (1982); (e) G.M.J. Schmidt, Pure Appl. Chem., 27, 647 (1971); (f) M.D. Cohen, Angew. Chem., Int. Ed. Engl., 14, 386 (1975); (g) J.R. Scheffer, Acc. Chem. Res., 13, 283 (1980); (h) M.D. Cohen and B.S. Green, Chem. Br., 9, 490 (1973); (i) I.C. Paul and D.Y. Curtin, Acc. Chem. Res., 6, 217 (1973); (j) J.M. McBride, Acc. Chem. Res., 16, 304 (1983); (k) G.R. Desiraju, Endeavour, 8 201 (1984); (l) J.M. Thomas, Pure Appl. Chem., 51, 1065 (1979); (m) M. Lahav, B.S. Green and D. Rabinovich, Acc. Chem. Res., 12, 191 (1979); (n) S.R. Byrn, "The Solid State Chemistry of Drugs," Academic Press, New York, 1982; (o) G.M.J. Schmidt, "Solid State Photochemistry," D. Ginsburg, Ed., Verlag Chemie, New York, 1976; (p) J. Trotter, Acta Cryst., B39, 373 (1983); (q) D.Y. Curtin and I.C. Paul, Science, 187, 19 (1975); (r) J.M. Thomas and W. Jones, "Reactivity of Solids," K. Dyrek, J. Haber and J.

Nowotny, Eds, Elsevier, 2, 551 (1980); (s) D.Y. Curtin and I.C. Paul, Chem. Rev., 81, 525 (1981); (t) G.R. Desiraju, Proc. Indian Acad. Sci., 93, 407 (1984); (u) V.M. Misin and M.I. Cherkashin, Russ. Chem. Rev., 54, 956 (1985).

40. M.D. Cohen and G.M.J. Schmidt, J. Chem. Soc., 1996 (1964).

41. (a) J.R. Scheffer, K.S. Bhandari, R.E. Gayler, and R.A. Wostradowski, J. Am. Chem. Soc., 97, 2178 (1975); (b) J.R. Scheffer, B.M. Jennings, and J.P. Louwerens, J. Am. Chem. Soc., 98, 7040 (1976); (c) J.R. Scheffer and A.A. Dzakpasu, J. Am. Chem. Soc., 100, 2163 (1978); (d) S. Ariel, S.H. Askari, J.R. Scheffer, J. Trotter, and F. Wireko, J. Am. Chem. Soc., submitted for publication.

42. A. Bondi, J. Phys. Chem., 68, 441 (1964). See also J.T. Edward, J. Chem. Ed., 47, 261 (1970).

43. (a) R.S. Becker, K. Inuzuka, and J. King, J. Chem. Phys., 52, 5164 (1970); (b) A. Devaquet and L. Salem, Can. J. Chem., 49, 977 (1971); (c) A Devaquet, J. Am. Chem. Soc., 94, 5160 (1972).

44. R. Bonneau, J. Am. Chem. Soc., 102, 3816 (1980).

45. D.I. Schuster, R. Bonneau, D.A. Dunn, J.M. Rao, and J. Joussot-Dubien, J. Am. Chem. Soc., 106, 2706 (1984).

46. N.J. Pienta, J. Am. Chem. Soc., 106, 2704 (1984).

47. (a) K. Wiesner, Tetrahedron, 31, 1655 (1975); (b) G. Marini-Bettolo, S.P. Sahoo, G.A. Poulton, T.Y.R. Tsai, and K. Wiesner, Tetrahedron, 36, 719 (1980); (c) J.F. Blount, G.D. Gray, K.S. Atwal, T.Y.R. Tsai, and K. Wiesner, Tetrahedron Lett., 4413 (1980).

48. For a thorough discussion of Wiesner's proposal, see (a) A.C. Weedon in "Synthetic Organic Photochemistry," W.M. Horspool, Ed., Plenum, New York, 1984, pp 96-99 and 131-134; (b) E. Piers, B.F. Abeysekera, D.J. Herbert, and I.D. Suckling, Can. J. Chem., 63, 3418 (1985).

49. K. Valenta and F. Grein, Can. J. Chem., 60, 601 (1982).

50. C.R. Jones and D.R. Kearns, J. Am. Chem. Soc., 99, 344 (1977).

51. (a) A. Weisz, M. Kaftory, I. Vidavsky and A. Mandelbaum, J. Chem. Soc., Chem. Commun., 18 (1984); (b) A Karpati, A. Rave, J. Deutsch, and A. Mandelbaum, J. Am. Chem. Soc., 95, 4244 (1973).

52. M. Kaftory and A. Weisz, Acta Cryst., C40, 456 (1984).

53. (a) W.K. Appel, Z.Q. Jiang, J.R. Scheffer and L. Walsh, J.

Am. Chem. Soc., 105, 5354 (1983); (b) J.R. Scheffer, J. Trotter, W.K. Appel, T.J. Greenhough, Z.Q. Jiang, A.S. Secco and L. Walsh, Mol. Cryst. Liq. Cryst., 93, 1 (1983).

54. (a) S. Ariel, S. Askari, J.R. Scheffer, J. Trotter, and L. Walsh, J. Am. Chem. Soc., 106, 5726 (1984); (b) S. Ariel, S. Askari, J.R. Scheffer, J. Trotter, and L. Walsh in "Organic Phototransformations in Nonhomogeneous Media," M.A. Fox, Ed., American Chemical Society, Washington, D.C., 1985, Ch. 15.
55. (a) S. Ariel, V. Ramamurthy, J.R. Scheffer, and J. Trotter, J. Am. Chem. Soc., 105, 6959 (1983); (b) J.R. Scheffer, J. Trotter, N. Omkaram, S.V. Evans, and S. Ariel, Mol. Cryst. Liq. Cryst., 134, 169 (1986); (c) S. Evans, N. Omkaram, J.R. Scheffer, and J. Trotter, Tetrahedron Lett., 26, 5903 (1985); (d) S.V. Evans, N. Omkaram, J.R. Scheffer and J. Trotter, Tetrahedron Lett., 27, 1419 (1986).
56. S.V. Evans, M. Garcia-Garibay, N. Omkaram, J.R. Scheffer, J. Trotter, and F. Wireko, J. Am. Chem. Soc., in press.
57. J.D. Dunitz, "X-Ray Analysis and the Structure of Organic Molecules," Cornell University Press, Ithaca, New York, 1979, pp 312-318.
58. S. Ariel, S. Evans, N. Omkaram, J.R. Scheffer, and J. Trotter, J. Chem. Soc., Chem. Commun., 375 (1986).
59. (a) W.A. Pryor, D.L. Fuller, and J.P. Stanley, J. Am. Chem. Soc., 94, 1632 (1972); (b) W.K. Stuckey and J. Heicklen, J. Chem. Phys., 46, 4843 (1967).
60. T.R. Darling, N.J. Turro, R.H. Hirsch, and F.D. Lewis, J. Am. Chem. Soc., 96, 434 (1974).
61. "Interatomic Distances Supplement," L.E. Sutton, Ed., Special Publication No. 18, The Chemical Society, London, 1965, p. S18s.
62. P.J. Wagner, B.P. Giri, J.C. Scaiano, D.L. Ward, E. Gabe, and F.L. Lee, J. Am. Chem. Soc., 107, 5483 (1985).
63. S. Mohr, Tetrahedron Lett., 3139 (1979) and 21, 593 (1980).

Chapter 2

TOPOTACTIC AND TOPOCHEMICAL PHOTODIMERIZATION OF BENZYLIDENECYCLOPENTANONES

CHARIS R THEOCHARIS AND WILLIAM JONES

1. INTRODUCTION

In 1969 it was reported that the uv irradiation of crystals of 2-benzyl-5-benzylidenecyclopentanone, 1, BBCP, yielded only one of the four possible photodimers - the centrosymmetric one (ref. 1); see Scheme 1 and Table 1. A similar result was subsequently reported for the bromo-derivative 2-p-bromobenzyl-5-benzylidenecyclopentanone, 2, BpBrBCP (ref. 2). It was further noted that the product from reaction of 2 consisted of single crystals of the photodimer.

The formation of well-ordered crystals without pronounced loss of crystallinity or morphological integrity is rare in uv induced [2+2] reactions. For example in the cinnamic acids, the solid state chemistry of which was crucial to establishing the ground rules for lattice controlled reaction (ref. 3), by far the major consequence of dimerization is the generation of an amorphous product. Whilst an amorphous organic product may in some contexts be of value, for possible utilization of solid-state chemistry a loss of crystallinity with conversion is expected to result in the loss of topochemical control and hence selectivity to a particular product.

For the cinnamic acid derivatives it is probable that the strong hydrogen bonding which exists in these solids does not allow sufficient relaxation of the product molecules within and around the so-called reaction cavity (refs. 4,5,6). In addition, the presence of large flexible side groups may increase the likelihood that reaction will occur without major changes in overall volume or shape by type of ballast or anchoring effect. Clearly, such anchoring groups are absent in the cinnamic acids.

Other systems also demonstrate pronounced changes in crystal integrity. Distyrylpyrazine, although yielding high molecular weight product undergoes significant crystal fragmentation and disruption during polymerization. Detailed transmission electron microscopy has revealed the extent to which dislocations, stacking faults and other imperfections are introduced during the reaction (ref. 5). Fig. 1 contrasts the difference in effect for BBCP and a related material dibenzylidenecyclopentanone, DBCP - see later. The retention

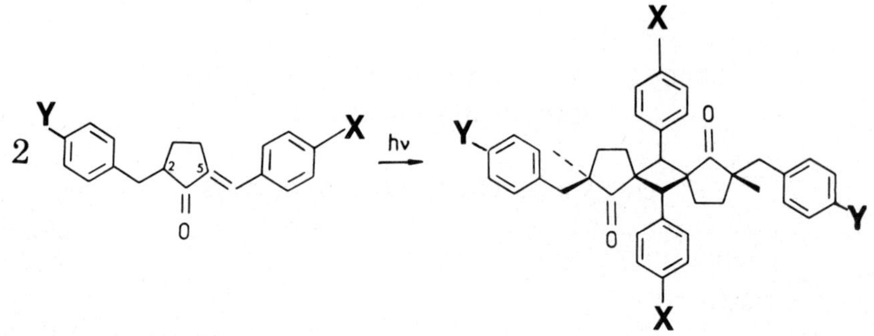

SCHEME 1

TABLE 1

Cell parameters and nomenclature for the various BBCP derivatives

	X	Y		Space group	a/Å α/°	b/Å β/°	c/Å γ/°	V/Å^3
1	H	H	BBCP	Pbca	31.302	10.784	8.867	2932
2	pBr	H	BpBrBCP	Pbca	34.222	10.923	8.427	3150
5	pCl	H	BpClBCP	Pbca	31.048	8.523	11.613	3073
6	oCl	H	BoClBCP	$P2_1/c$	14.399	11.602 93.00	9.228	1542
7	mBr	H	BmBrBCP	$P2_1/c$	12.950	27.166 101.80	9.243	3183
8	H	pCl	pClBBCP	$P2_1/c$	17.175	10.587 76.33	8.796	1554
9	H	Me	pMeBBCP	$P2_1/c$	17.341	10.675 102.54	8.736	1579
10	H	pBr	pBrBBCP	$P2_1/c$	17.458	10.556	8.854	1583
11	pBr	Cl	pClBpBrBCP	$P2_1/c$	17.522	7.906 91.19	11.888	1646
12	pBr	Me	pMeBpBrBCP	$P2_1/c$	18.381	11.209 94.46	8.285	1702

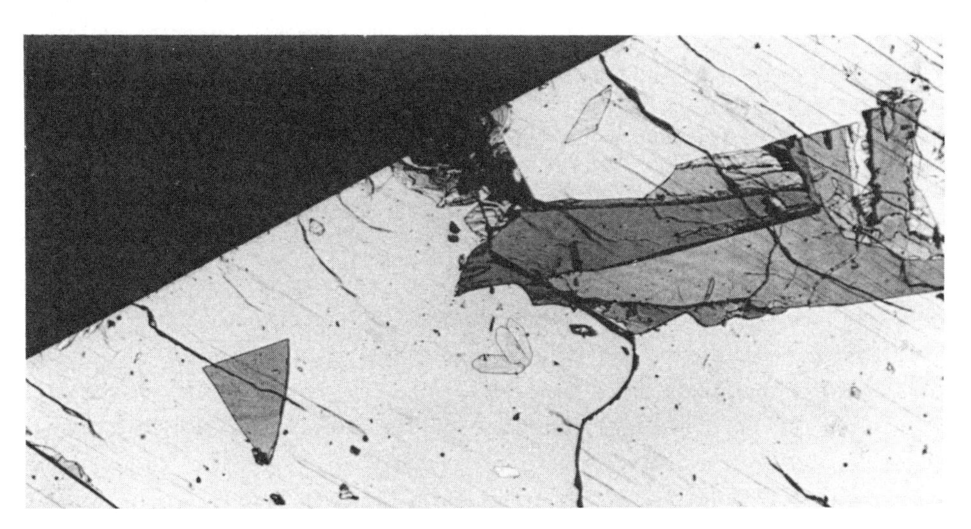

Fig. 1(a) Optical micrograph of single crystal of BBCP after dimerization.

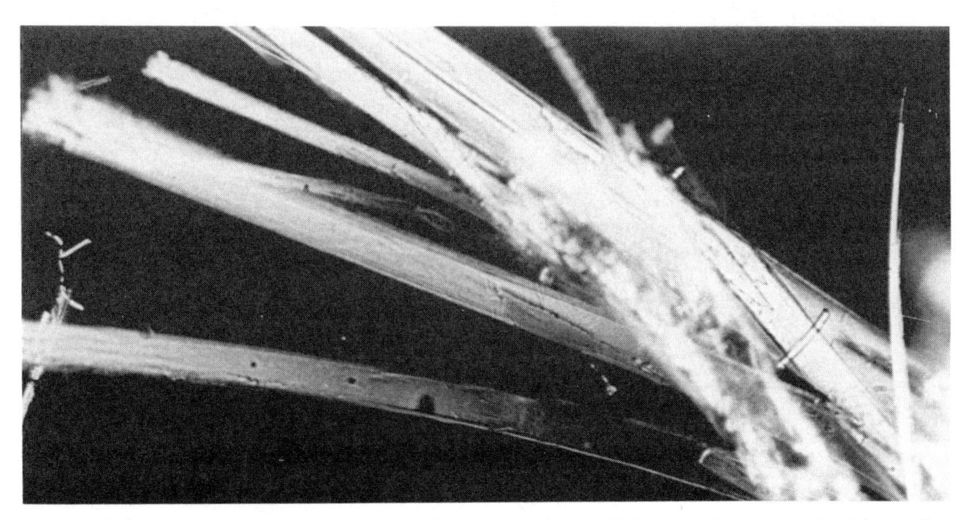

Fig. 1(b) Optical micrograph of DBCP after 10% reaction indicating the pronounced fracturing of the crystal which takes place.

of crystallinity throughout suggests that a quantitative yield of pure product may be obtained. In principle it is possible to locate by diffraction methods product molecules within the reactant lattice. The generation of only one photoproduct for each of BBCP and BpBrBCP as well as the formation of good crystals of product have led to a detailed study of this family of compounds. Numerous derivatives have been investigated and several aspects explored. Amongst these are: monitoring the course of the reaction by single-crystal diffractometry; crystal engineering; mixed-crystal formation and structural mimicry. These topics will be reviewed in this chapter.

2. IMPORTANT ASPECTS OF THE CRYSTAL STRUCTURE-REACTIVITY RELATIONSHIP.

2.1 Crystal packing.

The explanation for the generation of the centrosymmetric dimer in high yield, in most instances it approaches 100%, is readily obtained when the crystal structures of the starting compounds are examined (refs. 7,8). BBCP packs in the orthorhombic space group Pbca. The long molecular axis is parallel to the long cell axis and the arrangement of the molecules is such that nearest neighbours are related by a centre of symmetry with reactive double bonds separated by 4.17Å. A view of the structure is shown in Fig. 2.

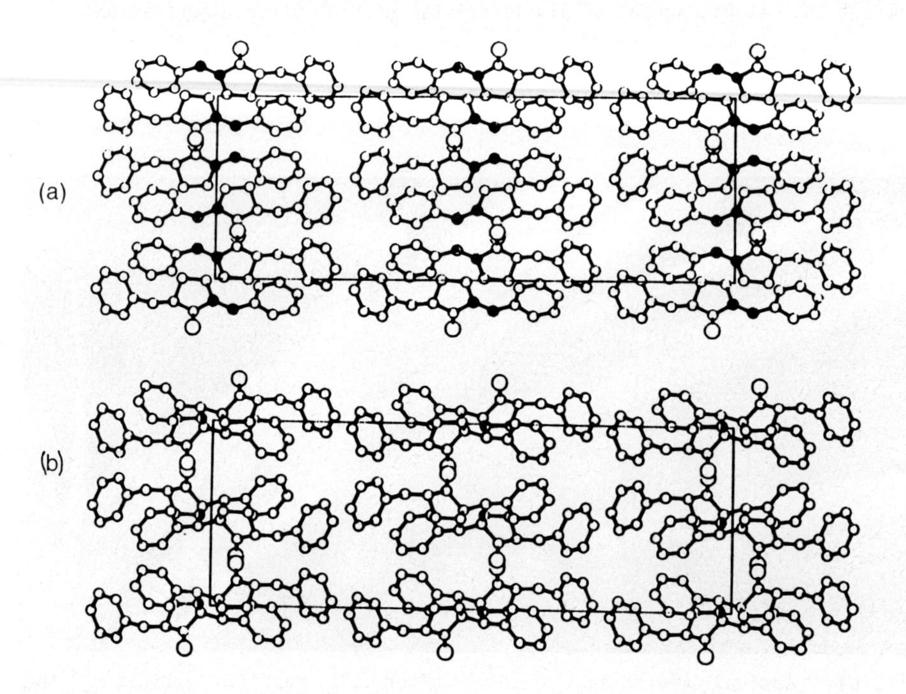

Fig. 2. View of the BBCP (a) monomer and (b) dimer crystal structures viewed along the c-axis.

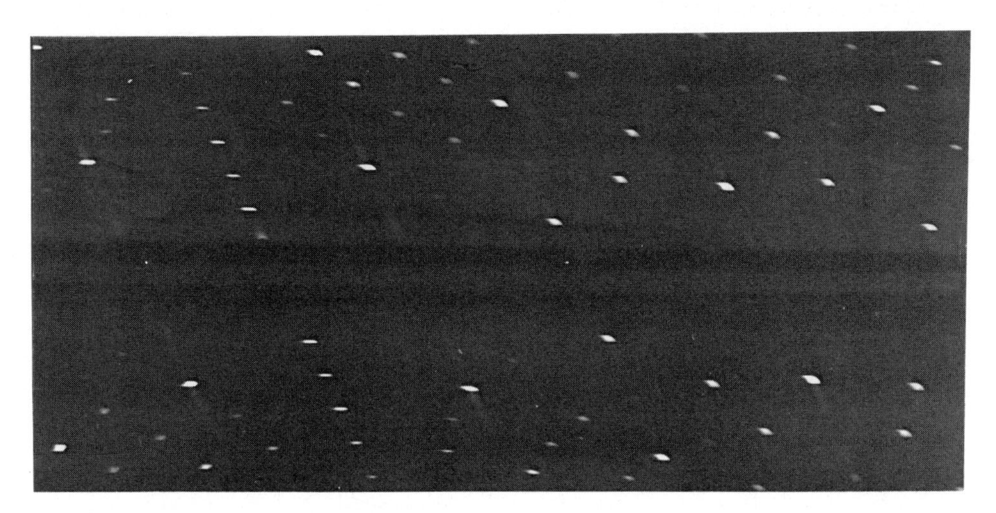

Fig. 3(a) Weissenberg photograph (hk0) of a BBCP crystal (CuK_α)

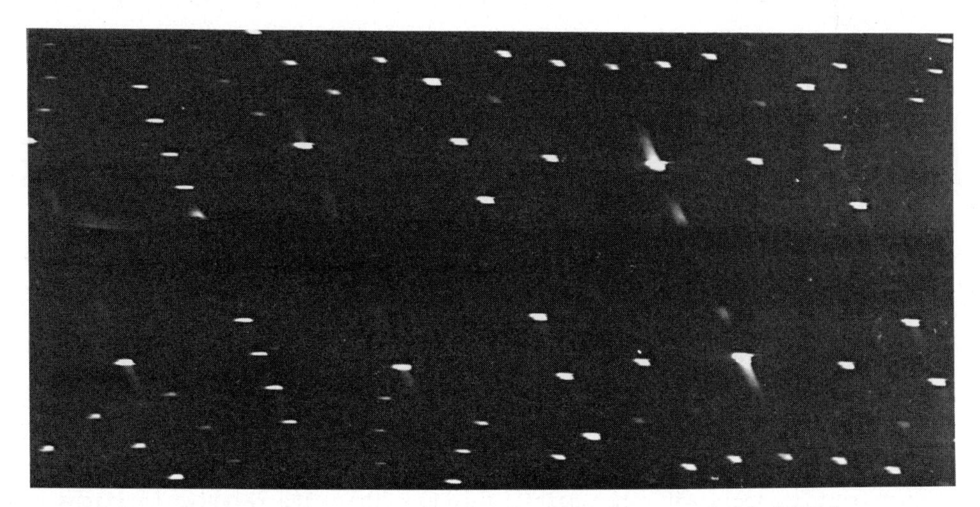

Fig. 3(b) Weissenberg photograph (hk0) of a BBCP dimer crystal (CuK_α)

Activation of molecules by uv irradiation leads to dimer formation. The structure of BpBrBCP is very similar, and also has incipient dimer pairs located within the crystal. For this derivative the reactive centres are separated by the slightly shorter distance (when compared with BBCP) of 3.8Å.

2.2 Crystallinity during reaction

That the crystallinity of the product is pronounced has been shown by conventional X-ray photographic techniques (refs. 8,9). Oscillation and Weissenberg photographs are shown, for BBCP, in Fig. 3. When a single crystal of BBCP is irradiated slowly and uniformly whilst mounted on a four-circle diffractometer, accurate cell parameter measurements show a

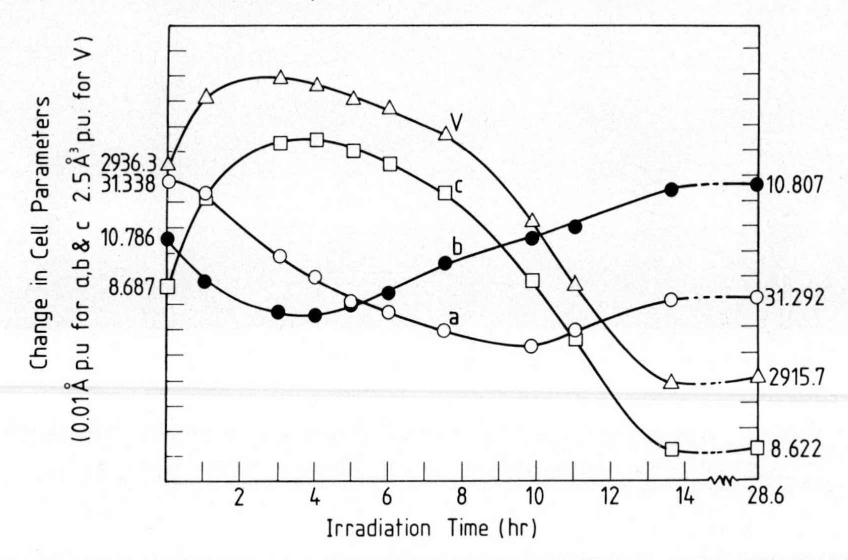

Fig. 4. Graph illustrating the changes in cell parameters which accompany the conversion of a single crystal of the monomer during reaction.

gradual and smooth conversion of monomer to dimer, as shown in Fig. 4. The space group remains Pbca with the cell parameters of the dimer very close to those of the monomer. The greatest change is 3% in the a-axis for BBCP and 7% for the b-axis in BpBrBCP. The unit cell orientation for the dimer phase is identical to that for the monomer, whilst it is noteworthy that for BpBrBCP the maximum change in cell parameters is along the line connecting the centres of the reacting double bonds i.e. the b-axis.

The crystal structure of the photodimer, solved from intensity data collected on an as-converted crystal (refs. 8,10), indicates that product dimer molecules occupy in an almost ideal manner crystallographic sites previously associated with incipient dimer pairs, see Fig. 2(b). The relative positions of the monomer and dimer molecules are shown in greater detail in Figure 5 and

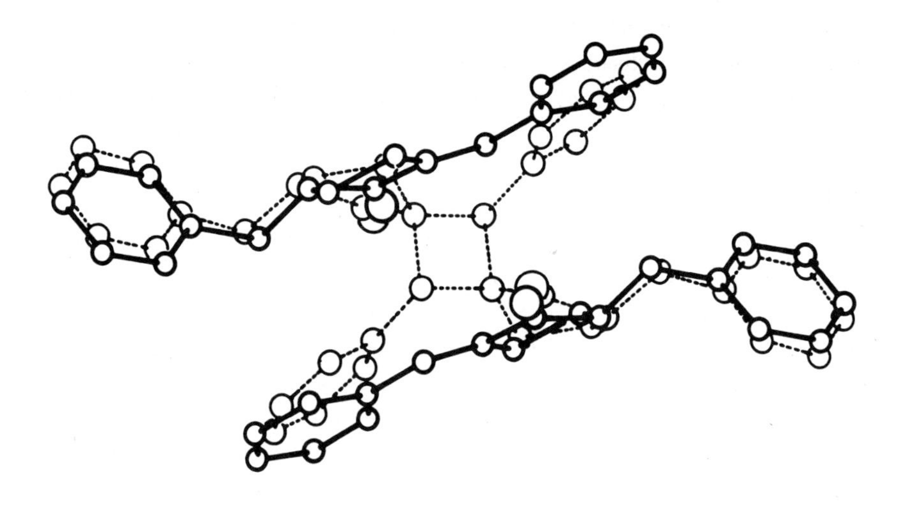

Fig. 5. Plot showing the incipient dimer pair and the corresponding dimer molecule (in dotted line). Note how the two monomer molecules occupy roughly the same volume as the dimer.

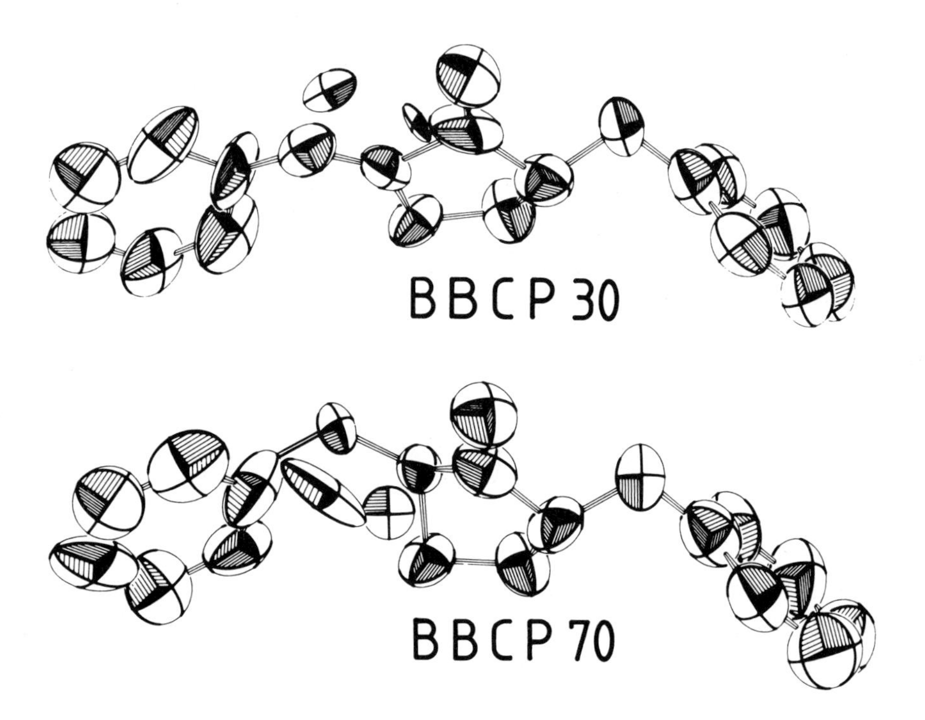

Fig. 6. ORTEP plots from the structure of a crystal containing 30% and 70% dimer. Both major and minor components are shown. The detached ellipsoids correspond to the C(5) and C(13) atoms of the minority component in each case.

Fig. 6. Rotation around the C(2)-C(6) bond allows the product molecules to relax after reaction without disruption of the lattice.

In passing it is perhaps appropriate to classify the possible types of reaction which may occur:

(i) in the cinnamic acids (ref. 3) there is overall retention of morphology but loss of crystallinity

(ii) in DSP and others (ref. 6) there is in general a retention of crystallinity but severe fragmentation

(iii) in DBCP (see later) there is fragmentation, but the product is amorphous, whereas in the dimerization of 9-cyanoanthracene (ref. 11) the product is crystalline but grows epitaxially upon the monomer surface. Incidentally the thermally induced monomerization of the corresponding dimer also yields crystalline product, but again as a separate phase growing topotactically from the dimer (ref. 12).

Systems which behave in a fashion similar to BBCP are the well studied diacetylene derivatives (ref. 13) and certain inorganic complexes which have been described by Sasada and his co-workers (ref. 14).

2.3 Least squares calculation of the atomic movements.

A simple mathematical model (refs. 15,16) aimed at rationalising the ease with which dimerization occurs has been developed. Two parameters are considered.

Firstly, for topochemical reactions the cooperative molecular rearrangement accompanying the reaction will determine whether a reaction is feasible and will be governed by factors including molecular repeat distances and proximity of those groups which react to yield the product. The root mean square displacement that encompasses these topochemical features is given by

$$R_1 = \sqrt{\frac{1}{n} \sum_{i}^{n} \left(t_{im} - t_{ip}\right)^2}$$

where n is the number of atoms in the molecule and t_i are the positional vectors of the i th atom from the origin for the monomer (m) and the product (p). Values outside a certain critical range would be expected to lead to incomplete reaction.

Secondly, the requirement that a reaction proceed cooperatively through the entire lattice to yield a single crystal product follows from the changes in unit cell parameter from reactant to product. A similar parameter to R_1 may be defined, yielding R_2. On the basis of such an analysis it was possible to rationalise (a) the enhanced reactivity of BpBrBCP compared to BBCP (R_1 = 0.68 and 0.72 respectively) and (b) the greater difficulty in converting BpBrBCP to a single crystal product compared with BBCP (R_2 = 1.27 and 0.06 respectively).

2.4 Flexibility of the molecular framework

The importance of the phenyl group in allowing the molecules to relax after reaction is seen in the solid-state reactivity of 2-benzylidene-γ-butyrolactone (3) (ref. 17). Uv irradiation leads to rapid dimerization, but the strain which is developed is sufficient that the crystals bend and eventually crack along [010].

Interestingly a comparison of the crystal structure and packing of 3 with the isoelectronic ketone, 2-benzylidenecyclopentanone (4) (ref. 17) reveals several important similarities and differences, noteworthy amongst which is the photostability of the latter. The separation distances for the centres of the olefinic bonds are 3.67 and 4.14Å respectively for 3 and 4. It has been pointed out that the generally accepted separation limit is ca. 4.2Å and that, therefore, the stability of 4 is surprising. However, consideration of orbital overlap rather than the mere separation of reaction centres offers some explanation. A full discussion of this point is deferred to Section 6.

Comparison of the behaviour of BBCP with that of DBCP (see later) shows that the anchoring group and its linkage to the main part of the molecule needs to be capable of altering it conformation during reaction. It is this ability that enables the group to remain essentially fixed whilst the reactive centres on the two monomers move towards each other, and which, therefore, helps maintain the overall size an shape of the reaction cavity, as shown in Figs. 5 and 6.

3. A COMPARISON OF REACTIVE AND UNREACTIVE MOTIFS

Numerous derivatives have been prepared involving substitution in the benzyl, benzylidene and cyclopentanone rings. As a result particular packing motifs have been identified and correlated with the observed reactivity or stability (refs. 18,19).

3.1 Substitution within the benzylidene ring.

BpBrBCP and BBCP pack in very similar arrangements; the difference in cell volumes arises from the elongation of the a-axis which is necessary to incorporate bromine with a slightly larger van der Waals radius (compare 1.95Å as against 1.20Å).

When chloro or methyl groups are placed in the same position, or if a substituent is placed in any other position within the benzylidene ring, the resulting packing motif is different to BBCP and BpBrBCP and the derivatives are photostable, see Fig. 7.

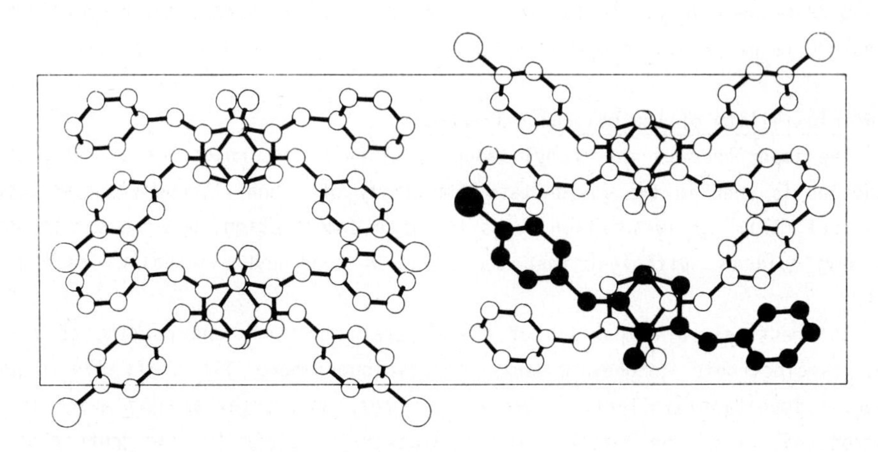

Fig. 7. Crystal structure of BpClBCP viewed along the b-axis.

The principal feature for reactive structures is the presence of incipient dimer-pairs related by a centre of symmetry. In general there is a short plane-to-plane distance indicative of an extremely strong interaction between the pi-conjugated systems of the molecules. Shortest intermolecular distances between double bonds within a pair are close to 4Å, e.g. 4.04 for BpBrBCP and 4.17Å for BBCP.

For the non-reactive crystals two principle packing motifs may be distinguished. The first is where nearest neighbour molecules (with respect to the exocyclic double bond) are related by a glide plane. The second is where nearest neighbours are related by a centre of symmetry.

Representative members of the first group (ref. 20) are BpClBCP (5) and BoClBCP (6). For BpClBCP the nearest neighbour molecules are related by a b-glide, the reactive centres being separated by 5.40Å. In BoClBCP nearest neighbour molecules are related by a c-glide, see Fig. 8, with a separation of reactive centres of 4.60Å. In all cases within this group the double bonds are not parallel. From Figs. 7 and 8 similarities between the two become clear. In both the molecules, when viewed along the shortest cell axis in each case, may be observed to form pairs of nearest neighbours with their long molecular axes crossed. The closest Cl....Cl distance for BoClBCP is 4.61Å for c-glide related molecules, whereas for BpClBCP the equivalent distance for molecules related by a centre of symmetry is 5.00Å. In both cases these are the nearest neighbour pairs; the differences arise from BoClBCP crystallizing in the $P2_1/c$ space group whereas BpClBCP crystallizes in Pbca.

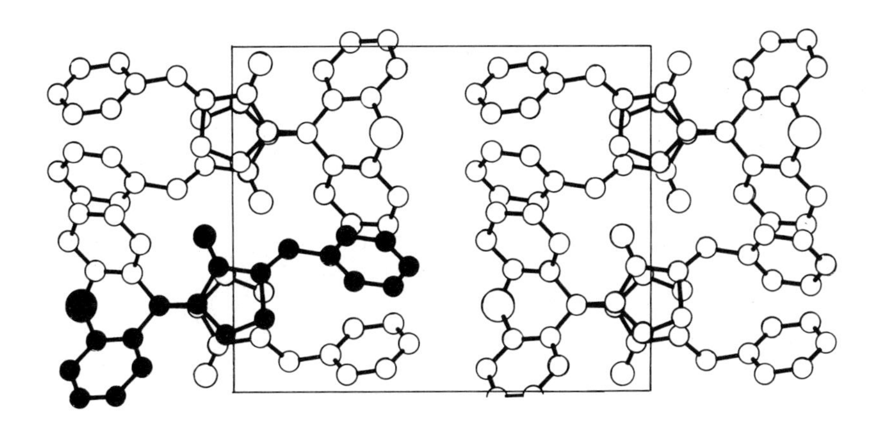

Fig. 8. Crystal structure of BoClBCP viewed along the c-axis.

The second group, that of nearest neighbour pairs related by centres of symmetry, is typified by BmBrBCP (7). For BmBrBCP there are two molecules within the asymmetric unit, Fig. 9, the two being related by a pseudo-centre of symmetry.

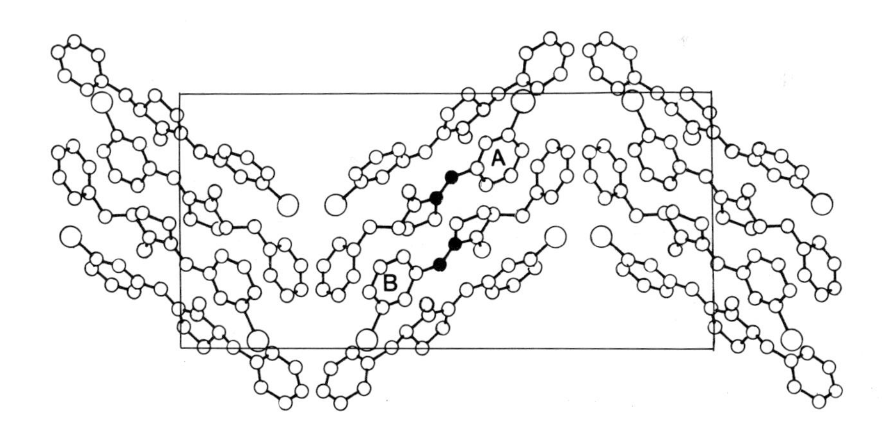

Fig. 9 Crystal structure of BmBrBCP viewed along the c-axis.

The molecules within the pair are distinguished by different distortions of the cyclopentanone ring and by the magnitude of the dihedral angle between the phenyl rings.

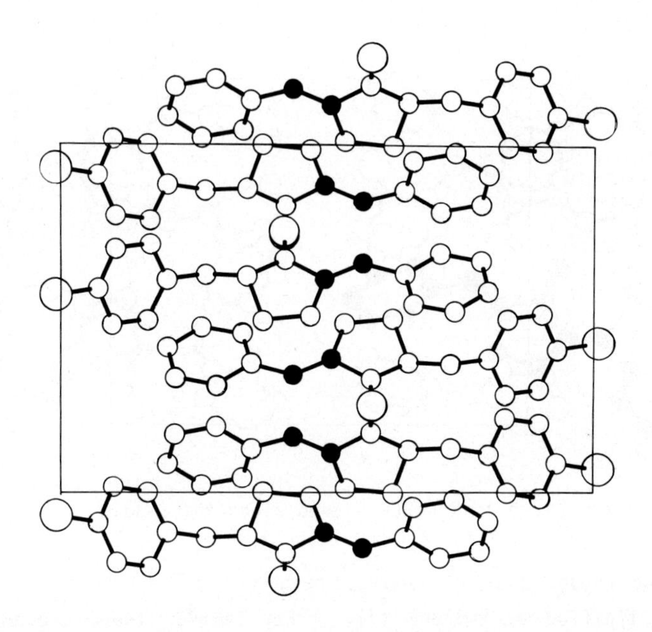

Fig. 10. The pClBBCP (8) structure viewed along the c-axis.

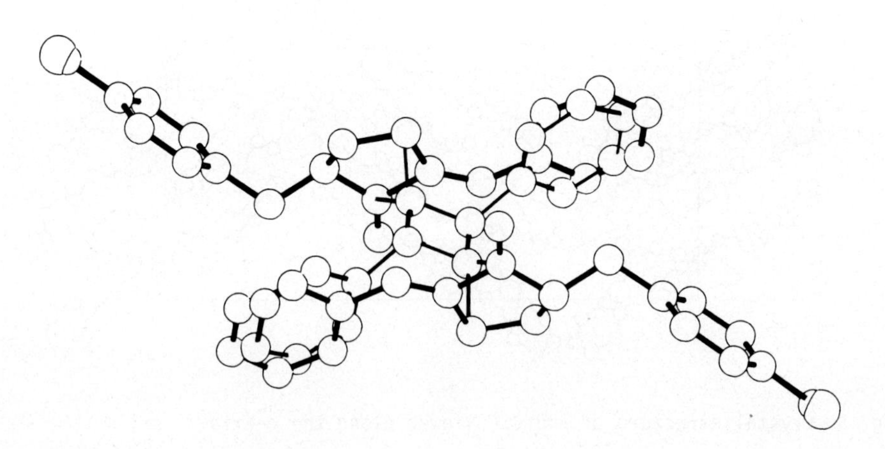

Fig. 11. Diagram illustrating the molecular structure obtained from an irradiated crystal of 8 after 20% conversion. Both monomeric and dimeric contributions for the benzylidene moiety are observed. In contrast the benzyl group refines to a single position.

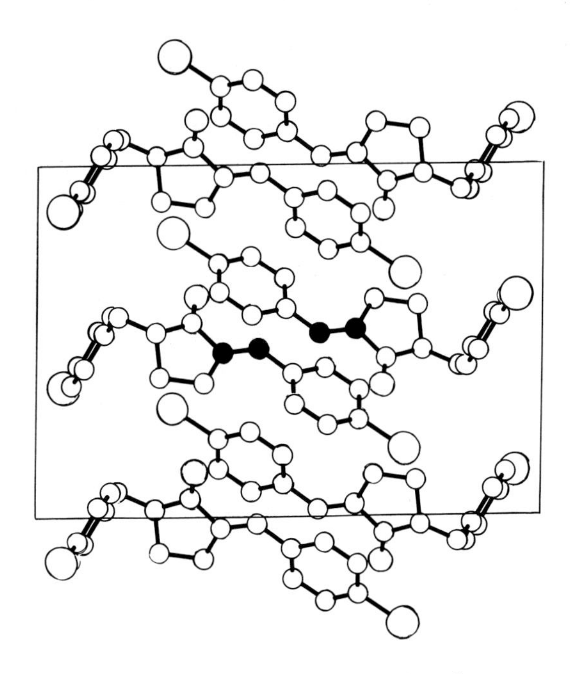

Fig. 12. The pClBpBrBCP structure viewed along the b-axis.

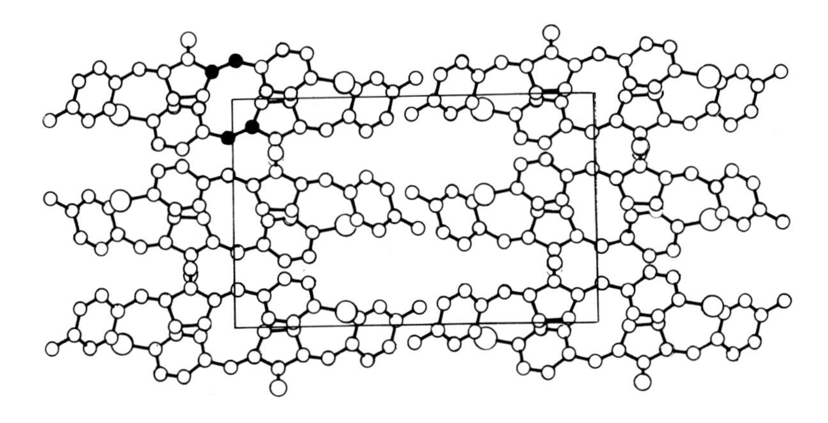

Fig. 13. The pMeBpBrBCP structure viewed along the c-axis.

4. COMPARISON OF pClBpBrBCP (11) AND pMeBpBrBCP (12)

Since BpBrBCP and pClBBCP are photoactive it might be expected that 11 might also be photoactive. This is not the case, however, and in 11 although nearest neighbour molecules are related by a centre of symmetry, the reactive centres are separated by 4.65Å (ref. 19). The structure for this compound is shown in Fig. 12. In this structure the dihedral angle subtended by the benzyl-phenyl group and the benzylidene-carbonyl moiety is 98^0. This is by far the largest encountered in this series of compounds. There is a significant difference in conformation about the C(2)-C(6) bond. In all the other BBCP derivatives the two hydrogen atoms attached to C(6) are trans to the C(2) hydrogen. In 11 the hydrogen atoms are guache. The peculiarity of this structure becomes more striking when it is compared with the methyl analogue 12. The structure of 12 is shown in Fig. 13.

There is a marked difference in conformation. What has been argued when enquiring of the origin of the large difference in structure of these two compounds is that an important role is played by the folding back of the p-chloro group in deciding the crystal packing. Detailed analysis of the non-bonded intermolecular contacts in the structure 11 reveals that the chloro-phenyl ring carbons are quite close to the bromine atom of the second molecule within the incipient dimer. The origin of this feature is ascribed to the electron withdrawing power of the chlorine atom leaving the phenyl ring carbons slightly postively charged. For the p-methyl derivative this interaction is less likely to be so important, and indeed the shortest contacts

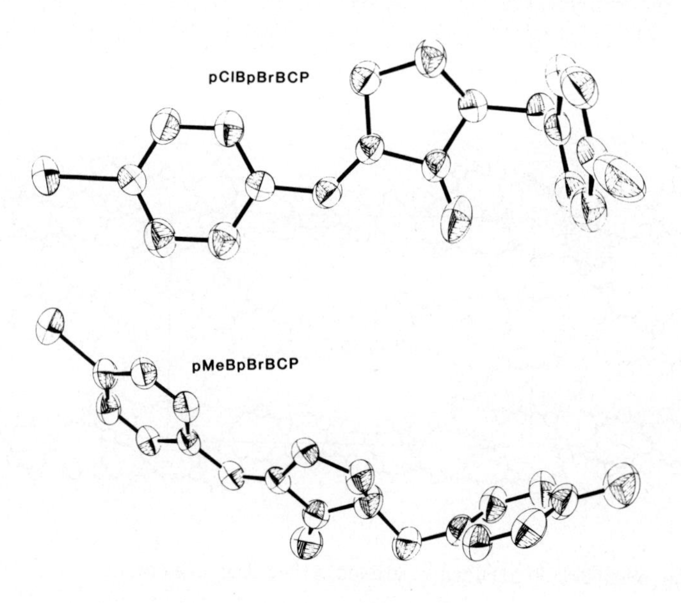

Fig. 14. ORTEP plots of pClBpBrBCP and pMeBpBrBCP.

for the ring carbon atoms and bromine within the pair are considerably shorter in 11 than in 12. The conformations of 11 and 12 are compared in Fig. 14.

Clearly from the viewpoint of crystal engineering in this series of compounds there is some support for a limited utilization of chloro/methyl exchange. In those cases, however, where electrostatic contributions may begin to dominate (e.g. in 11) an unambiguous prediction of structure may not be possible

5. MIXED CRYSTAL FORMATION AND STRUCTURAL-MIMICRY

Given that the crystal structures of 8 and 9 are isomorphous with very similar cell dimensions, it is not surprising that mixed crystals result from solutions of both in a chloroform/methanol mixture (ref. 22,23). In such crystals the components are randomly distributed over the lattice sites; there is no evidence for any superlattice formation. Varying composition for the crystallizing solution results in crystals whose cell dimensions are intermediate between those of the pure components, with the actual value dependent upon the ratio of the two within the crystal. Structural refinement yields the chloro/methyl ratio and suggests that a continuous substitution of chloro for methyl (and vice versa) is possible whilst retaining the same reactive packing motif. Fig. 15 show the packing diagram for the mixed crystal.

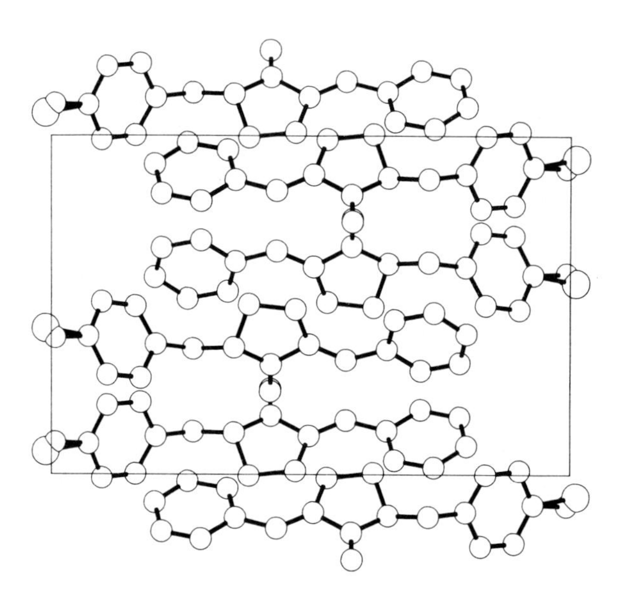

Fig. 15 Packing diagram along the c-axis of a pClBBCP/pMeBBCP mixed crystal.

For 11 and 12 when these were dissolved in an equal mixture of chloroform/methanol in a 9:1 ratio and the solution slowly evaporated, single crystals were obtained with cell dimensions which were slightly, but significantly, different from those of 12. X-ray intensity data and structural solution, phasing on the bromine position, showed that the atoms in the benzylbenzylidene-cyclopentanone framework were in a conformation very similar to that for 12. Further difference maps revealed both methyl and chloro substituents and showed that mixed crystals contained both compounds in a statistically averaged fashion.

The crystals were able to undergo topotactic and topochemical conversion upon irradiation to give asymmetrical cyclobutane products. For mixed crystals of 11 and 12 as well as 8 and 9, single crystals were studied during conversion. For example, a single crystal containing 66% of 9 and 34% of 10 was irradiated for 20 hours. After full data collection, the structure was solved by using as a starting model the monomer carbon framework, but excluding the atoms making up the exocyclic double bond, C(5) and C(13). These were located from difference maps at two sets of positions, the first corresponding to those for the monomeric crystal; the second corresponding to those of the product molecule. In subsequent difference maps, the whole benzylidene moiety (C(14)...C(19)) was resolved to two alternative positions; one corresponding to dimer and one to monomer.

6. THE SOLID STATE REACTIVITY OF DIBENZYLIDENECYCLOPENTANONE, 13.

The solid state reactivity of 2,5-dibenzylidenecyclopentanone, DBCP, (ref. 24) is of interest because by replacement of the C(2)-C(6) single bond present in BBCP by a double bond several features arise. (i) There is an additional reactive centre. (ii) The varied conformation open to BBCP and its derivatives is lost, see Fig. 16; DBCP is essentially a planar molecule.

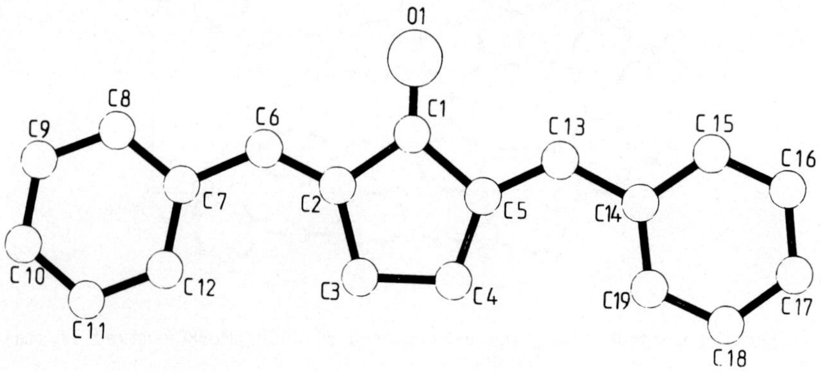

Fig. 16. Atom numbering scheme used for the DBCP molecule.

(iii) The removal of the chiral centre at C(2) will lead to an increased range of packing motifs - BBCP derivatives are synthesised in such a way that both optical isomers are generated and in the absence of spontaneous resolution crystallization into racemic space groups (i.e. Pbca and $P2_1/c$) takes place; for DBCP this will not necessarily be the case. (iv) Finally a new chiral centre may be subsequently introduced into the pentanone ring.

When single crystals of DBCP (recrystallized from chloroform/methanol mixture) are irradiated under a nitrogen atmosphere an amorphous product results. Analysis confirms that a number of products are produced, see Scheme 2. The major one (structure III in the Scheme) is that which might be expected to result from a topochemical reaction.

Within the parent crystal nearest neighbour molecules are related by translation along the b-axis, Fig. 17. The double bonds C(2)-C(6) and C(5')-C(13'), where ' indicates symmetry related molecules, lie in planes parallel to each other, but with the double bonds themselves not parallel, but rotated, as shown in Fig. 17, with respect to one another by 56^0.

This arrangement is not one which might have been considered conducive for a topochemical reaction since early work had suggested that a parallel relationship was necessary. However, the p_z orbital on C(6) is oriented in the direction of the corresponding orbital of C(5'). These are the orbitals which initially form the pi-component of the double bond and subsequently form part of the cyclobutane ring. It is probable, therefore, that upon photoexcitation interaction and overlap of these orbitals may occur with subsequent ring formation.

In addition to product III the product IV was also isolated from irradiated crystals. It has been suggested that this may result from the reaction between a DBCP molecule and the biradical V. Biradical V could be obtained within the perfect crystal by reaction of molecules related by a two-fold axis.

Observations by Kaupp and Zimmermann (ref. 25) on thin films of DBCP grown from methylene chloride or methanol solutions indicated that a third product was also produced, VI. No trace of this was observed from crystals obtained from chloroform/methanol. Indeed, within the structure shown in Figure 17 this product could only be produced by reaction between molecules related by the centring of the lattice, or between nearest neighbours after rotation of 135^0. Both of these processes are unlikely. X-ray studies of DBCP, however, recrystallized from methylene chloride indicated an additional phase which may be responsible for this product. Some evidence from NMR spectra supports this conclusion.

In DB(+)3MeCP nearest neighbour molecules are related by a two-fold screw axis, and the distance separating the carbon atoms of neighbouring double bonds, C(5)-C(13) and C(2')-(C6') is 3.87Å. This structure is shown in Figure 18.

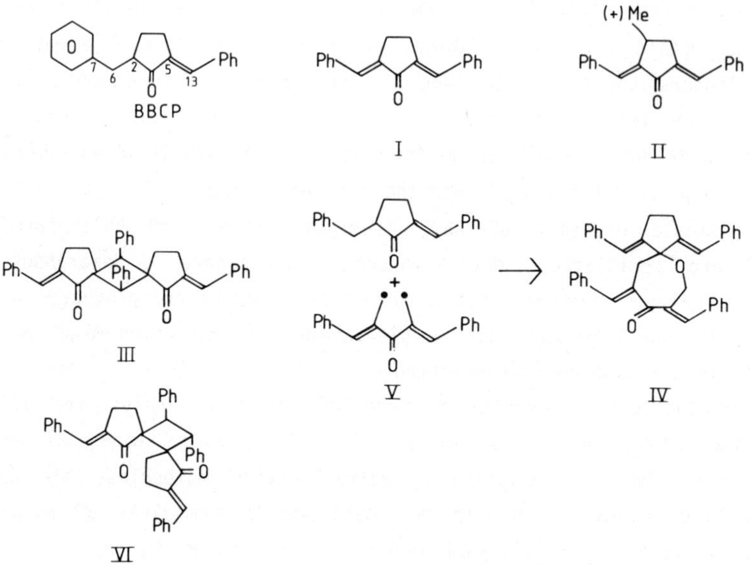

SCHEME 2

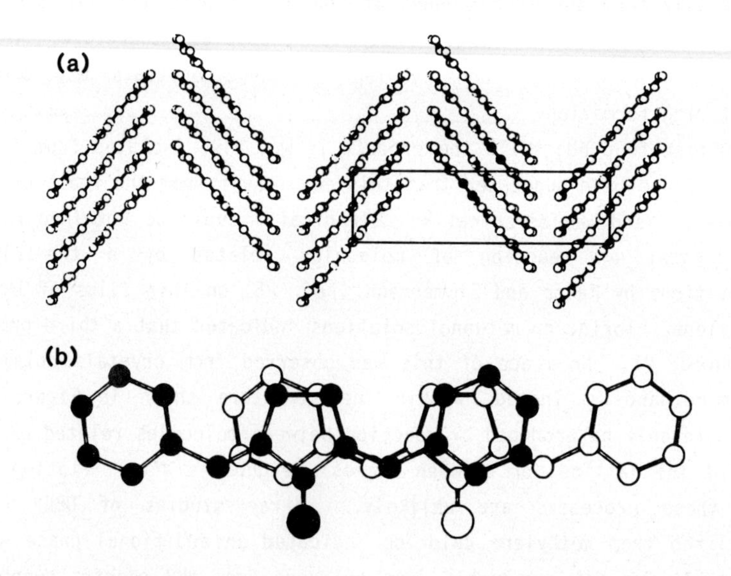

Fig. 17. DCBP structure viewed (a) along the z-axis and (b) at right angles to the mean molecular plane showing the geometry of the incipient dimer.

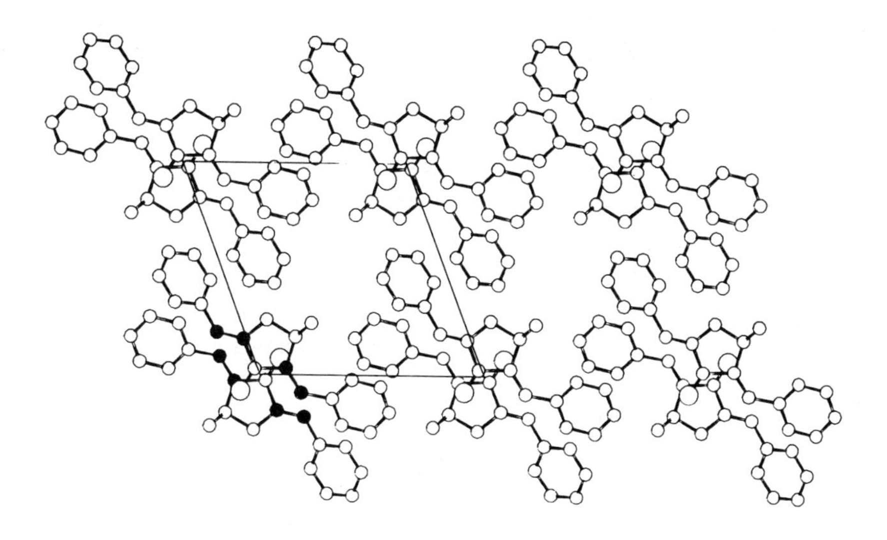

Fig. 18. Structure of DB(+)3MeCP viewed along the b-axis.

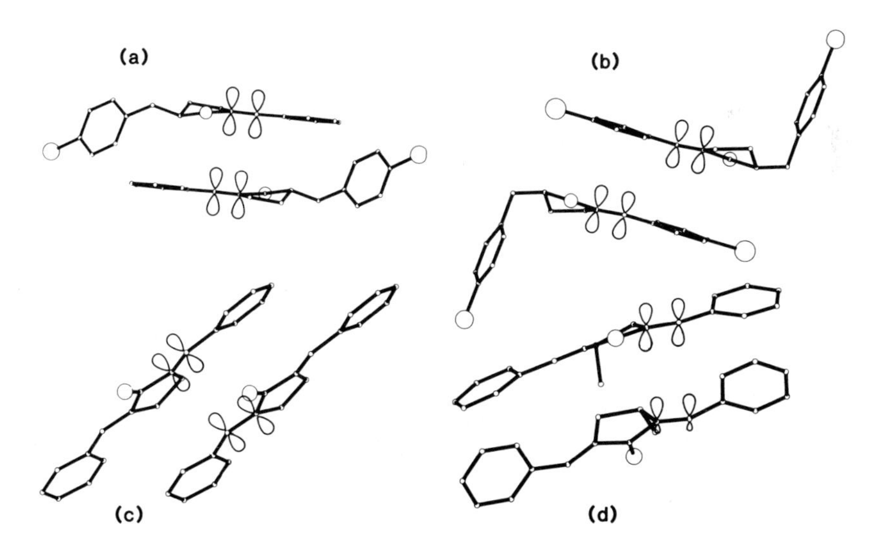

Fig. 19. Orbital overlap of potential incipient dimer pairs in (a) pClBBCP (b) pClBpBrBCP, (c) DBCP and (d) DB(+)3MeCP.

Although this distance is suitable for [2+2] cycloaddition, the crystal is photostable. This situation arises because the benzylidene groups to which these two bonds belong are not parallel. This prevents the necessary overlap of potentially reactive orbitals. No photoreactivity is therefore expected from these crystals and none is observed. This is in contrast to the case for DBCP, where the bond distance was similar, but overlap of reacting orbitals was possible in spite of the two bonds not being parallel, Figure 19.

7. THEORETICAL CONSIDERATIONS

Molecular orbital calculations within the MNDO approximation (ref. 26) were carried out using the benzylidene carbonyl-moiety (i.e. benzylidene methylketone) as a model for both BBCP, and to a lesser extent trans cinnamic acid. Since the moiety represents the photochemically active portion of the BBCP molecule, it was considered as an adequate model for the solid state photodimerization under investigation, since the nature of the excited state would be independent of the physical state of the sample. However, on determining the reaction coordinate certain geometric constraints were imposed upon the two reacting fragments, in order to model the situation that obtains in the solid state.

Thus, the ground state of the molecule was found to have a heat of formation of 10.86 kcal mol^{-1}. The maximum density for the HOMO was located on C(5), and for the LUMO on C(13), with the two lobes having the same phase. The geometry was in good agreement with that observed in the crystal structure of BBCP. The lowest excited singlet state was found to have a heat of formation of 42.60 kcal mol^{-1}, with similar disposition of the HOMO and LUMO as the ground state. The lowest excited state was found to be a triplet state with heat of formation of 42.30 kcal mol^{-1}. Again, maximum density for the HOMO was located on C(5), and for the LUMO on C(13), but the two contributions had opposite phases.

The very similar energies of the lowest singlet and triplet excited states mean that transition from the former to the latter is extremely facile. Furthermore, it is clear that the excited triplet state thus formed, would be vibrationally excited. This can be correlated with the so-called phonon assistance of the reaction previously inferred (ref. 27). The molecular orbital symmetry is such that reaction between two ground state molecules or between one in the ground and in the singlet excited state is not allowed. On the other hand reaction between a ground state molecule and one in the lowest excited triplet state is allowed. Thus, the facility of energy transfer between states is crucial to the reaction occurring under topochemical control. The lifetime of the triplet state for DBCP as measured from the phosphorescence in emission spectra at 77K (ref. 28) was only 200 μs. The brevity of the lifetime of the excited state (and hence the apparent absence of any energy hopping) may also be related to the shape of the excited state. This is almost identical to that of one half of the dimer species. This change in shape compared with the monomer has two consequences: firstly, it makes energy hopping less likely, and secondly it brings the reactive centres closer together. The speed of reaction is also related to the fact that the "transition state" is clearly more akin in shape to the product than the reactants. In contrast, the situation that exists in anthracene, where energy hopping does occur, the excited anthracene molecule has a shape very similar to the ground state molecule. The symmetry of the orbitals in the triplet excited and in the ground states indicates that both the "head" to "head" and "head" to "tail" reactions are intrinsically possible. Further, the cycloaddition has to be a non-conserted process, since only one pair of orbitals of the two involved are initially of the correct symmetry. Further studies along these lines are currently in progress.

REFERENCES

1 G.C. Forward and D.A. Whiting, J. Chem. Soc. (C), (1969) 1868-1873.
2 D.A. Whiting, J. Chem. Soc. (C), (1971) 3396-3398.
3 G.M.J. Schmidt, Pure Appl. Chem., 27 (1971) 647-678.
4 M.D. Cohen, Angew. Chemie Int. Ed. Engl., 14 (1975) 386-393.
5 W. Jones, J. Chem. Res. (S), (1978) 142-143.
6 H. Nakanishi, W. Jones, J.M. Thomas, M. Hasegawa and W.L. Rees, Proc. Roy. Soc. Lond., A369 (1980), 307-325.
7 W. Jones, H. Nakanishi, C.R. Theocharis and J.M. Thomas, J. Chem. Soc. Chem. Comm., (1980) 610-611.
8 H. Nakanishi, W. Jones, J.M. Thomas, M.B. Hursthouse and M. Motevalli, J. Phys. Chem., 85 (1981) 3636-3642.
9 H. Nakanishi, W. Jones and J.M. Thomas, Chem. Phys. Lett., 71 (1980) 44-48.
10 H. Nakanishi, W. Jones, J.M. Thomas, M.B. Hursthouse and M. Motevalli, J. Chem. Soc. Chem. Comm., (1980) 611-612.
11 W.L. Rees, M.J. Goringe, W. Jones and J.M. Thomas, J. Chem. Soc. Faraday Trans.II, 75 (1979) 806-809.
12 C.R. Theocharis and W. Jones, J. Chem. Soc. Faraday Trans. I, 81 (1985) 857-874.
13 G. Wegner, Angew. Chemie. Int. Ed. Engl., 20 (1981) 361-381.
14 Y. Ohashi, K. Yanagi, T. Kurihara, Y. Sasada and Y. Ohgo, J. Am. Chem. Soc., 104 (1982), 6353-6359.
15 N.W. Thomas, Ph.D. Thesis, "Structure and Reactivity in Organic Crystals", University of Cambridge, 1983.
16 S. Ramdas, H. Nakanishi, W. Jones and J.M. Thomas, Unpublished Results.
17 S.K. Kearsley and G.R. Desiraju, Proc. Roy. Soc. Lond., A397 (1985) 157-181.
18 W. Jones, S. Ramdas, C.R. Theocharis, J.M. Thomas and N.W. Thomas, J. Phys. Chem., 85 (1981) 2594-2598.
19 C.R. Theocharis, W. Jones, M. Motevalli and M.B. Hursthouse, J. Cryst. Spec. Res., 12 (1982) 377-389.
20 C.R. Theocharis, Ph.D. Thesis, "Studies of Oriented Organic Molecules". University of Cambridge, 1982.
21 N.W. Thomas, S. Ramdas and J.M. Thomas, Proc. Roy. Soc. Lond., A400 (1985) 219-227.
22 W. Jones, C.R. Theocharis, J.M. Thomas and G.R. Desiraju, J. Chem. Soc. Chem. Comm., (1983) 1443-1444.
23 C.R. Theocharis, W. Jones and G.R. Desiraju, J. Am. Chem. Soc., 106 (1984) 3606-3609.

24 C.R. Theocharis, W. Jones, J.M. Thomas, M. Motevalli and M.B. Hursthouse, J. Chem. Soc. Perkin Trans II, (1984) 71-76.
25 G. Kaupp and I. Zimmermann, Angew. Chem. Int. Ed. Engl., 20 (1981) 1018-1019.
26 E.L. Short and C.R. Theocharis, Submitted for Publication.
27 J. Swatkiewicz, G. Eisenhardt, P.N. Prasad, J.M. Thomas, W. Jones and C.R. Theocharis, J. Phys. Chem., 86 (1982) 1764-1767.
28 C. Brauchle and C.R. Theocharis, Unpublished Results.

Chapter 3

THE PREDICTION OF CHEMICAL REACTIVITY WITHIN ORGANIC CRYSTALS USING GEOMETRIC CRITERIA

Simon K. Kearsley

1 BACKGROUND TO TOPOCHEMISTRY

Topochemistry embraces the study of chemical reactions that occur within solids, and it is the nature of these reactions that is examined in this chapter. Most of these investigations have involved bond-forming reactions and isomerisations within organic molecular crystals. The key distinguishing feature of a topochemical reaction is the way the initial solid state environment, for instance the molecular packing, dictates both the possibility of reaction and the structure of its products. Reactions that have product molecules related to the geometric arrangement of the molecules in the reactant crystal structure would be an example of the topochemical phenomenon. Such studies of reactivity focus on only the initial geometric relationship of the molecules and it is only this that is considered to have a bearing on the eventual outcome of photo-excitation or heating. This is because reactions have been regarded simply as a manifestation of the topochemical phenomenon and efforts were primarily directed toward controlling and rationalising the structure of the crystalline environment for these molecules. The many examples of topochemical reactions (Schmidt 1967) indicate that the initial geometric configuration is of paramount importance and suggest that mechanism is merely secondary.

It is questioned whether the products of solid state reactions can be predicted or even rationalised solely in terms of the atom positions derived from crystal structure data. If this were so, there should be a perfect correlation between the geometric arrangement of the reactants within the lattice and the molecular geometry of the products. On the other hand, reactions may be considered as a function of molecular orbital interactions where a potential flow of electrons dictates the possible reaction products. It is emphasised that structures that have their component atoms defined by maxima in electron density (point atoms) will give little or no immediate indication of the distribution of electron density within a particular molecular orbital. The orbital orientation is implicit in the geometry of atomic positions; however, factorising the electron density into the occupied molecular orbitals can only be done with great difficulty. If reactions are dependent on molecular orbital overlap there can be no such absolute correlation. However, it is tempting to determine the extent to which geometric or topochemical criteria explain such solid state reactions.

For many reasons the [2+2] photocycloaddition is by far the most studied topochemical reaction (Schmidt 1967 and references therein) The orientational prerequisites for double bonds follow from the two most common symmetry elements in crystals, those of translation and inversion. Perhaps this is the reason such [2+2] reactions are prevalent. Double bonds contribute to areas of conjugation in molecules and such areas are generally planar. These planar regions elicit stacked packing modes in molecular crystals and thereby ensure most of the orientational needs for [2+2] reactions. For instance, for β-structures where the short crystallographic axis is around 4 Å, one can readily deduce that the reacting atoms are the same distance apart. Furthermore, the orientation of the reacting orbital lobes will be nearly optimal. (α-structures are ones where the reacting molecules are related by a centre of symmetry and γ-structures are all those that are unreactive.) The main attention of this chapter will be on examining such [2+2] reactions.

2 REACTIONS WITHIN A DIFFUSIONLESS ENVIRONMENT

In an attempt to delineate the extent to which topochemical principles are applicable to chemical systems other than molecular crystals, the following examples are examined. The discussion of these examples will centre on what specifically defines a topochemical reaction and the features that distinguish it from other types of reactions. Consider the following reactions:

(i) The [2+2] cycloaddition of benzylbenzylidenecyclopentanone (BBCP, Fig. 1) monomers to a dimer within the bulk of a single crystal (Thomas 1981) and the polymerisation of distyrylpyrazine (Nakanishi et al. 1979) are examples of reactions where there exists a direct relationship between the geometric configuration of the monomers in the lattice and the molecular structure of the product compounds. Furthermore, the reactions undergo a single-crystal to single-crystal transformation where there are also relationships between the crystal structures of monomer and product molecules (Nakanishi et al. 1979; 1980; 1981a).

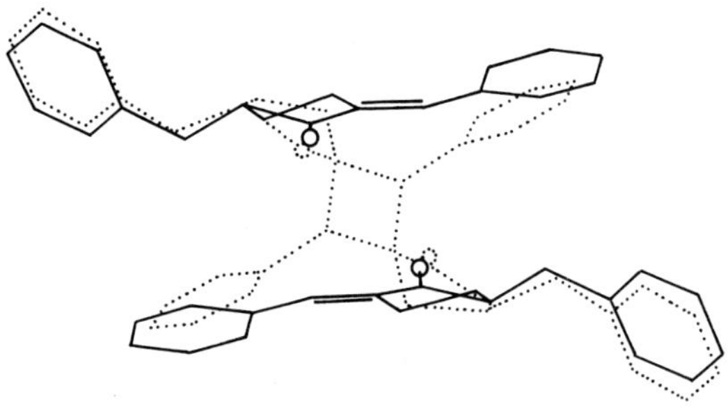

Fig. 1. The crystallographic configuration of two molecules of benzylbenzylidenecyclopentanone is depicted by the solid lines. Superimposed is the structure of the [2+2] dimer drawn with dotted lines (Thomas 1981).

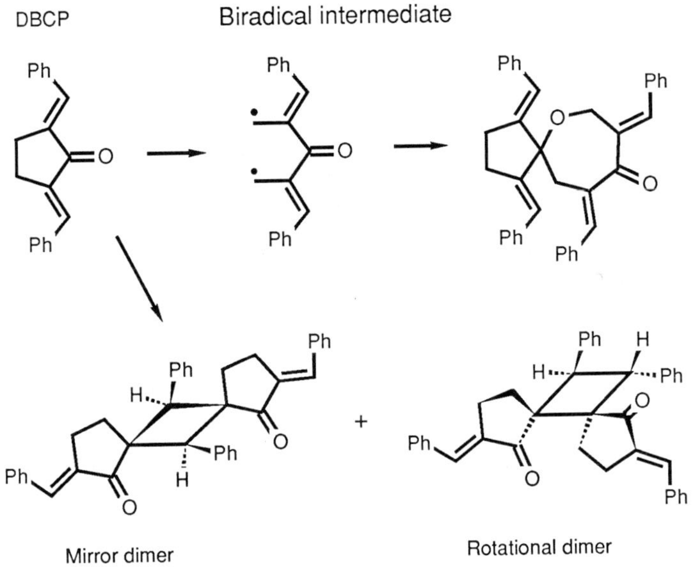

DBCP

Biradical intermediate

Mirror dimer

Rotational dimer

Scheme 1

(ii) The irradiation of dibenzylidenecyclopentanone (DBCP) in the crystalline state (Theocharis et al. 1983) yields several products (Kaupp and Zimmermann 1981); two of these products have been positively identified as having a stereochemistry which can be rationalised in terms of the reactant crystal structure. This is not so for the third major product which has a relative yield of 8%

(Scheme 1). There is no relationship between the product crystal structure and that of the starting material. As the reaction proceeds the crystal deteriorates due to the build-up of strain in the lattice.

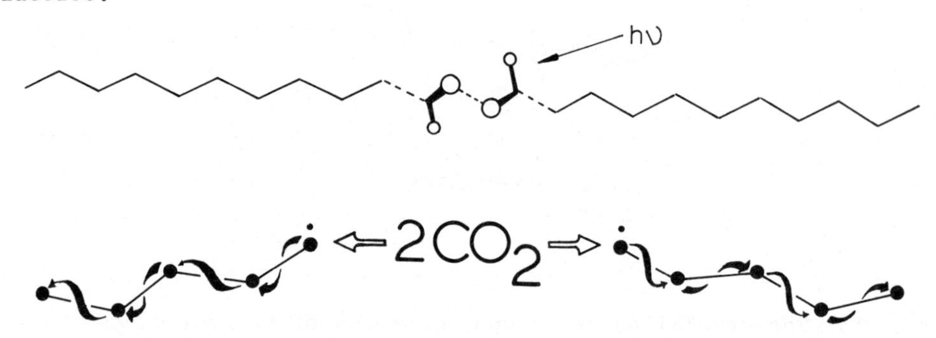

Fig. 2. The upper drawing of diundecanoylperoxide, viewed down the crystallographic two fold axis, shows which bonds are broken when the compound is irradiated with UV light. The lower drawing shows how the two CO_2 molecules generated by the reaction push the decyl radicals apart. The proposed motion of the aliphatic chain is such that each carbon atom in the chain replaces the adjacent atom by simultaneous rotation around and translation along the chain axis.

(iii) The photolysis of diundecanoylperoxide (McBride 1983) releases two molecules of CO_2 within the lattice. The strain generated by the presence of the CO_2 molecules is resolved by gross molecular movement on the part of the aliphatic chain. The postulated mechanism of movement involves a concerted-rotational motion about the aliphatic chain axis rather than an arbitrary buckling in the vicinity of highest strain area, and the motion has been determined to be a direct consequence of the surrounding molecules (Fig. 2). The final product is again related to the geometry of the reactant lattice once the CO_2 has diffused away from the reaction site.

(iv) The monomerisation of dianthracene and subsequent dimerisation in a rigid polymer matrix or in single crystals of dianthracene are diffusionless reactions (Tomlinson et al. 1972, Ferguson and Mau 1974). The ease with which dimerisation occurs is due to the arrangement of monomers held by the polymer or dianthracene matrix.

(v) Many examples of heterogeneous catalysis on crystalline surfaces show remarkable specificity. Breaking the catalytic process into diffusion and static stages, it is apparent that the catalytic steps occur when the substrate is fixed in an ordered fashion on the catalyst. Reference is being made to chemisorbed species where the thermodynamic aspects of the mechanism are important; that is, the catalytic environment supplies a new mechanism for a particular reaction in addition to the rate enhancement. Physisorbed processes require the catalyst more for energy transfer purposes; the topography of the catalyst environment is not as important since in many cases multiple layers of substrate form on the catalytic surfaces. More specifically, the nature of the surface topology and hence the binding sites decides the course of reaction. Key factors required are the structure of the catalytic surface and the thermodynamic binding energies of the reactants and products to the catalyst.

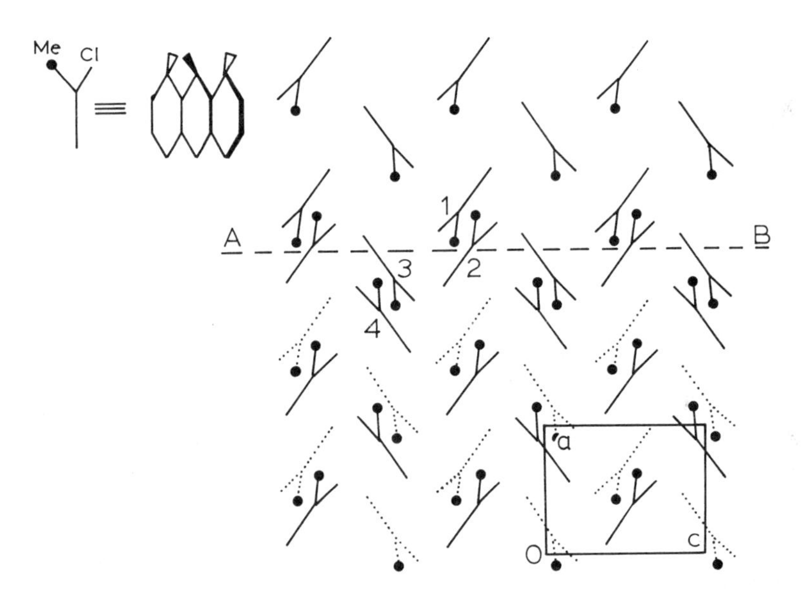

Fig. 3. A schematic depiction of a planar fault in 1,8-dichloro-9-methyl-anthracene (outline taken from Ramdas et al. 1977). The molecules drawn with solid lines are in the same plane. The ones that are dotted were in the same plane as the ones on the upper side of the fault line but have been displaced by 1/2[010] (into the page). The trace of the planar fault is shown by the dashed line AB. A displacement vector of 1/2[010] on the (100) plane brings the molecules 1 and 3 into close contact with molecules 2 and 4; the pairs 1/2 and 3/4 are incipient dimers.

(vi) In enzyme reactions, excluding the diffusion stages, the environment supplied by the protein backbone restricts the conformation and orientation of the substrates so that one specific reaction dominates. The enzyme is not a totally static medium and adjusts its conformation in subtle ways during reaction. Here, such cooperative effects during reaction are key factors that determine the pathway taken.

(vii) The reactions of anthracene (Cohen et al. 1971), 9-cyano-anthracene (Kawada and Labes 1970) and 1,8-dichloro-9-methylanthracene (Desvergne et al. 1975; Ramdas et al. 1977) occur at structurally defined dislocations in the crystal lattice (Fig. 3).

(viii) Mutagenesis by formation of thymine dimers in the DNA base sequence is a consequence of the fixed positional and orientational relationship of juxtaposed thymine bases attached to the sugar phosphate backbone (Haseltine 1983). The reactions in the crystalline state of thymine analogues and derivatives have been studied in detail by Frank and Paul (1973) and references therein.

2.1 Semantics of the solid state environment

If one considers thermodynamic rather than kinetic aspects of the aforementioned reactions, their mechanisms can be seen to have several features in common. The structure of the reaction site is well defined on a molecular scale. The reactants are "set" by whatever mechanism into orientations that are favourable for reaction. A consequence of the last factor is that the entropy of activation for the reaction is significantly altered. One can postulate that these reactions occur with the minimum of atomic and molecular movement, which is a consequence of the geometric constraints "set" by the structure of the reactant environment. The topochemical postulate states that: reactions in solids occur with the minimum of atomic or molecular movement (Cohen and Schmidt 1964). The extent of the applicability of the topochemical postulate depends upon how "solid" is defined and to what degree an atomic or molecular movement is considered to be minimal.

If the definition of a solid is taken to be a diffusionless environment, then minimal movement of atoms and molecules is implied since there are no random translational motions of the reactant. The only motion occuring arises as a consequence of the changing conformation of the molecules and rehybridisation of the atoms during reaction. Seven of the above examples fit into this definition of environment. Note that the active site of an enzyme can be considered to be the diffusionless environment for the substrate.

Reactions on catalysts are excluded from the list of examples if "in solids" is specified since the reaction in most cases occurs at the phase interface. If solids are also defined as having periodic properties, then topochemistry takes on a much narrower meaning. Crystallographic restrictions exclude polymers and enzymes (examples iv, vi and viii). Reactions that are not a result of the bulk periodicity in the solid, but rather the aperiodic environment of defects are similarly excluded. It can also be argued that mechanisms at defects are due primarily to kinetic phenomena.

The proposed recoiling mechanism in photolysed molecules of diundecanoylperoxide (Fig. 2) is not considered to be within the definition of minimum atomic or molecular movement since the atoms that are eventually bonded move away from each other at the initial stages of reaction. The exclusion of the above reaction makes it difficult to determine what is to be considered minimal atomic and molecular movement; for instance, for the dimerisation products generated by reaction of DBCP monomers (Scheme 1), one can imagine that they occur with atomic movements of similar magnitude to the peroxide.

The topochemical postulate is intrinsically very broad. However, traditionally, topochemistry has been restricted to reactions which only include (i) and possibly (ii). It should be mentioned that the definition of topochemistry has been developed crystallographically as a result of studies of reactions in categories (i) and (ii). However, reactions that were excluded because the reaction environment was not defined crystallographically or the movement of participating atoms was considered too large can still be explained by the topochemical postulate.

2.2 Energetics of reactions in the organic solid state

Some of the differences which distinguish reactions in solids from those in fluids include extreme specificity and quantitative yield of the product. We turn our attention to a comparison of the thermodynamics and kinetics of reaction processes in solids and fluids for an explanation.

Figure 4(A) shows a general potential energy profile of reactants going to products in fluids and solids. The diagrams suggest that the energy liberated by reactions in solids is not as great as for fluids; this is because the product molecules do not have the same shape as the cavity left by the reactant species and therefore will require energy to deform the original environment. Furthermore, the activation energy will probably be greater in solids by similar reasoning. From the above, which indicates that solid state

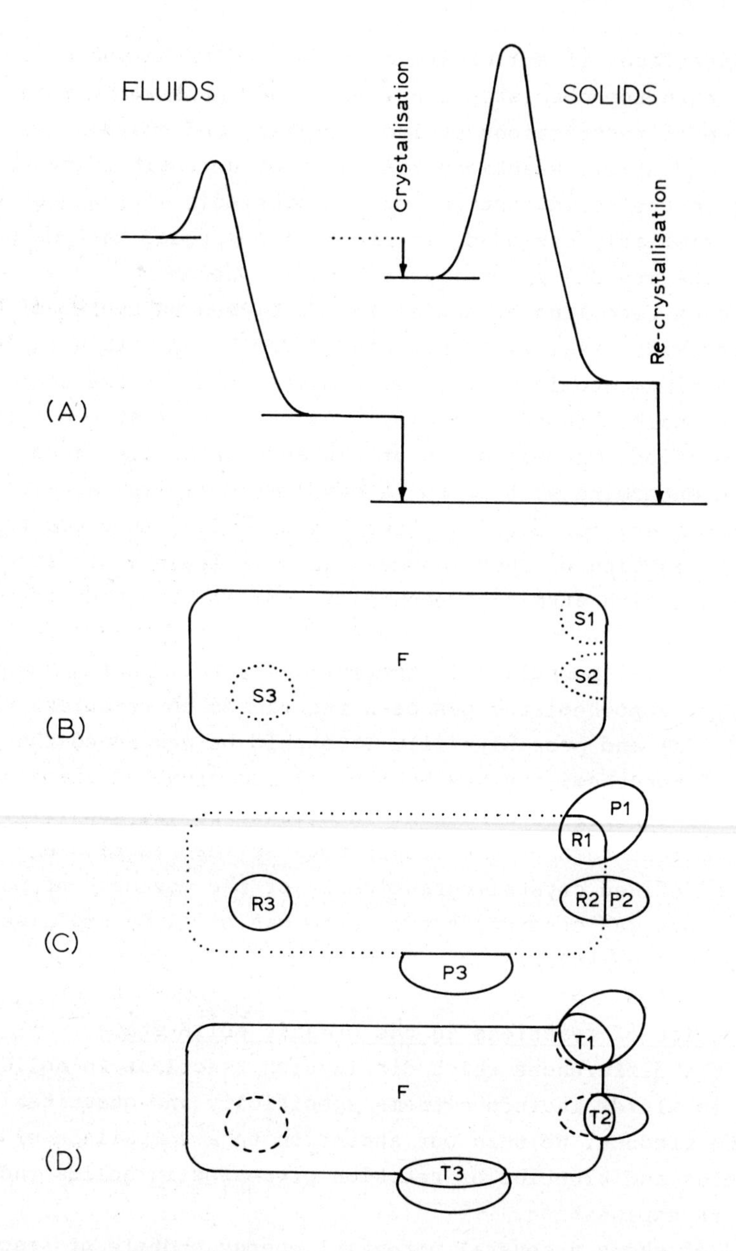

Fig. 4. (A) Comparison of the reaction coordinate for a reaction occuring in fluid or solid states. Schematic drawings (B, C and D) show how the entropy of activation changes for fluids and various solid phases. For full details about these diagrams refer to section 2.2 of the text.

reactions are not as enthalpically viable as reactions in fluids, it is clear that this is not an adequate explanation for the types of reactions seen in solids.

There are two important kinetic aspects to consider, those of the translation and orientation of the reactant molecules. That is, how far do the reactant molecules have to travel to reach one another and will they collide favourably. In solids, reactant molecules are at their most concentrated; therefore the solid phase constitutes the limiting case for kinetic parameters dependent on concentration of reactants. More importantly, it is the orientation of the reactant molecules that differs in fluids and solids. Fluids have a random collisional probability associated with the rate of reaction, whilst in solids the reactant molecules are pre-oriented.

Figure 4(B) shows schematically the accessible intermolecular configurations of the reactant molecules on the reactant potential surface. Some of these configurations correspond to solid state arrangements (S1, S2 and S3). These portions of the potential surface will invariably be deep, narrow wells. The accessible intramolecular configurations (R1, R2 and R3) for the corresponding solid state intermolecular configurations are shown in Figure 4(C). Adjacent to some of these are the product molecule configurations P1 and P2. For instance, region R1 connects region P1 via a smooth change in configurational geometry and may do this in the solid if the region R1 overlaps the region S1 of the intermolecular configurational surface. Note, product P3 cannot be produced in the solid state as there are no appropriate solid phase configurations. In contrast, solid phase S3 is stable as there are no viable intramolecular routes to products without substantially changing the intermolecular geometry. Combining the two surfaces in Figure 4(D) we see that there are small channels (T1, T2 and T3) that will lead to product molecule configurations (P1, P2 and P3). The small channels correspond to transition state configurations of the reactant molecules. In fluids the accessible area on this potential surface is vast but effectively bounded for finite concentrations; this is denoted in Figures 4(B) and 4(D) by the oblong region marked F. Therefore the probability of the reactant molecules having transition state geometries is reduced. In the solid state, the accessible configurational area is extremely small (areas S1, S2 and S3); if these areas include the transition state configurations then the probability of reaction is very high since the areas S and T are comparable.

The salient difference between topochemical reactions and reactions in fluid media is usually the difference in the entropy of activation; this is related to the ratio of the areas F:T and S:T in Figure 4(D). The "all or none" feature of topochemical reactions is then explicable. The accessibility of both T1 and T2 is feasible in fluids, but only one, the other or neither are accessible in solids depending on the phase (S1, S2 or S3).

Thermally activated reactions in solids with accessible transition states will always proceed to an equilibrium. The [2+2] cycloadditions are thermally disallowed, but if the molecules are shunted into excited state manifolds, then reactions can ensue. For photochemical reactions there are competing processes, foremost of which are internal conversions to other forms of energy not contributing to the reaction. In solution cinnamic acids do not have time to dimerise as their excited states are not long lived enough to allow for favourable collisions, while in the solid they are pre-oriented and collide effectively. Thus the solid medium provides a route to product molecules not normally accessible in solution and an inherent selectivity; for instance, in the solid, 1-chloro-anthracene leads only to the centrosymmetric dimer whilst in solution both isomers are produced; see Thomas et al. (1977) and references therein.

It becomes apparent even on thermodynamic grounds there is still little basis for distinguishing among the reactions in the eight examples given in the previous section; the only difference is the manner in which they attain their static environment; in some cases, (e.g. (v), (vi) and (vii)) kinetic effects are paramount. Therefore, this chapter treats as topochemical only reactions in which the molecular structure of the product can be explained in terms of the perfect lattice geometry of the reactant with electronic and kinetic effects playing seemingly secondary roles.

3 MECHANISM OF REACTION

Little is known in general about the way in which these solid state photoreactions proceed. The reasons for this are due to the problems in probing the solid state experimentally and identifying what the molecule is doing during reaction. However, McBride (1983) has demonstrated that ESR and FTIR can be used to give information about the reaction centre. Electron microscopy has substantiated mechanisms of propagation and the role of crystalline imperfection on reactions (Jones and Thomas 1979). Relationships between possible reaction coordinates and local breakdown in packing density have also been investigated (Gavezzotti 1983).

Principles applicable to the fluid state can be shown to be useful in the discussion of the likely events during a solid state reaction. Such principles are those of "product stability" and "least motion" - either electronic or nuclear (Hine 1977). Hine (1972) has combined these principles and has stated that: "The making and breaking of bonds will tend to take place in separate steps except where the formation of a sufficiently unstable intermediate may be avoided by combining several such bond changes into one concerted step." As a corollary, there will be many reaction paths available but depending upon the chemical environment only a few will be preferred. In the solid state, it is suggested that topochemical factors determine which reaction path is to be taken.

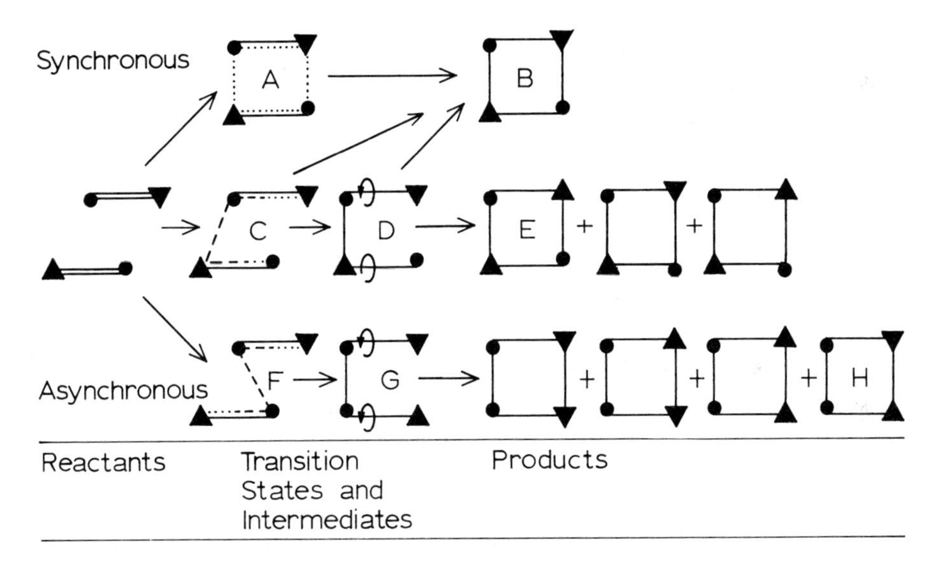

Fig. 5. Some possible [2+2] reaction mechanisms are depicted schematically. The orientation of the reactant double bonds is assumed to be fixed as they would be in solid state. The orientation of the triangles and circles denotes the regiochemistry and stereochemistry of the carbon atoms.

Figure 5 shows two extreme types of bond-making processes for a [2+2] cycloaddition, the concerted and the biradical route. There are also intermediate mechanisms which are the combination of each of the extremes to various degrees. Note that if prohibitively high energy is required to reach the transition state to the biradical intermediate, then the reaction is more likely to go synchronously.

As a result for concerted reactions there is intrinsically more regiochemical and stereochemical control. In a reaction that involves a biradical, more than one product could be expected because the intermediates can change their configuration. This is shown in Figure 5 where the products will have triangles placed at different corners of the squares. Note that the topochemical postulate asserts that reactions in solids occur with the minimum of atomic or molecular movement (Cohen and Schmidt 1964). One implication of this is that favoured reactions are the ones that disrupt the lattice to a lesser degree, hence this places a restriction on the type of mechanisms but does not suggest them. It is therefore distinct from the "least nuclear motion" hypothesis (Hine 1977) which does suggest reaction routes. Thus, because of the spatial control afforded by the crystalline environment, the reaction routes are most probably confined to: a -> b or c -> d -> b since by both these mechanisms there is least molecular movement over the entire reaction.

4 GEOMETRIC CRITERIA FOR REACTION OF DOUBLE BONDS

In general, the requirement for the reaction of double bonds is that their p(z) orbitals be nearly colinear. It is also preferable to have the double bonds (σ-bond component) in a parallel arrangement. This is achieved in solution by diffusion, such that at one particular instant the optimal configuration is met, and if the energy is available reaction will ensue. In the lattice, parallel configurations are generated by certain packing arrangements; where molecules are related by translation or inversion symmetry operations the double bonds will be parallel. The threshold value for maximal separation of double bonds for which reaction of the [2+2] type may occur is considered to be around 4.2 Å (Schmidt 1971). The distance of 4.2 Å refers to the distance between positions of the pre-bonding atoms. In cases where the double bonds are askew then both pre-bonding atom-atom distances are quoted. Geometric constraints can be formulated for other types of bond forming reactions in the solid state, although there are many more examples for the [2+2] cycloaddition.

Already, there has been some revision of the above distance constraint criterion (Theocharis et al. 1983; Gnanaguru et al. 1984), and although the reactivity of many compounds that undergo [2+2] cycloadditions in the solid state can be successfully rationalised on the above basis, there are exceptions. It is the compounds that

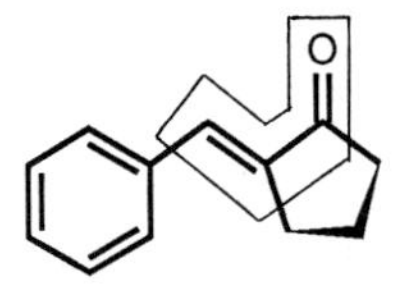

Benzylidenecyclopentanone (BCP)
(Kearsley & Desiraju 1985)

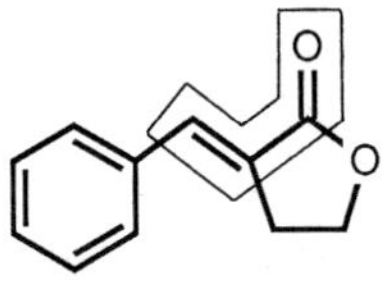

Benzylidenebutyrolactone (BBL)
(Kearsley & Desiraju 1985)

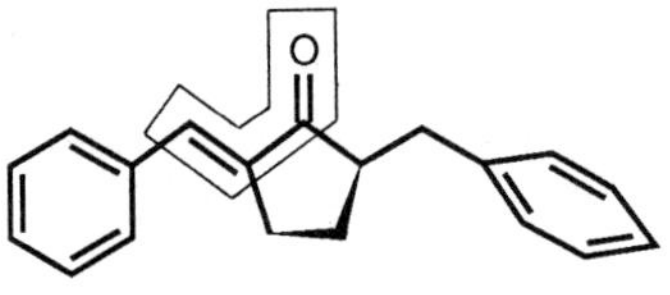

Benzylbenzylidenecyclopentanone (BBCP)
(Nakanishi et al. 1981)

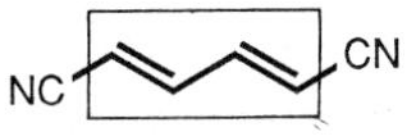

Muconodinitrile
(Filippakis et al. 1967)

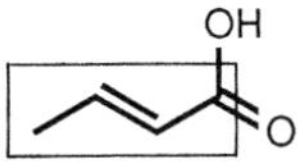

Crotonic acid
(Shimizu et al. 1974)

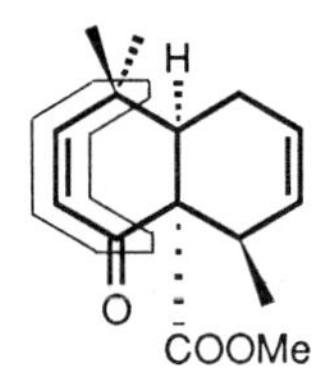

Napthalenone derivative (NAPTH)
(Ariel & Trotter 1984)

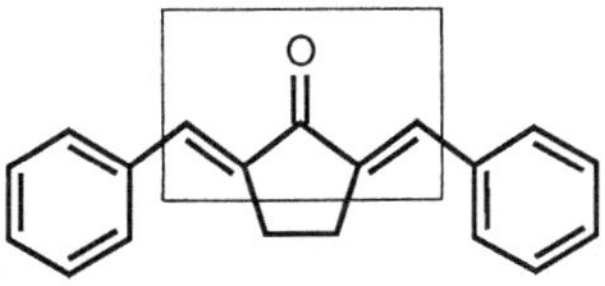

Dibenzylidenecyclopentanone (DBCP)
(Theocharis et al. 1983)

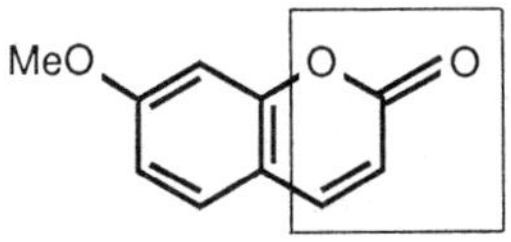

7-Methoxycoumarin (7MC)
(Gnanaguru et al. 1984)

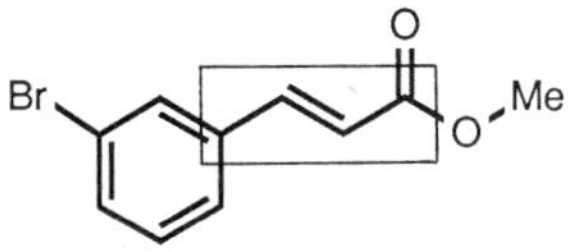

Methyl-m-bromocinnamate (mBrCA)
(Leiserowitz & Schmidt 1965)

Scheme 2

flout this topochemical rule that will be the main subject of the rest of this chapter.

4.1 Reacting orbitals

It has been pointed out by Frank and Paul (1973) that "the distance between reacting carbons is not necessarily the most important factor in permitting photochemical reactions and a sometimes more important factor is the orientation of the participating orbitals". Given a knowledge of the structure, a measure of the orbital configuration is implied by quoting relevant geometric data (Begley et al. 1978; Theocharis et al. 1982). The amount of orbital overlap can be quantified by elaborating on distances and angles between atoms and planes. Any unique combination of distances and angles which specifies the position and direction of potentially reactive orbitals (double bonds) is sufficient to predict photoreactivity, provided of course, that spatial proximity is the only criterion for reaction.

Figure 7. Legend
The following three pages show a schematic depiction of the orbital overlap for nine of the compounds in Table 1. These diagrams are constructed from the crystal data. Two projections are shown for each compound. The one on the left shows a view perpendicular to both the axis through the carbons connected by the double bond and the z-axes of p(z) orbitals. On the right the projection is along the z-axes of the reacting p(z) orbitals. For the projections on the left, an arc, more properly a spherical shell, shows the 4.2 Å boundary; this is drawn with the midpoint of the lower double bond as the centre of the sphere which contains the arc depicted. Scheme 2 shows the conjugated fragment of the molecule drawn for each compound; the plane defined by these atoms will be referred to as the plane of conjugation. The solid circles in the projections on the right show the intersection of the plane of conjugation with the spherical shell. Thus the radius of this circle is a function of the plane separation and shows the 4.2 Å limit for this separation. The dashed lines indicate the van der Waals envelopes for these fragments. See section 4.2 for details on how the p orbitals are drawn.

An simple numerical description for the overlap of the reacting orbitals is suggested. This formulation is still only dependent on the coordinate geometry of the atoms. This is the distance between the apices of the reacting p(z) atomic orbital lobes of carbon. The p(z) orbital of carbon is assumed to be participating in π-conjugation and not in any σ-bonding. The apex of the p(z) orbital lobe was found by a geometric construction deduced from the atomic con-

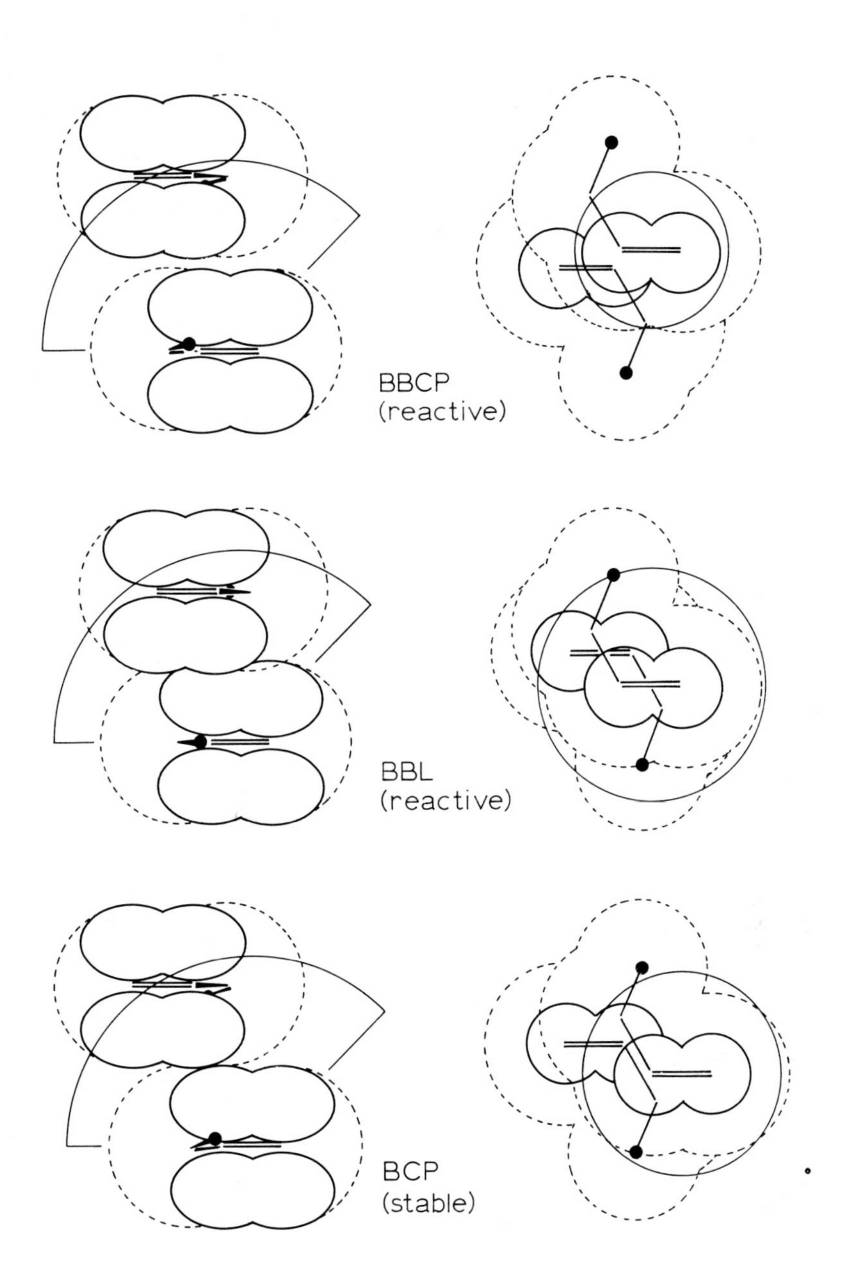

Fig.7

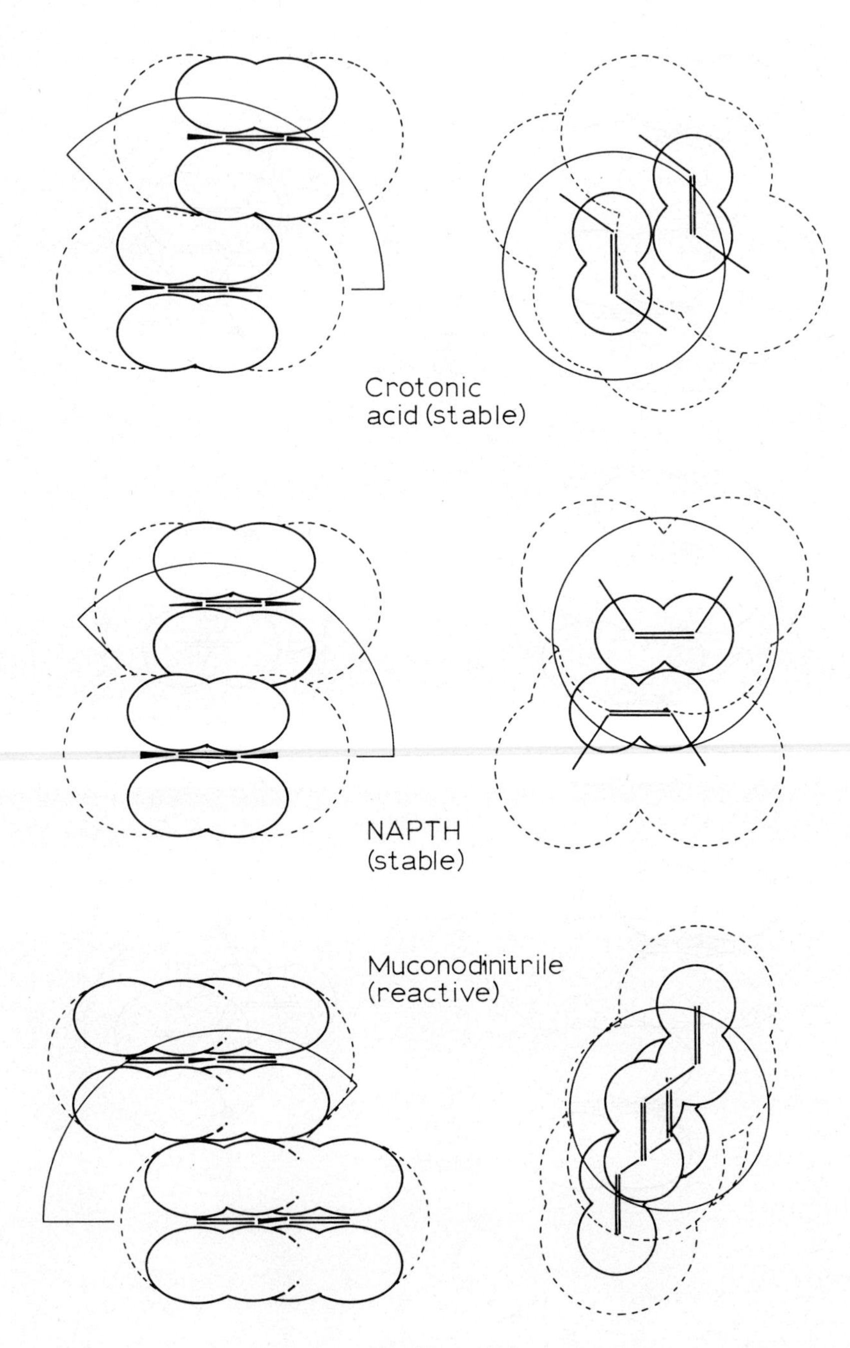

Fig.7 continued

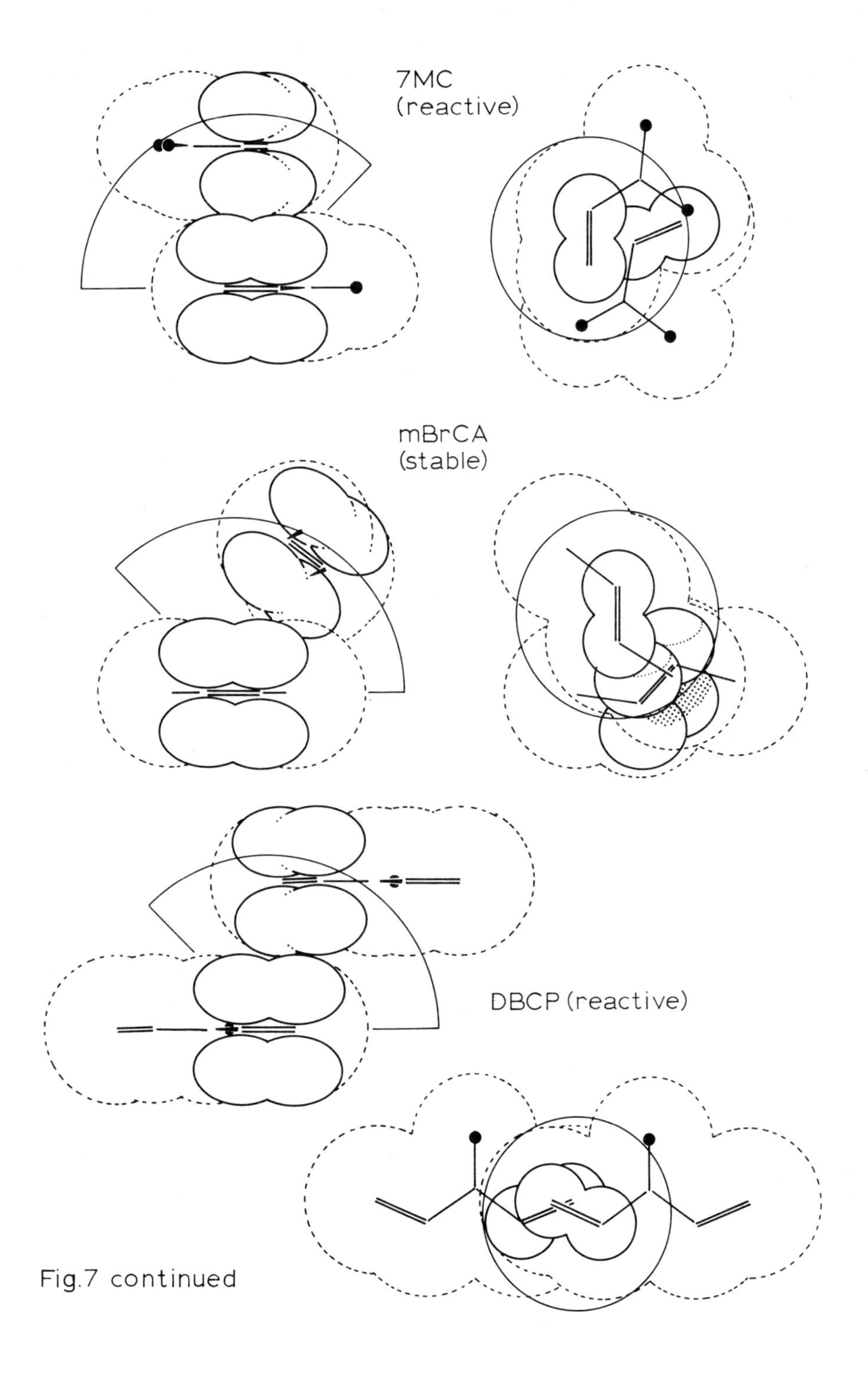

Fig.7 continued

nectivity (Figure 6B). The p orbital lobe apex is chosen as that point along the p(z) orbital axis that intersects the van der Waals surface of the atom. If the point T is the apex of the p atomic orbital and T' is the corresponding point on the opposing pre-bonding atomic orbital, then the smaller the distance T-T', the more parallel and colinear is the alignment of the reacting p orbitals. Note that this measure of orbital overlap can be applied to other types of reactions provided that rotation of participating orbitals during reaction is not necessary. Table 1 gives geometric information about various compounds including the ones drawn in Figure 7. It is not supposed to be a comprehensive list of compounds that undergo [2+2] cycloadditions in the solid state, but serves to demonstrate the topochemical anomalies and especially interesting reactions.

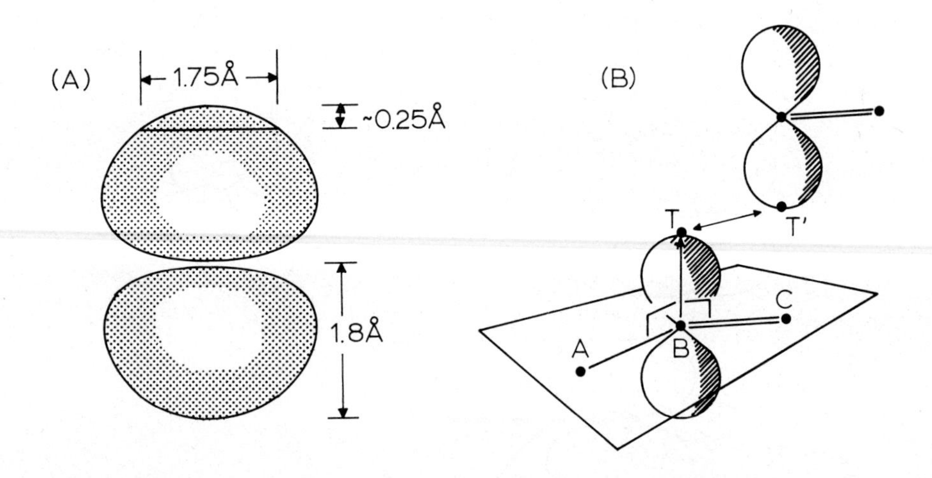

Fig. 6. (A) Drawing showing the shape of a p orbital; the contour encompasses 95% of the electron density. This is scaled so that the apex of this orbital coinsides with the van der Waals radius of carbon. The top 1/4 Angstrom cap of the orbital has an approximate diameter of 1.75 Å, this relates to the small circles drawn in the plane projections of the orbital shown in Figure 7. (B) The position T is constructed by using the plane normal through point atoms A, B and C. The plane normal is given a magnitude of 1.8 Å (the van der Waals radius of carbon). Thus the distance between T and T' of the reacting orbitals is a description of the amount of orbital overlap and to some extent their orientation with respect to each other.

4.2 Exact size and shape of orbitals

Electron density maps for the atomic orbitals of hydrogen have been calculated. Contours on these maps signify the size of the orbital containing certain percentages of the electron density. These orbitals can be scaled by multiplying by a screening constant to get the size for a carbon atom p orbital. Such a p orbital is drawn in Figure 6A; the single contour represents 95% of the electron density. For practical purposes the orbitals should be drawn to the van der Waals radius; this corresponds to 90-95% of the electron density, and partially justifies the drawing scheme in Figure 7.

It may be the case that the disposition of reacting orbitals is more critical in some directions than in others. Furthermore, the choice of van der Waals radius is in general quite arbitrary. This will have a bearing on the numbers shown in Table 1 and the drawing scheme in Figure 7. It should be borne in mind that this measure of topochemical reactivity is suggested only as an improvement and is certainly not definitive. But, it will be shown to be of more diagnostic value than other geometric measures of reactivity.

4.3 Parallel double bonds

The reactivity of BBCP, BBL, BCP and NAPTH compounds (Fig. 7 and Scheme 2) can be rationalised visually in terms of the amount of orbital overlap; the first two compounds are reactive because the p orbitals overlap substantially more than the others. From Table 1 it can be seen that all the above compounds are within the 4.2 Å limit. The double bond in NAPTH is isolated from any other conjugated part of the molecule and can be considered to be near the limit of lateral displacement for parallel double bonds. It therefore suggests that those compounds which are reactive and have lateral displacements greater than this that are probably assisted in reaction by other factors. Comparison of compounds that are reactive and have double bonds within 4.2 Å shows that the upper limit on the SUM (see Table 1 for the definition of SUM) of the orbital overlap has a value around 3.5. Thus, compounds that do not show the expected reactivity need additional or alternative rationalisation.

Begley et al. (1978) have shown that the relative rates of the photochemical [2+2] reaction of BBCP derivatives (in KBr pellets) correlate with the geometric configuration of the "embryo cyclobutane ring". This correlation is somewhat misleading; there can be

TABLE 1

Geometric information is given about various compounds including the ones drawn in Figure 7. The plane separations are those computed from the plane defined by the maximal moment of inertia through the flat conjugated fragment of the molecule. In the case of skew double bonds a rough estimate of the plane separation is taken as the perpendicular distance of the midpoint of one double bond from the plane defined by the other. Lobel is the distance T-T' between orbital lobes that will form a bond; Lobe2 is the distance between the other pair of reacting orbitals. The orbital overlap SUM is defined as the sum of the distances between reacting orbital lobes 1 and 2 (SUM = Lobel + Lobe2). To distinguish between the atom-atom distance criterion and the orbital overlap SUM, the latter is specified without units in the text. The distance between bonding atoms is given after the corresponding T-T' interaction if they differ substantially as they would do for skew double bonds. The starred (*) T-T' interactions flag the oxetan oxygen-carbon overlap. The lateral displacement is the distance between the double bonds measured along the axis perpendicular to both the carbon double bond axis and the z-axes of the p(z) orbitals. All distances are measured in Angstroms.

Compound	Separation	Distance	Lobel	Lobe2	SUM	Reaction	Lateral	Remark
Compounds depicted in Figure 7 and Scheme 2.								
BBCP	3.84	4.12	1.56	1.56	3.12	yes	0.11	parallel
BBL	3.35	3.67	1.47	1.47	2.94	yes	0.80	parallel
BCP	3.55	4.14	2.15	2.15	4.30	no	0.75	parallel
Crotonic acid	3.37	4.06	2.28	2.24	4.52	no	1.86	parallel
NAPTH	3.34	3.79	1.79	1.79	3.58	no	1.67	parallel
Muconodinitrile								
trans	3.57	3.96	1.71	1.71	3.42	yes	0.67	parallel
centre	3.57	3.64	0.76	0.76	1.52	yes	0.53	parallel
7MC	3.43	-	1.23 (3.63)	2.33 (4.14)	3.56	yes	-	skew(70)
mBrCA	3.39	-	1.14 (3.48)	2.05 (4.41)	3.19	no	-	skew
DBCP mirror	3.52	-	0.93 (3.72)	0.93 (3.72)	1.86	yes	-	skew(55)
DBCP rotational	3.52	-	0.46 (3.59)	2.06 (4.14)	2.52	?	-	skew(125)
Benzylbenzylidenecyclohexanone (Nakanishi et al. 1981b)								
	3.57	3.79	1.30	1.30	2.60	yes	0.98	parallel
Benzyl-p-bromobenzylidenecyclopentanone (Nakanishi et al. 1981a)								
	3.54	3.80	1.27	1.27	2.54	yes	1.26	parallel
p-Chlorobenzylbenzylbenzylidenecyclopentanone (Theocharis et al. 1981)								
	3.80	4.03	1.37	1.37	2.74	yes	0.07	parallel

TABLE 1 continued

Compound	Separation	Distance	Lobe1	Lobe2	SUM	Reaction	Lateral	Remark
p-(X)benzyl-p-bromobenzylidenecyclopentanone (Theocharis et al. 1982)								
X = methyl	3.62	3.92	1.48	1.48	2.96	yes	1.45	parallel
X = chloro	3.62	4.65	2.90	2.90	5.80	no	1.82	parallel
Benzyl-m-bromobenzyidenecyclopentanone (Theocharis 1982)								
	3.59	4.36	2.42	2.42	4.84	no	-	parallel
DBCP chiral derivative (Theocharis et al. 1983)								
	3.39	-	2.09(4.29)	1.23(4.11)	3.32	no	-	skew
	3.88	-	1.25(3.83)	1.94(4.28)	3.19	no	-	skew
7-Chlorocoumarin (Gnanaguru et al. 1984)								
	3.40	4.45	2.87	2.87	5.74	yes	-	parallel
Diphenyloctatetraene (Drenth and Wiebenga 1955)								
	-	-	2.26(4.04)	2.65(4.79)	5.91	no	-	skew
	-	-	2.16(3.96)	2.58(4.73)	4.74	no	-	skew
1-Methylthymine (Hoogsteen 1963)								
	3.36	3.81	1.75	1.75	3.50	yes	0.42	parallel
1,1'-Trimethylenebisthymine (Frank and Paul 1973)								
intra	3.19	-	1.52(3.55)	1.65(3.45)	3.17	no	1.53	skew
inter	3.63	-	0.86(3.63)	0.82(3.75)	1.68	yes	0.84	parallel
Acridizinium salts (Kearsley 1983)								
Perchlorate	3.46	4.26	2.40	2.40	4.80	yes	-	parallel
Bromide	3.66	3.86	1.46	1.46	2.92	yes	-	parallel
Muconic acid (Bernstein and Leiserowitz 1972), molecules A and B								
trans A	3.36	3.79	1.77	1.77	3.54	yes	1.95	parallel
centre A	3.36	3.78	1.75	1.75	3.50	no	1.68	parallel
trans B	3.20	3.79	2.07	2.07	4.14	?	1.91	parallel
centre B	3.20	4.30	2.90	2.90	5.80	no	0.69	parallel
Dimethylmuconate (Filippakis et al. 1967)								
	3.06	4.29	3.08	3.08	6.16	no	1.08	parallel
Monomethylmuconate (Rabinovich and Schmidt 1967)								
trans	3.40	3.88	2.01	2.01	4.02	yes	0.27	parallel
centre	3.40	-	1.57(3.76)	1.48(3.63)	3.20	no	1.54	parallel
trans	3.48	3.88	1.69	1.73	3.42	yes	0.24	parallel

(continued)

TABLE 1 continued

Compound	Separation	Distance	Lobe1	Lobe2	SUM	Reaction	Lateral	Remark
Distyrylpyrazine (alpha; Sasada et al. 1971) (gamma; Nakanishi et al. 1976)								
gamma	3.80	-	2.62(4.19)	2.59(4.36)	5.21	no	1.56	skew
alpha	3.48	3.94	1.89	1.89	3.78	yes	1.32	parallel
4-(2-Carboxyvinyl)-cyanocinnamic acid dimethyl ester (Nakanishi and Sasada 1978)								
	3.31	3.96	2.17	2.21	4.38	yes	1.94	parallel
	3.34	3.96	2.07	2.18	4.25	yes	1.86	parallel
Formylcinnamic acid (Nakanishi et al. 1985)								
	3.39	4.83	3.48	3.40	6.88	yes	2.53	parallel
Dicinnamoylbenzene (Nakanishi et al. 1984)								
	3.72	-	1.20(3.90)	1.21(3.95)	2.41	yes	0.90	skew
	3.74	-	1.30(4.09)	1.13(3.97)	2.43	yes	1.25	skew
Crotonamide (Shimizu et al. 1974)								
	3.49	3.98	1.93	1.93	3.86	no	1.66	parallel
3-Nitro-4-hydroxycinnamate (Hanson 1975)								
	3.39	3.87	1.89	1.89	3.78	no	0.17	parallel
3,4-methylenedioxycinnamic acid (Desiraju et al. 1984)								
	3.39	3.80	1.74	1.74	3.48	yes	0.22	parallel
3,4-dimethoxycinnamic acid (Desiraju et al. 1984)								
	3.15	4.04	2.56	2.56	5.12	yes	2.48	parallel
Complex of 3,4-dimethoxy and 2,4-dinitro cinnamic acids (Sarma and Desiraju 1985)								
	3.49	-	1.58(3.84)	1.64(3.76)	3.22	no	0.69	parallel
2,5-dimethyl-1,4-benzoquinone (Rabinovich and Schmidt 1964); stack A and B:								
endo-butane A	3.46	4.01	2.05	2.04	4.09	yes	1.29	parallel
exo-butane A	3.46	4.03	2.09	2.09	4.18	no	1.26	parallel
endo-butane B	3.46	4.01	2.05	2.04	4.09	yes	0.26	parallel
exo-butane B	3.46	4.65	3.11	3.11	6.22	no	2.30	parallel
oxetan 1 A	3.46	-	0.49(3.49)	0.38(3.48)*	0.87	yes	0.03	parallel
oxetan 2 B	3.46	-	1.02(3.61)	1.09(3.61)*	2.11	yes	1.03	parallel
oxetan 3 B	3.46	-	1.53(3.77)	1.49(3.74)*	3.02	no	1.56	parallel

TABLE 1 continued

Compound	Separation	Distance	Lobe1	Lobe2	SUM	Reaction	Lateral	Remark
2,6-dimethyl-1,4-benzoquinone (Rabinovich and Schmidt 1967)								
endo-butane	3.55	3.98	1.81	1.82	3.63	yes	0.33	parallel
exo-butane	3.55	4.54	2.83	2.91	5.74	no	2.19	parallel
oxetan 1	3.55	-	1.05(3.72)	1.02(3.69)*	2.07	yes	0.94	parallel
oxetan 2	3.55	-	0.94(3.69)	1.01(3.64)*	1.95	no	0.92	parallel
Benzoquinone (Trotter 1960)								
oxetan	3.17	-	1.05(3.33)	2.05(3.78)*	3.10	no	-	skew
endo-butane	3.27	-	2.00(3.70)	2.75(4.19)	4.75	no	-	skew
exo-butane	3.27	-	1.48(3.48)	3.24(4.50)	4.72	no	-	skew
Duroquinone (Rabinovich et al. 1967)								
exo-butane	3.46	-	0.81(3.56)	2.72(4.41)	3.53	no	-	skew(90)
endo-butane	3.46	-	1.96(3.98)	2.01(4.01)	3.97	• no	-	skew(90)
Acrylic acid (Higgs and Sass 1963)								
dimer	3.15	-	1.59(3.50)	3.01(4.33)	4.60	no	-	skew(67)
polymer	3.15	-	1.59(3.50)	C3-C3 biradical				
polymer	3.15	-	2.08(3.75)	C2-C2 propagation				
Thienylacrylic acid (Block et al. 1967)								
beta form	3.51	3.91	1.97	1.97	3.94	yes	1.43	parallel
gamma form	3.50	-	1.30(3.72)	1.31(3.73)	2.61	no	C5=C3 to C4=C2	
Furylacrylic acid (beta form) (Filippakis and Schmidt 1967)								
dimer	3.50	3.84	1.61	1.61	3.22	yes	1.24	parallel

no simple correlation between kinetics and ground state overlap derived from the geometry of point atoms. They do not adequately explain the one exception to the trend that reaction rates are faster when there is greater orbital overlap. Apparently the iodo derivative has a structure isomorphous to the bromo derivative with a distance of 3.88 Å between double bonds. The bromo derivative has a relative rate far greater than BBCP (correlating with the SUM in Table 1; 2.54 and 3.12 respectively) but the reaction for the iodo derivative was slower than BBCP. It is unlikely that the actual bond-forming process is going to be the rate determining step. Effects of steric interactions on relaxation of environment, phase separation and population of defects will also have a bearing on the rate.

4.4 The necessity for a parallel double bond configuration

For a concerted reaction the reacting atom pairs ought to be comparable distances from one another so that the reaction may proceed smoothly. It has been demonstrated that some non-parallel bonds configurations react within the topochemical limit of 4.2 Å.

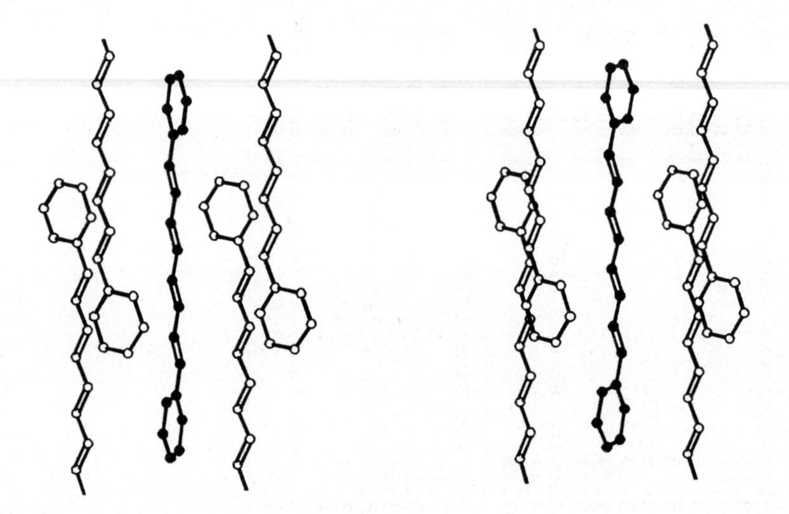

Fig. 8. Stereo diagram of light stable 1,8-diphenyloctatetraene showing surrounding glide related molecules. The conjugated sections of the molecules are inclined about 70 degrees to each other (Drenth and Wiebenga 1955).

Non-parallel double bond configurations can arise in two ways: the first is skewness in the plane of conjugation and the second is skewness in the direction of the double bonds and the p(z) orbital plane. Glide and screw symmetry operators that produce non-parallel configurations invariably involve both types of skewness as demonstrated by diphenyloctatetraene (Fig. 8) and m-bromo-methylcinnamate (mBrCA, Fig. 7).

The apex-apex distance as a measure of orbital overlap suggests that the octatetraene derivative will not react; the SUM far exceeds the value of 3.58 observed for the unreactive NAPTH. The rare double dimerisation of dicinnamoylbenzene (Nakanishi et al. 1984) closes both double bonds on each end of the molecule. The orbital overlap SUM also suggests this should happen; the reacting p orbitals, although skew, are substantially more colinear than mBrCA or the polyene.

The cases of light-stable mBrCA (Schmidt 1971) and 2,5-dibenzylidene-3-(+)-methylcyclopentanone (a DBCP derivative, Theocharis et al. 1983) are not so easily explained. The orbital overlap SUM being less than 3.5 suggests that they should react. However, for a concerted reaction the arrangement is extremely poor and more importantly, the overlap in the z-direction for the less optimally positioned p orbital is far less than in other compounds. The z-component of the apex-apex distance is 0.3 and 0.7 for BBCP and mBrCA respectively (Fig. 7). For mBrCA the angle between the planes containing the double bonds is 44°. Clearly there is a possibility for one of the p orbital lobes to interact to form a biradical intermediate. But seemingly the intermediate is unable to twist to close the cyclobutane ring. Not finding a viable alternative reaction route (e.g. polymerisation) the radical would collapse back to the ground state. This suggests an improvement on the T-T' distance criterion; for instance the z-component in the p orbital frame of reference could be given more weight, but, for double bonds skewed only in the plane of conjugation, this would not matter.

Double bonds askew in only the plane of the conjugated system can be achieved by having two or more distinct double bonds within the same molecule reacting between intermolecular units (e.g. DBCP), or two molecules within the asymmetric unit that stack in a plane-to-plane manner as in the case of 7-methoxycoumarin (7MC, Ramasubbu et al. 1982). DBCP has already been cited as an example of the condition of totally parallel configuration not being necessary (Theocharis et al. 1983). The SUM of the apex-apex distances that give the observed mirror dimer is 1.86 suggesting extremely good overlap as Figure 7 corroborates, even though the angle between the double bonds is 50°. The interesting feature of this compound is that it gives three products which have been identified by Kaupp and Zimmermann (1981), as depicted in Scheme 1.

The second most abundant dimer is unusual in that the explanation for its formation is through a radical mechanism. The intermediate biradical is stabilised by conjugation with surrounding p orbitals of the carbonyl and two exo double bonds. There is extremely good overlap of p orbitals of one of the broken methylene chain lobes with that of the p orbital of a neighbouring carbonyl oxygen. Therefore, this can be considered a topochemical reaction. The rehybridisation of the reacting carbons is effectively independent of the motion of the orbital that is forming the bond. That is, the p(z) orbital movement in bond formation is not coupled to any rotation needed for alignment.

Once the first bond is formed, probably giving a biradical intermediate, the other orbitals involved in completing the reaction must rotate and align themselves as part of the motion of bond formation. This can work for or against a reaction in the solid state, as the excited geometry may or may not be favourable topochemically; mBrCA could be an example of the latter. In DBCP, once this initial bond formation is accomplished there is also the need for some gross molecular rearrangement for the seven membered ring to be formed.

What is not explained is the presence of a third product dimer of DBCP which has rotational symmetry (Scheme 1). A concerted reaction can certainly be ruled out because the arrangement of the reacting atoms precludes this. A biradical intermediate would allow the eventual rotation product to be formed. Comparing the orbital overlap SUM, 2.52, suggests that this reaction is well within the 3.5 limit. Therefore this too can be considered to be a topochemical reaction. Note, although it is fortuitous that the SUM of this product is less than that for BBCP, only one lobe contact is really important, the one which produces the probable biradical with overlap SUM of 0.42; this corresponds to the route f -> g -> h in Figure 5. Objections to this may be that the rotation of the double bonds through 130° is unlikely, but the massive molecular re-orientation the molecules have to undergo to close the seven membered heterocyclic ring is just as great a movement and is observed.

For 7MC the double bonds are even more askew; the angle is about 70°. The dimer which was isolated and characterised crystallographically (Ramasubbu et al. 1982) corresponds to the preformed asymmetric pair when the molecules are rotated through 70° (Fig. 7). The orbital overlap parameters in Table 1 correspond to this product. For one lobe there is excellent contact whilst the other is marginal with a SUM of 2.33. This poor contact may not have a bearing on the initial stages of reaction, as in the case for DBCP. The reaction is topochemical in the sense that the reactant lattice dictates the formation of the biradical intermediate. However, minimal molecular movement after this event cannot be realised. Both DBCP and 7MC are flat planar molecules which form dimers that are very different in shape. As reaction proceeds, crystallinity is lost, and closure is made possible as the environment around the intermediate changes (this assumes some sort of cooperativity between molecules and intermediates).

5 CHOICE OF REACTION PATHWAY

Refering back to the discussion on the difference between solid and fluid state reactions, different products and hence reaction paths would be expected if the area of allowable configurations for a particular solid phase included two or more transition state configurations. Also, the ratio of products would be determined by the relative accessibility of these transition state configurations (i.e. ratio of the areas T1 and T2 in Fig. 4).

5.1 Alternative [2+2] reactions

Reactions have already been encountered that produce more than one product in the solid state. On the other hand, where there is a possibility of more than one reaction route there are examples in which only one route is selected. The photopolymerisation of trimethylenebisthymine (TMT) is an example of the latter (Frank and Paul 1973). In the fluid state it is accepted that in most cases intramolecular reactions are kinetically favoured over intermolecular reactions, provided the steric factors are comparable, and this is certainly the case for TMT. In the solid state they suggest that the preference for polymerisation between molecules over intramolecular dimerisation is due in part to molecular adjustments, caused by neighbouring molecules reacting, enhancing intermolecular overlap of adjacent unreacted molecules. The orbital overlap SUM also indicates that this is so (see Table 1 and Fig. 9). Neither suggestion satisfactorily explains why the intramolecular [2+2] product is not seen without involving non-topochemical arguments (vide infra).

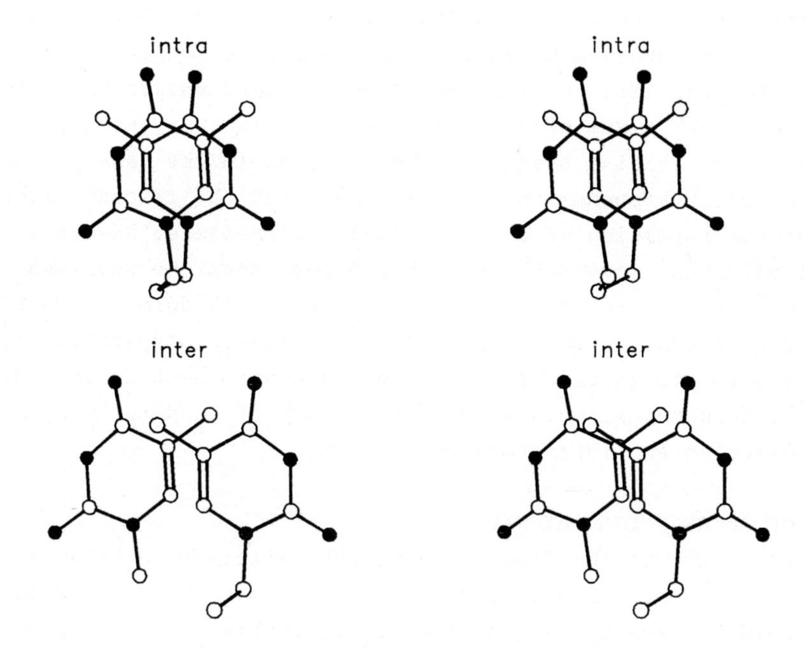

Fig. 9. Stereo diagram of trimethylenebisthymine showing overlap between rings (Frank and Paul 1973).

The topochemistry of the p-benzoquinones has been studied by Rabinovich and Schmidt (1967) and shows several interesting features. The crystal structure of 2,5-dimethylbenzoquinone contains two molecules in the asymmetric unit; each asymmetric unit assembles into its own column of molecules (stacks A and B in Fig. 10) by crystallographic translation. Examining the orbital overlap SUMs one would expect that each stack has its own topochemistry. Certainly for the 2,5- and 2,6- derivatives the photoproducts observed, which include [2+2] cycloadditions to the carbonyl group forming oxetans, can be rationalised by the orbital overlap scheme (Fig. 10). However, the 2,3-dimethyl derivative remains a complete anomaly. According to its crystal structure, where pairs of molecules are related by centres of symmetry, one would expect dimers of centric symmetry; however, the NMR (Cookson et al. 1961) shows dimers of mirror symmetry.

Muconic acid and some of its derivatives have the possibility of forming dimers with double bonds related by translation or centres of symmetry. (Even if the centre is not present, the product dimer

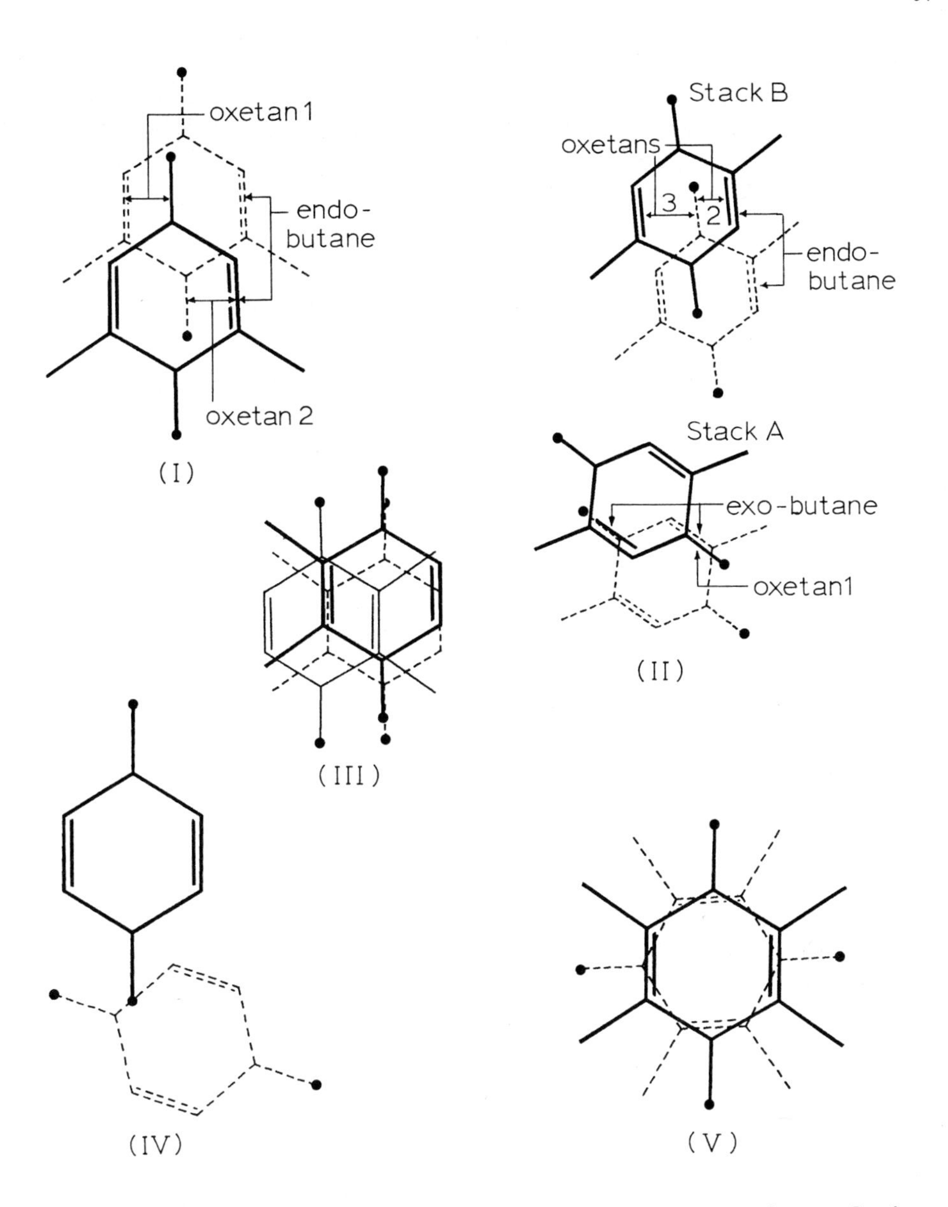

Fig. 10. The nearest neighbour overlap of 5 p-benzoquinone derivatives is shown. The 2,6- (I) and 2,5- (II) dimethyl derivatives are both reactive forming cyclobutanes and oxetans (Rabinovich Schmidt 1967; 1964 respectively). Table 1 quantifies the overlap for some of the expected topochemical products. Three molecules are shown for the 2,3-dimethyl derivative (III) (Rabinovich 1967). The narrow lined molecule can react with either inversion related molecules (bold and dotted lined molecules) to form centric dimers; however, this is not seen. Benzoquinone (IV) (Trotter 1960) and duroquinone (V) (Rabinovich et al. 1967) both polymerise in the solid state; the contacts for forming the [2+2] dimer are seen to be very poor.

can have a centre of symmetry.) The orbital overlap for the latter is more favourable; however, the centric product is only seen in the dinitrile derivative. Furthermore, and in contrast to TMT, muconodinitrile shows the same propensity toward both trans and centric products (Schmidt 1971) even though the orbital overlap favours the centric dimer (see Fig. 7 and Table 1). The crystal structure of muconic acid, like 2,5-dimethylbenzoquinone, is comprised of two molecules in the asymmetric unit forming stacks A and B by translation. The orbital overlap for stack B is not conducive to photoreactivity; therefore, a maximal theoretical yield of 50% is expected for the product. Lahav and Schmidt (1967) never investigated this topochemical aspect of muconic acid as it was clear that other polymeric products were being formed.

5.2 Polymerisation route

Generally, the formation of oligomers and polymers is broken down into three steps: initiation, propagation and termination. Only the initiation, that is the making of the first bond, can be rationalised topochemically. Propagation will depend upon the relationship of the dimer intermediate to the surrounding monomer environment; this is not expected to be the same as the original monomer-monomer relationship. Seemingly polymeric routes become favourable when [2+2] dimeric routes are ruled out; this is the explanation given for the reactivity of p-benzoquinone and duroquinone (Fig. 10; Rabinovich and Schmidt 1967).

Bamford et al. (1963) have shown that acrylic acid polymerises in the solid state. The structure shows nearest neighbour molecules with double bonds 3.87 Å apart and approximately 67° askew. The sum of the orbital overlap (Table 1) implies it is unlikely to react in a [2+2] fashion (cf. with the SUM of Coumarin, 3.57). Hirshfeld and Schmidt (1964; Schmidt 1967) have proposed that interaction of the carbons with short contacts in acrylic acid leads to formation of an open chain dimer biradical which initiates polymerisation. The atoms involved are described in Table 1 and depicted in Figure 11. Unlike the probable biradical dimer intermediate in DBCP, this biradical intermediate does not rotate to form a cyclobutane ring. The spatial proximity of another neighbouring reactive site on another molecule changes the reaction route to polymerisation.

Where propagation can compete with the [2+2] ring closure, polymers are seen along with the dimeric products; this is the case for the trans,trans-muconic acid derivatives. Lahav and Schmidt (1967)

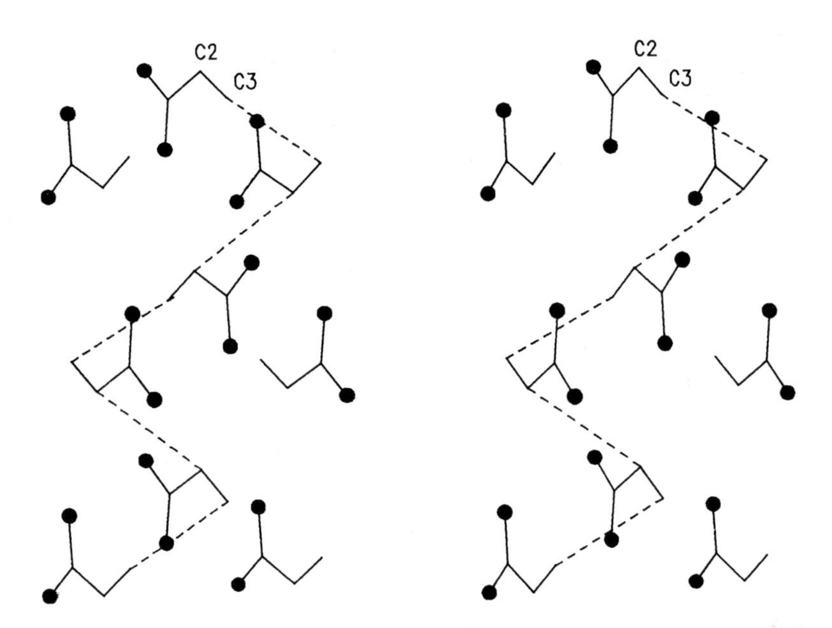

Fig. 11. Stereo diagram of acylic acid showing possible polymerisation routes for molecules in the (010) plane (Higgs and Sass 1963). Carbon atom C3 bonding to C3' in an adjacent molecule is expected to initiate polymerisation. The direction of propagation along [100] is dictated by whether C3 is pointing up or down in the diagram. Propagation along [001] would be random without cooperative effects between polymer and monomer molecules.

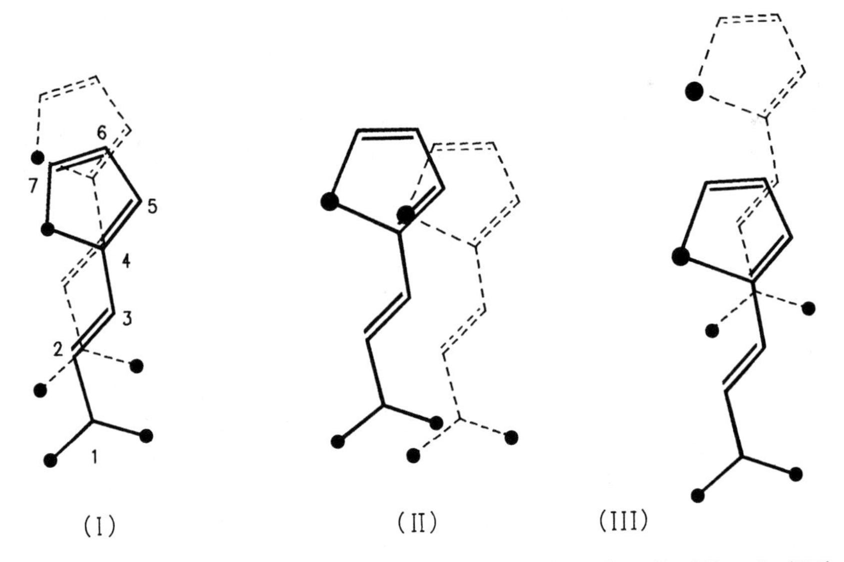

Fig. 12. Nearest neighbour configurations for furyl (I), β-(II) and γ-thienyl (III) forms of acrylic acid (Filippakis and Schmidt 1967; Block et al. 1967). The atom numbering scheme for (I) applies to both (II) and (III).

have also studied the photoreactions of triene systems (heterocyclic analogues of cinnamic acids). The β-form of thienylacrylic acid only dimerises; it has a lobe SUM of 3.94, greater than would be expected for a reactive compound. In contrast the γ-form only photopolymerises and the furyl derivative does both. The γ-thienyl derivative is unable to dimerise topochemically because its overlap SUM is greater than 5. The obvious short contacts in both polymerising compounds are between translated molecules along the shortest axis, these are depicted in Figure 12. The manner in which these compounds polymerise depends upon the relative chemical activity of the atoms that are in nearest intermolecular contact. It was suggested by Lahav and Schmidt (1967) that the reason why the β-thienyl only dimerises and does not polymerise in the same manner as the furyl derivative is due to the activity of the furyl oxygen. The furyl oxygen increases the activity of the endo-cyclic double bonds making alternative reactions possible. Also they suggest that possible polymerisation routes would occur between stacks of molecules. If one assumes that the conformation of the initial diradical (on the monomer molecule) is the same as that of the monomer molecule, then there is no overlap of consequence between radical orbitals and conjugated systems in adjacent stacks of molecules. Similarly for the muconic acid derivatives the orbital overlap is negligible between stacks of molecules. Lahav and Schmidt (1967) rationalise the absence of polymerisation for the cis,cis-muconic acid by the lack of appropriate inter-stack contacts. However, Figure 12 shows that the molecular overlap of the β-thienyl conjugated system is marginal compared to that of the furyl and γ-thienyl counterparts and hence could explain the lack of polymerisation.

6 MECHANICAL COOPERATIVITY

6.1 Steric effects

Bond formation may be feasible topochemically but rigidity of the surrounding lattice molecules will have a bearing on how the reaction proceeds. Although the chemical potential for bond formation is an order of magnitude greater than the existing nonbonding forces that hold the lattice together, these nonbonding forces restrict the diffusion necessary for bonding reactions and so are responsible for the greater activation barrier for reaction in solids. In fact Ariel et al. (1984) have rationalised the inactivity of NAPTH on the grounds of steric inhibition (see reacting molecules B and C in Fig. 13). They corroborate this finding by simulating the motion of the reacting molecules and see a rapid development of repulsive interactions between molecules A,C and B,D (Fig. 13). However, they assume the reacting molecules are rigid and one could imagine that the so called "steric compression" would be relieved if the intramolecular geometry were allowed to change; there is plenty of torsional freedom in the carbomethoxy group and the conformation of the rings could also be adjusted. Other explanations of the anomalous stability of NAPTH such as initial orbital overlap (Table 1 and Fig. 7) or lack of vibrational assistance are just as plausible.

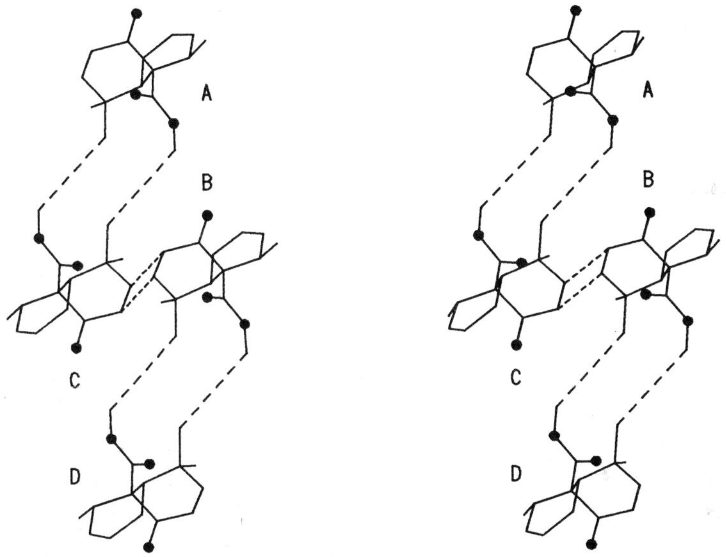

Fig. 13. Stereo packing diagram of napthalenone derivative (Ariel and Trotter 1984); see text for details.

The conformation of the product molecules is rarely the same as that of the reactants and as a consequence the lattice never remains undistorted. Attempts to quantify the amount of distortion generated by reaction have involved defining areas of major or minor steric interaction and has led to the concept of the "reaction cavity" (Cohen 1975). The strain in the lattice manifests itself in varying degrees. Where the lattice is severely distorted, the deformation energy will be released by the formation of defects, particularly dislocations; this will produce a new environmental relationship between molecules. Once a reaction has commenced, there can be cooperative effects of a mechanical nature arising from steric interactions from parts of the molecule remote to the reacting centre. As a consequence of reaction, i.e. the changing geometry of the molecule, these remote fragments are forced to interact with the coordinating molecules which may enhance or deter further reaction. An example of mechanical cooperativity enhancing reaction, which has been extensively researched, is the polymerisation of polymorphs and compounds similar to distyrylpyrazine (Jones and Thomas 1979; Nakanishi et al. 1980; Braun and Wegner 1983). Note that the overlap SUM for the α-form of distyrylpyrazine is 3.78, larger than that for the unreactive NAPTH (3.58); this would imply that there is positive mechanical cooperativity between the intermediate oligomers and the monomer lattice.

Other aspects related to the above are the pathways in which the reaction proceeds through the crystal. A random dimerisation of molecules within a perfect monomer lattice is in effect introducing defects (impurity molecules). At one point the solubility limit of the defect molecules will be reached and the dimers will recrystallise forming two separate phases. Reactions where reactants and products "appear" to form a solid solution are termed homogeneous. The definition is qualified because there is no absolute way of imagining a truly homogeneous reaction; i.e., a reaction where the product molecules do not perturb the reactant lattice in any way. Heterogeneous reactions are ones in which the product formation is nonrandom, i.e., reactions that start at defects in crystals or propagate at the phase interface between reactants and products. (In short, one phase develops into two.) Many reactions may start homogeneously and then proceed via a heterogeneous mechanism once the solubility limit of one phase in the other is exceeded. In general "homogeneous" refers to the limit in which heterogeneity is detectable. This limit is dependent on the size of the sample taken and where the reaction takes place in the crystal. This is not an easy quantity to define or measure. On a relative scale the degree of homogeneity/heterogeneity is indicative of the degree of disruption to the lattice caused by reaction.

6.2 Reactions in a translational stack

Theoretical maximum yields have been computed for reacting molecules that are related by crystallographic translation (Flory 1939, 1953; Cohen and Reiss 1963; Desiraju and Kannan 1986). The results differ slightly depending on the precise mathematical model used;

86% was calculated by Flory and 82% by Desiraju and Kannan. Yields significantly different to this imply strong cooperative effects. The excited monomer molecule which is unable to dimerise because it is sandwiched between dimers in the translational stack will seek other reaction pathways. This could account for the polymerisation products in the p-benzoquinones, muconic acid derivatives and heterocyclic trienes.

Other possibilities are that cooperative effects, originating from the primary initial reaction (for instance topochemical dimerisation), lead to an environment that is chemically advantageous for alternative reaction routes. Such may be the case for DBCP where the product distribution is indicative of steric effects enhancing routes to minor products (Kaupp and Zimmermann 1981).

7 EXCITED STATE GEOMETRIES

There is little structural information immediately accessible about the geometry of the excited molecules in the solid state. The reason for this is the very short lifetimes and the dilution of the excited molecules within the ground state matrix. In solution, excited states will often have quite different equilibrium geometries when compared to that of the ground state. For instance, in one particular excited state, formaldehyde has a pyramidal structure with a dipole moment different from that of the ground state. Furthermore, excited frontier molecular orbitals (e.g. LUMO) may have sufficiently different electronic configurations from that of the ground state orbital (HOMO) so that overlap for reaction between neighbouring molecules may become critically dependent on this. Excited state electronic configuration may change to the extent that overlap now becomes unfavourable. This may be so, even though the ground state configuration dictates a possible topochemical reaction.

The cis-cinnamic acids or the β-substituted-trans-cinnamic acids isomerise in the solid state, probably going through the orthogonal π-bond arrangement of typical olefins in preference to feasible topochemical [2+2] cycloadditions. This alternative reaction was rationalised in terms of the relatively non-planar conformation of these acids (Schmidt 1971). It is pointed out that there are examples of twisted molecules undergoing [2+2] reactions, for instance trans-cinnamide (Leiserowitz and Schmidt 1969). It must therefore be important to know to what extent the geometry is allowed to change during the course of the reaction.

7.1 Degree of electronic overlap

The electron population of the orbitals must be considered. (Orbitals do not physically exist unless they are filled with electrons.) To get a true picture of the electronic profile of the reactive site, knowledge of the electron density is required.

In the ground state of the molecule, the outermost molecular orbitals make the contacts between adjacent molecules; these molecular orbitals can be involved with intermolecular bond forming reactions. Although the sum total of all the electron density may be approximately spherical (for instance, in carbon this is partially due to the two core electrons being localised at the atom sites), individual molecular orbitals, which determine reactivity, are very directional and rarely localised at the atom nuclear sites. Ohno et al. (1983) deduced the profile of electron density of the frontier molecular orbitals, those that are in most cases intrinsically more reactive, by the experimental technique of Penning Ionisation Electron Spectroscopy. Two things become apparent. (i) Molecular orbitals, notably the highest occupied molecular orbital (HOMO), extend beyond the classical van der Waals radii of the individual atoms. Not only are the junctions between intersecting van der Waals spheres given a higher proportion of the electron density (representing the bonding electrons) but also other positions on the molecule not involved in bonding (such as lone pairs of electrons). (ii) The amount of exterior electron density can be a measure of the relative activity of a specific molecular site.

As the molecule attains the excited state the profile of the electron density may change. For instance, the π^* orbital for ethene, which in the excited state contains an electron (assuming a π to π^* transition), has a substantially different electron density distribution compared to the bonding orbital of the ground state (Fig. 14). As a consequence, the overlap between the filled anti-bonding orbital and the ground state orbital has changed (providing that the excited state atomic geometry is the same).

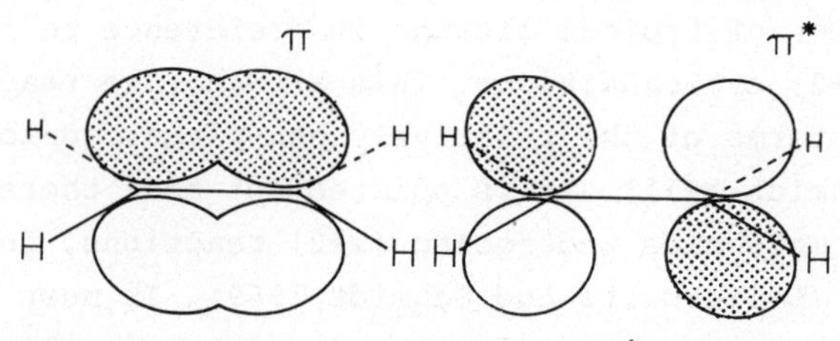

Fig. 14. Schematic depiction of π and π^* orbitals of ethene taken from Jorgensen and Salem (1973).

7.2 The whereabouts of the excitation energy

Regardless of the shape of molecular orbitals, the excitation energy must reside within the vicinity of the topochemically arranged functional groups. Desiraju and Sarma (1983) have rationalised the inactivity of a donor-acceptor complex of 3,4-dimethoxy and 2,4-dinitro cinnamic acids (SUM of 3.22) in terms of the excited electronic configuration being localised solely upon the nitro groups.

There is also an intermolecular counterpart to this, where excitation energy is allowed to travel (as excitons) through the crystal. These excitons may get trapped at defects or impurities within the crystal lattice rather than fostering topochemical reactions in perfect regions of the crystal lattice.

7.3 Excimers

Where conjugated molecules pack in stacks or dimer pairs, there is the possibility of excimer formation. The term "excimer" is used to describe a complex between two excited state monomers to distinguish it from the excited state of a dimer. Since most of the compounds studied are flat planar molecules with conjugation, the likely excimers are of the π-π type. Excimers in solids can be identified by structureless fluorescence at wavelengths longer than that which was absorbed (Forster 1969; Michel-Beverle and Yakhot 1977). Ferguson and Mau (1974) have concluded from fluorescence spectroscopic studies done at various temperatures that an excimer state is involved as an intermediate in the formation of dianthracene. It is mentioned here since in the solid state the formation of excimers is under topochemical control. The topochemical limitations are not so stringent for the formation of excimers as they are for the formation of a dimer, because the charge transfer interaction is less localised in comparison to the configuration of bond-making atoms.

As in solution, reactions tend to follow the route with the least energy barrier, and if the transition state can be lowered by transient molecular bonding, then the mechanism will not always be determined by the geometric factors at the bonding sites. For instance the Diels-Alder reaction between cyclopentadiene and maleic anhydride is geometrically selective for the endo product. This is because there is also bonding between excited state orbitals on other atoms in the transition state (dotted lines in Fig. 15). Thus parts of the molecule remote to the reacting site might have a role to play in drawing the molecules into a reacting configuration within the solid state.

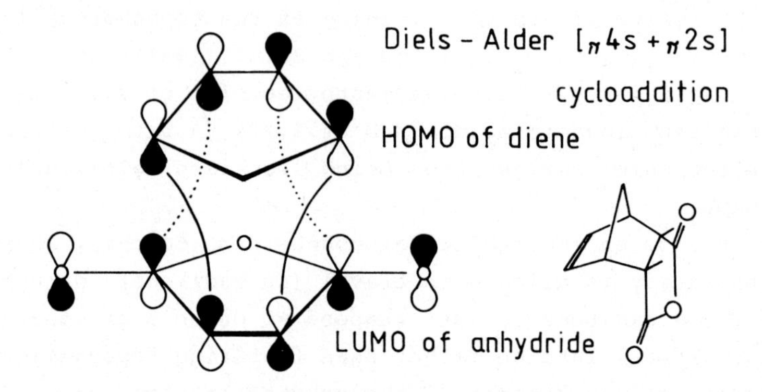

Fig. 15. Transient bonding in the transition state geometry of a Diels-Alder cycloaddition explains the preference for the endo product.

It is stressed that if reaction is a consequence of an excimer intermediate, the reaction itself cannot be considered topochemical with respect to the ground state lattice environment. This is because excimers generally have different geometric configurations from the monomer pair which are not predictable from topochemical rules and therefore neither is the reaction. Calculations have been carried out by Warshel and Huler (1974) on the equilibrium configuration of the pyrene excimer. They did this by constructing a semi-empirical excimer force field and then optimising the excimer configuration while simultaneously minimising the lattice forces. The resultant plane separation was, as in solution, about 3.2 Å (cf. with crystallographic separation of 3.5 Å). But, unlike in solution, the excimeric pair is displaced from total overlap; the coordinating molecules in the lattice cause this effect. The chemical potential for the formation of excimers is not as great as that for bond formation, hence excimer formation is more dependent on the rigidity of the surrounding lattice.

Some of the compounds in Table 1 dimerise with atom-atom contacts far exceeding 4.2 Å. In particular the apex-apex distance for the [4+4] photodimerisation of acridizinium perchlorate (Kearsley 1983) is of the same magnitude as that for the m-bromo derivative of BBCP (Theocharis 1982) and far exceeds that of BCP. The latter two compounds are stable but the perchlorate reacts in a fashion similar to that expected and observed for acridizinium bromide (Table 1). Even though the apex-apex contact of the bonding orbitals suggests that the perchlorate ought to be light stable, its reactivity can be rationalised if pre-bonding intermediates are considered. The intermediate could foster a more favourable topochemical arrangement so that dimerisation is made possible. In the case of the perchlorate the necessary intermediate is likely to be an excimer. Also for 7MC and 7-chlorocoumarin there is a possibility of excimer interactions which could augment pre-bonding overlap. For BBCP derivatives and NAPTH compounds their π-conjugated fragments do not overlap to any great extent, thus reaction intermediates of this sort are not expected.

If the geometry of an excimer is sufficiently different and its formation takes precedence over immediate bonding reactions, then there is the possibility that this intermediate could prevent the reaction of a previously feasible topochemical bonding arrangement. This could explain the lack of intramolecular dimerisation in trimethylenebisthymine; Figure 9 shows that an intramolecular excimer would decrease overlap between the rings. Excimer intermediates could be envisioned for some of the other planar conjugated molecules. (In Table 1: monomethylmuconate, dicinnamoylbenzene, 3,4-dimethoxycinnamic acid, formylcinnamic acid, 2,5-dimethylbenzoquinone and γ-thienylacrylic acid.)

8 THERMAL ASSISTANCE

8.1 Phonons

The concept of molecular vibrations, intra and intermolecular, assisting reactions as part of the mechanism cannot be ignored, especially if these influence the orbital overlap before the possible onset of reaction. The existence of quantised lattice vibrations, or phonons, implies a cooperative effect between particles (Kittel 1971). For molecular crystals the intermolecular forces are weak and it may be that the vibrations in one area of a crystal are not coupled to vibrations in another area. Such crystal vibrations may be responsible for solid state reactivity of certain organic compounds. However, Prasad and Swiatkiewicz (1983) go on to say that upon excitation, electronic-phonon or phonon-phonon couplings are able to deform the lattice locally. It is this local deformation, especially if it is along the possible reaction coordinate, that assists photochemical reactions. Mirsa and Prasad (1982) have also suggested that such polaron (combination of an electron and its strain field in the lattice) mechanisms are an alternative to proposed excimer intermediates.

Rationalisation of solid state phenomena in terms of phonons, polarons and their interaction with the lattice and electrons is valid for metals, inorganic salts and covalent solids where the bonding forces are strong. However, for molecular crystals where only nonbonding forces hold the molecules together, the coupling between various dynamic effects is going to be weak and the disappearance and emergence of vibrational modes less open to interpretation.

8.2 The degree of vibrational overlap

Dwarkanath and Prasad (1980) have established to some extent that organic solid state photochemical reactions are thermally assisted. Swiatkiewicz et al. (1982) found that BBCP did not react at liquid nitrogen temperatures when irradiated with UV of wavelengths greater than 340nm whilst at room temperature the dimer was obtained as usual. This implies that for BBCP, static overlap at 77K between potentially reactive orbitals is insufficient for a topochemical reaction and that dynamic effects dependent on temperature do assist reaction. Comparing BBCP with the other reactive compounds in Figure 7 shows that the overlap is similar when viewed perpendicular to the plane of conjugation, but there is a substantial gap between the van der Waals envelope for BBCP whilst no such gap exists for the others.

If the molecule absorbs UV light and only fluoresces from an excited state of lower energy, then the excess energy goes into heating the crystal. The extra vibrational energy, and hence enhanced amplitude of motion of atoms and molecules, may also be responsible for assisting reactions in the immediate neighbourhood. For the dimerisation of BBCP to be thermally assisted, the vibration needed is seen intuitively to be one in which the two molecules of the incipient dimer are brought closer together, especially in the z-direction of the p orbitals. Such modes of distortion may include translational vibrations of the molecule itself; various bending or coupled translation-libration (rocking) modes can be envisaged to bring the reacting orbitals into bond-forming proximity.

A very crude approximation to the change in the amplitude of vibration for BBCP was made by computing the difference in total potential energy of the lattice with respect to two oscillating molecules (Kearsley 1983). That is, the model is of a pair of reacting molecules related by centres of symmetry oscillating in concert. The direction of vibration is normal to the plane defined by the benzylidenecyclopentanone fragment (similar fragment depicted in Scheme 2). The energy was computed by summation of all the atom-atom pair-potentials over only the coordinating molecules (Kitajgorodskij 1973; Williams 1972). Data points were derived about the crystallographic equilibrium separation at 0.025 Å intervals; these are the circles in Figure 16.

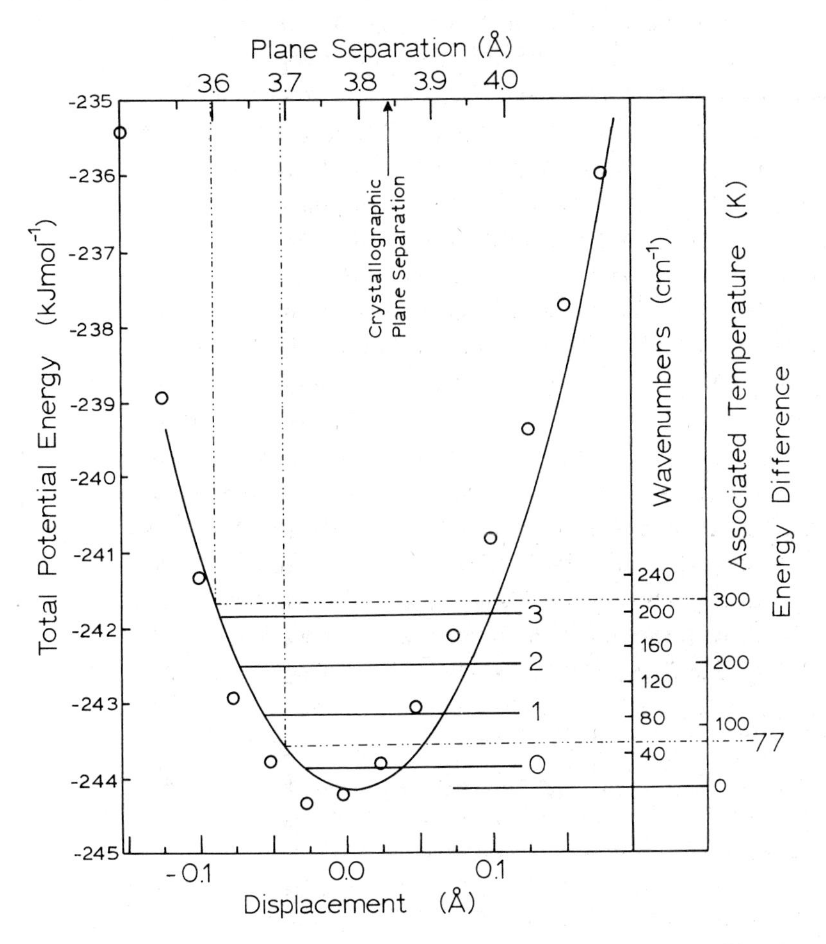

Fig. 16. Variation of potential energy with respect to displacement from the equilibrium crystal structure for oscillating incipient dimer pairs of BBCP molecules; see text for details.

A parabola was fitted to the points that gave the least sum-squared deviation (14 points), the solid line representing the least squares curve. Thus for a maximum displacement of 0.15 Å the oscillating pairs are treated harmonically. (The displacement coordinate has its origin corresponding to the minimum of the fitted parabola.) The force constant was determined as 284 kJ/mol/A^2 (twice the coefficient of the quadratic term describing the energy of the oscillator). Using the mass of the molecule gave the fundamental frequency of vibration as 1.66 THz (equivalent to 55 wavenumbers) with a zero point energy of 0.33 kJ/mol.

If one considers the oscillator classically then at liquid nitrogen temperature (77K), the maximal amplitude of vibration is around 0.1 Å and at room temperature (300K) around 0.2 Å. It may be the case that 0.1 Å difference is enough to create the extra overlap needed for reaction. In quantum mechanical terms the population density at certain energy levels must be assessed at various temperatures. It is sufficient for this approximation to say that at 300K the population density is more classical (level 3) and thus the probability of finding the molecule at a closer separation is greater, whilst at 77K near the first quantum level, the greatest probability is at the equilibrium point of the crystal structure. No effort was made to relax the vibrating pair in the lattice, so the energy values will be high and the profile of the energy well too narrow.

One can conclude that the vibrational amplitude of the molecules will increase by 0.05 Å thus periodically reduce the plane separation by at least 0.1 Å. This sort of molecular vibration is comparable to that expected of organic molecules; Kitajgorodskij (1973) quotes amplitudes in the region of a few tenths of an Angstrom and angular librations of 2-3 degrees.

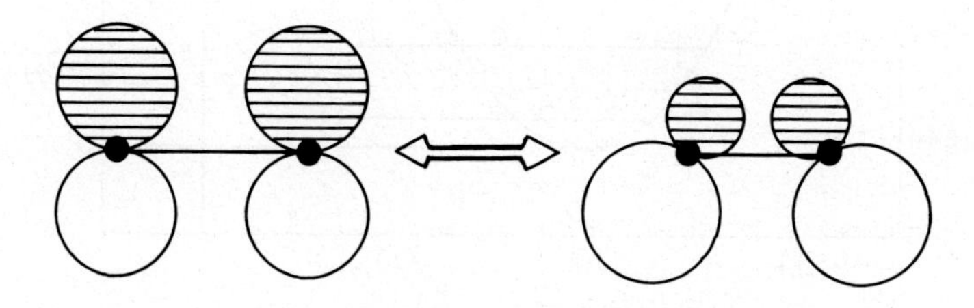

Fig. 17. Schematic depiction of the correlation of electron density with vibrations of the molecule.

The above calculation was done to show what sort of change in vibration amplitude for molecules is expected for a 200 degree increase in temperature. Lattice modes, dispersion curves and other dynamical quantities can be deduced from the static crystal structure. However, just as there is little immediate information about the molecular orbital configurations from the geometry of the molecules, knowledge of correlation of electron density with vibrational structure is also unobtainable from geometry. These can be coupled to lattice vibrational modes. Figure 17 shows how the orbitals (schematic and exaggerated) can be deformed as the molecules oscillate. Overlap will be even harder to assess because the profile of the electron density is continuously changing within the time scale of a vibration. (The Born-Oppenheimer approximation applies to time scales for electronic transitions around 10^{13}-10^{16}s while molecular crystal vibrational energies have values around 10 to 200 cm^{-1}; equivalent to a time scale about three orders of magnitude less than an electronic transition.)

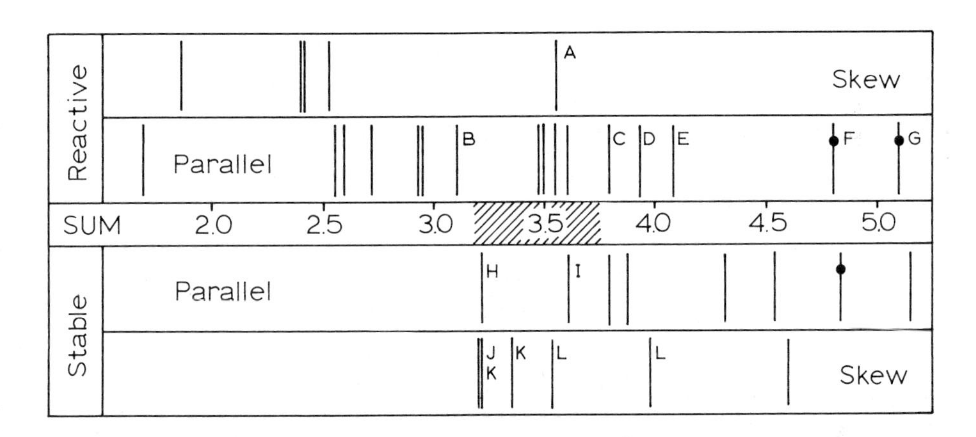

Fig. 18. Summary of the overlap SUM for most of the molecules in Table 1. All the lines show compounds that have atom-atom distances less than 4.2 Å except the lines with the dark circles. Compounds of special interest are labelled with letters:
A: 7-Methoxycoumarin acid (7MC)
B: Benzylbenzylidenecyclopentanone (BBCP)
C: α-Distyrylpyrazine
D: β-Thienylacrylic acid
E: 2,5-Dimethylbenzoquinone
F: Acridizinium perchlorate
G: 3,4-Dimethoxycinnamic acid
H: Complex of 3,4-dimethoxy and 2,4-dinitro cinnamic acids
I: Napthalenone-carbomethoxy derivative (NAPTH)
J: m-Bromocinnamic acid (mBrCA)
K: Chiral DBCP derivative
L: Duroquinone

9 CONCLUSION

It is apparent from the above discussion that explanations in addition to the topochemical rules will be necessary to rationalise the likelihood and course of reactions within single crystals. It is seen that merely stating distances between reacting atoms does not show why these molecules should, or do, participate in reactions. The fact that so many [2+2] cycloadditions seem to obey the 4.2 Å criterion is incidental because orbital configuration and overlap is to a large extent implicit in molecules that are assembled in stacks where each molecule is related to another by either translation or inversion symmetry operations. For other reactions, such fortuitous correlations do not occur.

An attempt to quantify the degree of overlap geometrically by taking the distances between the apices of hypothetical atomic orbitals was successful in explaining anomalies not resolved by measuring atom-atom distances. This is summarised in Figure 18. It can be seen that this geometric construction is more discriminating than measuring atom-atom distances. The hatched area (Fig. 18) indicates a delimiting range between reactive and stable apex orbital SUMs. Compounds not rationalised by the geometric overlap scheme (compounds C to G, H and perhaps I) must entail non-topochemical explanations for reactivity. Such processes which modify the topochemical reactivity pattern are: cooperative mechanical assistance, thermal assistance, excited state reaction intermediates (excimers), diversion into secondary reactions as in the case of competing polymerisation routes and intermolecular steric hindrance. These processes will manifest in varying degrees of importance for different compounds and even in different crystal phases of the same compound.

Reactions develop from the interaction of molecular orbitals as they are excited; this is intimately coupled to the mechanical interaction between molecules and the lattice. To predict reactions with any sort of certainty, both electronic and steric effects must be quantified; this cannot be done by inspection of the point atom description of the crystal structure alone. This does not mean that the topochemical rules are not justified, but put in a realistic perspective, they are very crude. Thus the topochemical delimiters (whether atom-atom or apex-apex distances) are an incomplete, albeit useful tool for rationalising the occurrence of solid state reactions.

Acknowledgement. This work was supported by National Science Foundation Grant DMR-8203662.

REFERENCES

Ariel, S., Askari, S., Scheffer, J.R., Trotter, J. and Walsh, L. 1984 J. Amer. Chem. Soc., 106, 5726.
Ariel, S. and Trotter, J. 1984 Acta Cryst., C40, 2084.
Bamford, C. H., Eastmond, G. C. and Ward, J. O. 1963 Proc. Roy. Soc., 271, 357.
Begley, M. J., Mazid, M. A. and Whiting, D. A. 1978 Acta Cryst., A34, 896.
Bernstein, J. and Lesiserowitz., L. 1972 Isr. J. Chem., 10, 601.
Block, S., Filippakis, S. E. and Schmidt, G. M. J. 1967 J. Chem. Soc. B, 233.
Braun, H-G, and Wegner, G. 1983 Makromol. Chem., 184, 1103.
Cohen, E. R. and Reiss, H. 1963 J. Chem. Phys., 38, 680.
Cohen, M. D. and Schmidt, G. M. J. 1964 J. Chem. Soc., 1996.
Cohen, M. D., Ludmer, Z., Thomas, J. M. and Williams, J. O. 1971 Proc. Roy. Soc., A324, 459.
Cohen, M. D. 1975 Angew. Chem. Int. Ed. Engl., 14, 386.
Cookson, R. C., Cox, D. A. and Hudec, J. 1961 J. Chem. Soc., 4499.
Desiraju, G. R. and Sarma, J. A. R. P. 1983 J. Chem. Soc. Chem. Comm., 45.
Desiraju, G. R., Kamala, R., Kumari, B. H. and Sarma, J. A. R. P. 1984 J. Chem. Soc. Perkin Trans. II, 181.
Desiraju, G. R. and Kannan, V. 1986 Proc. Indian Acad. Sci. (Chem. Sci.), 96, 351.
Desvergne, J.-P., Thomas, J. M., Williams, J. O. and Bovas-Laurent, H. 1975 J. Chem. Soc. Perkins II, 84.
Drenth, W. and Wiebenga, E. H. 1955 Acta Cryst., 8, 755.
Dwarkanath, K. and Prasad, P. N. 1980 J. Amer. Chem. Soc., 102, 4254.
Ferguson, J. and Mau, W-H. 1974 Mol. Phys., 27, 377.
Filippakis, S. E. and Schmidt, G. M. J. 1967 J. Chem. Soc. B, 229.
Filippakis, S. E., Leiserowitz, L. and Schmidt, G. M. J. 1967 J. Chem. Soc. B, page 305 (Muconodinitrile), page 290.
Flory, P. J. 1939 J. Amer. Chem. Soc., 61, 1518.
Flory, P. J. 1953 "Principles of Polymer Chemistry", Cornell Univeristy, Ithaca, New York.
Forster, T. 1969 Angew. Chem. Int. Ed. Engl., 8, 333.
Frank, J. K. and Paul, I. C. 1973 J. Amer. Chem. Soc., 95, 2324.
Gavezzotti, A., 1983 Mol. Cryst. Liq. Cryst., 93, 113; J. Amer. Chem. Soc., 105, 5220.
Gnanaguru, K., Ramasubbu, N., Venkatesan, K. and Ramamurthy, V. 1984 J. Photochem. 27, 355.
Hanson, A. W. 1975 Acta Cryst., B31, 1963.
Haseltine, W. 1983 Cell, 33, 13.
Higgs, M. A. and Sass, R. L. 1963 Acta Cryst., 16, 657.
Hine, J. 1972 J. Amer. Chem. Soc., 94, 5766.
Hine, J. 1977 'Advances in Physical Organic Chemistry', ed. V. Gold and D. Bethele, 15, 1.
Hirshfeld, F. L. and Schmidt, G. M. J. 1964 J. Polym. Sci., A-2, 2181.
Hoogsteen, K. 1963 Acta Cryst., 16, 28.
Jones, W., and Thomas, J. M. 1979 Prog. Solid State Chem., 12, 101.
Jorgensen, W. L. and Salem, L. 1973 'The organic chemist's book of orbitals', Academic Press N. Y. and London.
Kaupp, G. and Zimmermann, I. 1981 Angew. Chem. Int. Ed. Engl., 20, 1018.
Kawada, A. and Labes, M. M. 1970 Mol. Cryst. Liq. Cryst., 11, 133.

Kearsley, S. K. 1983 Ph.D. thesis submitted at the University of Cambridge, U.K.
Kearsley, S. K. and Desiraju, G. R. 1985 Proc. R. Soc. Lond., A379, 157.
Kitajgorodskij, A. 1973 'Molecular Crystals and Molecules', Academic Press, N. Y.
Kittle, C. 1971 'Introduction to Solid State Physics', Wiley and Sons, chapter 5.
Lahav, M. and Schmidt, G. M. J. 1967 J. Chem. Soc. B, page 312 (Muconic acid), page 239.
Leiserowitz, L. and Schmidt, G. M. J. 1965 Acta Cryst., 18, 1058.
Leiserowitz, L. and Schmidt, G. M. J. 1969 J. Chem. Soc., (A), 2372.
McBride, J. M. 1983 Mol. Cryst. Liq. Cryst., 96, 19; Acc. Chem. Res., 16, 304.
Michel-Beyerle, M. E. and Yakhot, V. 1977 Chem. Phys. Letters, 49, 463.
Mirsa, T. N. and Prasad, P. N. 1982 Chem. Phys. Letters., 85, 381.
Nakanishi, H., Ueno, K. and Sasada, Y. 1976 Acta Cryst., B32, 3352.
Nakanishi, H. and Sasada, Y. 1978 Acta Cryst., B34, 332.
Nakanishi, H., Parkinson, G. M., Jones, W., Thomas, J. M. and Hasegawa, M. 1979 Isr. J. Chem., 18, 261.
Nakanishi, H., Jones, W., Thomas, J. M., Hasegawa, M. and Rees, W. L. 1980 Proc. R. Soc. Lond. A369, 307.
Nakanishi, H., Jones, W., Thomas, J. M., Hursthouse, M. B. and Motevalli, M. 1981a J. Phys. Chem., 85, 3636.
Nakanishi, H., Theocharis, C. R. and Jones, W. 1981b Acta Cryst., B37, 758.
Nakanishi, H., Hasegawa, M. and Mori, T. 1984 Acta Cryst., C40, 509.
Nakanishi, H., Hasegawa, M. and Mori, T. 1985 Acta Cryst., C41, 70.
Ohno, K., Mutoh, H. and Harada, Y. 1983 J. Amer. Chem. Soc., 105, 4555.
Prasad, P. N. and Swiatkiewicz, J. 1983 Mol. Cryst. Liq. Cryst., 93, 25.
Rabinovich, D. and Schmidt, G. M. J. 1964 J. Chem. Soc., 2030.
Rabinovich, D. 1967 J. Chem. Soc. B, 140.
Rabinovich, D. and Schmidt, G. M. J. 1967 J. Chem. Soc. B, page 127 (2,6-Dimethyl-1,4-benzoquinone), page 144 (photochemistry of p-quinones), page 286 (Monomethyl muconate).
Rabinovich, D., Schmidt, G. M. J. and Ubell, E. 1967 J. Chem. Soc. B, 131.
Ramasubbu, N. Row, T. N. G., Venkatesan, K., Ramamurthy, V. and Rao, C. N. R. 1982 J. Chem. Soc. Chem. Comm., 178.
Ramdas, S., Thomas, J. M. and Goringe, M. 1977 J. C. S. Faraday II 73, 551.
Sarma, J. A. R. P. and Desiraju, G. R. 1985 J. Chem. Soc. Perkin Trans. II, 1905.
Sasada, Y., Shimanouchi, H., Nakanishi, H. and Hasegawa, M. 1971 Bull. Chem. Soc. Jpn., 44, 1262.
Shimizu, S., Kekka, S., Kashino, S. and Haisa M. 1974 Bull. Chem. Soc. Jpn., 47, 1627.
Schmidt, G. M. J. 1967 'Reactivity of the Photoexcited Organic Molecule', Interscience: London, 227.
Schmidt, G. M. J. 1971 Pure and Applied Chem., 27, 647.
Swiatkiewicz, J., Eisenhardt, G., Prasad, P. N., Thomas, J. M., Jones, W. and Theocharis, C. R. 1982 J. Phys. Chem., 86, 1764.
Theocharis, C. R., Nakanishi, H. and Jones, W. 1981 Acta Cryst., B37, 756.

Theocharis, C. R. 1982 Ph.D. thesis submitted at the University of Cambridge, U.K.
Theocharis, C. R., Jones, W., Motevalli, M. and Hursthouse, M. B. 1982 J. Cryst. and Spec. Res., 12, 377.
Theocharis, C. R., Jones, W., Thomas, J. M., Hursthouse, M. B. and Motevalli, M. 1983 J. Chem. Soc. Perkin II, 71.
Thomas, J. M., Morsi, S. E. and Desvergne, J.-P. 1977 'Advances in Physical Organic Chemistry', ed. V. Gold and D. Bethele, 15, 63.
Thomas, J. M. 1981 Nature, 289, 633.
Tomlinson, W. J., Chandross, E. A., Fork, R. L., Pryde, C. A. and Lamola, A. A. 1972 App. Optics, 11, 533.
Trotter, J. 1960 Acta Cryst., 13, 86.
Warshel, A. and Huler, E. 1974 Chem. Phys., 6, 463.
Williams. D. E. 1972 Acta Cryst., A28, 629.

Chapter 4

PHONON SPECTROSCOPY OF ORGANIC SOLID STATE REACTIONS

P. N. PRASAD

I. INTRODUCTION

Technological applications of solid state reactions are more recognized now than ever before. A recent symposium on the solid state polymerization was a witness of the growing importance of reactions in condensed phase whereby novel products of potential application could be synthesized [1]. One example is a group of polymers called polydiacetylenes which can only be synthesized by reactions in the solid state [2]. Proposed applications of this group of polymers range from temperature indicators and lithography to nonlinear optical devices. The term "crystal engineering" has been introduced to emphasize the importance of the highly selective reaction pathway which solid state chemistry offers [3,4]. An exciting prospect has been the introduction of chirality by using solid state chemistry [5]. Solid state reactions are also significant to pharmaceutical industries. Since most drugs are marketed in the solid dosage form, environmental effect on the chemical stability of solid materials are of significant consequence.

There are many distinct merits which a solid state reaction offers in contrast to that of conventional solution chemistry:

1. Solid state reactions are highly selective. Therefore, the products obtained by solid state chemistry are fewer in number than those produced from the same starting material in the liquid state. In many cases the product of a solid state reaction is different from that obtained from the same reactive molecule in the liquid phase. An example is the photopolymerization of 2,5-distyrylpyrazine which in the solid state yields a crystalline polymer upon photoreaction [6]. In contrast, the product obtained from polymerization in solution is an amorphous polymer.
2. There are examples of reactions such as thermal rearrangement of methyl-p-dimethyl aminobenzene sulfonate which proceed much more rapidly in the solid state than in the liquid state [7].

3. Since solid state reactions are specific to the crystalline environment, different reactivity may be exhibited by different polymorphic forms. An example is provided by the photodimerization of trans-cinnamic acid where the α-modification yields α-truxillic acid but the β-modification produces β-truxinic acid [3].
4. The chiral environment of a reaction site in a crystal can be used to produce an "absolute" asymmetric synthesis with quantitative enantiomeric yield [5]. Thus, a quantitative chiral synthesis can be conducted in the absence of any external chiral agents.
5. Solid state reactions can also be conducted in condensed monolayer and successively built multilayer Langmuir-Blodgett films to produce highly ordered films of products for various electronic and optic applications. Diacetylene carboxilic acids have been polymerized in the form of Langmuir-Blodgett films to produce ordered monolayer and multilayer polydiacetylene films [8].
6. Recently, it has been recognized that electroactive polymers can be produced by using gas-solid interface reaction with AsF_5 [9,10]. The advantage of this method is its simplicity and the ability to polymerize and dope simultaneously.

X-ray diffraction has been the most widely used technique in the past for the study of solid state reactions. This method, however, yields only space and time-averaged results. Consequently, the details of reaction dynamics is lost in such averaging. In order to utilize the full potential of solid state chemistry in crystal engineering of novel materials, it is essential that we understand the reaction dynamics which incorporates the dynamic nature of the lattice. Phonons are low frequency molecular motions (see below) and, therefore, representatives of the dynamic behavior of the lattice. Therefore, phonons serve as excellent probes to study the reaction dynamics.

The method of phonon spectroscopy of solid state reactions developed in this authors laboratory has been found to provide valuable information on reaction dynamics [11-18]. Phonon spectroscopy can conveniently be used to investigate the following aspects of reaction dynamics in solids:

(i) The reaction mechanism. Reactions in solids have been described in terms of either a homogeneous or a heterogeneous mechanism [3]. A homogeneous mechanism involves a single phase transformation where the reactant and the product would form a solid solution. A heterogeneous mechanism involves two separate phases. In the latter case, the reactant and the product segregate in distinct phases.

(ii) The existence of any lattice intermediates.

(iii) The role of low frequency molecular motions. In this regard, the concept of phonon-assisted reactions has been introduced by the research group of this author.

Phonons in organic crystals can be investigated by any of the following techniques: (i) low frequency laser Raman spectroscopy, (ii) electronic spectroscopy at cryogenic temperatures, (iii) X-ray diffraction thermal parameters analysis, (iv) Far-IR spectroscopy, (v) Brilluoin scattering, and (vi) the recent technique of phonon echo [19]. In the investigation of solid state reactions, the first three techniques have been found to be particularly useful [17,18].

This review focuses mainly on the work done by the research group of this author using this novel approach of phonon spectroscopy to investigate the reaction dynamics [11-18]. In sections 2 and 3, the basic concepts of phonon motions and electron-phonon coupling in organic solids are reviewed. Section 4 deals with a discussion of the reaction mechanism and lattice control of a reaction. It also introduces the principles based on which phonon spectroscopy can be used to study reaction mechanisms, lattice control of the reaction, and existence of any lattice intermediate. In section 5 the concept of phonon-assisted reactions is introduced and discussed. Sections 6 and 7 present selected examples of thermal and photochemical reactions. Section 8 presents some new results which demonstrate a novel application of nonlinear laser spectroscopy for the study of photochemical reactions in condensed phase.

2. NATURE OF PHONON MOTIONS IN ORGANIC SOLIDS

The nature of phonon motions in organic solids is described in detail in the review by Venkataraman and Sahni [20]. However for the sake of clarity for those readers who are not familiar with this subject, some basic concepts are reviewed here.

Because of the large difference between the valence force field and the crystal field, distinct molecular units are clearly identifiable in organic solids. Therefore, molecular motions in organic solids can be described by two sets of displacements: (i) displacement within the molecule labelled by internal coordinates $Q_{in\alpha}$ with i, n, and α labelling the degrees of freedom, the unit cell, and the site within the unit cell and (ii) rotational and translational displacements $R_{in\alpha}$ of the entire molecule. The total potential energy V which consists of the valence force field and the crystal field can be expanded in a Taylor series with respect to displacement around the equilibrium position like this [21]:

$$V = V_o + 1/2 \, \Sigma_{i,j} \left(\frac{\partial V}{\partial \xi_i \partial \xi_j}\right)_0 \xi_i \xi_j + 1/3! \, \Sigma_{i,j,k} \left(\frac{\partial^3 V}{\partial \xi_i \partial \xi_j \partial \xi_k}\right)_0 \xi_i \xi_j \xi_k$$

$$+ 1/4! \, \Sigma_{i,j,k,l} \left(\frac{\partial^4 V}{\partial \xi_i \partial \xi_j \partial \xi_k \partial \xi_l}\right)_0 \xi_i \xi_j \xi_k \xi_l \qquad (1)$$

where term $\xi_i = Q_{in\alpha}$ or $R_{in\alpha}$. The first order term vanishes at the equilibrium point. The third and fourth order terms describe the cubic and quartic anharmonic interactions. In harmonic approximation, only terms up to second derivative are retained. Therefore, in harmonic approximation

$$V_h = V_o + \Sigma_{i,j} \left(\frac{\partial^2 V}{\partial \xi_i \partial \xi_j}\right)_0 \xi_i \xi_j \qquad (2)$$

In general, the displacements ξ_i and ξ_j can be internal ($Q_{in\alpha}$) or external ($R_{in\alpha}$). However, based on the large difference between the valence force field and the crystal field, one usually makes the so called "rigid molecule approximation" in which the internal displacements $Q_{in\alpha}$ are decoupled from the external displacements $R_{in\alpha}$. Therefore, the potential V_h now can be separately defined for internal and external vibrations as follows

$$V_h^{internal} = V_o + 1/2 \, \Sigma_{ijk,jn'\alpha'} \left(\frac{\partial^2 V}{\partial Q_{in\alpha} \, \partial Q_{jn'\alpha'}}\right)_0 Q_{in\alpha} \, Q_{jn'\alpha'} \qquad (3)$$

$$V_h^{external} = V_o + 1/2 \, \Sigma_{in\alpha,jk\alpha} \left(\frac{\partial^2 V}{\partial R_{in\alpha} \, \partial R_{jn'\alpha'}}\right)_0 R_{in\alpha} \, R_{jn'\alpha'} \qquad (4)$$

The solution of vibrational problem defined by potential (3) is what describes the motion within the molecules i.e. internal vibrations. These motions are usually in the range 100 - 3000 cm^{-1}. Under crystal field interaction the same internal vibrations of molecules at different sites may couple to produce a vibrational exciton band which is characterized by a wave vector k that describes the phase relation between displacements at different sites. The dispersion of ω_k vs k is called the exciton band width and typically it is very small (generally 5 cm^{-1} or less). This dispersion determines the extent of delocalization of vibrational motion in the crystal. Internal vibrations in organic solids, therefore, are fairly localized.

The solution of the vibrational dynamics matrix utilizing the potential of equation (4) gives the low frequency external vibrations. These vibrations are derived from hindered translations and rotations of the entire molecule and are called phonons. Again, intermolecular coupling leads to a dispersion of the phonon frequency ω_k vs the wave vector k. This dispersion for phonon motions is large, hence phonons in organic solids are highly delocalized [22]. For a crystal containing Z number of molecules per unit cell there are 6Z number of phonon branches (each having ω_k vs k dispersion, i.e. consisting of a closely spaced phonon levels of different k vectors), 3 of them are acoustic phonon branches because at k=0 ω_k=0 for them. Also, for them the group velocity $\frac{d\omega_k}{dk}$ for low value of k is the acoustic speed, hence the name acoustic phonons. Other (6Z-3) phonon branches have $\omega_k \neq 0$ at k=0. They are the optical phonon branches. In an optical excitation, one generates phonons of k=0, because the wave vector of light is very small. Therefore, only optical phonons can be generated. These k=0 optical phonons can be studied by optical spectroscopy such as Raman spectroscopy and Far-IR spectroscopy where they appear in the low frequency (10-200 cm^{-1}) spectral region. If the molecule is at a centrosymmetric site in the crystals, then hindered rotations (librations) and translations do not couple for k=0 (also called the center of Brilluoin zone). Under this situation, phonons observed in the Raman spectra correspond to librational motions of the molecule and have g-symmetry; those observed in the Far-IR spectra are derived from hindered translational motions of the molecules. Often terms such as librational phonons and translational phonons are used to characterize decoupling between these two types of external molecular motions. As an example, naphthalene crystallizes in a $P2_1/C$ space group with two molecules per unit cell each occupying a centrosymmetric site. Therefore, there would be 12 phonon branches in total; 3 are acoustic branches (which at least at k=0 are derived from translational motions of the molecule). The remaining 9 branches are optical phonons. At k=0, six of them are librational phonons which appear in the Raman spectra [23]; the three translational phonons are the u-phonons which can be probed by Far-IR spectroscopy.

For the study of solid state reactions, laser Raman spectroscopy has been found to be very useful. This point has been emphasized and discussed in detail in an earlier review [17]. The advantages are derived mainly from the physical nature of the Raman process. It is a scattering process which in a laser excitation mode permits microsampling, space and time resolution, convenient detection of low frequency vibrational spectra, and ease of sample preparation. For details, readers are referred to the earlier review [17].

Some information regarding averaged librational and translational phonon motions can also be obtained from the rigid-motion analysis of thermal parameters of X-ray diffraction studies. Using the thermal parameters, one can calculate the root mean square amplitudes of librations and translations along the principal axes of the molecule. The averaged librational and translational phonon frequencies (average value of any crystal splittings) can be calculated from these amplitudes using Cruickshank's formula [24]. A nice example of the applicability of this method in obtaining phonon frequencies and relating them to those observed in the Raman spectra is provided by the work by Chen and Prasad [25]. An advantage of this method is that it may provide insight into the nature of molecular displacement associated with a specific phonon mode (frequency). This information can be of significant value in understanding the molecular mechanism of reaction dynamics in solids.

3. ELECTRON-PHONON COUPLING IN ORGANIC SOLIDS

The electronic excitations in organic solids are described in the framework of Frenkel excitons [26]. Here a brief description is presented using the formalism of second quantization which is very convenient for discussing exciton-phonon coupling. For details readers are referred to the review by Hochstrasser and Prasad [27]. In the second quantization formalism the electronic hamiltonian in real lattice space is given as

$$H^f = \Sigma_n (\Delta E^f + D_n^f) B_n^+ B_n + \Sigma'_{n,m} M_{nm}^f B_m^+ B_n \quad (5)$$

This equation for simplicity assumes one molecule per unit cell, where n labels the lattice position. In the above equation ΔE^f is the excitation energy of an isolated molecule in the excited state f which in a pure crystal is independent of n. The term D_n^f is the static crystal field contribution to the site energy at site n. In other words

$$D_n^f = \langle nf|V_{cryst}|nf\rangle - \langle ng|V_{cryst}|ng\rangle \quad (6)$$

Therefore, the energy $\Delta E^f + D_n^f$ is the excitation energy of site n. The operators B_n^+ and B_n are, respectively, the creation and annihilation operators for site n, hence $B_n^+ B_n$ terms describe excitation localized at site n. The term M_{nm}^f is the excitation exchange between sites n and m through crystal field interactions i.e.

$$M_{nm}^f = \langle nf,mg|V|ng,mf\rangle \quad (7)$$

This term leads to delocalization of the excited electronic state throughout the crystal in form of an electronic exciton band which consists of closely spaced exciton levels each characterized by a wave vector k.

In hamiltonian (5), both interactions D_n^f and M_{nm}^f are dependent on the intermolecular separation. Since during phonon motions the intermolecular separation changes, the electronic hamiltonian (5) is modulated by phonons which is the source of electron-phonon coupling. In the adiabatic approximation, the electron-phonon coupling can be calculated by expanding H in a Taylor series in the phonon motion coordinates (librational and translational displacements) R. For small displacements and in the linear coupling limit the electron-phonon interactions derived from the D and M terms can be written as [27]:

$$H_D^f(R) = \sum_{n,m} B_n^+ B_n \, [P_{nm}(R) D_m^f(R)] \tag{8}$$

$$H_M^f(R) = \sum_{n,m}' B_n^+ B_m [P_{nm}(R) M_{nm}^f(R)] \tag{9}$$

In the above equations the operator $P_{nm}(R)$ is given by

$$P_{nm}(R) = \sum_j \left[R_n^j \left(\frac{\partial}{\partial R_n^j} \right)_{R=Ro} + R_m^j \left(\frac{\partial}{\partial R_m^j} \right)_{R=Ro} \right] \tag{10}$$

The index j runs over various intermolecular displacement modes (phonon co-ordinates). Then the electronic hamiltonian becomes

$$H^f(R) = H^f(R_o) + H_D^f(R) + H_M^f(R) \tag{11}$$

The two electron-phonon interaction terms described by equations (8) and (9) have different physical meanings. The interaction $H_m^f(R)$ describes the modulation of electronic delocalization (exciton band) by phonon motions. In other words, it leads to scattering of an exciton level k into another exciton level k' by creating (or destroying) a phonon of wave vector (k-k'). The term $H_D(R)$ describes the variation of the site energy with the phonon motion. It describes the displacement of the molecular equilibrium position (R_o) at site n with excitation to electronic level f. This interaction then leads to a local lattice deformation in the excited state which is operative in a region around the excited state molecule. This term, therefore, tends to localize the excitation energy and reduce the exciton band width. Since the Franck-condon factor for a vibronic transition is determined by the extent of displacement of the excited state equilibrium position, the electron-phonon coupling strength $H_D(R)$ which leads to the displacement of the equilibrium position of the phonon co-ordinate in the excited state, determines the strength of the phonon side bands in the electronic spectra.

In the limit $H_D(R) >> H_M(R)$, the electronic excitation is described by a localized picture and the electronic spectra will exhibit a prominent phonon side band on the high energy side of the zero-phonon line (pure electronic transition) at liquid helium temperatures. The relative strength of the phonon side band with respect to the zero-phonon line, therefore, can be related to the strength of the electron-phonon coupling H_D. For a detailed quantitative description readers are referred to the review by Hochstrasser and Prasad [27].

As H_D increases in relation to H_M, the electronic excitation becomes more localized due to excited state equilibrium displacement. A prominent but structured phonon side band appears along with the zero phonon line in the liquid helium spectra. In the limit of $H_D >> H_M$, the excitation is localized, the ratio of intensity of phonon side bands to zero-phonon lines becomes larger than unity and processes involving more than one phonon may give rise to the phonon side band. In this limit, the excited state has a fairly distorted local geometry due to local lattice relaxation. However, the individual molecules still retain their identity in terms of the vibronic features (involving intramolecular vibrations) in the electronic spectra. This is the case of formation of a polaron which is an electronic excitation dressed with phonons due to local lattice relaxation. The excitation can still hop around from site to site as a localized unit. In this process of energy transfer, as the electronic excitation moves from site a to another site b, the equilibrium displacement at site a disappears because it returns to the ground state, while a local lattice deformation builds at site b which now is in the excited state.

In the limit of severe dynamic distortion, an excited state molecule may deform sufficiently to come close to another molecule which is in the ground state and form an excited state dimer geometry. This excited state dimer is called an excimer. It should be pointed out here that what we have discussed is a dynamic mechanism (electron-phonon induced) for the formation of an excimer. There are cases where the excimer formation is promoted by a defect site which provides a suitable geometry for an excited state dimer formation. The manifestation of an excimer state formation is that the electronic spectra even at liquid helium temperature are very broad and featureless due to large lattice distortion; a large stokes shift (gap) between the adsorption maxima and the emission maxima is observed.

4. REACTION MECHANISM AND LATTICE-CONTROL

Phonon spectroscopy can conveniently be used to determine the reaction mechanism and the existence of any lattice intermediate as a consequence of lattice control of reaction in a solid. Raman phonon spectra can be monitored as a function of reaction progress to obtain this information. The basic principle of phonon probing of reaction mechanism is that phonon motions are determined by crystal field and symmetries. Therefore, they exhibit a large manifestation of change in crystalline interactions and crystal symmetries. Furthermore, for most organic solids the important intermolecular interactions are short range, hence phonon motions are particularly sensitive to the local organization of molecules and probe the immediate (nearest neighbor) structure of a given crystalline site. If this local environment changes by a solid solution formation, or by formation of a lattice intermediate, the phonon spectra consequently exhibit a change.

Our work on phonon bands in a multicomponent system provided the basis for developing the concept of phonon probing of solid state reactions. Our work on phonons in multicomponent systems is discussed in detail in two earlier reviews [17,22]. Here we summarize only the important results for the sake of clarity with an example of a binary system which may exemplify the reactant and the product. A binary system can be classified in three categories: (i) a solid solution, (ii) a segregated two phase physical mixture, and (iii) a new phase different from that of either component.

Our work has shown that phonon bands in a binary solid solution show a monotonic shift in frequency from the phonon spectrum of one component to that of the other component as the concentration of the solid solution is changed [17,22]. The shift in frequencies is also accompanied by a broadening of the phonon transition. The frequency shift can be visualized as a consequence of the perturbation on the local structure caused by random mixing of the two components at the molecular level in a solid solution. One does not observe an increase in the number of phonon bands (one set due to one component and the other set due to the second component) because of the highly delocalized nature of phonons. In a simple minded picture one can simply say that a phonon motion propagates as a highly delocalized wave over both components because the perturbation due to differences between them is small compared to the delocalization effect. Consequently, the binary solid solution acts as an amalgamated system for phonon bands and only one effective phonon set shifted in frequency represents both components. This behavior has been labelled as the amalgamation limit [22,23]. The broadening of phonon transition is simply the result of substitutional disorder which scatters the phonon waves.

In the case of a two-phase physical mixture there is no mixing at the molecular level and no perturbation of the local structure results. As a result, no perturbation on the phonon spectra are observed; the observed spectra of a binary system in this case are simply a superposition of that of individuals components. If the two components form a new crystalline phase at a specific concentration (because of either a lattice intermediate or a crystalline complex formation), the local environment has a different symmetry and crystal field. Consequently, the phonon spectrum observed is different from that of either component.

Phonon spectra can also be used to determine the nature of order in the lattice. The presence of disorder, for example due to strain produced by the product formation, leads to temperature independent broadening of phonon transitions in the phonon spectra. By investigating the extent of temperature independent broadening of the phonon spectra one can investigate the extent of order in a lattice.

Now we come back to the application of phonon spectroscopy specifically to solid state reactions. In the case of a homogeneous reaction, a solid solution forms between the reactant and the product. The phonon spectra, in such a case, should show a monotonic change accompanied by broadening of the transitions as the reaction progresses. In the case of a heterogeneous reaction, the phonon spectrum of the product superimposes on that of the reactant, their relative intensities changing with the advancement of the reaction. In case where the reaction proceeds through an intermediate crystalline phase, a different phonon spectrum will be observed with the reaction progress. Raman spectroscopy is applicable to all solid state reactions and, therefore, more general. Unfortunately, Raman spectroscopy is limited in its sensitivity, and is very useful in probing the conversion of > 1%.

In cases where the reactant and/or the product show electronic emission under the reaction conditions, the emission spectra can also be used to derive information on the reaction mechanism [14]. A homogeneous mechanism will create a homogeneous internal strain field as the reaction proceeds. This strain field will give rise to a shift of the emission maximum and a broadening of the spectral profile. Emission spectroscopy is highly sensitive and more suitable for the very initial stage of the reaction. The limitations of emission spectroscopy are the lack of detailed information regarding the chemical nature of the rearrangement, its dependence on the crystal quality, and the complication due to energy transfer. Therefore, emission spectroscopy can provide valuable information regarding the mechanism of a process at the initial state (< 1%) where the Raman method is not sensitive. After the initial state and throughout the conversion range (> 1%), Raman spectroscopy is the preferred technique for the investigation of a reaction.

A solid state reaction may involve a lattice intermediate. The possibility of a lattice intermediate is more when a reaction is intermolecular in nature. Examples of such reactions are intermolecular rearrangements and intermolecular aggregation (such as polymerization). The use of Raman spectroscopy permits us to investigate the possible existence of both lattice and molecular intermediates. The reason is that, in the same experimental arrangement, we can obtain both the Raman phonon spectra and the intramolecular vibration spectra as a function of the reaction progress. Although ultra-short time resolutions can easily be achieved in Raman spectroscopy, often one needs time resolution of only several seconds. The reason is solid state reactions are usually slow which need time scales of minutes to hours. The Raman phonon spectra are used as probes to investigate the occurrence of any lattice intermediates. The intramolecular vibrations give information on the existence of any molecular intermediates. A temperature dependence analysis of the phonon line widths of the product in the Raman spectra can be used to derive information on the disorder-induced line width. From this information one can get insight into the nature of order and strain in the product lattice.

5. CONCEPTS OF PHONON-ASSISTED REACTIONS

We have introduced the concepts of phonon-assisted reactions which suggests that both photochemical and thermal reactions in solid state can be assisted by phonon motions [11-18]. The concept of phonon-assisted chemical reactions bears a direct analogy to that of a phonon-assisted physical transformation in solids. Important phonon interactions can be divided in two categories: (i) anharmonic phonon-phonon interactions and (ii) electron-phonon interactions.

As discussed in section 2, vibrations in an organic solid are treated in the framework of harmonic approximation under which second derivative of the crystal potential with respect to displacements in equation (1) are retained. The anharmonic potential can then be written as

$$V_A = 1/3! \sum_{i,j,k} \left(\frac{\partial^3 V}{\partial \xi_i \partial \xi_j \partial \xi_k}\right)_0 \xi_i \xi_j \xi_k + 1/4! \sum_{i,j,k,l} \left(\frac{\partial^4 V}{\partial \xi_i \partial \xi_j \partial \xi_k \partial \xi_l}\right)_0 \xi_i \xi_j \xi_k \xi_l \qquad (12)$$

Depending on the nature of the displacement co-ordinate ξ, the anharmonicity can be divided in three categories: (i) external lattice anharmonicity for which all displacements ξ in equation (12) are external librational and translational displacements R (phonon co-ordinates); (ii) internal anharmonicity for which all displacements ξ are intramolecular motions Q; and (iii) mixed mode anharmonicity which involves both R and Q. The anharmonic phonon-phonon interactions important in relation to thermal reactions are derived from the external lattice anharmonicity. Manifestations of the anharmonic phonon-phonon interactions are temperature dependent broadening and frequency shift of phonon transitions. Expressions for the line width and the frequency shift derived from anharmonic phonon-phonon interactions can be found in reference 28. The frequency shift is towards lower values as temperature increases.

In most organic solids, as the temperature increases a phonon band broadens and shift in frequencies to lower values but the width is still small compared to the frequency. As the frequency shifts to lower values, the amplitude of a phonon motion increases. In cases where anharmonic phonon-phonon interactions are strong for a specific phonon mode, a large shift in phonon frequency towards zero value accompanied by a rapid broadening is observed as temperature increases. This behavior is called mode-softening and the phonon motion is then overdamped which leads to a large amplitude motion along the phonon co-ordinate.

In many systems, mode-softening creates a lattice instability which gives rise to a phase transition. In a reactive system, large amplitude displacements, owing to mode-softening, are analogs of molecular collisions in a gas phase. Thus mode-softening can be expected to assisted in reactivity. This possibility has led us to propose the new concept of phonon-assisted reactions [11]. The role of phonon mode-softening in a thermal reaction can easily be investigated by Raman spectroscopy if the soft-mode is a Raman-active optical mode. The mode-softening can be observed as a gradual shift of the phonon frequency to the zero value accompanied by a rapid broadening of the transition with the increase of temperature. Some information about the co-ordinate of the soft mode can be derived from the thermal motion anisotropy obtained from the thermal parameter analysis of the x-ray diffraction studies as discussed in section 2.

The electron-phonon interactions, discussed in section 3, can give rise to phase-transitions such as metal-insulator transitions. Electron-phonon interactions can also assist photochemical reactions in solids [12,13,18]. Photochemical aggregation reactions, such as dimerization or polymerization reactions, can be assisted by the occurence of strong electron-phonon interaction in the reactive electronic state. This strong electron-phonon interaction creates a local lattice-deformation in the reactive (excited) electronic state. The deformation traps the electronic excitation and, at the same time, it may provide a local preformation of the product lattice if the distortion is along the reaction co-ordinate. Both these features assist a photochemical aggregation reaction. Depending on the strength of the electron-phonon coupling, one may observe the formation of a polaron or an excimer. As discussed in section 3, the formation of a polaron does not lead to the loss of the identity of the monomer in the excited state, but simply the excitation is localized by local lattice-deformation [13]. The excimer formation requires a severe distrotion of the local structure which leads to an excited state dimer. It may also be pointed out, that the polaron mechanism is a purely dynamic effect which can occur even in a defect-free lattice. In contrast, the excimer formation can occur either by a dynamic effect due to strong electron-phonon coupling or by a static effect due to sites deformed by the presence of defects as discussed in section 3.

The information on the formation of a polaron or an excimer is derived from the low temperature electronic absorption and emission spectra of the reactive crystals. As discussed in section 3 a strong electron-phonon coupling in the reactive state manifests itself as a very strong phonon-side band in the liquid helium temperature spectra.

6. EXAMPLES OF THERMAL REACTIONS

(i) Thermal rearrangement of methyl-p-dimethyl aminobenzene sulfonate.

The thermal rearrangement of methyl-p-dimethyl aminobenzene sulfonate (abbreviated as MSE) in the solid state yields p-trimethylammmonium benzene sulfonate (abbreviated as ZWT). This reaction serves as a beautiful example where phonon spectroscopy has proved to provide valuable information on the reaction dynamics. This reaction can be represented as follows [7]:

$$(CH_3)_2N-C_6H_4-SO_3-CH_3 \longrightarrow (CH_3)_3\overset{+}{N}-C_6H_4-SO_3^-$$

MSE ZWT

The reaction involves the transfer of an ester methyl group from the oxygen atom of the MSE molecule to the nitrogen atom. Kuhn and Ruelius [29] reported this reaction occurring at room temperature and proposed an intermolecular migration of the methyl group from the SO_3-CH_3 group of one molecule to the $(CH_3)_2N$ group of a neighboring molecule.

Sukenik et al [7] conducted a thorough investigation of this rearrangement reaction by using X-ray diffraction studies, deuteration effect study, and field desorption mass spectrometry. They established the intermolecular nature of the methyl group migration. Their kinetic study confirmed the previous work in that the rate of this methyl rearrangement reaction increases significantly with an increase in temperature. At a temperature of 81°C, only 14° lower than the melting point of MSE, it takes only 20 minutes to convert nearly half of MSE into ZWT, while to acheive the same extent of conversion into the product takes a week at room temperature.

The reaction rate drops rapidly when MSE is heated to the melting point. Also, this reaction does not occur in solution, MSE can be stored in the solution phase for a long time. Hence the thermal rearrangement of the methyl group is controlled by a specific orientational (topochemical) effect in the solid state. The X-ray diffraction study of MSE confirms this interpretation [7]. The MSE crystal structure reveals a column structure in which the molecules are stacked along the b-axis, but are tilted around molecular axes so that neighboring molecules are not parallel to each other. Nevertheless, such an arrangement causes the distance between the carbon atom of the ester methyl groups and the nitrogen atom of the adjacent molecule to be 3.54Å. According to the topochemical principle, this close alignment of the intermolecular reaction centers is ideal and provides a straightforward explanation for increased reaction rates for this reaction in the crystal phase.

As the reaction involves a methyl group transfer, the effect of this small molecular rearrangement in the Raman spectra of intramolecular vibrations is found to be mild. One of the spectral region which shows a more pronounced effect of this rearrangement is the ~760 cm^{-1} mode. This can be a vibration of the methyl group. The Raman phonon spectra have also been studied as a function of the reaction progress to determine the mechanism of the reaction [11]. Figure 1 shows the phonon spectra of MSE, a partially converted sample, and ZWT.

The phonon spectrum of the partially converted sample can easily be seen to be a simple superposition of those of MSE and ZWT. This feature is found for samples of the entire conversion range. Therefore, a heterogeneous mechanism is found for the entire conversion range of the thermal rearrangement reaction of MSE. Also, this study provides no evidence for any lattice or molecular intermediate for this reaction.

The thermal reaction is intermolecular and, therefore, it involves a cooperative transformation. Both the thermal nature and the cooperative transformation make this reaction suitable to test the applicability of the model of a phonon-assisted reaction which involves mode-softening. Figure 2 shows the phonon spectra of pure MSE at three different temperatures [11].

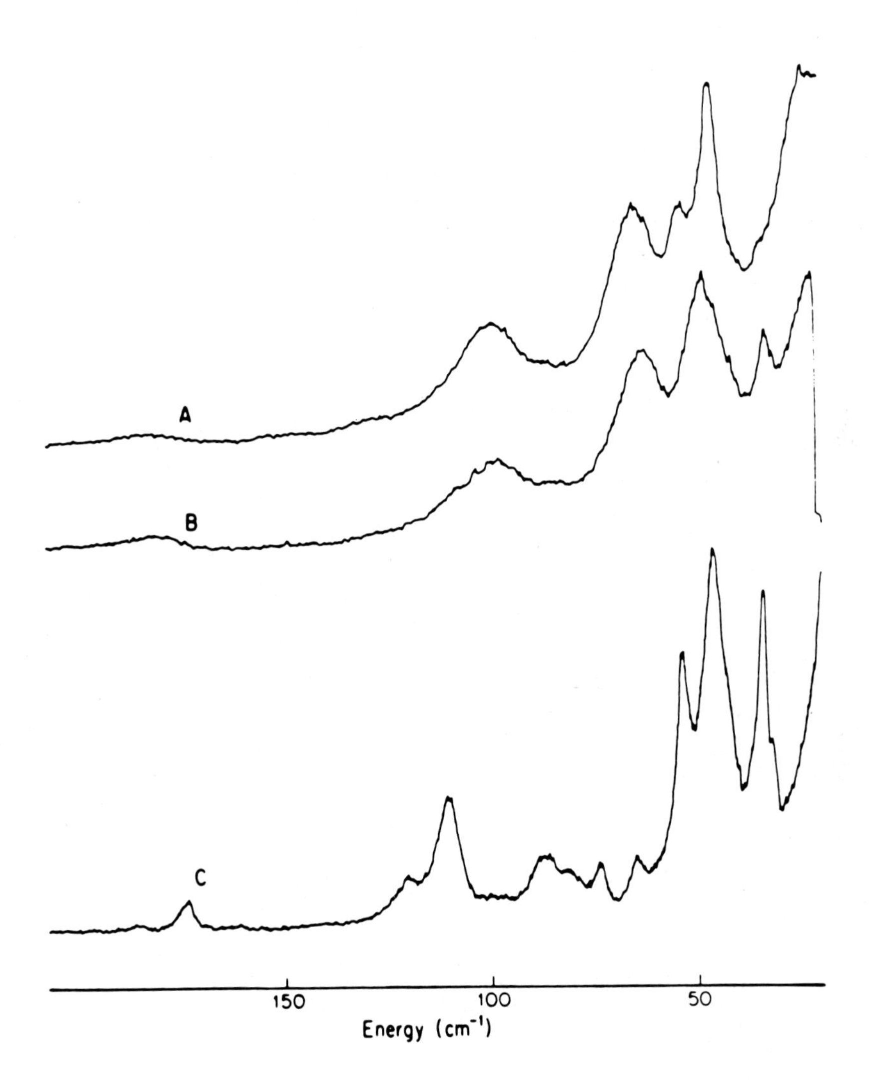

Fig. 1. The Raman phonon spectra of (A) MSE, (B) a partially converted sample, and (C) ZWT at 152 K. (Reproduced from Ref. 11.)

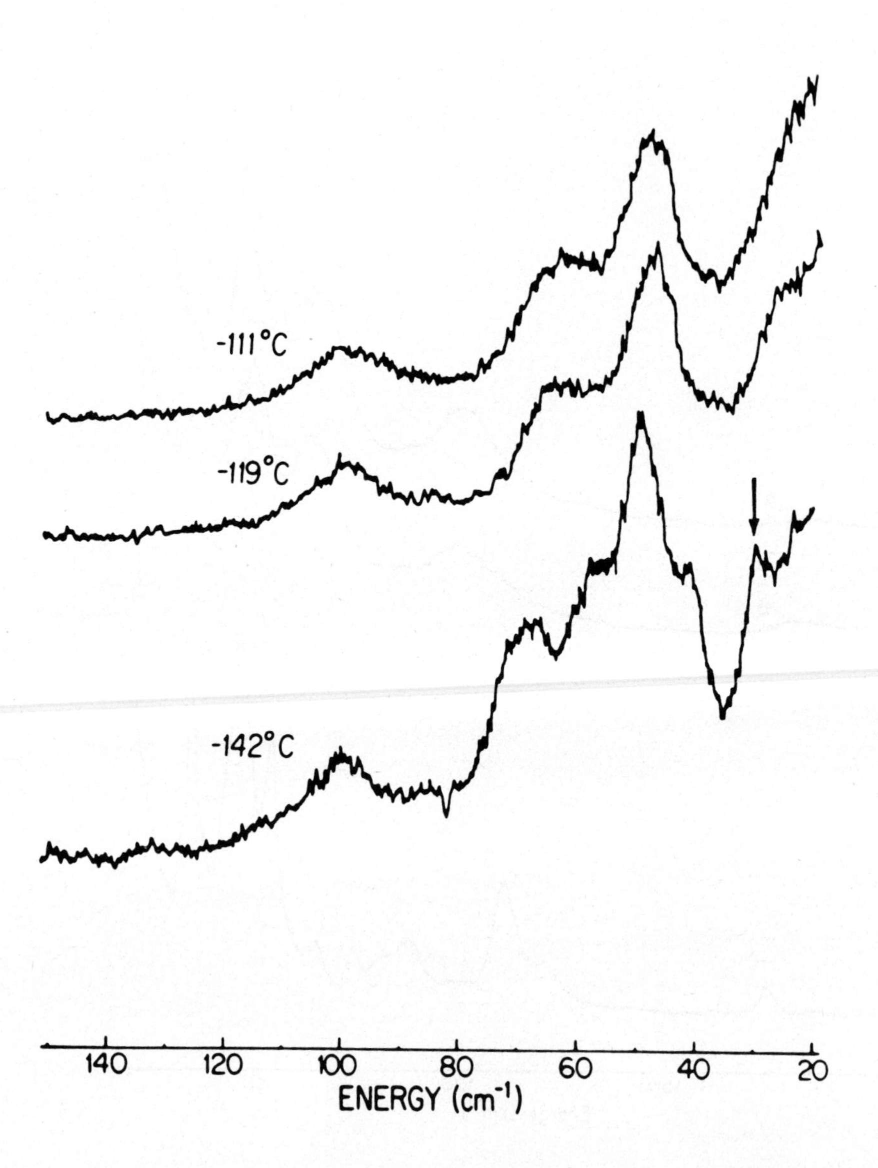

Fig. 2. The temperature dependence of the Raman phonon spectra of MSE. The soft mode is indicated by an arrow. (Reproduced from Ref. 11).

As the temperature increases, the bands shift to lower frequencies and broaden. This effect is more pronounced for the band at 27 cm^{-1} which is indicated in the figure by an arrow. By 162 K this band broadens considerably and shifts to merge with the Rayleigh wing. This mode is indentified as a soft mode, and this observation clearly lends support to a dynamic model of reactivity assisted by phonons in the lattice.

For further confirmation of the mode-softening and a possible identification of the molecular nature of the over-damped mode, we used the rigid-body motion analysis of the thermal parameters of the room temperature x-ray diffraction study. A thermal-motion analysis (TMA) program was used to calculate the components of the libration (L) and the translational (T) tensors with a least-square fit of the published thermal parameters [7] of all non-hydrogen atoms of the molecule. The librational frequencies were calculated by the method of Cruickshank [24], using the appropriate eigenvalues of the L-tensor and the corresponding moments of inertia. Figure 3 shows the results of this calculation.

The amplitude of libration along the L_1 axis is the largest. The corresponding librational frequency calculated by Cruickshank's formula is 23.5 cm^{-1}. On the other hand, the frequency of the 27 cm^{-1} soft mode at room temperature is 18 cm^{-1}. Considering the simplifying assumptions involved in the use of Cruickshanks formula, the agreement between the two values is reasonable. In other words, L_1 libration appears to be the soft-mode.

Figure 3 also shows the orientation of the two neighboring molecules of the stack along the b-axis together with their librational principal axes. The L_1-axis is perpendicular to the b-axis. The dashed lines refer to the methyl sulfonate group and the neighboring dimethyl group positions during the L_1 libration. It can be seen that during this libration the two reactive groups come close to each other. Therefore, a mode softening of the L_1 libration can assist this rearrangement reaction. We conclude from this thermal motion analysis, in combination with the temperature effect on the Raman phonon spectra, that this reaction is phonon-assisted.

(ii) Thermal rearrangement of dimethyl-3,6-dichloro-2,5-dihydroterephthalate

The thermal rearrangement reaction of dimethyl-3,6-dichloro-2,5-dihydroxyterephthalate leads to the conversion of a yellow crystal to a white crystalline isomer. The reaction rate again is highly dependent on temperature. This rearrangement process has been extensively investigated by Curtin and co-workers using the x-ray diffraction technique [30,31].

In solutions, this compound also exhibits yellow and white forms [32]. A yellow solution is obtained in weakly interacting solvents such as

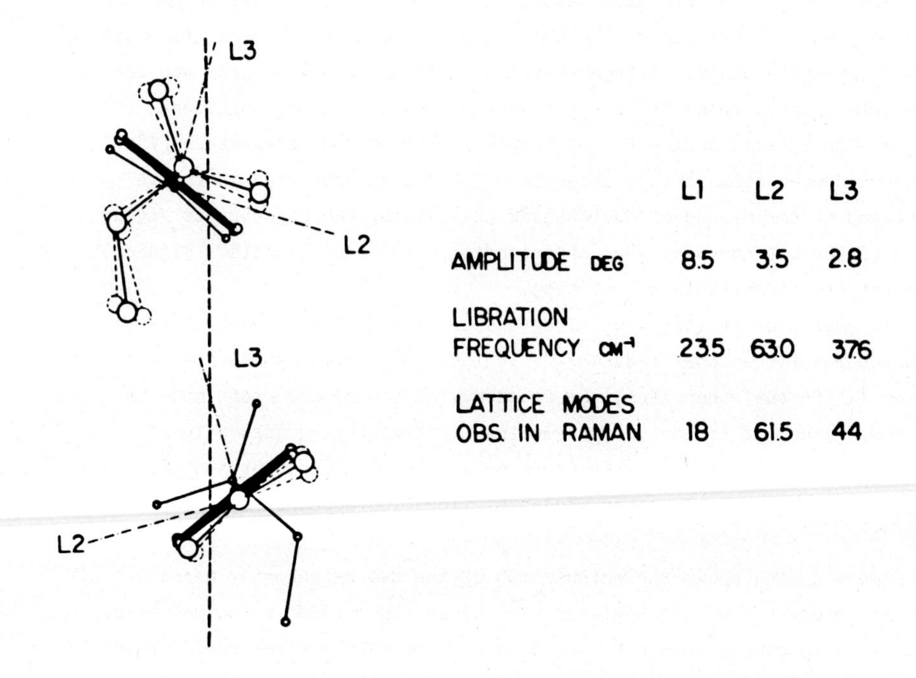

Fig. 3. The projection of two neighboring molecules of MSE, in the plane containing the b-axis as the vertical axis. The phonon frequencies and amplitude calculated from the thermal parameters are displayed along with the frequencies observed in the Raman spectra. (Reproduced from Ref. 18).

chloroform. On the other hand, a white solution results in a solvent like methanol, which forms strong hydrogen bonds. The spectroscopic studies of these solutions reveal an equilibrium between two limiting structures [32]. The two forms differ in the nature of the hydrogen bond. The Y-form(yellow)

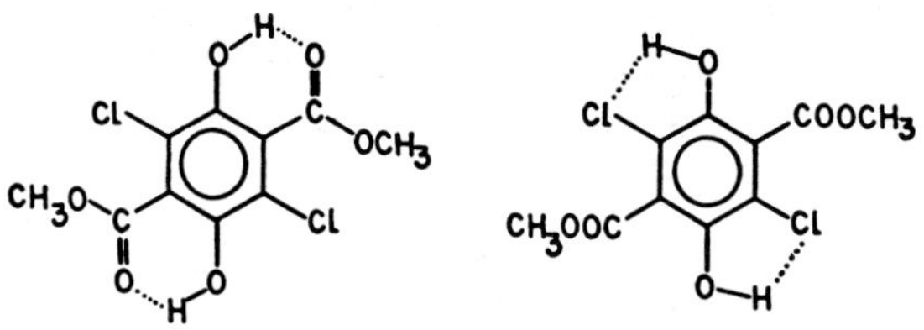

Y W

involves a hydrogen bond to the carbonyl oxygen atoms at each end of the molecule. The W form (white) has two intramolecular hydrogen bonds to the chlorine atoms. The molecular structures of the Y and W forms in the solid state differ in the conformation of the carboxymethyl groups with respect to the benzene ring. They are coplanar in the Y form but are distinctly out of the plane of the benzene ring in the W form. In the solid state, the Y form still contains the hydrogen bonding, which is predominantly intramolecular and of a chelate type. A detailed structural analysis [28,29] of the W crystalline form, however, suggests a bifurcated hydrogen bond involving an intermolecular O-H•••O bond and an intramolecular H•••Cl bond. Since there are differences in both the molecular conformation and the nature of the hydrogen bond, the two forms have been classified as chemical isomers [30,31].

Although both isomers crystallize in the same space group, $P\bar{1}$, the unit cell of the Y isomer contains one one-half molecule per asymmetric unit while that of the W isomer contains two one-half molecules per asymmetric unit. In other words, the conversion of the Y form to the W form leads to a doubling of the unit cell. A closer inspection of the spatial arrangement of the W structure reveals that it can be generated from the Y structure if a considerably large reorientation of the molecule is assumed. Specifically, a reorientation of nearly 180° of half of the molecules around the Cl-Cl axis is required in addition to the conformational change [30,31]. This feature suggests that the reaction may be a good candidate to be phonon assisted.

A search for any lattice or molecular intermediate by Raman study of this rearrangement reveals that when the Y-isomer crystal is converted to the W-isomer crystal at high temperature, the modes associated with the hydrogen-bond in the W-isomer do not show up immediately [14]. Once the crystal is cooled and left for some time, these modes appear. This observation suggests that, in the conversion of the Y-isomer to the W-isomer, the development of the intermolecular hydrogen bond is probably the last process involved in the rearrangement mechanism.

Because the crystals of the Y-isomer, fortunately, emit at the conversion temperature, both the electronic emission spectra and the Raman phonon spectra were used in conjunction to investigate the reaction mechanism [14]. The emission spectra were recorded as a function of time at 383 K and results are shown in Figure 4.

The results of the emission studies were interpreted as follows [14]. During the initial phase of rearrangement, the reaction process creates a homogeneous internal strain-field, which gives rise to the shift of the emission maximum and a broadening of the spectral profile. (c.f. curves 1 and 2 in Figure 4). This homogeneous internal strain may arise either from the formation of the W-isomer randomly distributed as a solid solution in the Y-isomer lattice, or from an intermediate structure which converts to the W-isomer later. As the rearrangement, at this stage, is too small to be probed by Raman spectroscopy, we cannot conclusively establish the exact nature of the rearrangement. In either case, the results of the study indicate that the rearrangement, at this initial state involves a homogeneous mechanism.

The distortion of the reactant lattice, created by further rearrangement, is severe enough to break the lattice and crack the crystal. However, despite the crystal fracture, the emission spectra show that the strain created within the reactant lattice still exists. (c.f. curves 2 and 3). Perhaps, a small phase separation of the W-isomer product may already occur at this state of the rearrangement. The process of phase separation of the W-isomer is evident when the Raman phonon spectroscopy becomes adequately sensitive to the conversion. At this conversion, the observed phonon spectra are found to be superposition of the unperturbed phonon bands of the Y-and the W-isomers which indicate a heterogeneous behavior. Therefore, the thermal rearrangements of the Y-isomer in the solid state is homogeneous only at the very initial stage of concentration, which is not detectable by the sensitivity limit of Raman phonon spectroscopy.

We used the rigid motion analysis of the thermal parameters in conjunction with the Raman phonon spectra to make assignments of the phonon bands. The temperature dependence study of the phonon spectra of the Y-isomer did not show any phonon mode-softening. In this case, our study suggests that the rearrangement is defect-controlled, and not phonon-assisted.

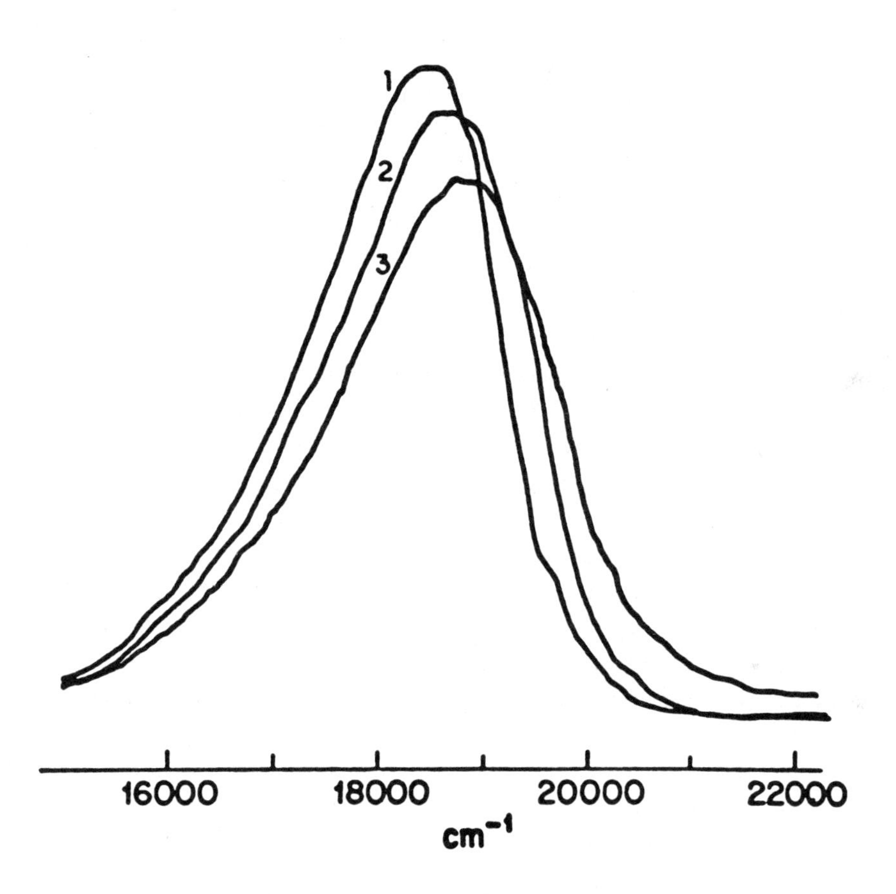

Fig. 4. The emission spectra of the Y-isomer single crystal, monitored as a function of time, after raising the temperature of the crystal to 383 K. The spectra 1, 2, and 3 were obtained after 5, 100, and 175 minutes respectively. (Reproduced from Ref. 14).

(iii) <u>Thermal rearrangement of bis (0-Iodobenzoyl) peroxide into 1-(2'-Iodobenzoyloxy)-1, 2-benziodoxolin'-3-one</u>.

The thermal rearrangement of bis(0-iodobenzoyl) peroxide (Structure I shown below) in solid state yields 1-(2'-iodobenzoyloxy)-1,2-benziodoxolin-3-one (Structure II shown below) (33,34).

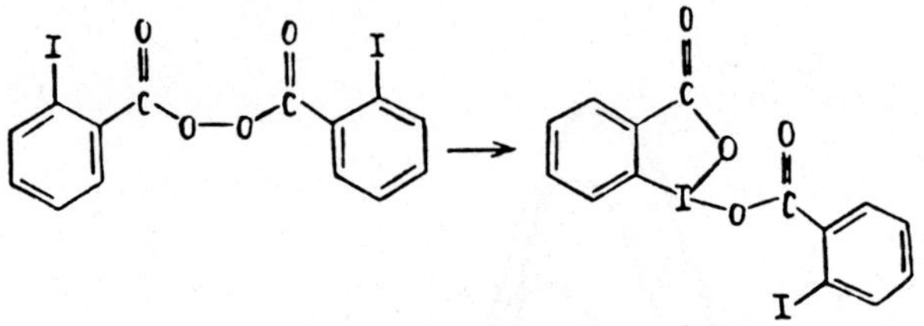

I II

This reaction has been suggested to take place by a free radical mechanism involving the rupture of the peroxide bond. The reaction can be greatly accelerated by heating the reactant crystal. The same rearrangement also takes place in several solutions. This specific thermal rearrangement, in solid state, exhibits a number of interesting features [33,34]: (i) The reaction has been classified topotactic, because it yields a crystal of the product in which the molecules assume preferred orientations relative to crystallographic directions of the reactant structure. (ii) The product has been found to exist in two polymorphic crystalline forms α and β, but the rearrangement below the $\alpha \rightarrow \beta$ transformation temperature (110°C) has been suggested to produce only the α-modification of the product. (iii) Although, the rearrangement reaction is intramolecular, the formation of the product lattice from the reactant lattice also involves a significant molecular reorientation. (iv) The oriented product produced by this rearrangement may be degraded further by two different, competing and consecutive reactions to produce 0-iodosobenzoic acid and 0-iodobenzoic acid.

The phonon spectroscopy was used to obtain the mechanism of reaction and to investigate if the reaction is phonon assisted [15]. The phonon spectra were monitored during the rearrangement. As the rearrangement progresses, the phonon bands of the product appear to gain intensity while those of the reactant decrease in intensity. During the intermediate conversion range, the observed phonon spectra are simply a superposition of the unperturbed phonon bands of the reactant and the product. This observation indicates a heterogeneous mechanism for this thermal rearrangement reaction.

To investigate the possibility of phonon-mode softening, which may assist the rearrangement, a temperature dependence study of the phonon spectra was conducted [15]. Except for a small shift in the peak positions and spectral broadening, no evidence for a phonon-mode softening is observed. From this result, it may be assumed that the process is not phonon-assisted. In view of the fact that the rearrangement reaction is primarily intramolecular, it is not surprising that the process is not phonon-assisted.

7. EXAMPLES OF PHOTOCHEMICAL REACTIONS

(i) Four-center type photopolymerization.

Four-center type photopolymerization in the crystalline state describes the reaction of a group of compounds containing a conjugated diolefinic group which has been extensively investigated by Hasegawa and co-workers [6]. By photopolymerization, the crystal of the monomer is converted into crystals of a linear polymer containing cyclobutane rings. Two specific examples of this class of photoreactive materials are 2,5-distyrylpyrazine (abbreviated as DSP) and 1,4-bis (β-pyridyl-2-vinyl)benzene (abreviated as P2VB). Of these two, DSP has been more extensively studied because it is more reactive. Phonon spectroscopy has successfully been used to study the reaction mechanism of DSP [12,16]. In addition, the concept of phonon-assisted reaction appears to be applicable to the photopolymerization of DSP. It also explains the enhanced photoreactivity of DSP compound to P2VB [18].

First we discuss DSP itself. The photopolymerization of DSP can be represented as follows:

2, 5-Distyryl pyrazine   Poly-2, 5-distyryl pyrazine

Some of the very interesting features of this photoreaction are listed below:

(i) Two crystalline modifications of 2,5-distyryl pyrazine are known [35]. One polymorphic form (the γ-form) is obtained in needle shapes by sublimation. This polymorphic form does not exhibit photopolymerization. The other form (the α-form) is obtained as plates from solution. It is this plate form which undergoes a rapid polymerization whereby the bright yellow crystals of 2,5-distyryl pyrazine turn white upon polymerization.

(ii) The solid state polymerization yields a polymer with high crystallinity, but in solution or molten state the polymerization results in an amorphous oligomer [6].

(iii) The polymerization process exhibits an interesting wavelength dependence [6,35]. When excited with the light of wavelength shorter than 4000 A, a high molecular weight polymer is produced. On the other hand, when irradiated with the light of wavelength longer than 4000 A, 2,5-distyryl pyrazine (the reactive polymorphic form abbreviated as α-DSP) converts only into an oligomer (pentamer on the average). This effect can be attributed to the shift of the electronic transition to higher energy with the growth of the polymer chain. As the chain increases to greater than five monomeric units, the energy required for further polymerization (excited state energy of a five unit oligomer) exceeds that of 4000 A light.

(iv) The temperature dependence of the photopolymerization of α-DSP indicates that more favorable conditions exist at temperatures far below rather than near the melting point of the monomer crystal [6]. The explanation offered is that a temperature much lower than the melting point is needed to maintain the rigidity of the crystal lattice of the reactant so that the crystalline polymer may be formed.

(v) An X-ray diffraction study of the crystal structure reveals that the principal features of the molecular arrangement in the monomer crystal remain unaltered in the polymer [36]. Therefore, the polymerization proceeds by minimum movements of the nuclei. This led to an earlier belief that the photopolymerization of α-DSP proceeds by a homogeneous mechanism whereby the reactants and the products form a solid solution [36].

Although the molecular rearrangement in α-DSP is small, a single crystal develops parallel cracks upon irradiation. These cracks develop further with the increase of the exposure time. As a result, the original crystal does not remain a single crystal after photopolymerization. However, the product still contains highly oriented polymer fibers.

This reaction is ideally suited for the investigation of the lattice control of the reaction mechanism because a stepwise photoreaction can be used to stabilize the intermediate oligomer structures. In our work, the 476.5 nm line of an argon-ion laser was used to produce oligomers [16]. At this wavelength the monomers absorb but not the oligomers. Consequently, the 476.5 nm excitation produces only finite chain oligomers by photoreaction. Further photoreaction of these oligomers can be carried out with a 355 nm (3rd harmonic) or 417 nm (Raman shift of 355 nm) wavelength from a pulse Nd Yag laser.

Figure 5A shows changes in the phonon spectra as the monomer is converted to the oligomer by the 476.5 nm excitation; Figure 5B shows changes in the phonon spectra when the oligomer is converted to a long chain polymer by excitation with the 355 nm excitation. First, the phonon spectra of the monomer, the oligomer, and the polymer will be compared. One can see that the phonon bands observed in the oligomer spectra are considerably broader than those observed in both the monomer and the polymer spectra. This result indicates the presence of a considerable amount of disorder in the oligomer lattice caused, possibly, by the lattice strain due to a substitutional disorder (presence of dimer, trimer, etc.). In contrast, the phonon spectra of the polymer contain sharp features indicating a highly ordered polymer lattice. Therefore, the process of conversion of the oligomer to the polymer involves ordering of the product lattice.

The phonon spectra monitored as a function of oligomerization (Figure 5A) reveal a continuous change in the frequency and the intensity of various bands. These changes, as discussed in section 4, are representative of a homogeneous reaction mechanism. In the same Raman study we also monitored the intramolecular vibrations as a function of the reaction progress. Our result indicates that at the start of the reaction, although the intramolecular vibration spectra show only a mild change, the phonon spectra change drastically. Since the intramolecular modes are more sensitive to chemical changes, this result indicates that there is an initial induction period in which a small amount of the product formed produces a substantial lattice rearrangement to facilitate further formation of the product.

The Raman phonon spectra monitored during the conversion of the oligomer to the polymer (Figure 5B), show that this process starts initially as a homogeneous process which also involves a considerable degree of lattice rearrangement. Then it turns heterogeneous (as there is no shift of phonon frequencies between spectra 4, 5 and 6) with a phase separation accompanied by a gradual ordering of the polymer product lattice.

Now we come back to the comparison of the photoreactivity of α-DSP and P2VB. Both α-DSP and P2VB are isomorphic [6]. In the past, the crystal packing has been used to demonstrate the application of topochemical principles which utilizes a static lattice picture and relates the reactivity to the separations of the reactive centers. The distance between the reactive double bonds of the neighboring molecules in α-DSP is 3.939Å. This value for P2VB is shorter (3.910Å). Therefore, one would expect P2VB to be more reactive than α-DSP. The contrary is observed [6]. This failure

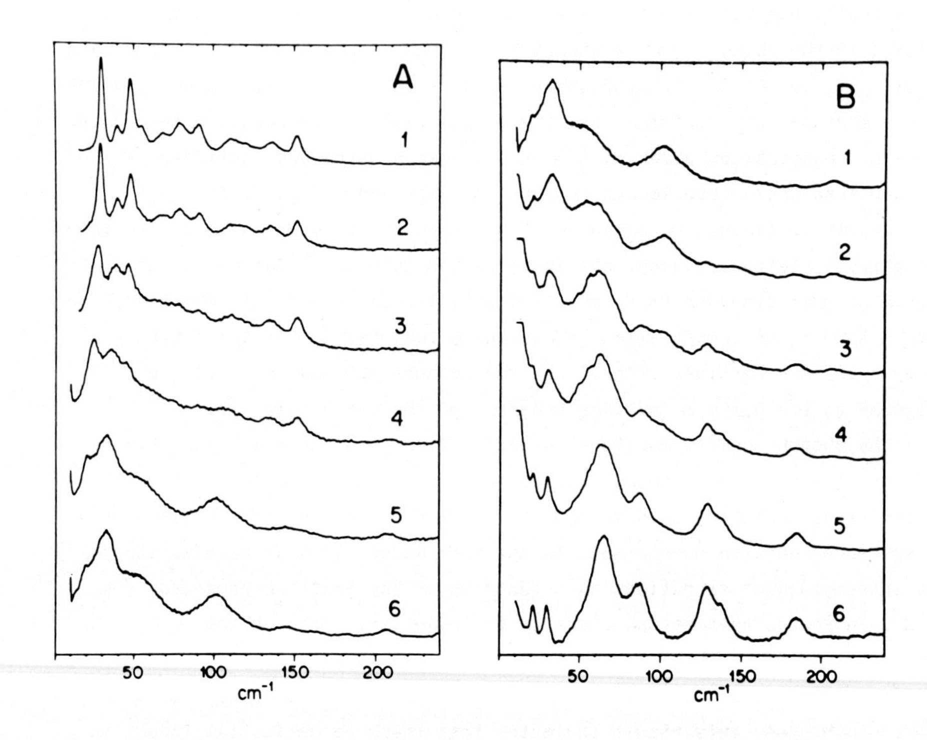

Fig. 5A. Raman phonon spectra monitored during the photo-oligomerization of a single crystal of 2,5-distyrylpyrazine at room temperature. The monomer spectrum and the spectrum of a quantitatively converted oligomer crystal are labeled 1 and 6, respectively.

Fig. 5B. Raman phonon spectra monitored as the oligomer of 2,5-distyrylpyrazine converts to polymer. The oligomer spectrum and that of the final polymer product are labeled 1 and 6, respectively.

may be because the dynamic nature of the lattice is not considered. In other words, electron-phonon coupling may be different for these two systems. The phonon-assisted process may manifest more in the case of α-DSP.

To examine if the differences in their reactivities could be related to the differences in the electron-phonon interaction in the excited state we studied the low temperature electronic spectra of these two systems. Figure 6 compares the electronic absorption and the emission spectra of the α-DSP and P2VB crystals at 4.2 K [18].

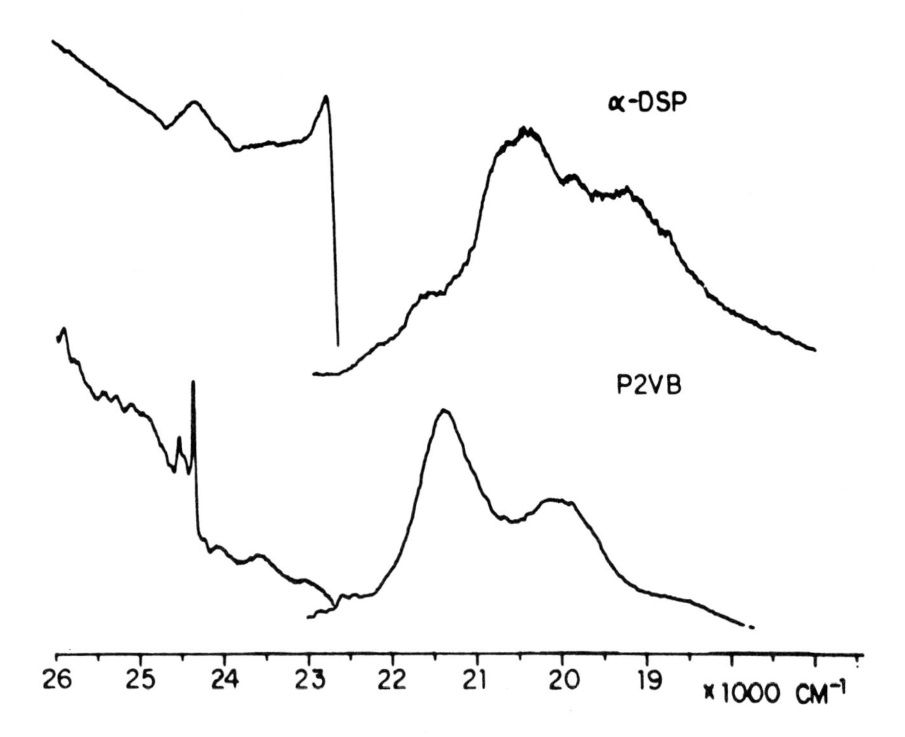

Fig. 6. Electronic absorption and emission spectra of a-DSP and P2VB, at 4.2 K are compared. (Reproduced from Ref 18).

The absorption spectra of α-DSP consist of considerably broad bands (FWHH > 100 cm^{-1}) compared to those of P2VB. This width is not due to disorder, because the phonon spectra of α-DSP, at this temperature, reveals a highly ordered lattice. We, therefore, conclude that this large width is due to a strong electron-phonon coupling in α-DSP. Hence, the photopolymerization in α-DSP is assisted by a strong electron-phonon coupling. In contrast, the electron-phonon coupling in the reactive excited state of P2VB is not so strong.

Comparing the emission spectra, both α-DSP and P2VB show broad excimeric emission. Therefore, the photopolymerization in both cases appear to be preceeded by excimer formation. The occurance of strong electron-phonon coupling in α-DSP (as inferred from the absorption spectra) suggests that the excimer formation is due to intrinsic dynamic process of local lattice relaxation because of this strong coupling with phonons. In absence of such strong electron-phonon coupling in P2VB, we assign the excimer formation in this case to arise from the presence of defects. Therefore, the photopolymerization in α-DSP is assisted by the intrinsic process of strong electron-phonon coupling. In contrast, the photopolymerization in P2VB is suggested to be defect controlled.

(ii) Photodimerization of 2,6-Dimethylbenzoquinone.

The various p-benzoquinones photodimerize in the solid state to yield various types of photodimers [37,38]. Among the p-benzoquinones, the 2,6-dimethyl derivative seems to photodimerize most readily. The time of exposure needed for extensive dimerization of this specific system depends on the size of the crystal, and can vary from 30 min to 2 h. The solid state photodimerization of 2,6-dimethyl-p-benzoquinone has been extensively studied by Robinovich and Schmidt [38] using the X-ray diffraction technique. They demonstrated the applicability of topochemical principle for this photodimerization. This reaction has also been investigated by IR and UV spectroscopy, which have provided useful information regarding the nature of the reaction [37,39]. There are two possible products that can be formed by the photodimerization of 2,6-dimethyl-p-benzoquinone. These products are (1) an oxetan and (2) a cage dimer. The reaction scheme forming these two products is represented as follows:

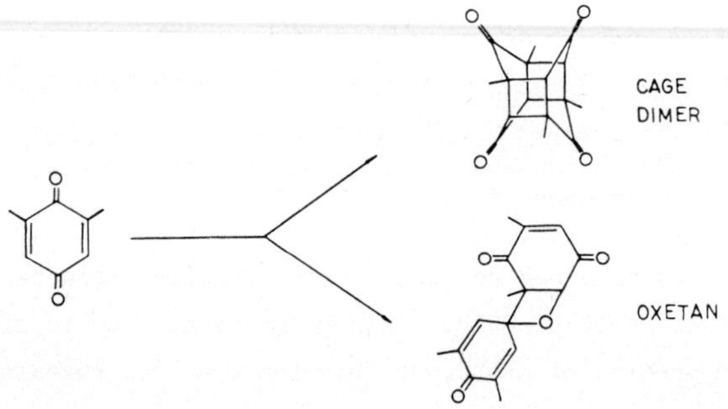

CAGE DIMER

OXETAN

Our Raman study of this reaction supports the formation of an oxetan structure [13]. The phonon spectra, obtained as a function of the advancement of the photodimerization, show the following behavior. The phonon bands in a slightly dimerized crystal, when compared with that of the monomer crystal, show a shift toward lower frequencies. Some new phonon features also appear. These new features grow in intensity with further progress of photodimerization. At this stage, no further frequency shift of the monomer phonon band is observed. These results have been interpreted as

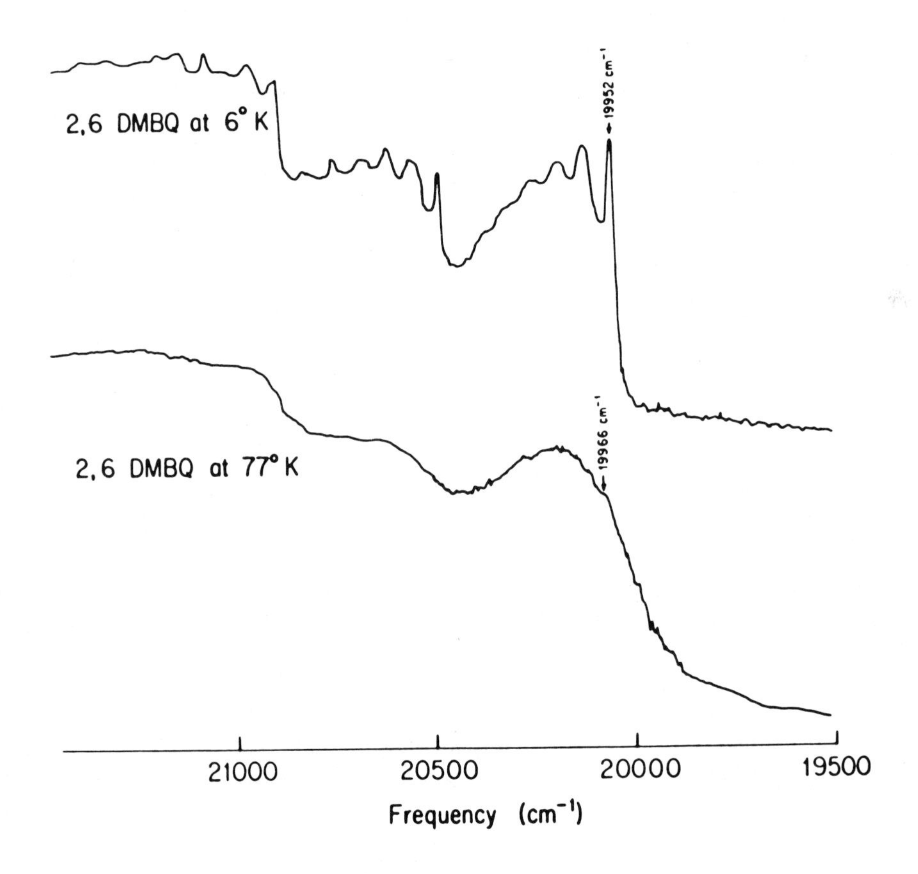

Fig. 7. Electronic absorption spectra of 2,6-dimethyl-p-benzoquinone at 6 K and 77 K. (Reproduced from Ref 13).

follows: (a) At the initial stage of the reaction, it proceeds by a homogeneous mechanism (b) As the concentration of the product increases with the reaction progress, the lattice distortion leads to the product phase separations. The reaction then turns heterogeneous [13].

In order to investigate the phonon-assistance of the photodimerization process, the electron-phonon coupling in the photoreactive state ($^1n\pi^*$) was studied [13]. The electronic absorption spectra of the monomer crystal were obtained at cryogenic temperatures and are shown in Figure 7. At 6 K, a sharp zero-phonon transition at 19952 cm^{-1} is accompanied by a strong phonon side band which appears to fit into a progression of a 65 cm^{-1} phonon. The phonon features Broaden by 77 K, confirming that they are due to phonon side band. This study is interpreted in terms of a selective and strong electron-phonon coupling with a specific phonon mode which gives rise to the formation of a polaron. It, therefore, is in support of the concept of a phonon-assisted photodimerization process in the crystalline phase. A selective and strong electron-phonon coupling with a specific phonon, the co-ordinate of which is the reaction co-ordinate, can create a lattice distortion bringing the reactant molecules in a configuration to initiate the photoreaction (preformation of a dimer configuration).

(iii) Photodimerization of 2-Benzyl-5-benzilidene Cyclopentanone.

The compound 2-benzyl-5-benzilidene cyclopentanone (abbreviated as BBCP) belongs to a class of photoreactive solids which contain a benzylidene cyclopentanone or a benzylidene cyclohexanone nucleus [40,41]. The photodimerization of the BBCP crystal can be illustrated (in terms of structural modes) as follows [40,41]:

(6)

A B

In the above illustration, (A) represents a pair of monomers in which the hatched circles represent the carbon atoms of the alkene group. Structure (B) represents the photodimer which now contains a cyclobutane ring. The photodimerization of BBCP is a single-crystal-to-single-crystal transformation which has been extensively investigated by Jones and Thomas and co-workers [40,41]. By using the X-ray technique, they found that the axes of the monomer lattice and the dimer lattice coincide. The lattice parameters of BBCP change only by 0.7% in the course of reaction. This uniquely small strain can allow a homogeneous mechanism of transformation of the monomer to dimer without the destruction of the single crystal.

The Raman phonon spectra were monitored as a function of the reaction progress [12]. The shift and broadening of the phonon transition implying a homogeneous reaction mechanism for the dimerization of BBCP is observed. The electronic absorption spectrum of the BBCP monomer single crystal at 4.2K is broad. This broad absorption again has been interpretted to arise from a strong electron-phonon coupling implying that the photodimerization process is phonon-assisted. No detectable emission from the monomer single crystal is observed, which suggests that the nonradiative energy-dissipation processes are highly efficient. This behavior is also consistent with a strong electron-phonon coupling in the BBCP crystal.

(iv) Polymerization of Diacetylenes.

Various derivatives of diacetylenes can be polymerized in the solid state thermally, photochemically or by using an X-ray radiation to produce a crystalline polymer according to the following scheme [42].

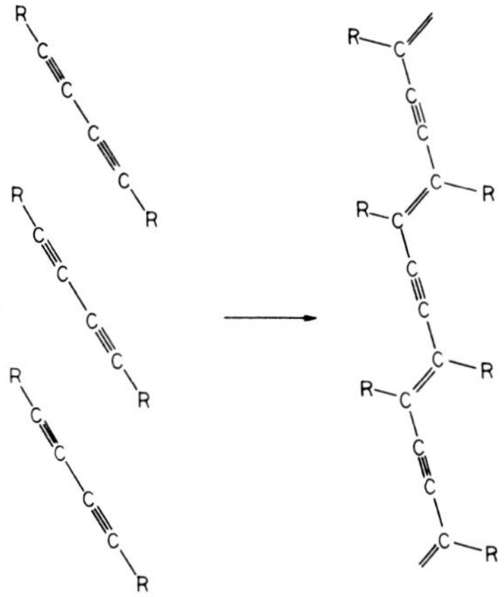

The solid state polymerization of diacetylenes of the general formula $R_1-C\equiv C-C\equiv C-R_2$ has been investigated [42-44] for many systems which differ in the substituent groups R_1 and R_2. The monomer crystal polymerizes thermally as well as by the action of ultraviolet or higher energy radiation. All these methods of solid state polymerization produce polymers which consist of well-aligned chains. The product is frequently a crystalline material with structural perfections similar to that of the monomer crystal. In other words, the monomer single crystal transforms into a single crystal of the polymer. This reaction also occurs in monolayer (Langmuir-Blodgett) films [43].

The solid state polymerization of diacetylenes is a reaction most extensively investigated by spectroscopic techniques [44]. The Raman spectroscopy has provided extremely useful information even at a very low conversion (<<1% polymerization) because of the resonance enhancement of the product polymer band. The pure diacetylene monomer is supposed to be white, but the polymer has strong absorption bands in the visible spectral range owing to extended conjugation. One generally uses a laser line in the visible region (for example, 5145Å = 19440 cm^{-1} lines of an argon ion laser) for Raman excitation. This excitation can provide a resonance condition for the polymer bands. The electronic-absorption maximum shifts continuously to lower energy as the polymer chain grows with the progress of the reaction. Therefore, the exact resonance condition varies as the reaction progresses.

We have studied the polymerization of a number of diacetylenes both in the form of a crystal and in Langmuir-Blodgett films by using Raman spectroscopy [8,17,45,46]. In the crystalline form, the phonon spectra monitored as a function of reaction progress shows a simple solid solution behavior in the low polymer conversion range [17]. Therefore, a homogeneous mechanism is established. In the higher polymer conversion range, a more complicated evolution of the phonon spectra is observed, indicative of lattice rearrangement.

For Langmuir-Blodgett films of para-toluene sulfonate polydiacetylenes (commonly known as PTS), the observed phonon spectra indicate the film to be multilayers where the phonon modes are determined by short range interactions [45]. Polarized Raman spectra were used to probe orientational effects resulting from preferred physical orientations within the film. The result indicates the presence of local anisotropy which varies from one location to another within the same film. Two other diacetylenes were studied in Langmuir-Blodgett films: (i) a diacetylene fatty acid $(CH_3)-CH_2)_{11}-C{\equiv}C-C{\equiv}C(CH_2)_8COOH$, and (ii) a diacetylene commonly known as 4-BCMU. In both cases monolayer film formation was observed [8,46]. The Raman spectra of these monolayer and successively built multilayer Langmuir-Blodgett films were successfully obtained by using the method of Raman optical wave guide. This method has been described in detail in reference [8]. For both these diacetylenes, the Raman spectra consist of the resonance enhanced double and triple bond stretching vibrations (intramolecular mode of the polymer) which permitted us to study the reaction and obtain information on the π-electron conjugation of the polymer backbone. However, no phonon spectra were observed. It appears that monolayer and relatively few layer thick Langmuir-Blodgett films do not have the sensitivity for Raman phonon spectroscopy. One may have to go to significantly thick Langmuir-Blodgett films (by successive multilayer deposition) to use phonon spectroscopy for the study of reaction mechanism and dynamics.

8. NONLINEAR SPECTROSCOPY OF SOLID STATE REACTIONS

Nonlinear spectroscopy involves simultaneous interactions of more than one photons to generate an excitation in the material. Nonlinear spectroscopic studies can provide valuable information on the dynamics of solid state reaction. Recently, we have started this novel approach to investigate the reaction dynamics.

The dynamics of photoreactive systems was investigated by two-photon absorption processes. The photopolymerizable α-DSP system was investigated by two-photon absorption. For this process a dye-laser, pumped by the second harmonic of a Nd-Yag laser, was used to provide a tunable wavelength range from 600 to 700 nm. Even at room temperature, a considerable amount of α-DSP fluorescence at the blue shifted wavelength is observed indicating significant two-photon absorption. However, no photoreaction is observed to take place even for a long irradiation time of several hours. This result in conjunction with the observation in our lab that the photopolymerization of α-DSP is greatly reduced at lower temperature leads us to propose that the reaction is biexcitonic. In other words, the initial step is started by deformed excitations (eg polaron, section 3) hopping and coming next to each other. The kinetic scheme will be

$$A + h\nu \longrightarrow A^*$$
$$A^* + A^* \longrightarrow \text{photoreaction}$$

Such a mechanism would be consistent with the absence of reaction by a two-photon absorption process, because such a process does not produce polarons in high concentration and, therefore, reduce biexcitonic processes. The temperature effect is simply a manifestation of reduced polaron hopping at lower temperature.

9. CONCLUDING REMARK

Phonon spectroscopy is a valuable approach for studying the reaction mechanism and reaction dynamics in solid state. It is hoped that this review will stimulate more interest for the use of this powerful technique in the investigation of solid state reaction. The concept of phonon-assisted reaction is exciting but needs to be further developed both by theoretical and experimental work. Nonlinear spectroscopy appears to be a novel approach for studying the reaction dynamics of photochemical processes.

ACKNOWLEDGEMENT

This work was supported by the Air Force Office of Scientific Research through contract number F49620855C0052.

REFERENCES

1 Symposium on Solid State Polymerization and the Structure and Properties of Polymers Produced by Lattice Controlled Processes. American Chemical Society 191st National Meeting, April 13-18, 1986.
2 D. Bloor, R. Chance and M. Nighan, Polydiacetylenes, Proceedings of a NATO Conference, Boston 1984.
3 J. M. Thomas, S. E. Morsi, and J. P. Desvergne, Adv. Phys. Org. Chem. 15 (1977) 63.
4 D. Y. Curtin and I. C. Paul, Chem. Rev. 8 (1981) 525.
5 L. Addadi and M. Lahav, J. Am. Chem. Soc. 100 (1978) 2838.
6 M. Hasegawa, Y. Suzuki, M. Nakanishi, and F. Nakanishi, Prog. Polym. Sci. Jpn., 5 (1973) 143.
7 C. N. Sukenik, J. A. P. Bonapace, N. J. Mandel, P. Y. Law, G. Wood and R. G. Bergman, J. Am. Chem. Soc. 99, (1977) 851.
8 R. Burzynski, P. N. Prasad, J. Biegajski and D. A. Cadenhead, Macromolecules 19 (1986) 1059, and reference therein.
9 L. W. Shacklette, H. Eckhardt, R. R. Chance, G. G. Miller, D. M. Ivory, and R. H. Baughman, J. Chem. Phys. 73 (1980) 4098.
10 R. Burzynski and P. N. Prasad, J. Polym. Sci. Polym. Phys. Ed. 23 (1985) 2193.
11 K. Dwarakanath and P. N. Prasad, J. Am. Chem. Soc. 102 (1980) 4254.
12 J. Swiatkiewicz, G. Eisenhardt, P. N. Prasad, J. M. Thomas, W. Jones and C. R. Theocharis, J. Phys. Chem. 86 (1982) 1764.
13 T. N. Misra and P. N. Prasad, Chem. Phys. Lett. 85 (1982) 381.
14 J. Swiatkiewicz and P. N. Prasad, J. Am. Chem. Soc. 104 (1982) 6913.
15 K. Dwarakanath, R. Burzynski, and P. N. Prasad, Mol. Cryst. Liq. Cryst. 100 (1983) 31.
16 J. Swiatkiewicz and P. N. Prasad, J. Polym. Sci. Polym. Phys. Ed. 22 (1984) 1417.
17 P. N. Prasad, J. Swiatkiewicz and G. Eisenhardt, Appl. Specty. Rev. 18 (1982) 59.
18 P. N. Prasad and J. Swiatkiewicz, Mol. Cryst. Liq. Cryst. 93 (1983) 25.
19 J. Swiatkiewicz and P. N. Prasad, J. Phys. Chem. 88 (1984) 5899.
20 G. Venkataraman and V. C. Sahni, Rev. Mod. Phys. 42 (1970) 409.
21 P. N. Prasad and R. Kopelman, J. Chem. Phys. 58 (1973) 5031.
22 P. N. Prasad, Mol. Cryst. Liq. Cryst. 52 (1979) 63.
23 P. N. Prasad and R. Kopelman, J. Chem. Phys. 57 (1972) 863.
24 D. W. Cruickshank, Acta Cryst. 9 (1956) 1005.
25 F. P. Chen and P. N. Prasad, J. Chem. Phys. 66 (1977) 4341.
26 A. S. Davydov, "Theory of Molecular Excitons (Translated by B. Dresner), Plenum Press, New York, 1971.
27 R. M. Hochstrasser and P. N. Prasad in E. C. Lim (Ed), Excited States, Academic Press, New York, 1974, p. 79.

28 J. C. Bellows and P. N. Prasad, J. Chem. Phys. 70 (1979) 1864.
29 R. Kuhn and H. W. Ruelius, Chem. Ber. 83 (950) 420.
30 D. Y. Curtin and S. R. Byrn, J. Am. Chem. Soc. 91 (1969) 1865.
31 S. R. Byrn, D. Y. Curtin and I. C. Paul, J. Am. Chem. Soc. 94 (1972) 890.
32 D. Y. Curtin and S. R. Byrn, J. Am. Chem. Soc. 91 (1969) 6102.
33 J. Z. Gougoutas and J. C. Clardy, J. Solid State Chem. 4 (1972) 230.
34 J. Z. Gougoutas, J. Am. Chem. Soc. 99 (1977) 127.
35 M. Hasegawa, Y. Suzuki, and H. Nakanishi, J. Polym. Sci. Part A1, 7 (1969) 143.
36 W. Jones, J. Chem. Res. (1978) 142.
37 C. R. Cookson, D. A. Cox, and J. Hudec, J. Chem. Soc. (1961) p. 4499.
38 D. Robinovich and G. M. J. Schmidt, J. Chem. Phys. 13 (1967) 1444.
39 W. O. George, M. J. Leigh, and J. A. Strickson, Spectrochim. Acta 27A (1971) 1235.
40 W. Jones and J. M. Thomas, Prog. Solid State Chem. 12 (1979) 101.
41 W. Jones and J. M. Thomas, J. Chem. Soc. Chem. Commun. (1980) pp 610, 611.
42 G. Wegner, Pure Appl. Chem. 49 (1977) 443.
43 B. Tieke and K. J. Weiss, J. Colloid Interface Sci. 101 (1984) 128.
44 D. Bloor, F. H. Preston, D. J. Ando, and D. N. Batchelder in K. J. Ivin (Ed), Structural Studies of Macromolecules by Spectroscopic Methods, Wiley, New York, 1976, p. 91.
45 R. R. McCaffrey, P. N. Prasad, M. Fornalik and R. Baier, J. Polym. Sci. Polym. Phys. Ed. 23 (1985) 1523.
46 J. Biegejski, R. Burzynski, D. A. Cadenhead and P. N. Prasad, Macromolecules 19 (1986) 2457.

Chapter 5

FOUR-CENTER TYPE PHOTOPOLYMERIZATION OF DIOLEFIN CRYSTALS

Masaki HASEGAWA

1. INTRODUCTION

Since the decade of the 1950's, solid state polymerization had been studied mostly in terms of the chain reactions of vinyl and cyclic ether compounds initiated by γ-ray, uv light, catalysts, or heat. Among these reactions, the polymerization of cyclic ethers, e.g., trioxane, had been extensively investigated from the point of view of the crystal lattice controlled process, since these cyclic ether crystals afford highly crystalline polymers. For example, certain correlations between the molecular arrangement of the monomer and the growth direction of the polymer chain had been visualized in the polymerization of the trioxane crystal initiated by γ-ray irradiation (ref. 1). The term "topochemical polymerization" was used for the first time for the solid state polymerization of the trioxane crystal. In the reaction of a solid solution composed of 3,3-bis(bromomethyl)- and 3-bromomethyl-3-ethyl-oxetanes, the composition of the two monomers in the resultant copolymer is determined by the monomer ratio in the solid solution, not by the monomer reactivity ratio (ref. 2). However, no definite correlation had yet been demonstrated between the monomer and the polymer in the dimensions of the crystal unit cell in these solid state polymerizations.

In 1958 an article described that a brilliant yellow crystal of 2,5-distyrylpyrazine (DSP, 1) was, on exposure to uv irradiation, turned into a white insoluble polymeric substance with a melting point of 331-333°C (ref. 3). The present author had made the same observation (ref. 4). By a further study of this photophenomenon, it was concluded that the DSP crystal was photochemically converted to a linear high polymer crystal by repeated intermolecular [2+2] cyclodimerizations (ref. 5).

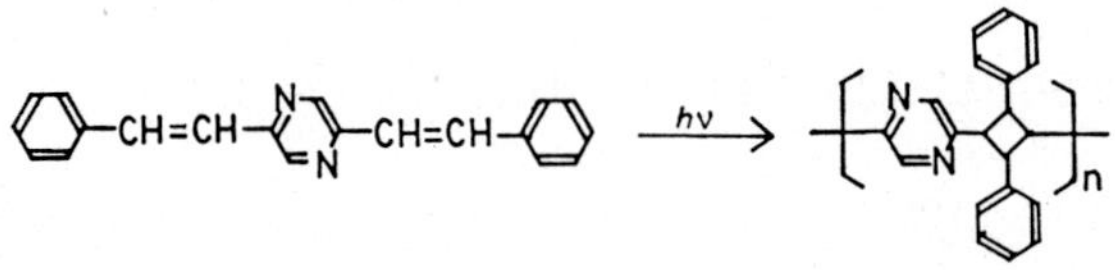

DSP Poly-DSP

The crystallographic studies have revealed that the polymerization of the DSP crystal follows the topochemical rule enunicated by Schmidt and his co-workers (ref. 6). In their topochemical concept, the reaction of several olefinic crystals tends to occur with a minimum of atomic and molecular motion. There are three types of crystal structures of photodimeric olefin molecules: the α-type crystal, in which the double bonds of neighboring molecules are arranged in parallel fashion and make contact at a distance of 3.6-4.1Å across a center of symmetry; the β-type, characterized by a lattice with one axial length of 3.9-4.1Å between translationally related molecules, and the γ-type, in which no double bonds of neighboring molecules are within 4.7Å. On the photoirradiation of cinnamic derivatives, for example, an α-type crystal gives a centrosymmetric dimer related to the α-truxillic derivative ($\bar{1}$-dimer), a β-type crystal gives a dimer of mirror symmetry related to the β-truxinic derivative (m-dimer), and a γ-type crystal is photostable.

The topochemical [2+2] photodimerizations of α-and β-type crystals will be illustrated for enone derivatives.

α-type crystal

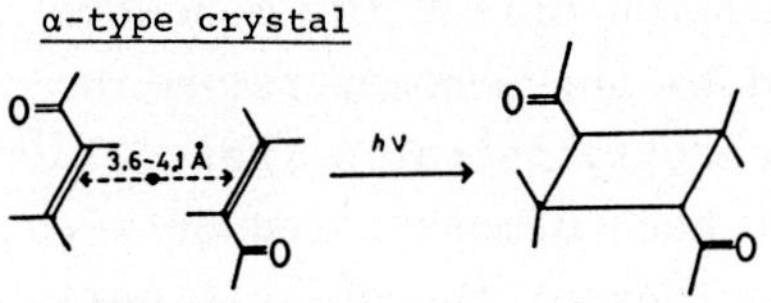

β-type crystal

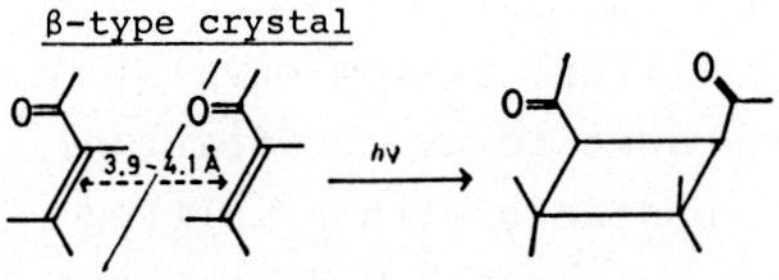

The reaction of the DSP crystal is a typical example of the extention of the topochemical [2+2] photodimerization of the α-type olefin crystal to the topochemical photopolymerization of the

α-type diolefin crystal (Fig. 1).

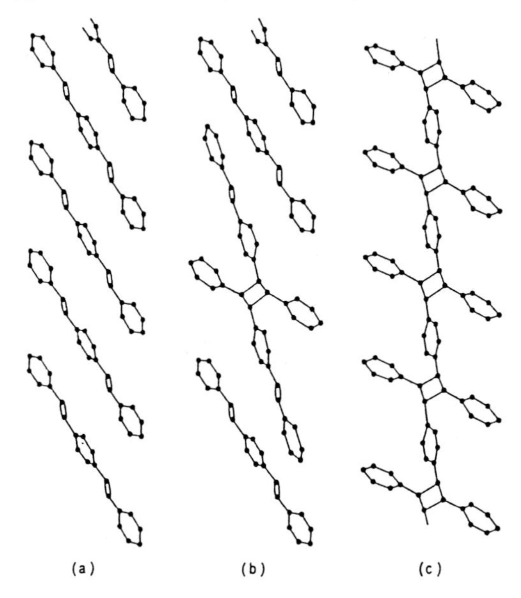

Fig. 1. Schematic illustration of the conversion of (a) monomer DSP (1) into (b) dimer and (c) polymer.

It was further demonstrated that the unit cell of the DSP crystal was strictly duplicated to that of the poly-DSP crystal, with retention of crystal symmetry, during the reaction, as shown in Table 1 (ref. 7).

TABLE 1
Crystal data of DSP (α and γ)[a] and poly-DSP

	Space group	a, Å	b, Å	c, Å	Z	Dx	C---C, Å[b]
DSP(α)	Pbca	20.638	9.599	7.655	4	1.244	3.939
poly-DSP	Pbca	18.36	10.88	7.52	4	1.257	---
DSP(γ)	$P2_1/a$	13.833	18.615	5.823	4	1.261	4.187
			(92.63)[c]				4.369

a A photoreactive α-type crystal of DSP is obtained from a benzene solution while a considerable portion of the photostable crystal

(footnotes continued on p.156)

(γ-type) is obtained by sublimation (ref. 8).
b Intermolecular carbon to carbon distance.
c β (°).

The topochemical photopolymerization of DSP was the first example of an organic reaction in which crystal-to-crystal transformation was visualized at the level of the crystal unit cell. The polymerization was named "four-center-type photopolymerization." The same type of crystalline state photopolymerization was discovered for a series of α,α'-bis(4-acetoxy-3-methoxybenzylidene)-1,4-phenylenediacelonitrile by Holm and Zienty, who published it as a patent in 1967 (ref. 10).

Since then, a great number of diolefin crystals have been found to photopolymerize to linear high molecular weight polymers (ref. 11).

Almost all of the reaction behavior of conjugated diolefin crystals and the configurations of the resulting product have been interpreted in terms of the topochemical principle of [2+2] photocyclodimerization (reactive double bonds are arranged in parallel fasion with a distance within 4.2 Å), while anomalous behavior of deviation from the accepted topochemical rule has been reported for several olefin crystals (ref. 12).

Optically active polymers and polymers with differently alternating substituents have been prepared by the four-center-type photopolymerization of unsymmetric diolefin crystals (refs. 13 and 14). In addition to several types of deviations from the topochemical principle, recent studies of the topochemical behavior of unsymmetric diolefin crystals have revealed a great variety of reactivities and photoproducts, which depend on a slight modification of the chemical structure or on the crystallization process of the monomer (refs. 13 and 15).

The solid state polymerization of diacetylene derivatives has been reported for many years (refs. 16 and 17). Hirshfeld and Schmidt suggested the possibility of a topochemical nature of the polymerization of the diacetylenedicarboxylic acid crystal (ref. 18). The lattice controlled polymerization mechanism of these diacetylene derivatives has been appreciated principally by the work of Wegner and his collaborators (ref. 19).

2. MOLECULAR DESIGN FOR PHOTOPOLYMERIC DIOLEFIN CRYSTALS AND POLYMER PREPARATION

Almost all of the photoreactive diolefin crystals have an α-type structure where the molecules are related with a center of symmetry, and are converted into a crystalline linear high polymer quantitatively on irradiation by uv ~ visible light with a wide range of reactivities. In the summary of preparative study of the four-center-type photopolymerization, molecular design for photopolymeric diolefin crystals was empirically proposed: the crystalline state [2+2] photodimerization of mono-olefins can be generally extended to the crystalline state photopolymerization of diolefins with a rigid linear structure consisting of conjugated double bonds separated by the 1,4-position of an aromatic ring (2,6-position of naphthalene in one case) (ref. 20). The reactive units are common between mono- and di-olefinic molecules, such as the stilbazole seen in the DSP series, the cinnamic acid in the p-PDA series (dialkyl 1,4-phenylenediacrylates), or the α-cyanocinnamic derivative, in the p-CPA series (dialkyl α,α'-dicyano-1,4-phenylenediacrylates).

Topochemically Photopolymeric Diolefins	Topochemical [2+2] Photodimerizations (Model reactions)
Ph-CH=CH-(pyrazine)-CH=CH-Ph	Ph-CH=CH-(pyridyl) $\xrightarrow{h\nu}$ cyclobutane dimer
DSP series	stilbazol
ROOCCH=CH-C$_6$H$_4$-CH=CHCOOR	Ph-CH=CHCOOH $\xrightarrow{h\nu}$ HOOC-cyclobutane-COOH
p-PDA series	Cinnamic acid
ROOC(CN)=CH-C$_6$H$_4$-CH=C(CN)COOR	(o-OCH_3)Ph-CH=C(CN)COOEt $\xrightarrow{h\nu}$ EtOOC, NC, CN, COOEt, CH_3O, OCH_3 cyclobutane dimer
p-CPA series	α-cyanocinnamic derivative

According to this molecular design, a great number of highly crystalline linear polymers have been prepared via the four-center type photopolymerization of conjugated diolefin derivatives.

These diolefins are prepared by aldol condensation between the corresponding aromatic methyl and formyl groups or by Witting's reaction.

Similar series of tetraolefins, such as 1,4-phenylene dibutadienoic acid, HOOC-CH=CH-CH=CH-C_6H_4-CH=CH-CH=CH-COOH, 1,4-phenylenebis(α-cyanobutadienoic) acid, HOOC(NC)C=CH-CH=CH-C_6H_4-CH=CH-CH=C(CN)COOH and their esters, have also been prepared and subjected to photopolymerization in the crystalline state (ref. 21).

However, since the polymerization of these crystals is strictly crystal-lattice-controlled, the same molecule sometimes behaves differently depending on its crystal structure modification. For example, the DSP crystal, crystallized from the benzene solution, is highly photoreactive while the sublimed DSP crystal is photostable (ref. 22).

Since all the monomers absorb the light of wavelength from the uv ~ visible region, polymerization proceeds by means of irradiation with a xenon or a high pressure mercury lamp; it sometimes proceeds very rapidly even on exposure to sunlight.

For the polymerization on a preparative scale, each monomer is dispersed as fine crystals in a suitable inert dispersant, such as water or water-ethanol, in a quartz flask and irradiated with an appropriate light source for a period of from several minutes to several tens of hours at room temperature or lower while being rigorously stirred. Several monomer crystals are orange ~ yellow but fade to opaque white during the reaction. Polymerization proceeds smoothly if the temperature is satisfactorily lower than the melting point of the monomer.

The potassium bromide-pellet method provides the most convenient way to detect a slight polymerizability at room temperature by means of IR spectroscopy.

Some typical examples of the polymerization conditions and the properties of the polymers are listed in Table 2.

Other molecules in which two olefin groups are connected by a flexible chain have been reported to polymerize into the cyclobutane polymers in several articles but these monomer crystals behave essentially differently from the photoreactive rigid diolefin crystals (refs. 24-26). For example, the polymethylene bismaleinimide derivative crystal gives an amorphous oligomer with a low conversion or a partially cross-linked polymer. The pentaerythritol tetracinnamate crystal reacts nearly quantitatively but gives an amorphous insoluble substance. The reported [2+2] photopolymerization behavior of polymethylene-dicinnamate crystals has not been confirmed at all by the present authors' trials under various experimental conditions. The details of these examples have been described in a review article (ref. 11).

One example of the outstanding, exceptional behavior of conjugated diolefin crystals is the reaction of the 1,4-dicinnamoylbenzene (1,4-DCB, 2) crystal, which has bis-chalcone type chemical structure (ref. 27). On photoirradiation, 1,4-DCB crystal does not give a linear high polymer, but gives, by repeating intramolecular "double" [2+2] photocyclodimerization, a tricyclic dimer, 21,22, 23,24-tetraphenyl-1,4,11,14-tetraoxo-2(13),12(13)-diethano[4.4] paracyclophane (4) in a high yield, plus a small amount of oligomeric substances.

The crystal structure of 1,4-DCB reasonably explains such anomalous topochemical photobehavior (Fig. 2).

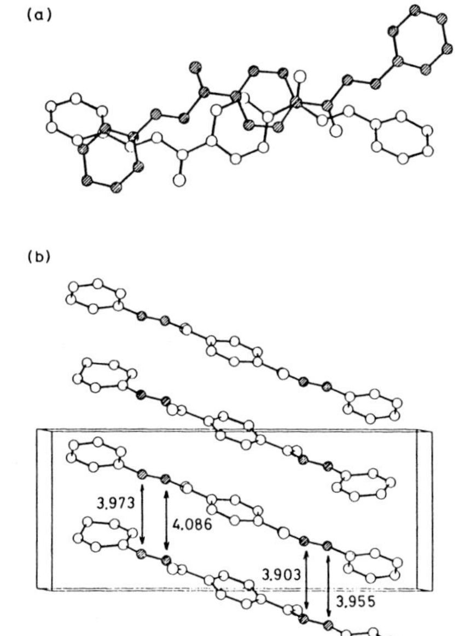

Fig. 2. The crystal structure of 2 projected onto (a) the plane of central benzene and (b) the (100) plane (unit; Å) (ref. 27).

TABLE 2

Polymerization conditions and properties[a]

Diolefin monomer	Reactivity	Reaction temp. (°C)	Polymer yield (%)	Decomp. point[b] (°C)	Reduced viscosity (dl/g)	Solvents for as-polym. polymer crystals	Ref.
General formula of polymers: $[-Ar-C_4(R)(R')(R')(R)-]_n$							
Ar: 1,4-pyrazylene R: phenyl, R': -H	very high	room	quant.[c]	339-343	1.0-10	conc. H_2SO_4 CF_3COOH (m-cresol, o-chlorophenol)[d]	5
Ar: 1,4-phenylene R: 2-pyridyl R': -H	low	room	quant.	340	0.3-2	(m-cresol, o-chlorophenol)[d]	5
Ar: 1,4-phenylene R: -COOMe, R': -H	high	room	quant.	415	>10	conc. H_2SO_4	23
Ar: 1,4-phenylene R: -COOEt, R': -H	high low	-25 room	quant. medium	347	1.4 0.16	conc. H_2SO_4	23
Ar: 1,4-phenylene R: -COO-Pr^n, R': -H	low	0-5	42	360	0.26	conc. H_2SO_4	23

TABLE 2 (continued)

Ar: 1,4-phenylene R: $COOC_6H_5$ R': -H	medium	room	85	420	0.90	aq. alkali (accompanied by hydrolysis)	23
Ar: 1,4-phenylene R: -COOEt, R': -CN	low	room	66	340	2.6	conc. H_2SO_4	20
Ar: 1,4-phenylene R: -COO-Pr^n, R': -CN	very high	room	quant.	335	3.0	conc. H_2SO_4	20
Ar: 1,4-phenylene R: -CH=C(CN)COOEt R: -H	high	room	quant.	245	1.3[η]	conc. H_2SO_4 CF_3COOH	21
General formula of the polymer:							
Ar: 1,4-phenylene R: -CN, R' : 3-methoxy-4-methoxycarbonylphenyl	high	room	quant.[c]	300	0.44[η]	conc. H_2SO_4	10[e]

[a] Results are always given not for the optimum conditions of polymerization but for a typical experimental run.

[b] Measured by the capillary method. The decomposition points are not strictly definable because almost all as-polymerized polymer crystals thermally depolymerize in the crystalline state, and the starting point of depolymerization is dependent on their molecular weight.

[c] A photostable crystal is obtained by recrystallization under specified conditions.

[d] Solvents in parentheses are the solvents used only for amorphous polymers.

[e] In the patent literature (ref. 10), a large number of crystals of this series of diolefin crystals have been reported to be photopolymeric.

In the crystal of 2, molecules are arranged not in α-type packing, but in δ-type packing, which is extremely rare for olefinic crystals. The reacting pair is arranged in a skewed position, and the distances between the intermolecular photoadductive carbons are 3.973 and 4.086 Å for one cyclobutane ring and 3.903 and 3.955 Å for the other. In a single crystal of 2 at the initial stage of photochemical reaction, both the intramolecular photocyclodimerization and intermolecular photopolymerization of diolefinic monocyclobutane 3 occur competitively to give a tricyclic dimer 4 as the main product and oligomers as the minor one.

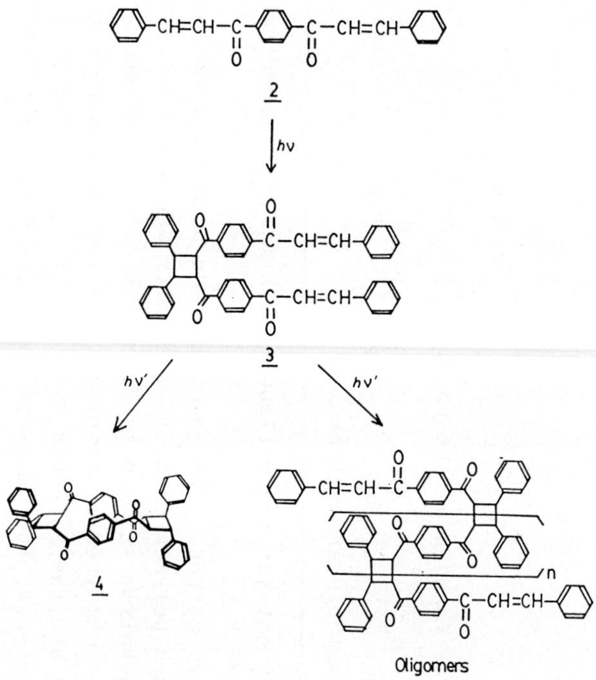

CH=CH–C(=O)–C(=O)–CH=CH

2

hν

C–CH=CH

C–CH=CH

3

hν′

hν′

4

CH=CH–C

C–CH=CH

n

Oligomers

Linear polymers derived from an α-type crystal of unsymmetric diolefins should have a "hetero-adduct" or "homo-adduct" structure, depending on their crystal structures, as is shown in the scheme.

a) Hetero-adduct polymer

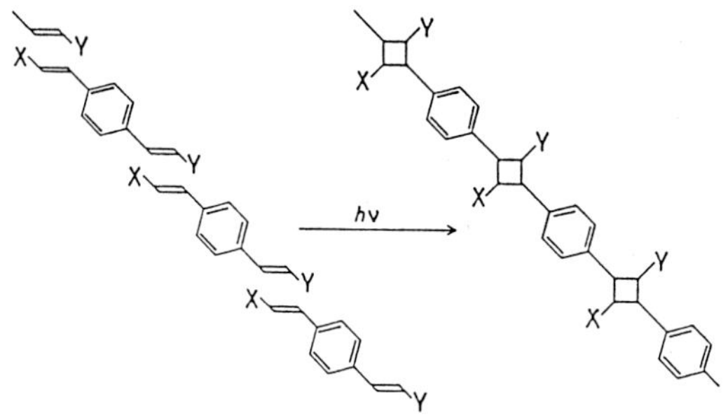

b) Homo-adduct polymer

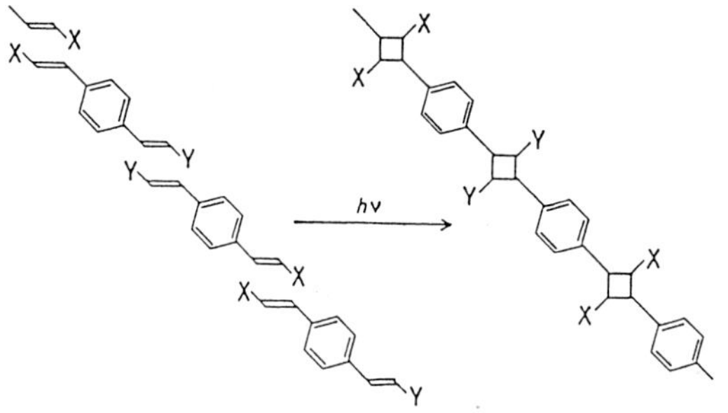

A few attempts to obtain photoreactive unsymmetric diolefin crystals have been carried out according to the molecular design described above. The first example of photoreactive unsymmetric conjugated diolefin crystals, with cinnamic and α-cyano-cinnamic units in a single molecule was reported by Addadi and Lahav. They succeeded in obtaining the optically active hetero-adduct type dimer and oligomers by means of the four-center-type photopolymerization of the chiral crystal of achiral unsymmetric diolefin molecules (refs. 13 and 28). The result is the first and only successful crystal engineering for "Absolute" asymmetric [2+2] photopolymerization, as is illustrated in the scheme below and in Table 3 (Group III). The molecular design in this study had been performed by combining the designs for the photopolymeric molecule and for the chiral crystal. Racemic oligomeric substances have been prepared by other workers from similar unsymmetric diolefin crystals (ref. 29).

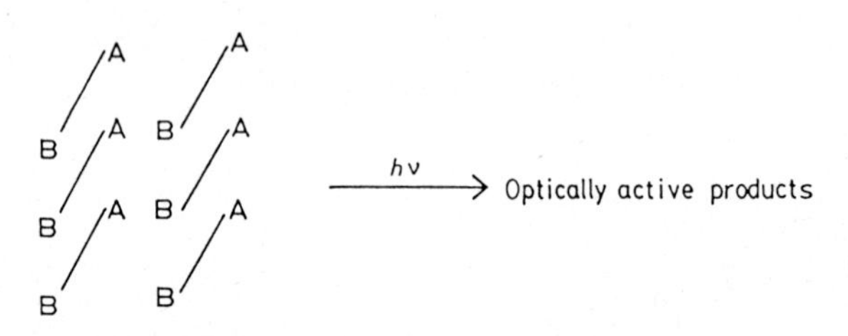

α-type chiral crystal

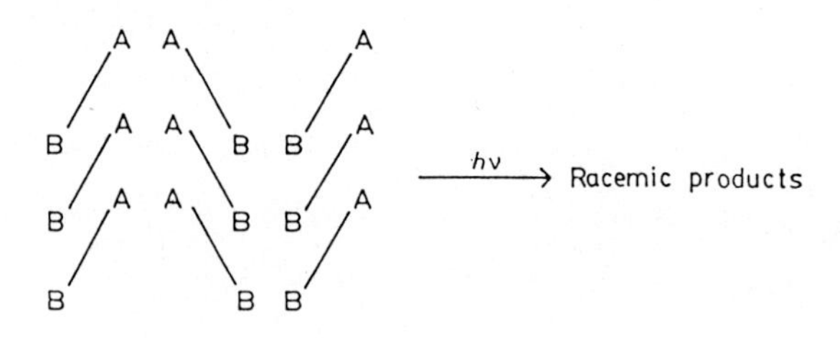

α-type achiral crystal

Recently a great variety of unsymmetric photoreactive diolefin crystals have been reported (refs. 14 and 15). These unsymmetric molecules have a general formula represented as (I)-(VII), where R and R' in Groups (I), (II), and (III) represent either hydrogen, alkyls, or phenyl. In Groups (IV) and (V), X is a hydrogen or cyano group, and Y represents either a cyano, alkoxycarbonyl, or benzoyl group. R in Group (VI), which is unsymmetric DSP, is either alkoxycarbonyl or chlorine. In group (VII), which are derived from the crystals in Group (I) and (IV), A, X, and R represent ethoxycarbonyl, hydrogen, and methyl respectively, or A is 4-pyridyl, and X and R are a cyano group and several alkyl groups.

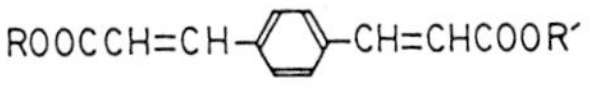

ROOCCH=CH-⟨⟩-CH=CHCOOR′

(I)

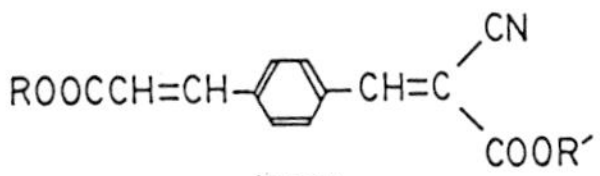

ROOCCH=CH-⟨⟩-CH=C(CN)(COOR′)

(II)

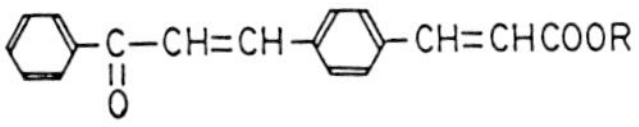

⟨⟩-C(=O)-CH=CH-⟨⟩-CH=CHCOOR

(III)

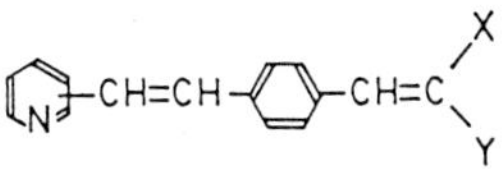

⟨N⟩-CH=CH-⟨⟩-CH=C(X)(Y)

(IV)

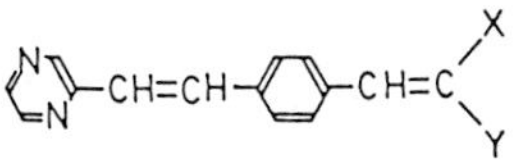

⟨N N⟩-CH=CH-⟨⟩-CH=C(X)(Y)

(V)

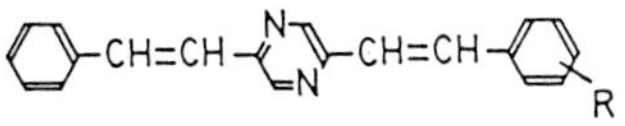

⟨⟩-CH=CH-⟨N N⟩-CH=CH-⟨⟩-R

(VI)

A-CH=CH-⟨⟩-[cyclobutane: X, COOR, ROOC, X]-⟨⟩-CH=CH-A

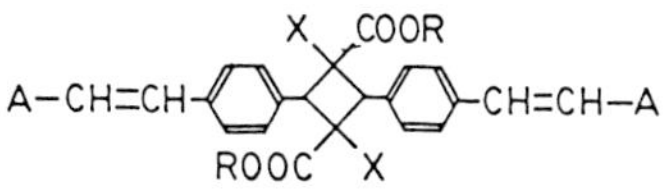

(VII)

Most of these unsymmetric diolefins in Groups (I)~(VI) are prepared in a way similar to that used in preparing symmetric diolefins. The diolefinic monocyclobutane derivatives in Group (VII) are derived by the topochemical photodimerization of the corresponding conjugated unsymmetric diolefin crystals in Groups (I) and (IV). These cyclobutane derivatives are the first examples of topochemically reactive non-conjugated diolefin crystals.

Many of these unsymmetric diolefins have an α-type packing crystal and, on photoirradiation, are converted into the corresponding crystalline or amorphous polymers, in which the cyclobutane has a head-to-tail hetero- or homo-adduct structure. A few reactions give photostable dimers which have a head-to-tail homo-adduct or head-to-head hetero-adduct structure (refs. 15(c) and 28(a)). The photoreactive dimers have been readily isolated at the intermediate stage of topochemical polymerization on photoirradiation by an ordinary high pressure mercury lamp, or by a xenon lamp in a few examples. Several examples of the photochemical behavior of unsymmetric diolefin crystals are shown in Table 3.

TABLE 3
Examples of photochemical behavior of unsymmetric diolefin crystals

Diolefins	Photoreactivity[a]	Photoproduct	Type of adduct[b]	Morphology	η_{inh}	Reference
Group I						
R : Me, R' : Et	++	Polymer	ht, homo	crystalline	3.12[c]	15(a)
Group II						
R : s-Butyl, R' : Et	+	Oligomers[d]	ht, hetero			13
R : Me, R' : 3-pentyl	+	Oligomers[d]	ht, hetero			28(a)
R : iso-Pr, R' : n-Pr	+	Dimer	hh, hetero			28(a)
Group III						
R : Et	+	Polymer				15(a)
R : phenyl	++	Polymer				15(a)
Group IV						
2-pyridyl, X : CN, Y : COOMe	++	Dimer	hh, hetero	amorphous		15(c)
4-pyridyl, X : H, Y : COOMe	++	Polymer	ht, hetero	amorphous	0.47	15(b)
4-pyridyl, X : CN, Y : COOMe	++	Polymer	ht, homo	amorphous	>0.13	14
4-pyridyl, X : CN, Y : COOEt	++	Polymer	ht, homo	amorphous	0.36	15(b)
4-pyridyl, X : CN, Y : $COOPr^n$	++	Dimer[e]	ht, homo	crystalline		14

TABLE 3 (continued)

Group V						
X : H, Y : COOMe	++	Polymer	ht, homo	crystalline	3.93	14
X : H, Y : COOEt	++	Polymer	ht, hetero	crystalline	8.19	14
X : CN, Y : COOMe	+	Polymer	homo	amorphous	0.24	14
Group VI						
R : 4-COOMe	++	Polymer	ht, hetero	crystalline	0.96	14
Group VII						
A : COOEt, X : H	++	Polymer	ht, homo	crystalline	0.21	15(a)
R : COOMe						
A : 4-pyridyl, X : CN[f]	++	Polymer	ht, homo	crystalline	0.87	14
R : n-Pr						

a ++ and + represent "highly reactive" and "reactive" respectively.

b ht and hh represent "head-to-tail" - and "head-to-head" -type cyclobutanes respectively.

c Acetone-insoluble part.

d The photoproducts from a single crystal are optically active.

e The dimer crystal (191 °C), as-prepared or recrystallized from ethyl acetate, is photostable.

f The photoreactive crystal is obtained by crystallization from a propanol solution.

3. POLYMERIZATION MECHANISM

The four-center type photopolymerization proceeds by way of a monomer crystal lattice controlled stepwise mechanism. In the monomer crystals, the relative position and orientation of reactive bonds are favorably aligned for product formation, and the principal features of molecular alignment remain unaltered during the reaction. Thus, the reaction obeys first-order kinetics. With increasing temperatures, the apparent reaction rate is accelerated below the melting point of the reacting crystals at the early stage, whereas the final conversion and the molecular weight of the final product are gradually diminished. For example, the diethyl 1,4-phenylenediacrylate crystal (mp 96 °C) photopolymerizes quantitatively to a crystalline high polymer at temperatures below 0 °C, including an extremely low temperature (4.2 K) (ref. 30), but above 25 °C partially cross-linked amorphous polymer is obtained in a poor yield (ref. 31).

In the temperature range where the reaction is strictly crystal-lattice controlled, the reaction rate under constant photoirradiation is assumed to be determined by the potential deviation of two olefin bonds from the optimal positions for the reaction and, therefore, to be dependent solely on the thermal motion of the molecules in the crystal. On the basis of the equation derived from such a reaction model, the kinetic behavior has been explained in terms of a temperature-dependent factor involving the product of the reciprocal and the exponential terms of temperature (ref. 32). Final yield will become a more significant indication, especially in the temperature range where the thermal motion is rather rigorous (refs. 31 and 33).

In conclusion, the polymerization of diolefin, which has an α-type packing crystal structure, readily affords a highly crystalline linear polymer in a high yield, as long as the polymerization temperature is sufficiently low to maintain the rigidity of the crystal lattice of the reactant.

Since the π-electron conjugation of diolefin monomers (A) is interrupted by the formation of a cyclobutane ring to produce a molecule larger than a dimer (B), the π-π^* electronic transition of B is shifted to a higher energy level than that of A. Therefore, an overall polymerization can, by controlling the wavelength for photoirradiation, be separated into two steps; the oligomerization starting from the monomer and the following polymerization to the high polymer.

The reaction scheme is as shown below.

$$A \xrightarrow{h\nu} A^*$$
$$B \xrightarrow{h\nu'} B^* \qquad \nu<\nu'$$
$$A^* + A \xrightarrow{k1} B \text{ (Dimer)}$$
$$A^* + B \xrightarrow{k2} B$$
$$B^* + A \longrightarrow B$$
$$B^* + B \longrightarrow B$$

A: $(R)(R')C{=}CH{-}Ar{-}CH{=}C(R')(R)$

B: $(R)(R')C{=}CH{-}Ar{-}[CH{<}C(R)(R')/C(R')(R){>}CH{-}Ar]_{n\geq 1}{-}CH{=}C(R')(R)$

A two-step mechanism due to two exited species, A* and B*, was visualized on the polymerization of a few monomer crystals. as is illustrated for DSP in Fig. 3.

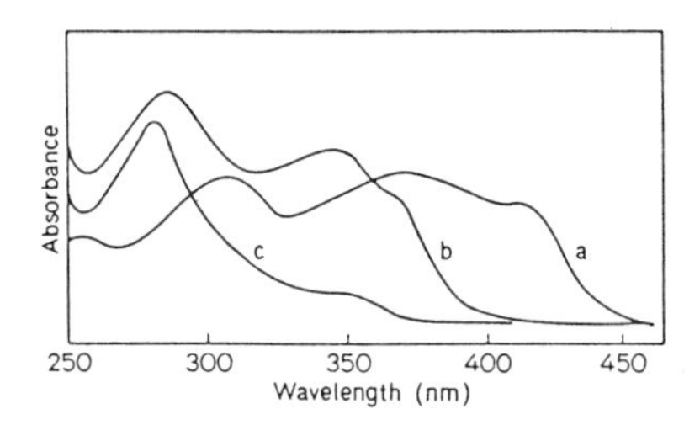

Fig. 3. UV spectra of (a) crystalline DSP, (b) DSP-oligomer, and (c) poly-DSP (ref. 34).

Upon the irradiation of the DSP crystal with 430 nm monochromatic light, the uv absorption spectrum changes and the maximum peaks are shifted to 350 nm and 290 nm of the DSP oligomer (B) from 420 nm and 370 nm of the DSP(A). On the further irradiation of the DSP oligomers at wavelengths shorter than 400 nm, crystalline poly-DSP with an absorption maximum at 280 nm is obtained (ref. 34).

As a result, in the four-center type photopolymerization, the molecular weight of the resulting polymer can be controlled by means of the reaction temperature and/or the wavelength of the irradiating light.

The photoquantum yields for the disappearance of olefin bonds are 1.2 and 1.6 for the overall oligomerization of DSP and for the overall polymerization of DSP oligomers respectively. Such a high quantum yield for single-photon stepwise reactions may reflect, partially at least, a high probability of effective collisions in the topochemical photopolymerization. On the other hand, the quantum yield is only 0.04 for the oligomerization of the 1,4-bis[2-(2-pyridyl)ethenyl]benzene crystal, although, in this crystal, the chemical structure and topochemical environment are extremely similar to those of DSP.

In the polymerization of the DSP crystal, the reaction rate of oligomerization between the excited monomer and the oligomers larger than the dimer ($A^* + B \rightarrow B$) is higher than the dimerization ($A^* + A \rightarrow B$). The different quantum yields for several monomer crystals have not been explained completely though the overlap of electronic orbitals in the monomer crystal, the volume changes in the crystal during the reaction, etc. seem to be related to the kinetics of the topochemical reaction.

In addition to the existence of two excitation species, A^* and B^*, it should be noticed that the topological environment varies, step-by-step, in the reacting crystal, resulting in a variation in the quantum yield of cyclobutane formation at each of the elementary processes. The step-by-step variation of the quantum yield is amplified in the reaction of several unsymmetric diolefin crystals that afford homo-adduct polymers (refs. 14 and 15).

In these unsymmetric crystals, two pairs of intermolecularly facing double bonds are placed differently in relative positions and orientations; accordingly, these pairs show a striking difference in topochemical reactivity. For example, ethyl methyl 1,4-phenylenediacrylate (5) (R and R' are ethyl and methyl respectively in Table. 3, Group I) photopolymerizes into a linear polymer (7) in the crystalline state. However, a large amount of the dimer (6) is formed exclusively and is accumulated at the intermediate stage of polymerization, even when without any control of the irradiating wavelength. The photoreactive dimer crystal (6) is easily isolaled from the reacting crystals. On the other hand, no other types of dimers are detected during the whole reaction process.

EtOOCCH=CH–⟨⟩–CH=CHCOOMe

5

↓

EtOOCCH=CH–⟨⟩–(cyclobutane, COOMe, COOMe)–⟨⟩–CH=CHCOOEt

6

↓

–[(cyclobutane, COOMe, COOMe)–⟨⟩–(cyclobutane, COOEt, COOEt)–⟨⟩]$_n$–

7

In the crystal of 5, the molecules are arranged in parallel and are related to the center of symmetry (ref. 35). The distances of the intermolecular double bonds are 3.891(6) Å between two methyl cinnamate groups (A in Fig. 4) and 4.917(6) Å between two ethyl cinnamate groups (B in Fig. 4). The distance between double bonds of ethyl and methyl cinnamate groups is approximately 6 Å (C in Fig. 4).

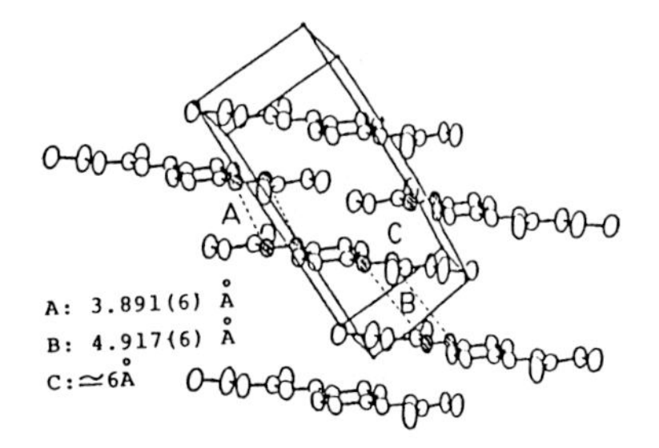

Fig. 4. Crystal structure of 5 viewed along the a-axis.

The selective dimer formation is doubtlessly due not to the selective excitation of the methyl cinnamate unit, but to different topochemical environment in the crystal 5.

In the topochemical reaction of ethyl α-cyano-4-[2-(4-pyridyl)ethenyl]cinnamate (8) (X is cyano and Y is ethoxycarbonyl in Table. 3, Group IV), one type of photoreactive dimer (9) is accumulated at the intermediate stage and then, on further photoirradiation, 9 is converted into a linear homo-adduct polymer (10). In the crystal of 8, every molecule is related to its neighboring molecules by two different centers of inversion along the stack direction. The two pairs of double bonds are separated by distance of 3.758(8) Å and 4.878(1) Å respectively (ref. 15 (b)), as is shown in the scheme below (Fig. 5).

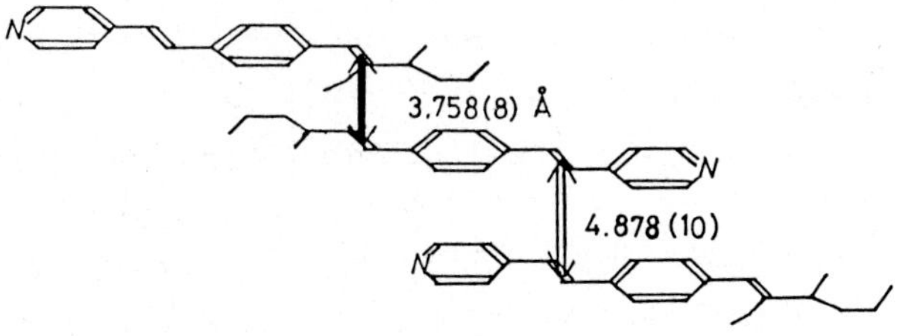

Fig. 5. Schematic drawing of the exclusive formation of the dimer 9 from 8.

As the former distance is within the conventional reactive distance, while the later is longer than the reactive one, the dimerization occurs exclusively between the former double bonds. In consequence, 8 gives a linear homo-adduct polymer 10 via the accumulation of the photoreactive dimer 9. No growing species with an odd-numbered degree of polymerization, such as a trimer or pentamer, have been detected at all in GPC curves, as is illustrated in Fig. 6 at the intermediate stages of the reaction of 8.

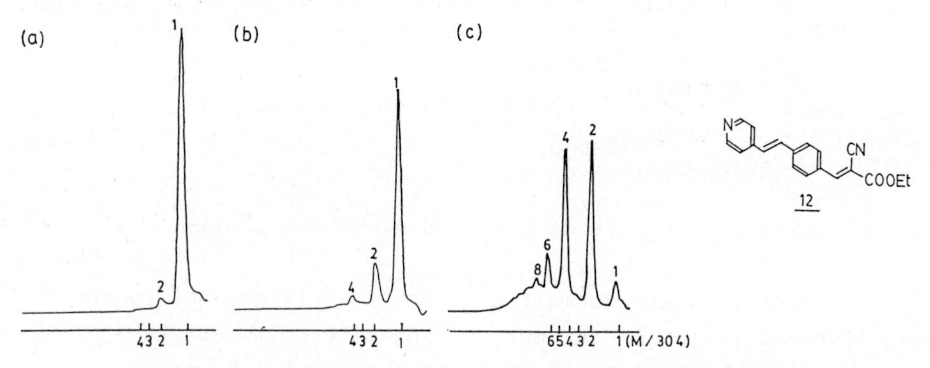

Fig. 6. GPC curves of the reaction of 8 irradiated for (a) 1h, (b) 3h, and (c) 10h.

The GPC study indicated that the monomer 8 reacts only with other monomer and that, after then, all the reacting species have an even-numbered degree of polymerization during the polymerization process.

"Even-numbered Polymerization Mechanism"

$$M + M \xrightarrow{h\nu} \text{Dimer (M-M)} \xrightarrow{h\nu} \text{Polymer } (M\text{-}M)_n$$

In addition to the even-numbered polymerization mechanism, all the reactive species in the crystal 8 should have pyridylethenyl groups at the growing terminal. In contrast to the reaction of 8, the GPC curves in the topochemical reaction of 5 show a small but definite peak of the trimer, indicating a small portion of the process of a reaction between the dimer (6) and the monomer (5).

Such an unusual behavior of the "even-numbered" chain growth should not be limited only to the reaction of unsymmetric diolefin crystals, but should be considered to be rather general for the topochemical reactions, at least to some degree. Therefore, the molecular weight distribution which was theoretically calculated is not applicable to the topochemical products.

4. CHARACTERISTIC PROPERTIES OF POLYMERS

The cyclobutane derivatives to which the chromophore groups are directly attached are generally cleaved into the original olefins photochemically or thermally, since these chromophores stabilize the intermediate radical species to make easy the thermal cleavage of cyclobutane, just as they absorb the photoenergy for cyclobutane photocleavage.

As is to be expected from such common behavior of cyclobutane derivatives, the polymers prepared by the four-center-type photopolymerization are depolymerized on uv photoirradiation or by heating into the monomer, mostly in extremely high yields. The polymers are readily photodepolymerized in a solution, while they are rather stable in the as-prepared crystal. All the crystalline polymers show no crystal melting point, but depolymerize thermally in the crystalline state into the crystalline oligomers. The deterioration from the high polymer to the oligomer crystals is a topochemical thermal depolymerization; this was confirmed by spectroscopic and x-ray diffraction pattern analysis and by the solution viscosity measurement of the products during the reaction. At the oligomer stage, these crystals melt and then, at higher temperatures, thermally depolymerize in a molten state into the monomer (ref. 36).

The outstanding characteristic behavior of these polymers is such that the thermal stability is gradually reduced with an increase in their molecular weight and that the thermal depolymerization occurs much more readily in solution than in the as-prepared crystalline state (ref. 37).

For example, when as-prepared poly-DSPs, for which the reduced viscosities are 3.0 and 8.4 dl/g, are kept at 290 °C for a long period, the longer polymer chain of 8.4 dl/g is broken down to a polymer with a reduced viscosity of 4.0 dl/g. On the other hand, the shorter polymer with the reduced viscosity of 3.0 dl/g keeps in original value after 10 h at the same temperature. No appreciable change is seen in either the x-ray pattern or DSC during the thermal treatment. The decreasing viscosity values almost level off after 10 h at values dependent on various specified temperatures (ref. 38). The same tendency of the temperature effect on the molecular weight is also seen in the case of the thermal depolymerization of these polymers in a solution. The average chain length at the level-off point in solution is much shorter than that in the crystalline state at the same temperature, indicating an enhanced thermal vibration in a solution (ref. 39).

Such an anomalous stability behavior has been interpreted in terms of the strain energy in these polymers, which had been accumulated step-by-step in the course of the topochemical process.

From unsymmetric diolefin crystals, the polymers with alternating hydrophobic and hydrophilic units in the polymer chain are prepared from the monomer in Groups IV and V of Table 3, as is shown in Fig. 7.

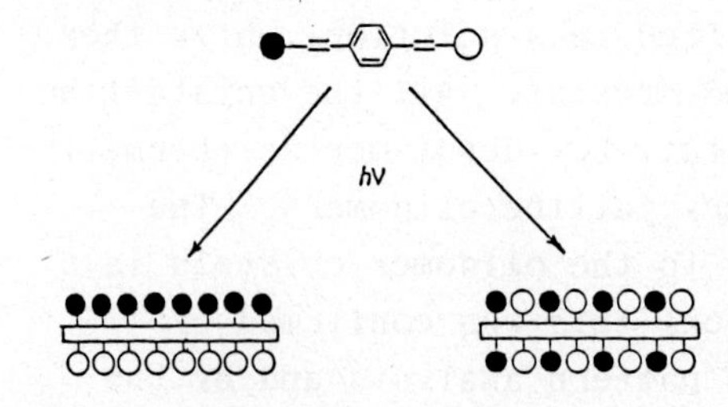

Fig. 7
(a) Hetero-adduct polymer ● Hydrophobic
(b) Homo-adduct polymer ○ Hydrophilic

REFERENCES

1 S. Okamura, K. Hayashi and M. Nishii, J. Polym. Sci., C-4 (1969) 839.

2 H. Watanabe, K. Hayashi and S. Okamura, J. Polym. Sci., B-1 (1963) 397.

3 C. F. Koelsch and W. H. Gumprecht, J. Org. Chem., 23 (1958) 1603.

4 F. Suzuki,Y. Suzuki and M. Hasegawa, Sen'i Kogyo Shikensho Kenkyu Hokoku, 72 (1965) 11.

5 (a) M. Hasegawa and Y. Suzuki, J. Polym. Sci., B5 (1967) 813. (b) M. Hasegawa, Y. Suzuki, F. Suzuki and H. Nakanishi, J. Polym. Sci., A-1, 7 (1969) 743.

6 (a) M. D. Cohen and G. M. Schmidt, J. Chem. Soc., (1964) 1969. (b) G. M. Schmidt, J. Pure Appl. Chem., 27 (1971) 647.

7 H. Nakanishi, M. Hasegawa and Y. Sasada, J. Polym. Sci., Polym. Chem. Ed., 10 (1972) 1537.

8 H. Nakanishi, K. Ueno and Y. Sasada, Acta Cryst. B-32 (1976) 3352.

9 H. Nakanishi, Y. Sasada and M. Hasegawa, Polym. Letts., 17 (1979) 459.

10 M. J. Holm and F. Zienty, US Pat. No. 3312688; CA, 67 (1967) 12151. J. Polym. Sci., A-1, 10 (1972) 1311.

11 The review articles of the four-center type photopolymerization: M. Hasegawa, Progr. Polym. Sci., Japan, 5 (1973) 143; Adv. Polym. Sci., 42, (1982) 1, Springer-Verlag; Chem. Rev., 83 (1983) 507.

12 For a review of anomalous topochemical results: M. Hasegawa, K. Saigo and T. Mori, ACS Symp. Ser., 266 (1984) 255.

13 L. Addadi and M. Lahav, Pure Appl. Chem., 51 (1979) 1269.

14 M. Hasegawa, Proceedings of 12th IUPAC Photochemistry Symp. July-Aug. (1986) Lisbon, Portugal; Pure Appl. Chem., 58 (1986) 1179.

15 (a) M. Hasegawa, S. Kato, N. Yonezawa and K. Saigo, J. Polym. Sci., Part C, Polym. Letts., 24 (1986) 513. (b) M. Hasegawa, H. Harashina, S. Kato and K. Saigo, Macromolecules, 19 (1986) 1276. (c) S. Kato, M. Nakatani, H. Harashina, K. Saigo M. Hasegawa and S. Sato, Chem. Lett., (1986) 847.

16 J. D.Dunitz and J. M. Robertson, J. Chem. Soc., (1947) 1145.

17 A. Seher, Ann, 589 (1954) 222.

18 F. L. Hirshfeld and G. M. J. Schmidt, J. Polym. Sci., A-2

(1964) 2181.
19 H. Bässler, H. Sixl and V. Enkelmann, Adv. Polym. Sci., 63, Springer-Verlag 1984 and the references cited therein.
20 F. Nakanishi and M. Hasegawa, J. Polym. Sci., Polym. Chem. Ed., 8 (1970) 2151.
21 H. Nakanishi, M. Suzuki and F. Nakanishi, J. Polym. Sci., Polym. Letts. Ed., 20 (1982) 653.
22 H. Nakanishi, M. Hasegawa and Y. Sasada, J. Polym. Sci. Polym. Chem. Ed., 10 (1972) 1537.
23 F. Suzuki, Y. Suzuki, H. Nakanishi and M. Hasegawa, J. Polym. Sci. A-1, 2 (1969) 2319.
24 N. Boeno, F. C. DeSchryver and G. Smets, J. Polym. Sci., Polym. Chem. Ed., 13 (1975) 201.
25 S. Watanabe and M. Hasegawa, Kenkyu Hokoku-Seni Kobunshi Zairyo Kenkyusho, 134 (1982) 33.
26 M. Miura, T. Kitani and K. Nagakubo, J. Polym. Sci., Part B, 6 (1968) 463.
27 M. Hasegawa, K. Saigo, T. Mori, H. Uno, M. Nohara and H. Nakanishi, J. Am. Chem. Soc., 107 (1985) 2788.
28 (a) L. Addadi, J. Van Mil and M. Lahav, J. Am. Chem. Soc., 104 (1982) 3422. (b) J. Van Mil, L. Addadi, E. Gati and M. Lahav, J. Am. Chem. Soc., 104 (1982) 3429.
29 (a) H. Nakanishi and Y. Sasada, Acta Cryst. B-34 (1978), 332. (b) F. Nakanishi, T. Tanaka, F. Miyagawa and H. Nakanishi, J. Chem. Soc. Japan, (1981) 412 (Japanese).
30 G. N. Gerasimov, O. B. Mikova, E. B. Kotin, N. S. Nekhoroshev and A. D. Abkin, Dokl. Akad. Nauk SSSR, 216 (1974) 1051.
31 H. Nakanishi, F. Nakanishi, Y. Suzuki and M. Hasegawa, J. Polym. Sci., Polym. Chem. Ed., 11 (1973) 2501.
32 M. Hasegawa and S. Shiba, J. Phys. Chem., 86 (1982) 1490.
33 G. R. Desiraju and V. Kannan, Proc. Indian Acad. Sci., 96 (1986) 351.
34 T. Tamaki, Y. Suzuki and M. Hasegawa, M. Bull. Chem. Soc. Japan, 45 (1972) 1988.
35 Crystal analysis was performed by S. Sato; to be published.
36 M. Hasegawa, H. Nakanishi, T. Yurugi and K. Ishida, J. Polym. Sci., Polym. Letts. Ed., 12 (1974) 57.
37 M. Hasegawa, H. Nakanishi and T. Yurugi, J. Polym. Sci., Polym. Chem. Ed., 16 (1978) 2113.
38 M. Hasegawa, H. Nakanishi and T. Yurugi, Polym. Letts., 14

(1976) 47.

39 M. Hasegawa, H. Nakanishi and T. Yurugi, Chem. Lett., (1975) 497.

Chapter 6

ORGANIC MOLECULES IN CONSTRAINED ENVIRONMENTS

JOHN M THOMAS AND KENNETH D M HARRIS

1. INTRODUCTION

It is well known that the reactivity of organic molecules embedded in their pure crystalline phases frequently contrasts with that of the same molecules in dispersed (dissolved or molten) states. There can, however, exist equal and even more striking contrasts between the behaviour of molecules incarcerated within a host solid and that of either the isolated, dispersed species or the corresponding species in its pure crystalline state. The host solids of interest in this context are of two types: (i) inorganic, generally aluminosilicate matrices consisting of channels, connected cages or sheets (Fig. 1); and, (ii) organic species typified by urea and thiourea that crystallize in the presence of molecules of an appropriate guest to yield so-called channel complexes (Fig. 2). Other organic molecules, notably tri-*ortho*-thymotide, apocholic acid, deoxycholic acid and perhydrotriphenylene, are also capable of forming well-defined channel complexes with a variety of organic guest species.

Compared with what is known about the solid-state chemistry of pure organic solids - see other chapters of this text - and about organic species intercalated by sheet silicates (refs. 1,2,3,4) - our knowledge of the behaviour of organic molecules incarcerated within organic hosts is sparse. Our intention here is to outline aspects of the chemistry of organic molecules constrained in one-dimensional channels within both inorganic and organic hosts, it being tacitly recognised that many of the key experiments in this field remain to be carried out. Nevertheless, it is not premature to focus on such systems: the growing interest in this area is a reflection of the realization that many issues, both of a fundamental and applied nature, are connected with it. Some obvious illustrative issues are the degrees of conformational and translational freedom of the guests, as well as the regularity or otherwise of the packing of these species within the channels, and the extent to which the host may be modified by, or respond to, the presence of the guest. Other considerations devolve upon the reactivity of the guest.

In this article we shall, so far as the inorganic hosts are concerned, focus on four specific materials, all aluminosilicates of the zeolitic kind: zeolite-L, silicalite-I, theta-1 and mordenite (Fig. 3). In particular, we shall discuss experimental and theoretical studies relating to the siting or reactivity of both small (e.g. benzene, pyridine) and large, long-chain (e.g. diundecanoyl peroxide) organic molecules inside the one-dimensional channels within these hosts. So far as the organic hosts are concerned, our discussion will be restricted to the inclusion compounds of urea and thiourea with some of the many guest species that they are capable of accommodating (Table 1).

2. EXPERIMENTAL AND THEORETICAL APPROACHES

No attempt will be made to assess critically the respective features of the numerous approaches which are available for the study of inclusion compounds. Suffice it to say that diffraction, spectroscopic, microscopic, thermochemical and computational procedures are available. It is relevant to recall briefly the primary merits of each of these methods.

2.1 X-ray diffraction

Seldom are well-defined, ordered single crystals of the organic guest: inorganic host complexes available, and often for the organic hosts there are difficulties. In the case of the zeolites, no example where a large, well-behaved single crystal of zeolite host capable of reversibly sorbing an organic guest has yet been found. The organic guest:organic host systems are generally better in forming reasonably acceptable single crystals of the inclusion compounds. But, by their very nature, these systems are not reversible. Removal of guest almost invariably leads to collapse of the host channel structure. However, while it is often possible to form moderately acceptable single crystals of the organic complex for x-ray diffraction studies, there is a tendency for the guest species to be either disordered or to form incommensurate repeats with respect to the host structure. In either situation structure determination is not straightforward. With the availability of ultra-high resolution powder diffractometers and powerful synchrotron sources, the prospects of being able to solve structures of the complexes by profile refinement are worth entertaining.

2.2 Neutron diffraction

The analysis of neutron powder diffraction data using the Rietveld powder profile method has already proved a boon in solving the structures of the compounds consisting of organic guests such as pyridine (ref. 5) or benzene (ref. 6) within zeolite-L. Newsam (ref. 7) has further demonstrated its viability by refining the structures of zeolitic complexes containing water.

2.3 Spectroscopic approaches

Multinuclear NMR has already been of value (ref. 3) for structural elucidation; but although this approach yields a wealth of valuable, sometimes unique, information, especially pertaining to the mobility and conformational freedom of the guest, it cannot, in general, match diffraction-based methods in arriving at quantitative structural information. Much more can, however, be expected of multinuclear NMR particularly when full advantage is taken of various specialized forms of the technique. Deuterium NMR, for example, as well as zero-field (ref. 8), dipolar dephasing (ref. 9) and two-dimensional NMR (ref. 10) are particularly promising, and have yet to realize their full potential for the study of inclusion systems.

Variable temperature ^{29}Si magic-angle-spinning NMR (MASNMR) affords interesting insights into the changes that occur to the framework structure of "hollow" siliceous zeolites. ^{29}Si MASNMR can also reveal changes in framework structure consequent upon the uptake of organic guests. Sometimes such changes are quite pronounced, and are apparently more readily investigated by NMR than by other spectroscopic methods or by diffraction techniques.

Infra red-, dielectric-loss-, u.v.- and photoluminescence spectroscopy have already played a useful role in affording information of the kind that we seek. The monographs on Inclusion Phenomena, edited by Davies, MacNicol and Atwood (ref. 11) deal with specific examples for the case of the organic hosts, and for the zeolitic hosts refs. 12 and 13 should be consulted.

Electron spin resonance spectroscopy (ESR) has also proved illuminating, as illustrated by investigations of isolated radicals and radical pairs formed by photolytic decomposition of long chain diacyl peroxides in a variety of different host matrices (see section 3.3) (ref. 14).

Specifically, when radical pairs are generated within a single crystal, the orientation of which with respect to the applied magnetic field is known, it is possible to deduce a considerable amount of information about the geometry of the radical pair from the magnitude of the zero-field splitting (ZFS) in the ESR spectrum, provided the spin distributions of the radicals are also known. For example, if alkyl radical pairs are located within a channel-type host, then alignment of the channel axis parallel to the magnetic field of the spectrometer, and subsequent measurement of the zero-field splitting, allows the component of the inter-radical vector along the channel direction to be determined (Fig. 4).

2.4 Microscopy

Optical microscopy is self-evidently of little value in probing detail at the atomic level, although it can occasionally indicate significant behavioural changes. For example, the collapse of an organic host (urea or thiourea) upon release of guest, is readily observed by even low-power optical microscopy. The morphology of thiourea or urea inclusion complexes is distinctly different from that of pure crystalline thiourea or urea.

High resolution electron microscopy is invaluable in probing the structure of the parent zeolitic hosts (see, for example, Fig. 5) as has been shown repeatedly in recent years (refs. 15,16,17). But, in view of the beam-sensitivity of organic molecules under electron irradiation, it is not possible to explore, in real-space, the structure of either the complexes of organic molecules within zeolites or any of the urea or thiourea inclusion compounds.

Recent high-resolution electron microscopic studies by Terasaki et al on the channel complexes formed between mordenite and selenium have yielded encouraging results. It is possible to distinguish those channels in the zeolite that are occupied by selenium from those that are not; and it transpires that domains of occupied channels separated by domains of unoccupied ones are a feature of this system.

2.5 Thermodynamic approaches

Thermodynamic measurements (for example, differential scanning calorimetry) are a useful guide to the occurrence of channel complex formation - see the work of Lahav, Leiserowitz and coworkers (ref. 18) - and to the breakdown of the complex by thermal energy. Moreover the magnitudes of the enthalpy and entropy changes serve as pointers to the relative gain or loss of translational, vibrational or rotational freedom of the guest and host components on breakdown or formation, respectively, of the channel complexes. Accurate thermodynamic studies, entailing careful measurement of heat-capacity changes as a function of temperature (ref. 19), are an invaluable guide to the manner in which the disposition and motion of the guest species in the urea inclusion compounds undergo change on thermal activation.

2.6 Computational procedures and computer graphics

Much more is known at present about the adaptation of these two approaches to the study of organic molecules retained within the channels of zeolitic hosts - to be precise highly siliceous variants thereof - than for the systems with an organic host. It is fortunate that experimental information exists (refs. 5,20) pertaining to the precise siting of pyridine or benzene inside zeolite-L so that the trustworthiness of the atom-atom calculations can be reliably assessed [such atom-atom calculations have also proved helpful for the study of the solid-state chemistry of pure organic crystals and structural defects therein (ref. 21)]. It is gratifying that, in all cases so far investigated both experimentally and computationally, the agreement between the directly observed siting of the guest species, or strength of binding, agrees well with that computed (refs. 22,23). These remarks apply to what is best described as the "static" situation: i.e. the location and orientation of an organic guest such as pyridine inside a zeolite-L host is computed at one temperature (i.e. the temperature at which the parameterized or semi-empirical constants and functions required for the computation are applicable).

Of late, progress has been made (ref. 24) in computational investigations of the influence of temperature upon the siting of the guest. Monte Carlo and molecular dynamics simulations (i.e. mathematical experiments) seem to be ahead of experiment (and *ab initio* theory) in arriving at the behavioural characteristics of organic species restricted within the channels of a host matrix. Yashonath *et al* (ref. 24) have already considered the accommodation of organic guests inside zeolitic hosts. Soon it ought to be feasible to tackle the organic guest:host systems such as those formed with urea or thiourea as host.

Computational studies on organic guests within zeolitic hosts were first initiated by the Russian worker Kiselev and coworkers (ref. 25), who showed that the first step entails the calculation of the interaction energy, $\phi_{(tot)}$, between the organic guest (usually a hydrocarbon) and the zeolite framework within which it is buried. Several approximations are made in their approach. First, the framework of the zeolite is assumed to be rigid and unperturbed by the presence of the guest species. Second, the guest is itself assumed to be rigid and present at low concentrations. And, third, the interaction energy is describable in terms of a summation based upon the well-known atom-atom procedure:

$$\phi_{(tot)} = \sum_{ij} \frac{B}{r_{ij}^{12}} - \frac{A}{r_{ij}^{6}} + \frac{Cq_i q_j}{r_{ij}}$$

where r_{ij} is the distance between atom i in the framework and atom j in the guest and the respective charges on these atoms are q_i and q_j.

In certain situations (see ref. 23) it seems adequate to employ the Lennard-Jones description of the atom-atom potentials:

$$\phi_{(tot)} = \sum_{ij} \frac{B}{r_{ij}^{12}} - \frac{A}{r_{ij}^{6}}$$

Kiselev *et al* derived values of the constants, A and B, from the reported polarizabilities of the atoms involved in the interaction. They and others (refs. 22,23,31) found that it is legitimate, in view of the details of the zeolitic structure, to ignore interactions between silicon atoms in the framework and atoms of the guest species.

Depending upon the particular molecule that is inserted as guest, appropriate values of B, A and C may be derived semi-empirically. Kiselev arrived at B, A and C for the case of methane in zeolite-Y by fitting his computed results to the experimentally determined enthalpies of adsorption.

3. RESULTS AND DISCUSSION

3.1 Silicalite-I and organic guests

We first deal with ^{29}Si MASNMR data pertaining to silicalite-I with and without sorbed organic species (ethanol, 1-propanol, n-decane or benzene) - see Fig. 6. The rich ^{29}Si MASNMR spectrum (refs. 26,27,28) indicates, *inter alia*, the relative ease with which the numerous crystallographically distinct tetrahedral silicon sites can be resolved in the parent host. The dramatic changes in the spectrum consequent upon the uptake of any of these guests signifies that structural modification of the silicalite does indeed take place when the inclusion compound is formed. Unfortunately, there is no easy route to the quantitative description of the magnitude of this structural change, although it is obvious that there are substantial changes (of the order of several degrees) in bond angles within the framework.

This result is noteworthy for two reasons. First, the time-honoured assumption (made by Kiselev and by others) that the parent host structure is invariant upon inclusion of guest is not really valid. Second, the ease with which secondary structural alterations of silicalite-I occur upon uptake of guest appears not to be a general feature of all zeolitic hosts. Parallel experiments (ref. 26) with mordenite, for example, show far less aptitude to change of spectral features upon uptake of the same guests. Furthermore, we have found (ref. 29) that temperature alone exerts a profound change upon the ^{29}Si MASNMR spectrum of silicalite-I when it is devoid of guest species. Moreover, x-ray diffraction studies, though less sensitive (compare Fig. 7(b) with Fig. 7(a)) than high resolution solid state NMR as a discriminating tool in this instance, confirms the ease with which the structure of the host silicalite-I changes as a function of temperature. In so far as we have been able to judge, silicalite-I seems to be exceptional, among the zeolites, in exhibiting this capacity to undergo flexural changes with change of organic environment.

It is appropriate to point out the similarity in behaviour of silicalite-I - which has exactly the same framework structure as the renowned and versatile Brønsted acid catalyst ZSM-5 - on the one hand with that of lysozyme on the other. Detailed crystallographic studies of the dynamic properties of lysozyme have been carried out by D.C. Phillips *et al* (ref. 30). The patterns of atomic displacements, deduced from the appropriate temperature factors, in crystals of both hen lysozyme and human lysozyme were derived from independent crystallographic structure refinement and were found to be broadly similar. The active site of lysozyme is located in a region of large atomic displacements; and Phillips et al opine that this protein mobility may play a significant part in biological activity. The point we wish to make - and it is, at this stage, not much more than a speculation - is that "intramolecular motion" of a kind is clearly a feature of the silicalite-I/ZSM-5 structure, which also has strong catalytic activity. This may amount to no more than a fortunate concomitance; but it is much more likely that the flexibility of the silicalite-I structure accounts in part for its efficacy as a versatile shape-selective catalyst. (A summarizing account of the shape-selective catalytic performance, and the computer graphic representation of the intracrystallite porosity of silicalite-I/ZSM-5 is given in ref. 31).

3.2 Computation of preferred siting of organic guests in zeolitic hosts

We illustrate the measure of success achieved using computational approaches by citing the results of Wright et al (refs. 5,20,33) on pyridine retained within the channels of (gallo) zeolite-L (prepared by Xin Sheng Liu (ref. 34)). Here there is good agreement between the computed position of the guest molecule expressed as a function of angle and distance - see Fig. 8 - and the siting of the pyridine molecule determined from Rietveld refinement of neutron powder diffraction data.

No experimental data are yet available that yield the siting of benzene inside the channels of silicalite-I or theta-1. But Ramdas et al (ref. 23) have already computed, using Lennard-Jones potentials, what might be expected. The computed results are shown in Fig. 9. It is seen that theta-1 offers greater lateral freedom to a benzene molecule situated in its channel than silicalite-I, a result not altogether unexpected in view of some distinct differences in shape selective catalytic behaviour of these two hosts.

The work of Yashonath et al (ref. 24) seeks to arrive at enthalpies of adsorption and numerous other thermodynamic and dynamic properties pertaining to small organic molecules incarcerated within the cavities and channels of zeolites. To date, a Monte Carlo calculation, invoking the well-known Metropolis algorithm, has been performed in the canonical ensemble (i.e. at constant volume and constant temperature) for methane inside a faujasitic zeolite (Si/Al = 3.0) containing some extra-framework Na^+ cations. Encouraging results have been obtained; and there is reason to suppose that, with the added power of molecular dynamics, greater insight into the behaviour of the constrained organic species will be forthcoming.

In more general terms, one may well enquire what photochemical bonuses could accrue from such computational work. Bearing in mind that photochemically assisted hydrogen tunnelling reactions can certainly take place (ref. 32) in organic crystals, it is probable that analogous processes may also occur in certain inclusion systems. Moreover, computational techniques should, in principle, reveal which sites within the host matrix should be "engineered" for accommodating say pyrazine or pyridine (with or without co-included methanol) in order to permit the initiation of favourable photochemical processes.

3.3 Studies of diacyl peroxides in urea and in zeolitic hosts

In order to investigate the influence of the environment of a constrained system on the chemical properties of the included (guest) species, it is desirable to design a probe which can be used within a variety of different host environments. In this respect Hollingsworth et al (ref. 14) have selected the photodecomposition of diacyl peroxides as such a "chemical probe" and have studied this reaction both within the diacyl peroxide/urea inclusion compounds and within zeolitic hosts (silicalite-I and ferrierite). The diacyl peroxides used in this study are shown in Fig. 10. Photolysis of these molecules leads to the following reaction:

$$RC(=O)OOC(=O)R' \xrightarrow{h\nu} R\cdot \; CO_2 \; CO_2 \; \cdot R' \longrightarrow \text{PRODUCTS}$$

The choice of this reaction was dictated, to a large extent, by the fact that several experimental techniques - including primarily ESR spectroscopy (see section 2.3) - may furnish detailed information regarding the behaviour of the radical pair produced.

Since urea is known to crystallize in a variety of different ways in its inclusion compounds (refs. 40,41,42) it is necessary, firstly, to establish the crystal structures of the diacyl peroxide/urea compounds. The host structure in UP/urea, determined from single crystal x-ray diffraction data, is shown in Fig. 11. The hydrogen-bonded array of urea molecules (the packing of which is consistent with the space group $P6_1$) contains parallel, linear, hexagonal channels within which the peroxide molecules are located. Although the packing motif of the urea molecules remains essentially invariant for each of the systems studied, the structural characteristics of the guest vary from one system to another. Thus in UP/urea, LP/urea and OP/urea it is apparent that some degree of three-dimensional ordering of the guest molecules exists whereas 6-BrHP/urea exhibits diffuse x-ray scattering consistent with a situation in which guest molecules are ordered only within each individual channel. For each of these diacyl peroxide/urea systems, the host and guest structures are incommensurate, at least along the channel direction. Solid state ^{13}C NMR experiments (employing the dipolar dephasing technique) indicate a high degree of mobility of the peroxide molecules in UP/urea and LP/urea at room temperature. Because of this feature, together with (a) the incommensurability of the host and guest, (b) partial disorder of guest molecules, and (c) the apparent decomposition of the peroxides in the x-ray beam, it has not proven possible to determine the location and structure of the peroxide molecules within the urea channels.

Two types of ESR experiments have been carried out to investigate the behaviour of the radical pairs generated upon photolysis of diacyl peroxide molecules included within the urea channel. First, as mentioned earlier, the magnitude of the zero-field splitting (ZFS) in the ESR spectrum of a photolyzed single crystal of the inclusion compound, orientated such that the channel axis is parallel to the applied magnetic field, yields the component of the inter-radical vector along that axis. Using this technique, it has been found (ref. 14) that the alkyl radicals in the radical pairs produced from UP/urea and 6-BrHP/urea are separated by more than 8Å at temperatures as low as 20K (Fig. 12). (This result should be compared with the corresponding distance of 6.06Å for decyl radical pairs produced in pure crystalline UP at 20K (ref. 43)). For comparison, the incipient radical centres in the parent peroxide molecule are separated by only about 5.7Å; see below

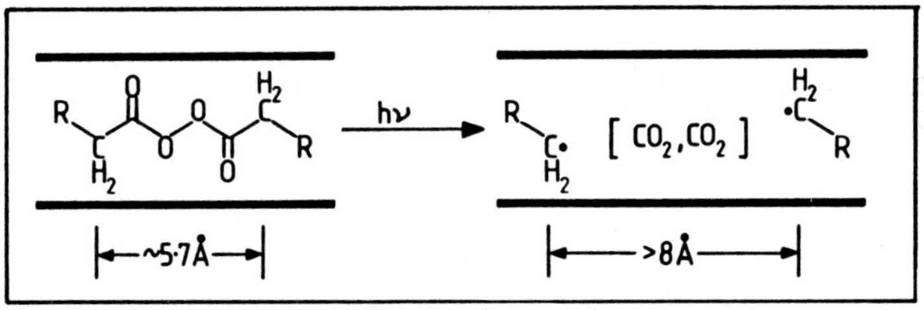

Separation of the components of the radical pair in this way is generally believed to occur as a consequence of the large pressures or "stresses" generated during this reaction (refs. 35,44). It is apparent that, in the inclusion system, the relatively loose packing of the peroxide molecules adjacent to the photolyzed peroxide is such that the recoil of the radicals in response to this stress can be readily accommodated within the channel.

In the second type of ESR experiment, the ESR signal due to the radical pair can be monitored as a function of time after photolysis to investigate the kinetics of subsequent reaction of the radical pair. The pairs thus produced within the urea channel have been found to possess exceptional kinetic stability: the half-life of the decyl radical pair produced in UP/urea is ∿17 min at 163K whereas that of the same radical pair produced in pure UP is only ∿1 min at 133K (ref. 43). The much greater kinetic stability of the radical pair in the inclusion compound can be attributed to the lack of a pathway for escape of the CO_2 molecules (either through the channel wall or along the channel) thus preventing collapse of the radical pair via recombination. The influence of substitution of the diacyl peroxide precursors on the persistence of the photogenerated radical pairs is exemplified by 4-MNP/urea and 6-BrHP/urea, (half lives: 42 min at 214K and 26 min at 191K respectively). In these cases the substituent effectively "anchors" the radicals within the channel, therefore providing an additional retarding influence on subsequent pair recombination.

Hollingsworth et al (ref. 14) also report that, when UP and 6-BrHP are incarcerated within silicalite-I, and when 6-BrHP is incarcerated within ferrierite (a zeolite structurally similar to silicalite-I), photolysis yields isolated radicals, rather than radical pairs, even as low as 20K. This observation may be due to a number of factors. For example, if the systems studied contain only low loading levels of peroxide then the local environment for each photolyzed peroxide molecule is probably that of an essentially "empty" channel. In such a case it may be that the radicals can diffuse sufficiently far apart, in response to the stress field generated at the reaction centre and/or by virtue of the surplus energy endowed by the reaction, that they behave as isolated radicals. It is also possible that, because both silicalite-I and ferrierite have intersecting channels (rather than unconnected, unidirectional ones as in the urea compounds), there is much latitude for the photogenerated radicals to find pathways by which they can migrate sufficiently far from one another so as to be uncoupled. In this respect, experiments with theta-1, which possesses non-intersecting channels, should prove instructive.

3.4 Reactions in thiourea inclusion compounds

In table 1 we listed some of the guest molecules which will form channel complexes with thiourea as host. This list refers to complexes prepared in the conventional manner - i.e. by crystallization from solution. Sergeev, Komarov and Zvonov (ref. 36) have, however, shown that a number of other inclusion compounds of thiourea can be formed if the components are co-condensed on to solid surfaces held at very low temperature and subsequently warmed up. We illustrate the nature of their ingenious approach by discussing the experiments they report on the solid-state reactivity of cyclopentadiene and maleic anhydride enclathrated within the channels of thiourea.

Typically, a glassy material is formed when vapours of the two potential guest species, cyclopentadiene and maleic anhydride, are co-condensed with the vapour of thiourea at 77K. Upon heating (to 130-160K), infra red spectroscopy indicates that a channel inclusion compound is formed. On further heating (to 240-270K) distinct changes occur in the infra red spectrum, and subsequent chromatographic analysis of the product formed shows that the Diels-Alder reaction (ca 90 per cent conversion) has taken place within the channels to yield a preponderance of the endo-isomer:

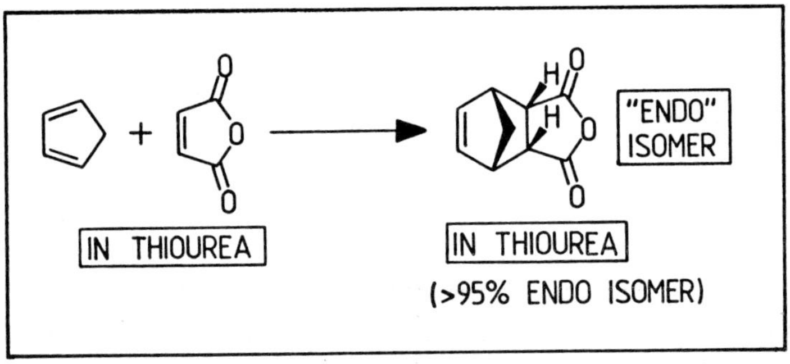

IN THIOUREA

"ENDO" ISOMER

IN THIOUREA
(>95% ENDO ISOMER)

Significantly infra red spectroscopy suggests that the framework remains unaltered during the course of the reaction. It is not easy (refs. 36,37) to obtain independent structural information, via x-ray diffraction, for the enclathrated anhydride and diene prior to and after the thermally stimulated reaction. From the nature of the product we may, however, infer that there is head-to-tail packing of the reactants along the channel axis.

It is pertinent at this stage to make reference to a system which provides perhaps the classic example of a molecule being constrained to behave differently, when present within a host matrix, than it does in its pure solid or liquid phases. Chlorocyclohexane exists, in liquid and vapour phases, in dynamic equilibrium between the "equatorial" and "axial" conformations shown below, with a predominance of the thermodynamically more stable equatorial conformer (refs. 45,46,47). Moreover, in the solid state, at sufficiently low temperature (e.g. -180°C) or high pressure (e.g. 30kbar), it exists only in this conformation (ref. 48). In complete contrast, however, when constrained within the channel in its inclusion compound with thiourea, chlorocyclohexane has been shown by infra red and Raman spectroscopy to take up the axial conformation, with no significant amount of the equatorial conformer present (refs. 49,50,51). Similar behaviour has been found for bromocyclohexane/thiourea and for the inclusion compounds of a number of other mono-, trans-1, 2-di-, and trans-1,4-di- substituted cyclohexanes in thiourea (refs. 52,53). Although significant amounts of the equatorial (or di-equatorial) conformer are present in some of the inclusion compounds reported in ref. 53, it is nevertheless true that, for each of these systems, there is a shift in the position of conformational equilibrium in favour of the axial (or di-axial) conformation upon enclathration by thiourea.

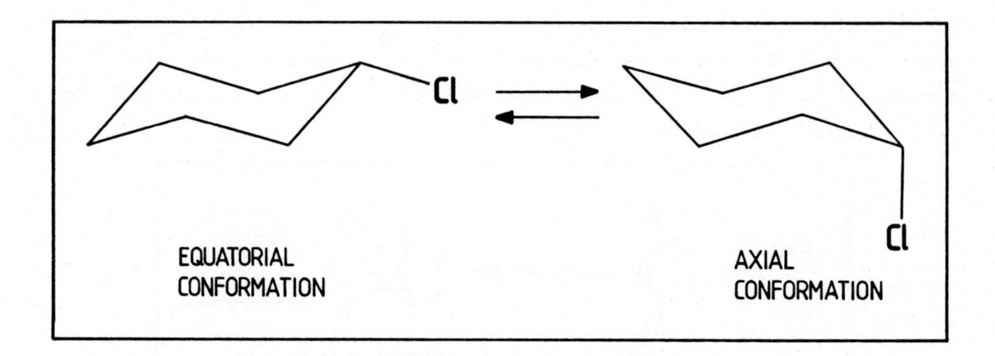

3.5 Future prospects

The study of solid state inclusion systems offers much scope for future experiment and computer simulation. By using zeolitic and other types (ref. 38) of porous, crystalline solids many novel catalytic and stoichiometric conversions of organic reactants can be effected, especially at relatively high temperatures (ca 1000K) at which the host matrix generally remains intact. Equally, as Baughman et al (ref. 39) demonstrated some years ago, certain unique organic solid state reactions can be carried out inside the channels of urea inclusion compounds, and the products isolated by subsequent dissolution of the host matrix.

So far as understanding the fundamentals of both photo- and thermally-stimulated reactions is concerned, the channels formed by organic and inorganic hosts offer much scope for novel experimental study. Nuclear magnetic resonance, photophysical and powder diffraction techniques are likely to prove particularly instructive. We also forecast that, by exploiting the techniques of molecular simulation and computer graphics, many important advances in our understanding of the behaviour of molecules in constrained environments should be forthcoming.

We thank our colleagues (mentioned on the references) for helpful discussions, and the British Petroleum Company p.l.c. for the award of a studentship to KDMH.

TABLE 1

A. Structural drawings of a selection of molecules which are known to form inclusion compounds with urea.

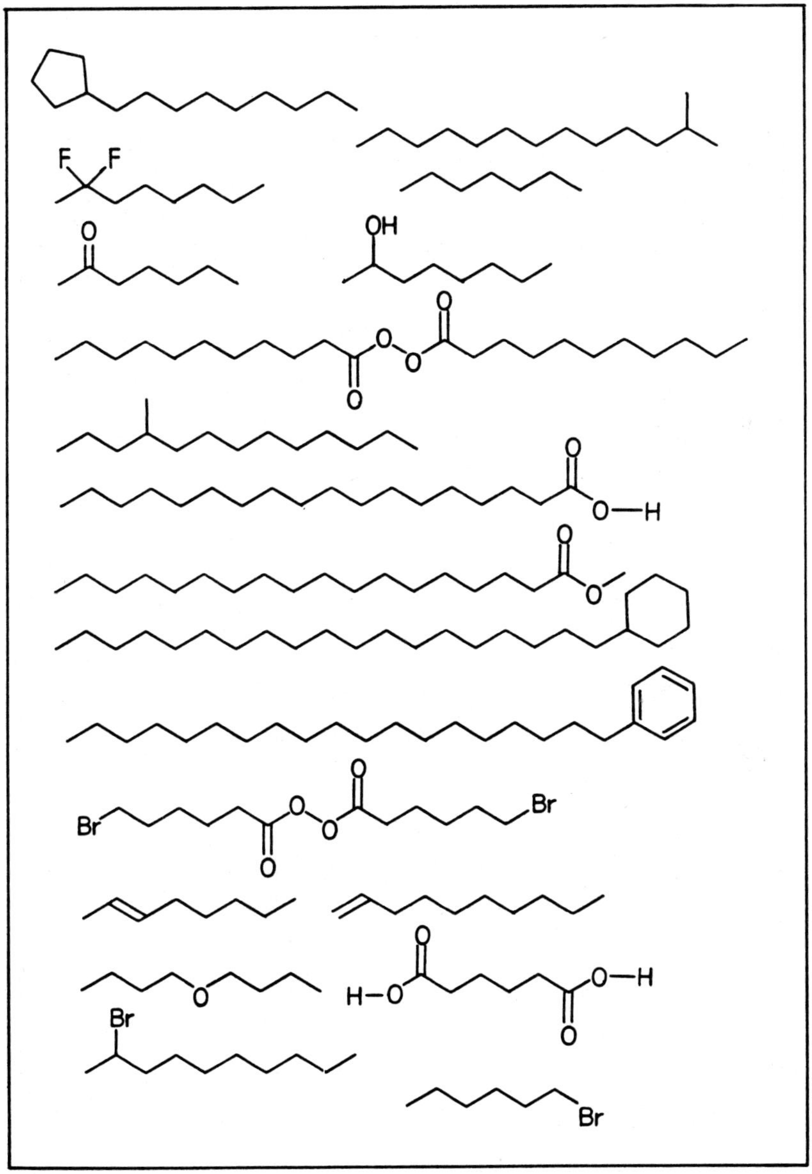

TABLE 1

B. Structural drawings of a selection of molecules which are known to form inclusion compounds with thiourea.

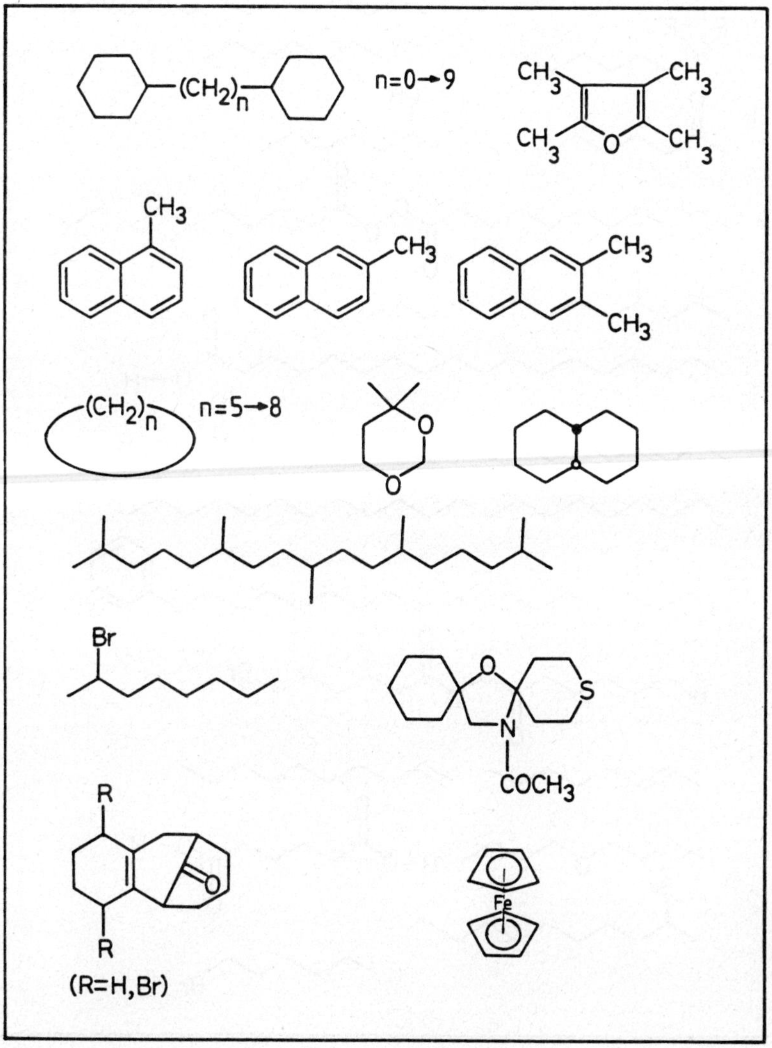

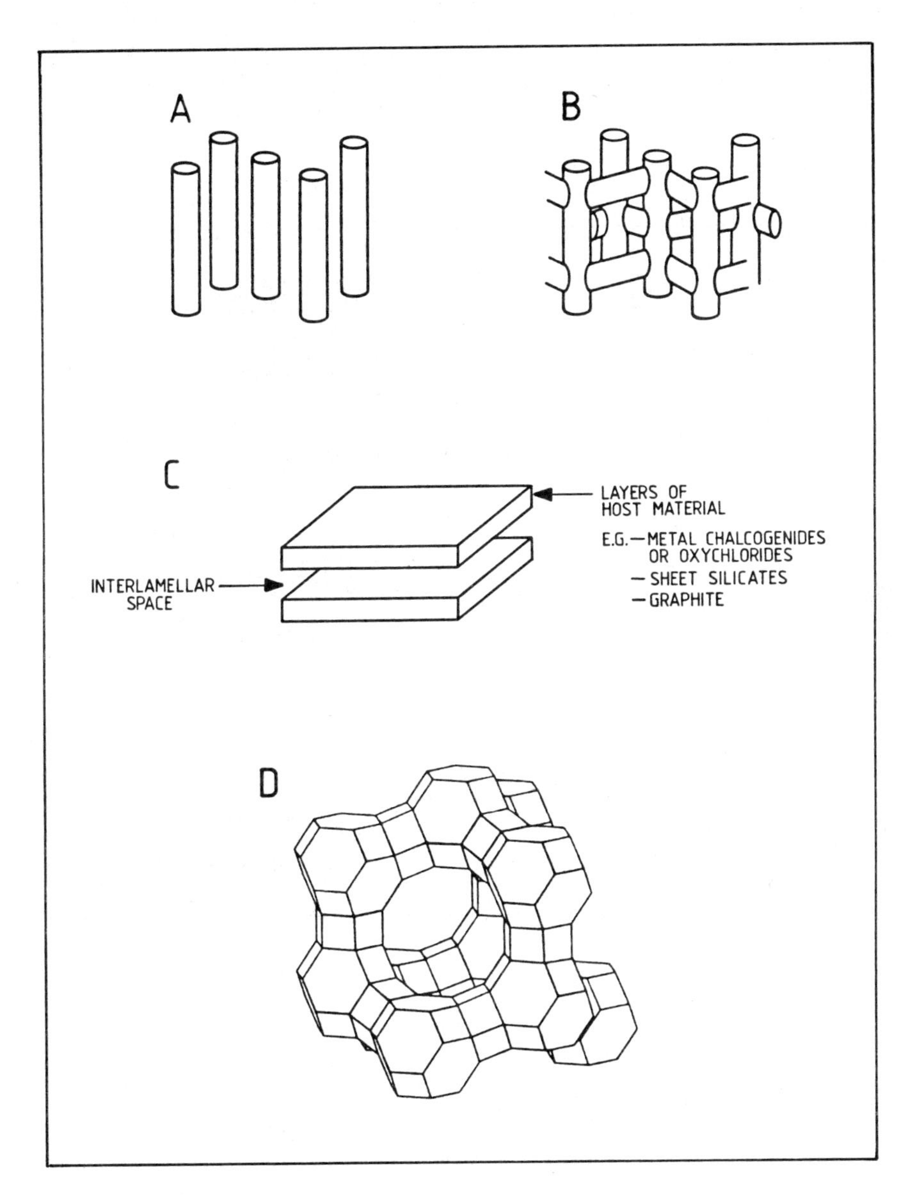

Figure 1

Schematic illustrations of various types of host solids capable of accommodating guest species.

A. Theta-1: a one-dimensional channel system.
B. Silicalite-I: a system containing intersecting channels.
C. Sheet structures that can accommodate guests free to move in two dimensions.
D. Faujasitic zeolites (i.e. zeolites X and Y) which contain three-dimensionally linked cavities.

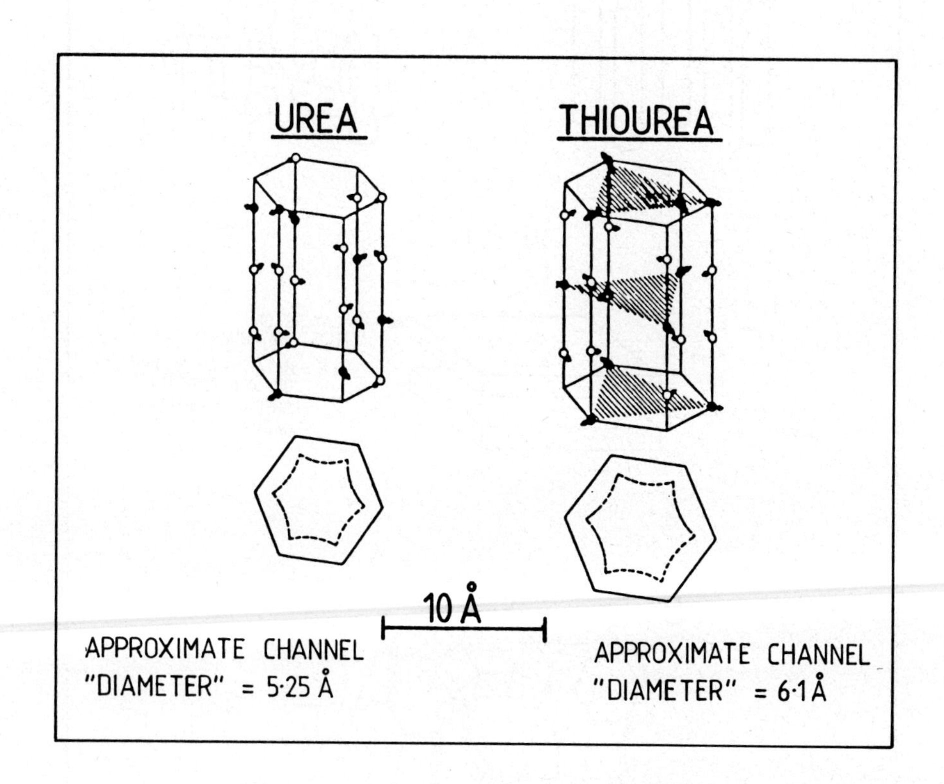

Figure 2

Schematic comparison of the host channel structure in urea and thiourea inclusion compounds.

(Modified from W. Schlenk, Ann. Chem., 573 (1951) 142).

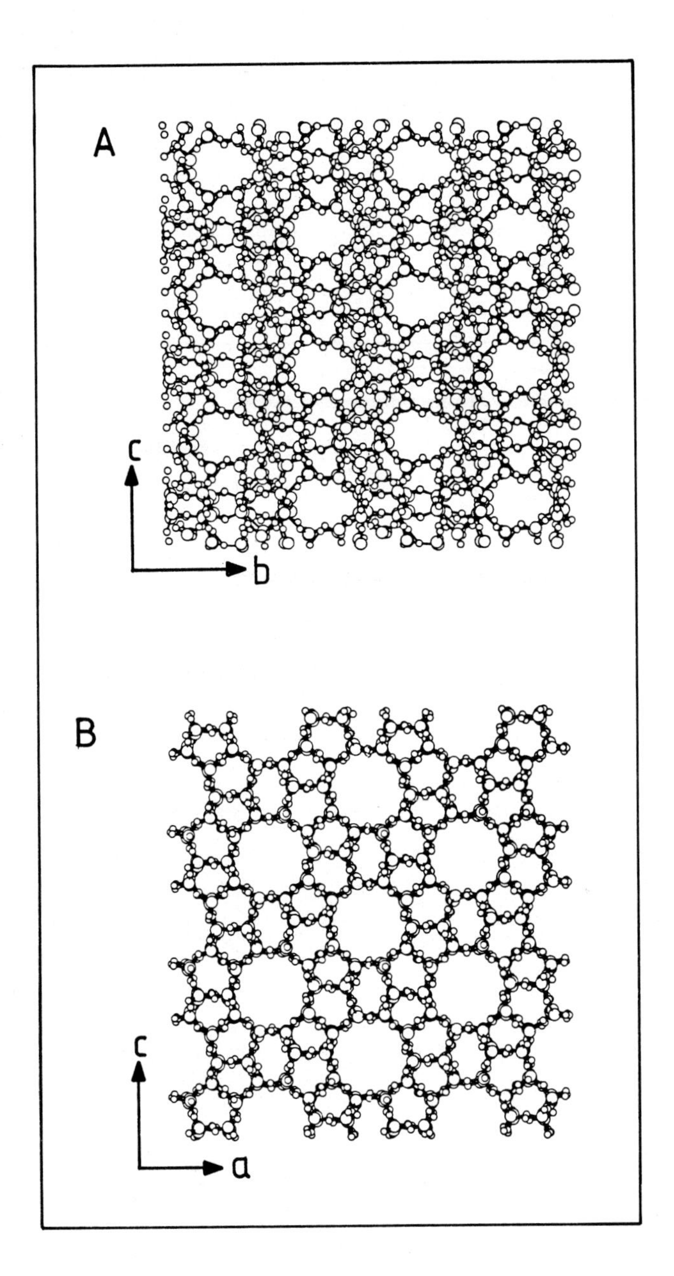

Figure 3

A. Structure of silicalite-I (ZSM-5) viewed along the crystallographic [100] direction.

B. Structure of silicalite-I (ZSM-5) viewed along [010].

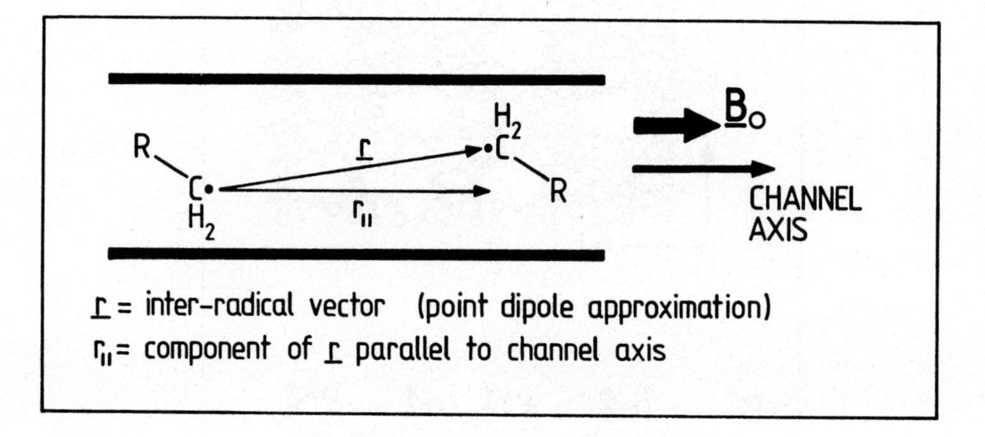

Figure 4

Illustration of the geometric relationships pertaining to the ESR experiment described in section 2.3.

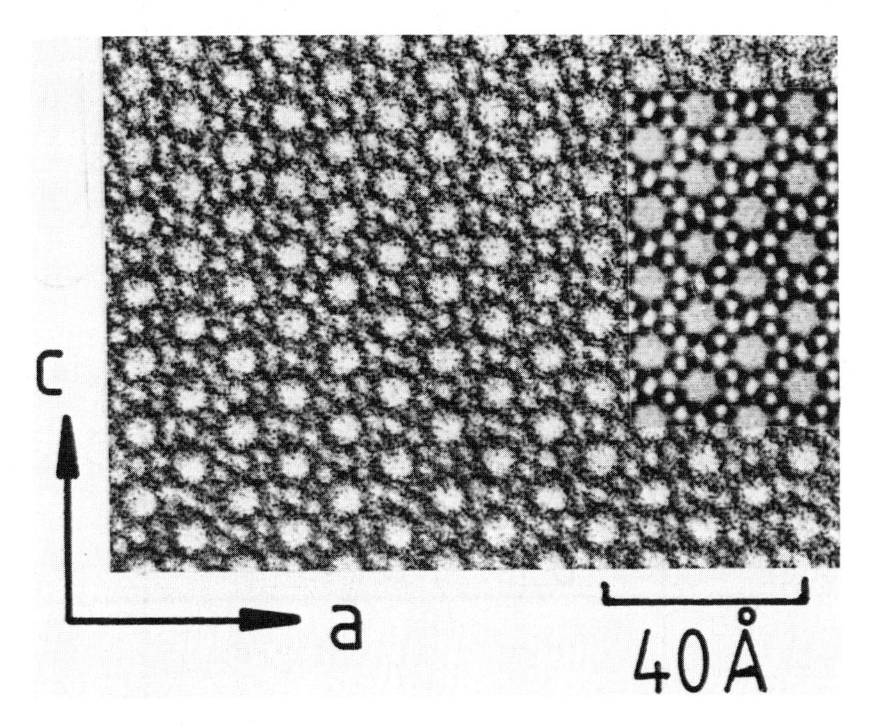

Figure 5

High resolution electron microscopic image of silicalite-I viewed along [010]. The main channels (of diameter ∿5.5Å) are clearly visible. (Compare with Figure 3B).

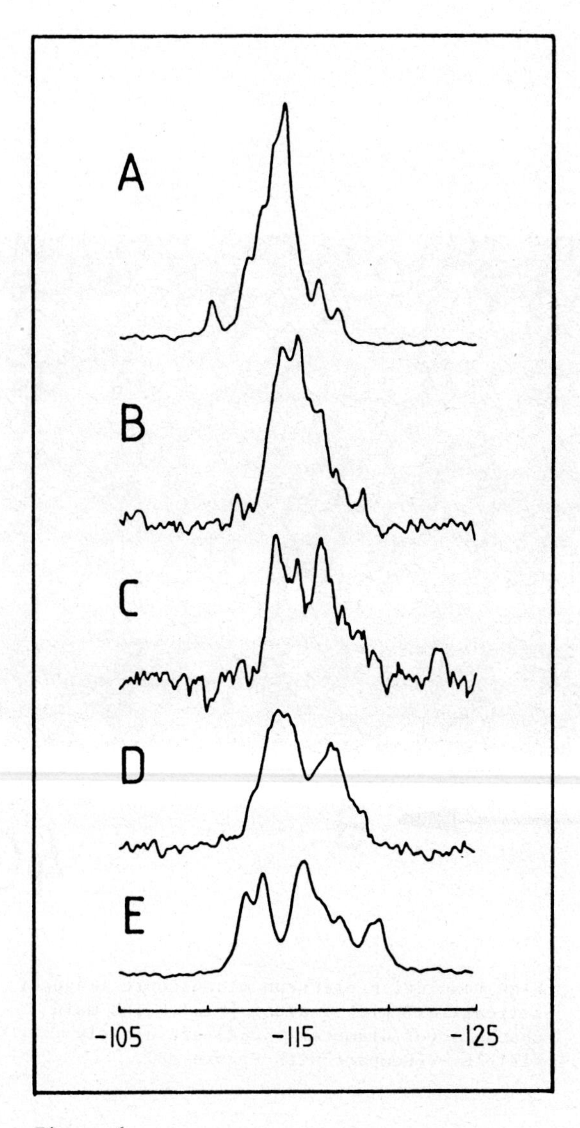

Figure 6

^{29}Si MASNMR spectra of silicalite-I (Si/Al > 4400) containing organic guest molecules.

A. parent host material
B. containing ethanol
C. containing 1-propanol
D. containing n-decane
E. containing benzene

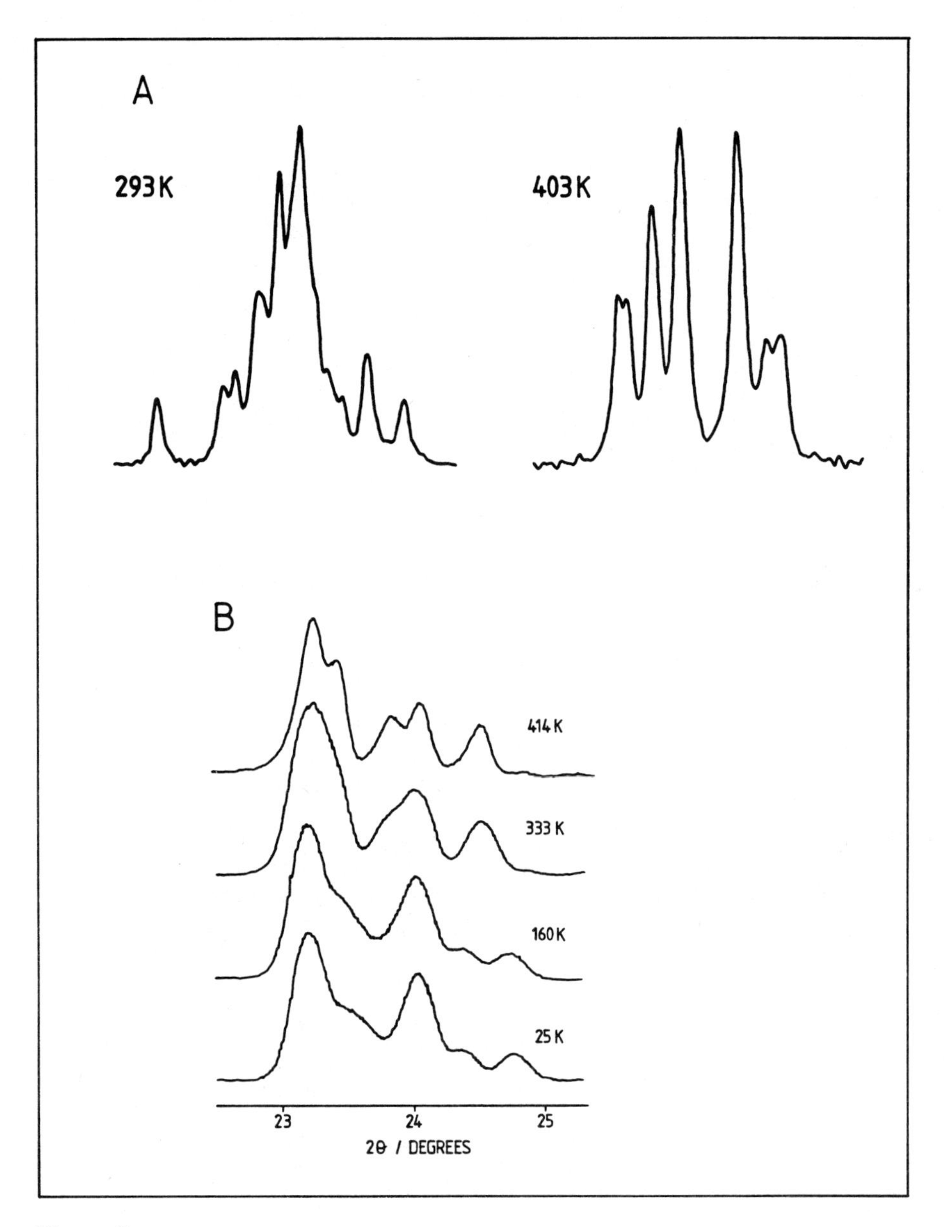

Figure 7

A. ^{29}Si MASNMR spectra of silicalite-I at two temperatures showing clear signs of structural differences.

B. X-ray powder diffractograms of silicalite-I at different temperatures, again showing evidence of structural change.

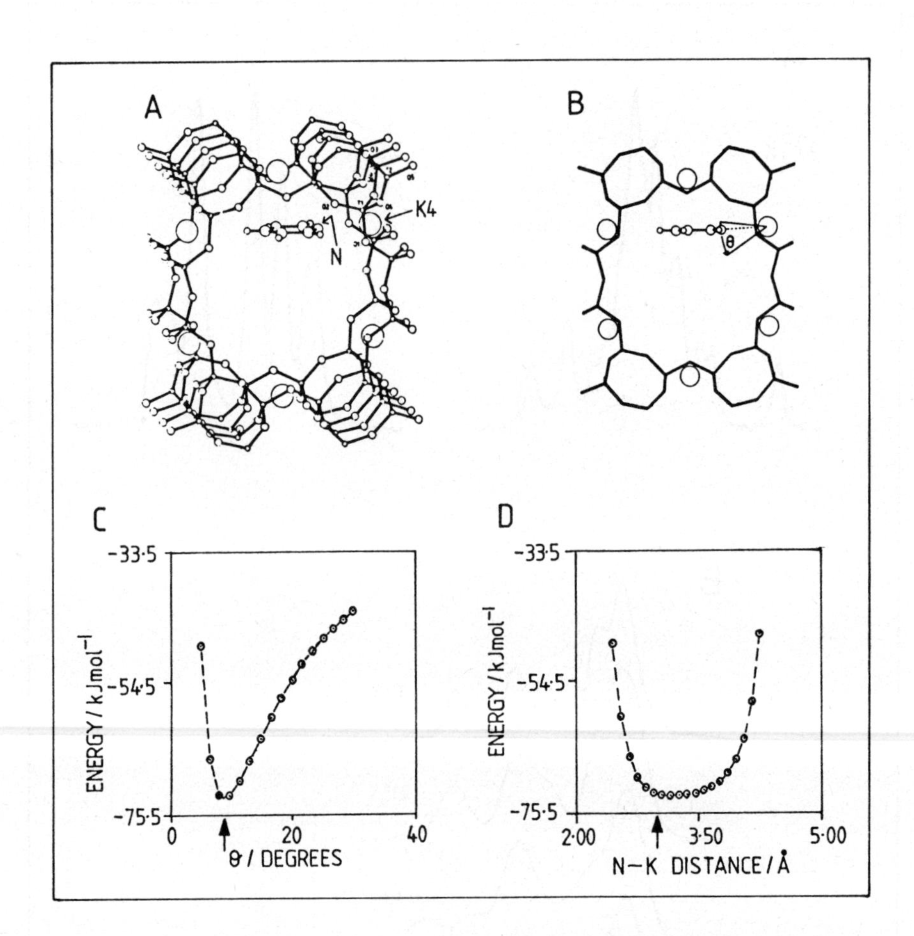

Figure 8

A. The location of pyridine in the main channel of potassium zeolite-L determined from neutron powder diffraction data.

B. Pyridine in zeolite-L, viewed along [001], showing the angle θ which was varied in the calculations, the results of which are shown in C.

C. Interaction energy between pyridine and zeolite-L as a function of the angle θ. The arrow indicates the experimentally observed angle.

D. The interaction energy as a function of the K(4) [zeolite-L]–N [pyridine] distance with θ fixed at 8.8°. The experimentally observed distance is indicated by the arrow.

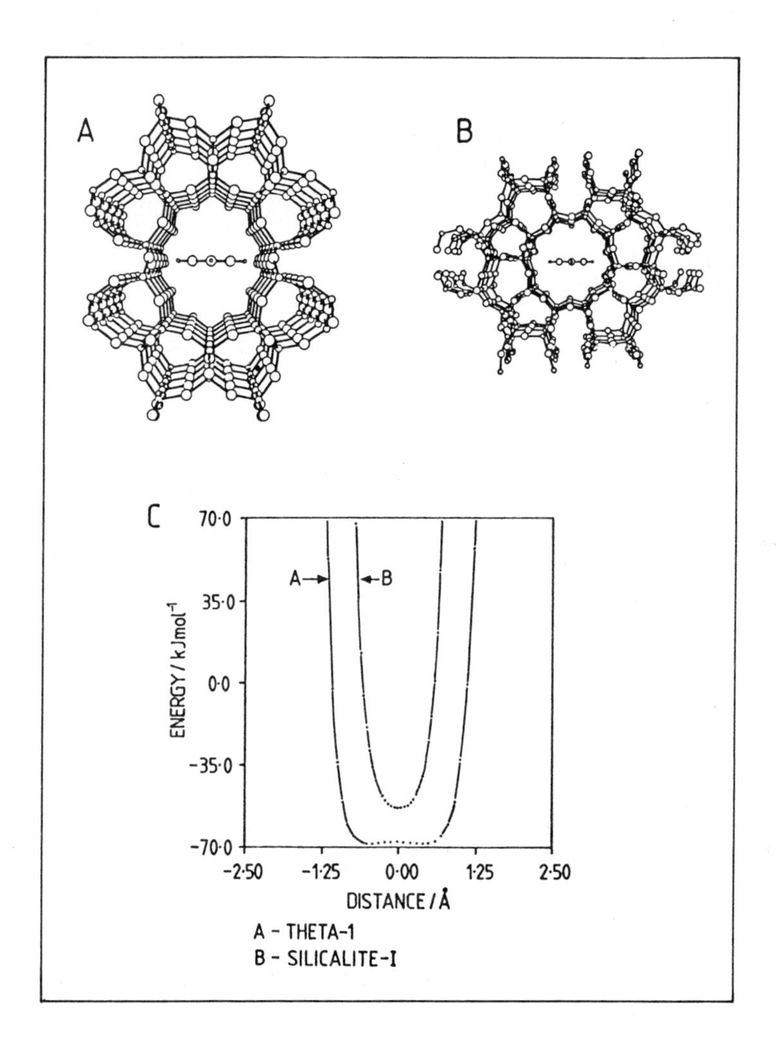

Figure 9

A. Illustration of a benzene molecule placed in the main channel of theta-1.
B. Benzene molecule in silicalite-I viewed down the straight channel.
C. Energy profiles for the benzene/theta-1 and benzene/silicalite-I systems as a function of the distance between the centre of the benzene molecule and the centre of the channel.

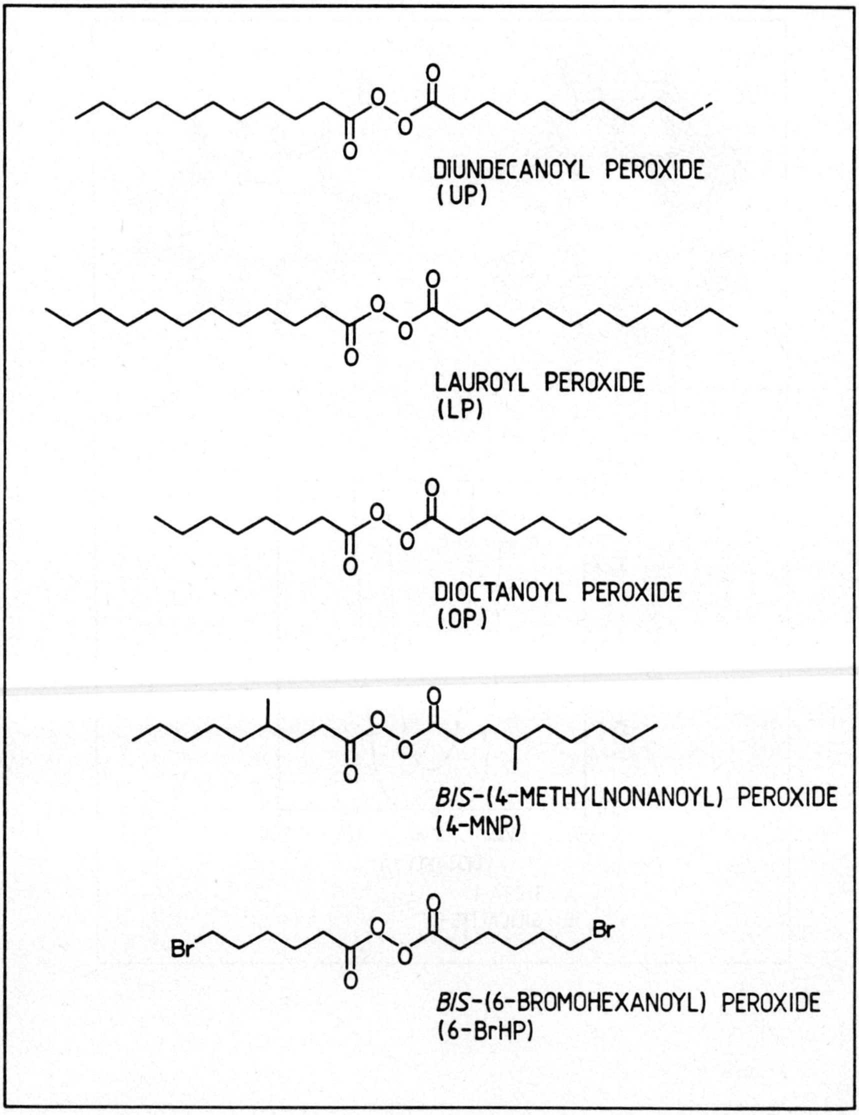

DIUNDECANOYL PEROXIDE
(UP)

LAUROYL PEROXIDE
(LP)

DIOCTANOYL PEROXIDE
(OP)

BIS-(4-METHYLNONANOYL) PEROXIDE
(4-MNP)

BIS-(6-BROMOHEXANOYL) PEROXIDE
(6-BrHP)

Figure 10

Structural formulae (and name abbreviations) of diacyl peroxides discussed in the text.

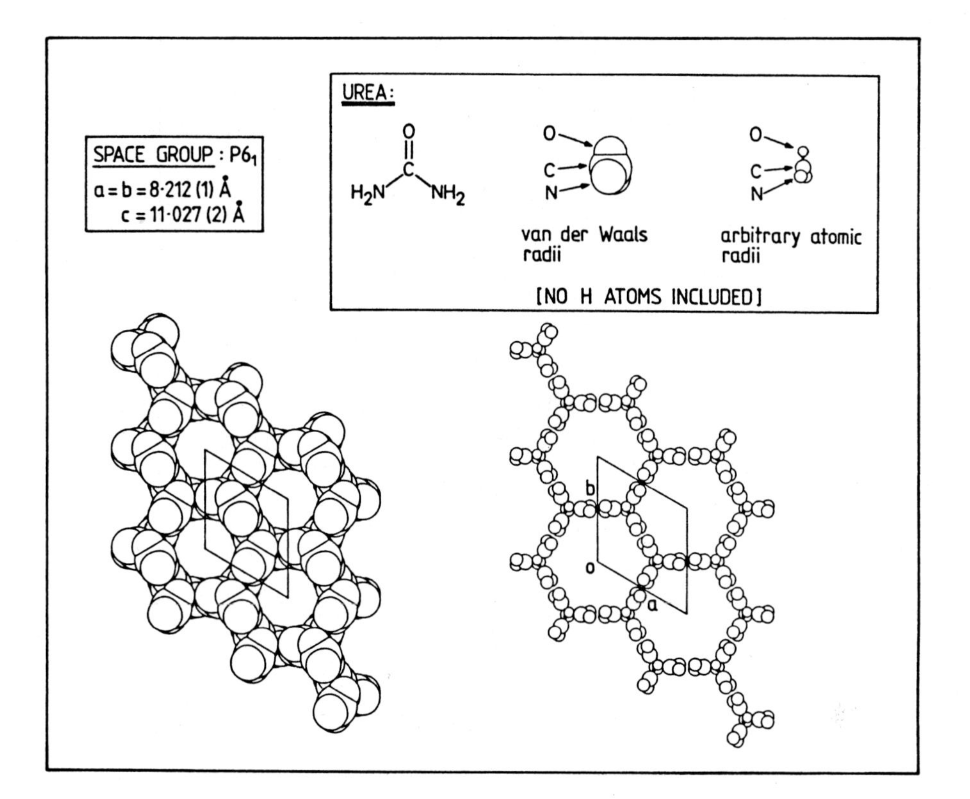

Figure 11

Two alternative representations of the host structure in UP/urea projected on to the crystallographic xy plane. The long-chain peroxide molecules (not shown in the diagram) are located in the channels parallel to the z-axis.

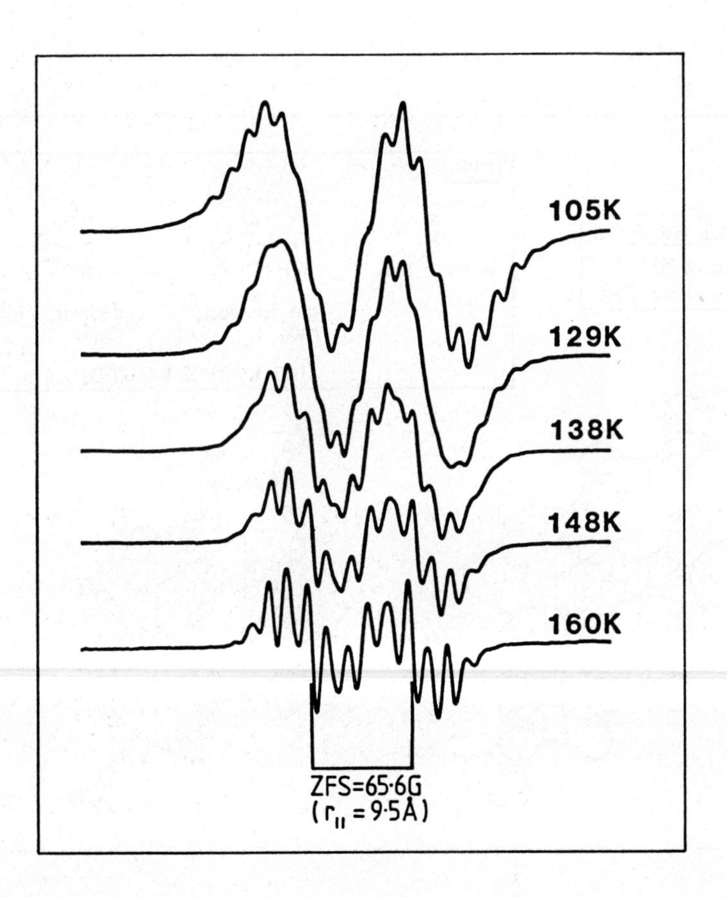

Figure 12

ESR spectra of decyl radical pairs generated by photolysis of a single crystal of UP/urea. With the crystallographic z-axis aligned parallel to the magnetic field, the zero-field splitting (ZFS) in the spectrum gives the component (r_{11}) of the inter-radical vector along the channel axis.

REFERENCES

1 R.M. Barrer, Zeolites and Clay Minerals, Academic Press, 1978.
2 J.M. Thomas, in M.S. Whittingham and A.J. Jacobson (Eds.), Intercalation Chemistry, Academic Press, 1982, 55.
3 J.M. Thomas and J. Klinowski, Adv. in Catalysis, 33 (1985) 199.
4 J.A. Ballantine, in R. Setton (Ed.), Chemical Reactions in Organic and Inorganic Constrained Systems, NATO ASI 165, Reidel, (1986) 197.
5 P.A. Wright, J.M. Thomas, A.K. Cheetham and A.K. Nowak, Nature, 318 (1985) 611.
6 C. Williams, Ph.D. Thesis, University of Cambridge, 1986; see also C. Williams, J.M. Thomas, A.K. Nowak and A.K. Cheetham, in preparation.
7 J.M. Newsam, J. Chem. Soc., Chem. Comm., (1986) 1295.
8 D.P. Weitekamp, A. Bielecki, D. Zax, K. Zilm and A. Pines, Phys. Rev. Lett., 50 (1983) 1807.
9 R.K. Harris, Nuclear Magnetic Resonance Spectroscopy, Pitman, 1983.
10 R. Benn and H. Gunther, Angew. Chemie Intl. Ed. Engl., 22 (1983) 350.
11 J.L. Atwood, J.E.D. Davies and D.D. MacNicol (Eds.), Inclusion Compounds, Academic Press, Vols. 2 and 3, 1984.
12 G.D. Stucky and F.G. Dwyer (Eds.), Intrazeolite Chemistry, ACS Symposium No. 218 (1983).
13 D. Olsen and A. Bisio (Eds.), Proc. Sixth Intl. Zeolite Conf., Butterworths, (1984).
14 M.D. Hollingsworth, K.D.M. Harris, W. Jones and J.M. Thomas, J. Inclusion Phenomena, in press.
15 J.M. Thomas and G.R. Millward, J. Chem. Soc., Chem. Comm., (1982) 1380.
16 J.M. Thomas, in G. Ertl (Ed.), Proceedings of the 8th Intl. Congr. on Catalysis, Verlag Chemie, 1 (1984) 31.
17 O. Terasaki, G.R. Millward and J.M. Thomas, in preparation.
18 M. Lahav, L. Leiserowitz, L. Roitman and C.P. Tang, J. Chem. Soc., Chem. Comm. (1977) 928.
19 N.G. Parsonage and R.C. Pemberton, Trans. Faraday Soc., 63 (1967) 311 and references therein.
20 P.A. Wright, Ph.D. Thesis, University of Cambridge, 1985.
21 S. Ramdas and J.M. Thomas, in M.W. Roberts and J.M. Thomas (Eds.), Chemical Physics of Solids and their Surfaces, Royal Society of Chemistry, London, 7 (1978) 31.
22 P.A. Wright, S. Ramdas, J.M. Thomas and A.K. Cheetham, J. Chem. Soc., Chem. Comm., (1984) 1338.
23 S. Ramdas, A.K. Nowak, A.K. Cheetham and S. Pickett, in preparation.
24 S. Yashonath, J.M. Thomas, A.K. Nowak and A.K. Cheetham, in preparation.
25 A.G. Bezus, A.V. Kiselev, A.A. Lopatkin and Pham Quang Du, J. Chem. Soc., Faraday Trans. II., 74 (1978) 367.
26 T.A. Carpenter, J. Klinowski and J.M. Thomas, unpublished work.
27 G.W. West, Aust. J. Chem., 37 (1984) 455.
28 C.A. Fyfe, G.J. Kennedy, C.T. de Schutter and G.T. Kokotailo, J. Chem. Soc., Chem. Comm., (1984) 541.
29 C. Williams, J. Klinowski, T.A. Carpenter and J.M. Thomas, unpublished work.
30 P.J. Artymiuk, C.C.F. Blake, D.E.P. Grace, S.J. Oatley, D.C. Phillips and M.J.E. Sternberg, Nature, 280 (1979) 563.
31 S. Ramdas, J.M. Thomas, P.W. Betteridge, A.K. Cheetham and E.K. Davies, Agnew. Chemie Intl. Ed. Engl., 23 (1984) 671.
32 D. Stehlik, private communication to J.M. Thomas, February 1986.
33 A.K. Cheetham, A.K. Nowak and P.W. Betteridge, Proc. Ind. Acad. Sci., (Chem. Sci.), 96 (1986) 411.
34 Xin Sheng Liu, Ph.D. Thesis, University of Cambridge, 1986.
35 J.M. McBride, B.E. Segmuller, M.D. Hollingsworth, D.E. Mills and B.A. Weber, Science, 234 (1986) 830.
36 G.B. Sergeev, V.S. Komarov and A.V. Zvonov, Dokl. Akad. Nauk. SSSR, 270 (1983) 139.

37 V.S. Komarov, W. Jones and J.M. Thomas, unpublished work; V.S. Komarov and W. Jones, in preparation.
38 J.M. Thomas, Nature, 322 (1986) 500.
39 R.H. Baughman, R.R. Chance and M.J. Cohen, J. Chem. Phys., 64 (1976) 1869; see also D.M. White, J. Amer. Chem. Soc., 82 (1960) 5678.
40 A.E. Smith, Acta Cryst., 5 (1952) 224.
41 Y. Chatani, Y. Taki, H. Tadokoro, Acta Cryst., B33 (1977) 309.
42 Y. Chatani, K. Yoshimori, Y. Tatsuta, Polym. Prepr., Am. Chem. Soc., Div. Polym. Chem., 19 (1978) 132.
43 B.E. Segmuller, Ph.D. Thesis, Yale University, 1977.
44 J.M. McBride, Acc. Chem. Res., 16 (1983) 304.
45 M. Larnaudie, Compt. Rend., 235 (1952) 154.
46 P. Klaeboe, J.J. Lothe, K. Lunde, Acta Chem. Scand., 10 (1956) 1465.
47 K. Kozima, K. Sakashita, Bull. Chem. Soc. Japan, 31 (1958) 796.
48 P. Klaeboe, Acta Chem. Scand., 23 (1969) 2641.
49 M. Nishikawa, Chem. Pharm. Bull., 11 (1963) 977.
50 K. Fukushima, J. Mol. Struct., 34 (1976) 67.
51 A. Allen, V. Fawcett, D.A. Long, J. Raman Spec., 4 (1976) 285.
52 K. Fukushima, K. Sugiura, J. Mol. Struct., 41 (1977) 41.
53 J.E. Gustavsen, P. Klaeboe, H. Kvila, Acta Chem. Scand., A32 (1978) 25.

Chapter 7

CLATHRATES

G. TSOUCARIS

"If we dissociate the elements, one might be unable to reassemble them again just as they were, and we have lost what we had at first: the whole with its details, its inner soul."

Arnold Schönberg, 1913

1 INTRODUCTION

Clathrate inclusion compounds are crystals comprising two different species, guest and host, whose association differs from that of other complexes or mixed crystals in two essential points: geometry and chemical bonding.

There is complete enclosure of the guest molecules in the cavities formed by a host framework or matrix. These cavities are isolated cages, channels or layers.

The host/guest association is not established by covalent or ionic bonds: van der Waals or dispersion forces are involved. The host/host association is assured by H-bonds and/or van der Waals bonds (with the exception of zeolites and other layer-type compounds, often considered as a class distinct from cage and channel clathrates).

Intuitive notions, like host/guest complementarity or mutual influence of hosts and guests, are introduced in the description of clathrates. These notions are similar to concepts which arise from biological sciences, such as antigen/antibody and enzyme/substrate complementarity.

Although the guest is not *stricto sensu* chemically modified, an important class of physical properties of the clathrate still depends rather on the host. As an example, the melting point is close to that of the pure host although the guest may be a practically perfect gas at room temperature (krypton enclosed in hydrates). This is a simple way of encapsulating the guest in voids formed by a three-dimensional crystalline framework. Conversely, bringing into solution or melting the clathrate generally leads to a complete dissociation of its elements, with a few notable exceptions, such as cyclodextrins and cryptates.

Let us consider the evolution of our knowledge about clathrates in relation to the history of chemistry and of crystallography. Molecules are invisible and chemistry is a science of modelization. The concept of molecular structure emerged in the nineteenth century through independent sources of observation and following different lines of thought: the invisible was not unknowable, points out H.M. Powell. Several decades before, the main developments of chemistry, mineralogy and crystallography (R.J. Hauy) were powerful precursors and contributors to the science of molecular structure. The next important step was accomplished in 1848 by Pasteur who proposed the tetrahedral bonding of carbon atoms in the tartaric acid molecule because he was able to obtain crystals of right-hand and left-hand configurations. Then Le Bel and van't Hoff set down the general principles of chemistry in space. The triumph of the molecular structure concept was so complete that purely ionic crystals were for a long time supposed to comply with this concept.

As late as 1927, the idea of ionic compounds was still strongly opposed by conservative chemists: "Prof. W.L. Bragg asserts that, in sodium chloride, there appear to be no molecules represented by NaCl. The equality in number of sodium and chlorine atoms is arrived at by a chessboard pattern of these atoms: it is a result of geometry and not of a pairing-off of these atoms . . . Chemistry is neither chess nor geometry, whatever X-ray physics may be . . . It is time that chemists took charge of chemistry once more and protected neophytes against the worship of false gods: at least teach them to ask for something more than chessboard evidence." (H.E. Armstrong, Nature, London, 120 478 (1927)).

Clathrates escaped the honour of such controversies, perhaps because they came either too early in the history of chemistry or after X-ray diffraction was recognized as the official photographer of molecular structures.

Chlorine hydrate was discovered by Davy <1> in 1811, by bubbling chlorine into cool water, and in 1823 Faraday <2> proposed the formula $Cl_2.10H_2O$ (in fact, $8Cl_2.46H_2O$ or $Cl_2.5.75H_2O$). F. Wöhler, who in 1828 synthesized urea in the laboratory, the first organic molecule, discovered in 1849 the quinol clathrates <3>, but at that time one could not understand their chemical nature. In 1886, Mylius came to the conclusion that the bonding was not of a known type, but that the host molecules were somehow able to lock in a volatile compound <4>. Finally, X-ray diffraction studies revealed the exact nature of these non-conventional chemical compounds (Palin and Powell in 1947 for quinol <5>, and Stackelberg in 1949 <6>, Claussen in 1951 <7>, and Pauling and March in 1952 <8> for hydrates). The term clathrate, coined by Palin and Powell, stems from the Greek 'κλαθρον', was transformed to clatra in Latin, and means 'bolt' or 'lock'.

Only a very small fraction of the large number of known inorganic and organic compounds is endowed with the ability of forming host lattices in clathrates. So far, over a hundred host molecules have been discovered. One is then confronted with several questions: can we foresee the characterization of the clathrate-forming hosts? Can we design in advance host molecules or, at least, can we modify existing hosts so that their inclusion properties are modified and 'modulated' according to predefined guests? What is the rôle of guest molecules on the stability of the clathrate? How easily is the clathrate formed and under which conditions does it decompose, thus liberating the guest?

In Section 2, an overall description is given of some clathrate frameworks, indicating how the guest can, to a certain extent, influence the structure of the host crystal.

Section 3 describes the chemical and thermodynamic nature of host/guest interactions, and more precise pictures of the association (i.e. position and dynamic behaviour of the guest molecule).

In Section 4, achievements and future possibilities of molecular structure design are examined.

In Section 5, we discuss properties of clathrates in relation to a wide range of applications in physics, chemistry, biology, medicine and agriculture.

In this introductory chapter, our aim is to illustrate the essential phenomena rather than to describe the different clathrate families. The number of publications on clathrates is already very large, probably over ten thousand. As a general reference, we use the three volumes of Inclusion Compounds <9>, and we particularly mention H.W. Powell's Introduction <10>.

2 CAVITY AND CHANNEL CLATHRATES

The main aim of this section is to describe prototypes of frameworks, focusing on the geometrical aspect of crystal structures.

2.1 Cage clathrates: hydrates

In these clathrates, the guest molecules are located within cages formed by the host framework. They are completely surrounded by the host molecules. The prototypes of cage clathrates are the hydrates described below.

'Holes' or voids commonly occur in crystal structures, and it is only the dimension of the voids that endows clathrates with very special properties. Thus, small octahedral and tetrahedral voids are found in cubic face centered (c.f.c.) metal structures, and can enclose hydrogen or carbon atoms. Solvent molecules are also often trapped in a crystal lattice. Clathrate voids may contain a variety of guest molecules, yielding a series of isomorphous crystals (Table 1).

Table 1

Hydrates

Type I	$6X.2Y.46H_2O$	(a ≅ 12 Å)
Ar, K, Xe, O_2, N_2, H_2S, CH_4, C_2H_2, C_2H_4, C_2H_6, CH_3CHF_2		
Type II	$8X.16Y.136H_2O$	(a ≅ 17 Å)
CH_3I, $CHCl_3$, $(C_2H_5)_2O$, cyclobutanone etc		
Types I and II		
COS, $(CH_3)_2O$, CH_3SH, cyclopropane, CH_3Br etc		
Type II with H_2S as the hilfgas		
butane, I_2, CCl_3Br, CCl_3NO_2 etc		

Hydrates - first discovered in 1811 - are characterized by a three-dimensional (3D) framework of water molecules closely analogue to that of ice. In ordinary hexagonal ice I_h and cubic ice I_c (Figures 1a and 1b), each H_2O molecule is tetrahedrally coordinated to four other molecules forming O-H..O hydrogen bonds. Thus each water molecule simultaneously donates and accepts two H atoms. There is a twofold disorder of the H atoms across each O...O edge, i.e. they are statistically distributed over the two models: O-H..O and O..H-O.

All these features are conserved in hydrates, but the spatial arrangement of second neighbours is different. Their basic structural component is a regular pentagon formed by O-H..O bonds. The regular solid generated by assembling such pentagons is the pentagonal dodecahedron (Figure 2). However, space cannot be completely filled in the periodical sense by such solids, although fivefold symmetry has been proposed in the recently discovered 'quasi-crystals'. How then are the elementary regular pentagons assembled to form a crystal? The key answer is the occurrence of other polyhedra (Figure 3) which, together with the regular 12-hedra, allow a periodical 3D arrangement. We shall describe briefly the most common hydrates of Types I and II <11>.

In Type I structures (diagram (a) of Figure 4), the host framework is composed of 12- and 14-hedra (diagrams (a) and (b) respectively of Figure 3) in the ratio of 1/3. The 14-hedra, shown in solid lines, have two hexagonal faces (shared with other 14-hedra) and twelve pen-

Figure 1a The arrangement of molecules in the ice crystal

There is one proton along each oxygen/oxygen axis, which is close to one or the other of the two oxygen atoms

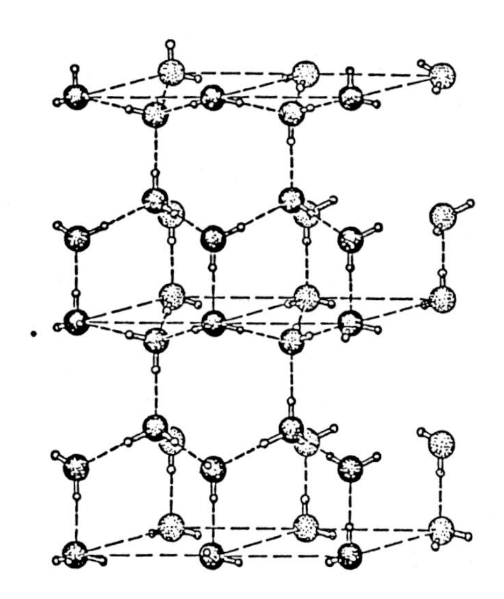

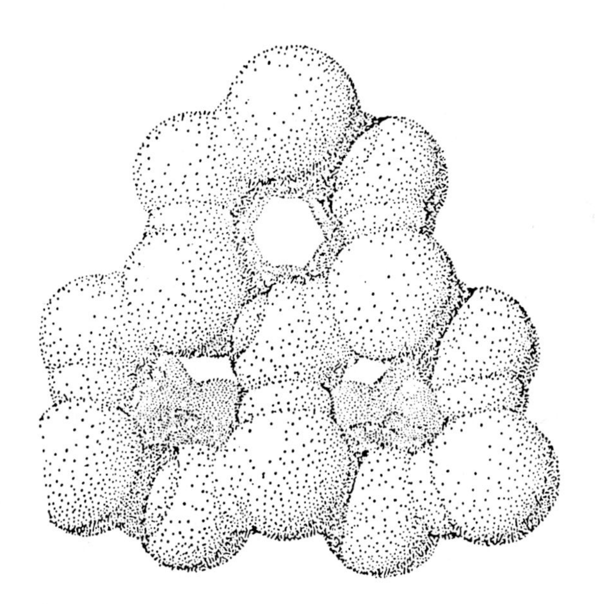

Figure 1b Representation of an ice (I) crystal showing the van der Waals radii of the atoms

(The view is down the c-axis illustrating the open 'shafts')

(Reproduced from Pimentel and McClellan, 1960, 'The Hydrogen Bond', Freeman, San Francisco)

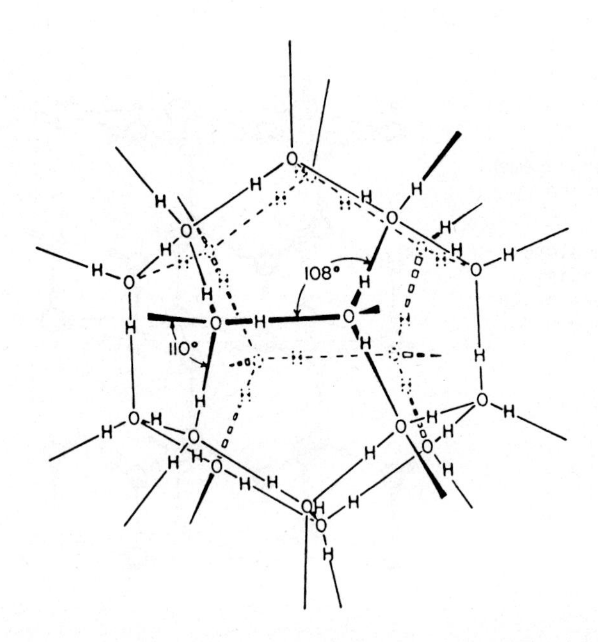

Figure 2 The $H_{40}O_{20}$ pentagonal dodecahedron

This shows one of the ordered arrangements of the hydrogen atoms. In the gas hydrates, the hydrogen atoms are twofold disordered across the O...O edge

(Reproduced from Inclusion Compounds, Vol.1, with permission from Academic Press, London, 1984)

tagonal faces. Eight of these faces are shared with 14-hedra, and the remaining four with other 12-hedra. As seen in (a) of Figure 4, the 12-hedra occupy the corners and centre (indicated by a cross) of the cubic cell. There are thus 8 × 1/8 + 1, i.e. two 12-hedra per cubic cell. The two 14-hedra, shown in solid lines, are located on one face of the cube, giving a 6 × 2/2, i.e. six 14-hedra per cubic cell.

We are now able to understand the otherwise puzzling stoichiometry of hydrates which, for Type I, are $X.5.75H_2O$ and $X.7.66H_2O$, as well as the intermediate stoichiometries. Indeed, the total number of H_2O molecules per cubic cell is 46, that of 12-hedra 2, and that of 14-hedra 6. Therefore, if we call X the molecular species occupying the 14-hedron cavities, and Y those occupying the 12-hedron cavities, the ideal stoichiometry for full occupancy is $6X.2Y.46H_2O$. Now, small guest molecules, up to the size of CH_3F or $H_3C\text{-}CH_3$, fit equally well into the 14- or 12-hedra sites. Thus, the stoichiometry is (X = Y): $8X.46H_2O$ or $X.5.75H_2O$. But larger molecules can

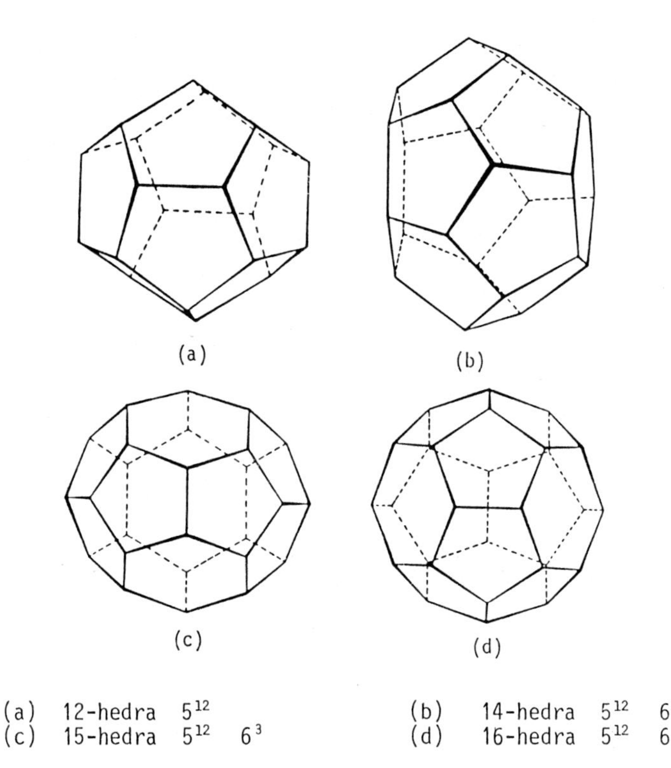

(a) 12-hedra 5^{12} (b) 14-hedra 5^{12} 6^{2}
(c) 15-hedra 5^{12} 6^{3} (d) 16-hedra 5^{12} 6^{4}

Figure 3 Voids in clathrate hydrates

(Reproduced from Inclusion Compounds, Vol.1, with permission from Academic Press, London, 1984)

fit only into the 14-hedra X sites, leading to $X.7.66H_2O$. The 12-hedra can be partially occupied and a non-stoichiometry is expected in the range $6X.46H_2O$ to $8X.46H_2O$. It should be noted that crystallization in the presence of air leads to partial inclusion of air in the 12-hedra. The structure of $6CH_2{=}CH_2.0.4$ air.$46H_2O$ was determined by neutron diffraction at 80 K.

In Type II hydrates (diagram (b) of Figure 4), the host framework is composed of 12- (site X) and 16-hedra (site Y) (diagram (d) of Figure 3). The unit cell formula is $8X.16Y.136H_2O$. In such hydrates, a new phenomenon has been discovered: the stabilization of a clathrate by an auxiliary, or 'hilfgas'. For example, n-butane is reported not to form clathrate hydrates, the molecule being too large <11>. However, in the presence of CH_4, n-butane does form a Type II hydrate, where it occupies the large 16-hedra cages (diagram (d) of Figure 3 and diagram (b) of Figure 4, and CH_4 occupies the small 12-hedra cages <12>. This stabilization phenomenon reveals interactions between the host lattice and the included molecules, and will be further described in the next section.

(a) Type I cubic Pm3n

$a \cong 12$ Å

Two face-sharing 14-hedra are shown with solid lines

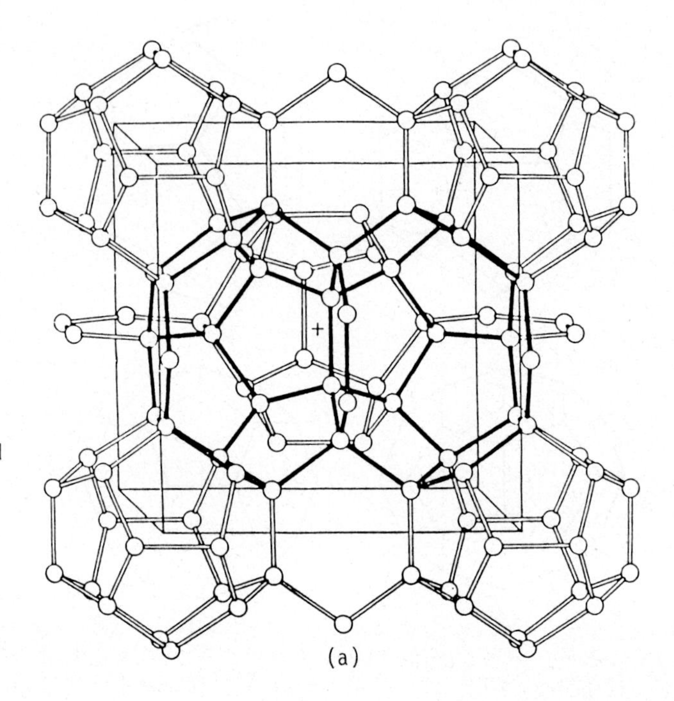

(a)

(b) Type II cubic Fd3m

$a \cong 17$ Å

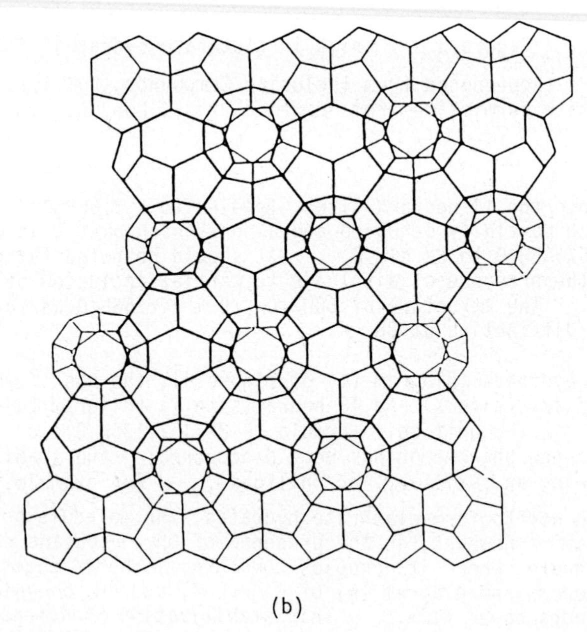

(b)

Figure 4 The host structures of hydrates

The hydrogen atoms are omitted

(Reproduced from Inclusion Compounds, Vol.1, with permission from Academic Press, London, 1984)

We shall now examine the influence of the chemical nature of a molecule upon its ability to form hydrates. The general criterion is a 'certain degree of hydrophobicity'. Although this statement may appear an etymological nonsense, it covers an important aspect of the physicochemical behaviour of water and of urea studied in the next section. Thus, molecules are not known to form hydrates if they contain either a single strong hydrogen bond-forming functional group (like amides and carboxylic acids), or several moderately strong hydrogen-bonding groups (like carbohydrates and polyalcohols). Yet acetone, alcohol, ethers and amines do form hydrates. The main host/guest interactions are van der Waals forces, and the guest is statistically or dynamically disordered. This set of observations usually indicates the ability of water to associate with molecules having little or no polarity, via van der Waals forces, in addition to the more usual H-bonds. The co-existence of van der Waals and H-bonding forces is relevant to the presence of an ice-like structure in liquid water, or the organization of the solvent water molecules around the side chains of proteins. For further discussion, see reference <11>. We only make mention here of the existence of alkylonium salt hydrates and amine hydrates involving other structures and more complex polyhedra than those of structures I and II.

Clathrasils <86>

A recent discovery of SiO_2 clathrates isomorphous to those of hydrates I emphasizes the overwhelming importance of geometrical factors over purely electronic ones in the formation of clathrates. Melanophlogite is a very rare mineral; it consists in a 3-dimensional host framework of $[SiO_4]$ tetrahedra including, in its natural form, CH_4, N_2 and CO_2. The reader would not be surprised by its composition: 46 SiO_2. 6 M^{14}. 2 M^{12} where M^{12} = CH_4 or N_2 are included in the 12-hedra, M^{14} = CO_2 or N_2, in the 14-hedra. We note the average Si-O-Si distance, ca. 3 Å, is comparable to that of O..H..O in hydrates, ca. 2.8 Å. The tedrahedra in SiO_2 being isomorphous to the OH_2 configuration, which is tetrahedral on average, it is not surprising to observe isomorphous clathrate structures despite the totally different electronic characters of the corresponding chemical bonds.

The mineral has been synthetized from aqueous solutions of silicic acid at 170°C. The presence of suitable guest molecules during crystallization is a necessary condition to stabilize the relatively open silica framework of melanophlogite. Obviously the crystallization takes place by condensation of the SiO_4 tetrahedra around these molecules. Once formed, the tetrahedral framework of melanophlogite does not collapse by heating up to ca. 1000°C even after part of the guest molecules have been driven out above 500°C.

The clathrate is the prototype of a whole family called clathrasils <87> containing much larger organic molecules such as piperidine and 1-aminoadamantane <88>.

2.2 From cavities to channels

In several clathrates, the three-dimensional framework of the host forms channels into which a succession of guest molecules fits. In contrast to the cavity clathrates, where there is a maximum dimension for the included guest, in channel clathrates the length of the guest must be greater than a certain minimum value. In some clathrates, the channels have variable cross-sections leading to structures containing cavities linked by narrow channels.

2.2.1 *Urea inclusion compounds*

It was discovered by chance <13> that urea $O{=}C(NH_2)_2$ forms channel clathrates in the presence of a great variety of long-chain molecules: hydrocarbons, alcohols, esters, aldehydes, ketones, mono- and di-carboxylic acids, amines, nitriles, thioalcohols, thioethers etc. The urea molecules are linked together by a continuous network of hydrogen bonds and form cylindrical channels of almost uniform cross-section <14> (Figures 5a and 5b).

Several features of the guest are of particular interest. The main chain must have at least six (non-hydrogen) atoms. An intuitive explanation is given in Section 3.2.1. A degree of substitution is allowed which increases with increasing guest length. Thus, 2-bromooctane, 2,2-difluorooctane, 2-octanol and 2-heptanone are included, whilst 2-bromohexane is not. Furthermore, 1-cyclopentylnonane forms a clathrate, whereas 1-cyclohexyloctane, 2,4-dimethyldodecane and 3-ethyldodecane do not, although all have 14 carbon atoms. Octadecylbenzene is included but benzene itself is not. It is only the degree of substitution and not the presence of a double bond that affects inclusion.

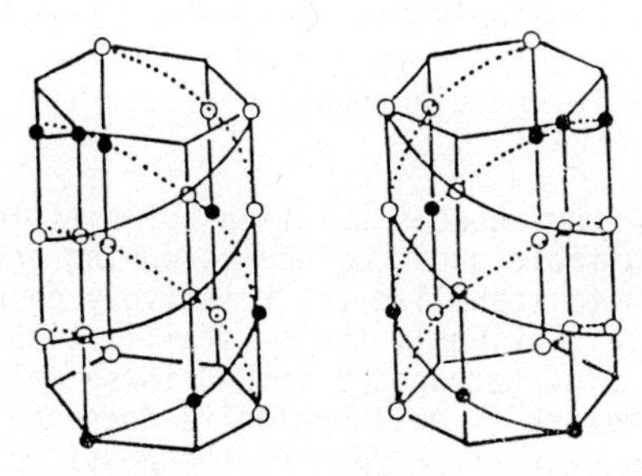

Figure 5a Idealized representation of urea arrangement in the hexagonal lattice

(Reproduced from Inclusion Compounds, Vol.2, with permission from Academic Press, London, 1984)

Figure 5b Cross-section of the urea.normal hydrocarbon inclusion compound

(Reproduced from M. Hagan, 1962, Clathrate Inclusion Compounds, Reinhold Publishing Corporation, New York)

2.2.2 *Variable-section channels: TOT clathrates*

Tri-o-thymotide is a host leading to a variety of architectural types: cage, channel and 'bottleneck' types <15,16,17>

Channel clathrates, crystallizing in hexagonal space groups P 6_1 and P 6_2 form in the presence of elongated molecules, such as those included in urea. In the presence of stilbene, a new type of cavity is observed. Channels run across the crystal structure but their section is not uniform. They consist of regions large enough to accommodate a benzene ring, linked by thin 'bottlenecks' (Figures 6a and 6b). In the presence of small molecules, such as benzene or 2-bromobutane, a cage-type clathrate crystallizes in space group P $3_1$2 which can be used for enantiomeric separation (Section 5).

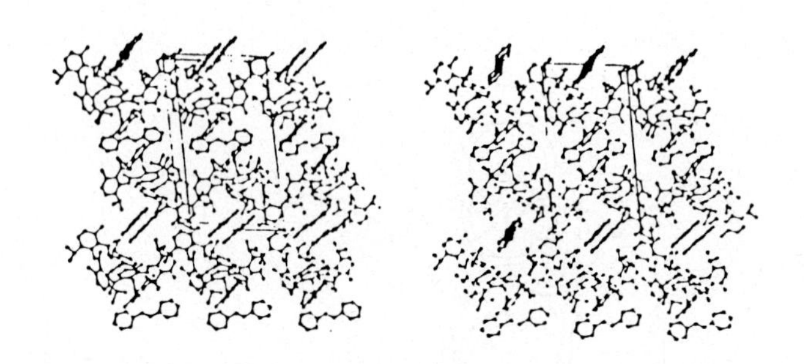

Figure 6a   Stereoscopic drawing of the
trans-stilbene.TOT clathrate structure viewed along the x-axis

(Reproduced with permission from
J.Am.Chem.Soc., 101 (25) 1979)

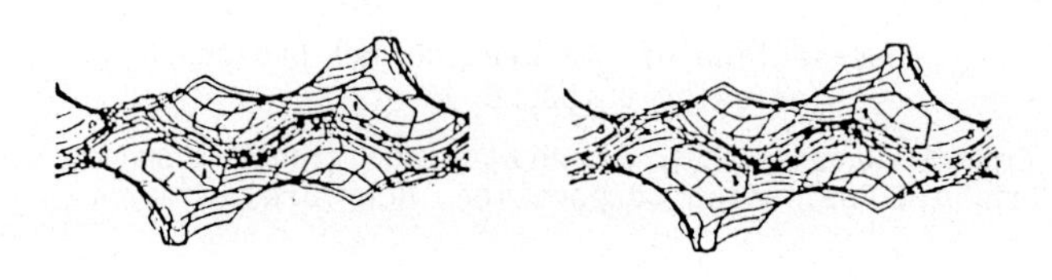

Figure 6b   Stereoscopic drawing of the cavity enclosing
trans-stilbene in the channel parallel to the a-axis
The contours of the volume accessible to stilbene atoms are drawn in sections
separated by 0.40 Å and parallel to the a-b plane
A point is considered as accessible to stilbene atoms if its distance to the
van der Waals envelope of TOT is larger than 1.5 Å

(Reprinted with permission from
J.Am.Chem.Soc., 101 (25) 1979)

Summarizing the architectural study of clathrates, we note that for certain host molecules a single cavity type is observed: cages for hydrates, channels for urea. Other hosts, such as TOT, form a variety of architectural types, depending on the guest as well as on crystallization conditions. In the hydrates and urea clathrates, the host/host bonding forces are primarily strong H-bonds. In TOT, only van der Waals forces are involved. This may account for the greater flexibility in the architecture of TOT clathrates.

2.3 Host/guest correlation: the quinol clathrates

So far we have considered a given host framework as a relatively rigid entity. However, the framework is endowed with a certain flexibility so as to adapt, in turn, to the demands of the guest. Thus, in hydrates of types I and II, the cell parameter may vary by up to 1 Å. However, because of disorder, it is not easy to give a clear picture of host/guest interactions. A prototype of fine tuning of the host architecture is provided by the quinol clathrates <18>.

Hydroquinone forms clathrates in space groups with trigonal symmetry. The clathrate is characterized by a hydrogen-bonded hexamer, where six hydroxyls belonging to six hydroquinone molecules form an almost regular hexagon which is the building block of the structure. The molecules point alternatively up and down (Figure 7). The floor and

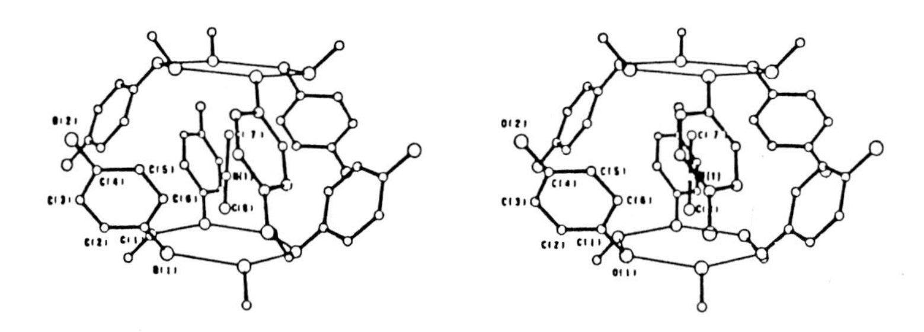

Figure 7 Stereoscopic drawing showing a CH_3NC guest molecule trapped inside a cage in the structure of hydroquinone, For clarity, all hydrogen atoms have been omitted

(Reproduced from Perkin Trans., 2 (1983) with permission from the Royal Society of Chemistry)

ceiling of the cage are formed by two such hexagons displaced along the hexagonal c-axis, the wall, by six benzene rings. The two hexameric units belong to two identical, but displaced, 3D interlocking networks. The link between them is the included molecule which establishes van der Waals bonds with both <24>. The scheme of this link (next page) became the emblem of several 'inclusion compound' textbooks.

The crystalline framework, called β-quinol, of space group $R\bar{3}$, is quite complicated and will not be discussed here. Pure hydroquinone forms a different structure called α-quinol. We focus our attention on the unit cell and space group variations induced by the guest (Table 2).

First - and this is exceptional - the empty clathrate has been crystallized. Its cell dimensions are indeed the smallest of the series. The symmetry around the centre of the cage is $\bar{3}$. The small H_2S molecule fits inside the cavity and its shape may account for the observed rotational disorder. With methyl alcohol, which is longer, such rotation is precluded. The space group changes to R3, the 3 axis now becoming polar. The C-O axis is inclined at an angle 35 degrees to the crystallographic c-axis, and a threefold orientational disorder is necessary to comply with the space group. The c-axis is increased with increasing guest size in the order HCl, CH_3OH, SO_2. An important change is observed with the longer CH_3CN and CH_3NC: it seems that the β-quinol framework has been stretched to the limit, since these clathrates decompose spontaneously, releasing the

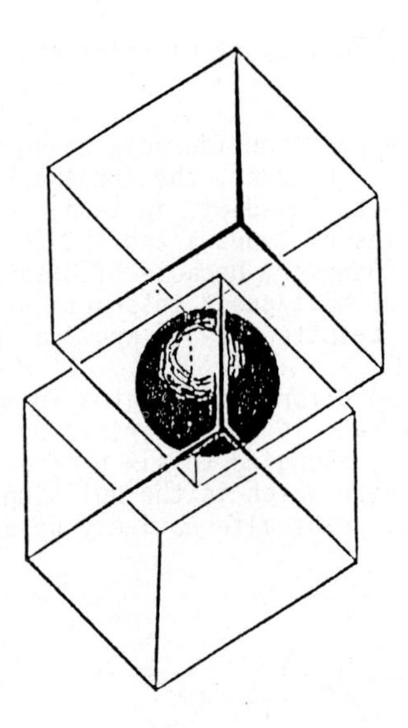

Table 2

Crystal data for β-hydroquinone clathrates

space group	lattice parameters[1]			guest	hexamer dimensions (O...O) (Å)	Ref.
	a	c	Z			
$R\bar{3}$	16.613 (3) Å	5.4746 (5) Å	9	none	2.678 (3)	<19>
	16.616 (3) Å	5.489 (1) Å	9 (host)	H_2S	2.696 (1)	<20>
R3	16.31 (5) Å	5.821 (1) Å	9 (host)	SO_2	2.727 (6) 2.733 (6)	<21>
	16.621 (2) Å	5.562 (1) Å	9 (host)	MeOH	2.653 (5) 2.779 (5)	<22>
	16.650 (1) Å	5.453 (1) Å	9 (host)	HCl	2.61 (1) 2.77 (1)	<23>
	15.946 (2) Å	6.348 (2) Å	9 (host)	CH_3NC	2.779 (6) 2.800 (6)	<24>
P3	16.003 (2) Å	6.245 (2) Å	9 (host)	CH_3CN	2.778 (mean)	<24>

[1]For $R\bar{3}$ and R3, the values of a and c given are referred to a hexagonal unit cell (α=β=90°, γ=120°).

guest molecule. The unit cell length of the isomorphous series increases considerably from 5.45 Å (HCl) to 6.35 Å (CH_3NC). One should also notice a concomitant increase in the length of the O-H..O bond in the hexagons. In the acetonitrile structure, a change in the Bravais lattice to P3 occurs, where sites which are equivalent in R3 are no longer so. Thus three crystallographically-independent cavities exist per P3 unit cell, instead of only one in R3, the overall structure being practically the same.

These facts show that the host framework can exhibit a certain adaptation to the steric demands of the guest. Another, perhaps more subtle, example of host/guest interaction is provided by the quasi-periodical variation of the unit cell parameters in TOT channel clathrates, as a function of guest length (Figure 8). In order to give a quantitative description of these interactions, we shall examine the thermodynamic properties of clathrates in the next section.

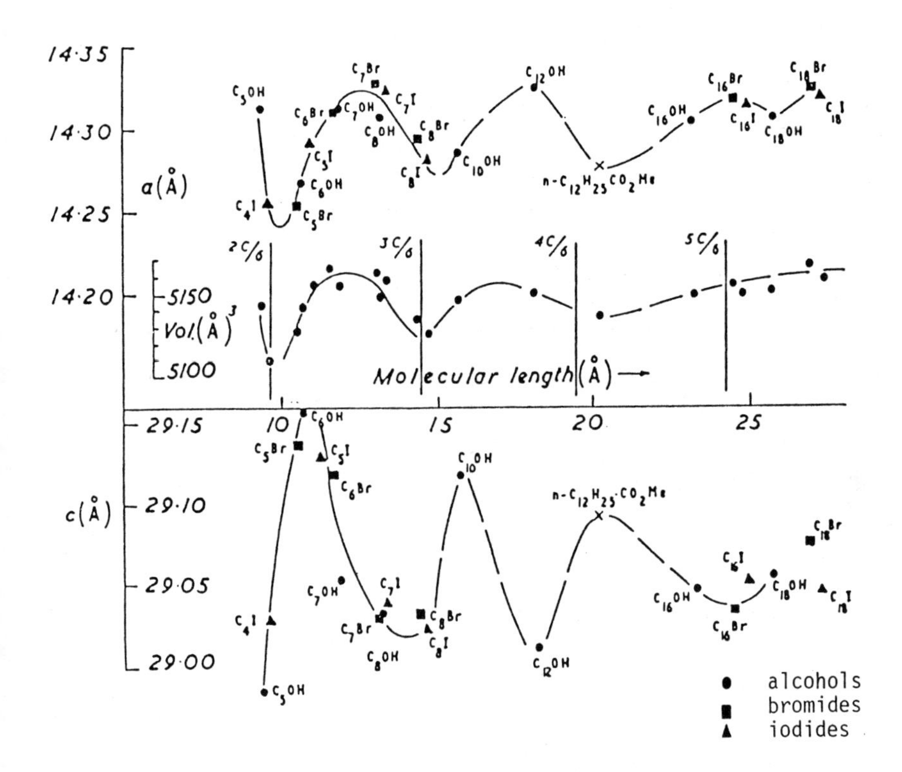

Figure 8 Variation of cell dimensions and volume with the length of the included molecule in the channel

(Reproduced from Inclusion Compounds, Vol.1, with permission from Academic Press, London, 1984)

3 THERMODYNAMIC STABILITY

3.1 Introduction

Although the general architecture involves mainly the host molecules, the nature of the guest molecule also plays an important rôle in clathrate formation, stability and properties.

In the simplified models described in Section 2, the rôle of the included molecule is limited to filling the cavities and increasing the stability of the clathrate by van der Waals type bonds to the atoms of the host. In the quantitative description of the thermodynamic properties of such a model, one neglects the interactions between guest molecules. However, a simple experiment on the hydrates shows, in a spectacular way, that the presence of included molecules in certain cavities may exert an influence on their inclusion in other cavities.

Indeed, the formation of Type II hydrates is limited by the size of the molecule to be included in the large 16-hedra cavity. n-propane (C_3H_8) forms a hydrate but the next homologous n-butane (C_4H_{10}) does not. However, in the presence of SH_2 or CH_4, n-butane can still be included. This is explained by the fact that SH_2 occupies the small 12-hedra cavities which leads to thermodynamic stabilization of the hydrate II framework which can now include n-butane, in spite of the energetically unfavourable steric hindrance. The small gas molecules helping stabilization are known as 'hilfgas'. This phenomenon shows that the stability of the entity 'host matrix/n-butane in 16-hedra cavities' is dramatically influenced by the presence or absence of included molecules in 12-hedra cavities, as a consequence of cavity/cavity interaction through the matrix. However, for most properties, these interactions can be neglected as a first approximation, although they must always be kept in mind.

Another question of great theoretical and practical importance is the 'permeability' of the cavities to the included molecule which can lead, depending on temperature and pressure, to a partial or total elimination of the guest. Such considerations are involved in the practical use of clathrates (high temperature behaviour, chromatography etc).

3.2 Thermodynamic studies

The fundamental thermodynamic characteristics associated with a chemical reaction are the variations of free enthalpy ΔG, enthalpy ΔH, and entropy ΔS. They are linked by:

$$\Delta G = \Delta H - T\Delta S \tag{1}$$

It is necessary to define the reaction conditions, i.e. temperature T, pressure P and the state of each reactant and product of reaction (gas, liquid or solid in a given crystalline form). These thermodynamic entities can be obtained from their values $\Delta G°$, $\Delta H°$, $\Delta S°$, at standard conditions of P = 1 atm and T = 298.3 K. $\Delta G°$ is linked to the mass law constant K through the relation:

$$\Delta G° = - RT \ln K \tag{2}$$

3.2.1 *Free enthalpy*

A chemical reaction occurs spontaneously if and only if $\Delta G < 0$. The value of $-\Delta G$ measures the 'chemical affinity' between the reactants. In a typical case:

$$(\text{host})_{\text{solution}} + (\text{guest})_{\text{solution}} \rightarrow \text{clathrate}$$

and we have:

$$\Delta G = G_{\text{clathrate}} - G_{\text{host}} - G_{\text{guest}} \tag{3}$$

Free enthalpy has a simple expression in the case of an ideal gas at pressure P. With certain clathrates (β-quinol, hydrates), it is possible to observe the reversible reaction:

$$(\text{host})_{\text{crystal}} + (\text{guest})_{\text{gas}} \rightleftarrows \text{clathrate}$$

and apply the ideal gas formula:

$$G_{\text{guest}} = G°_{\text{guest}}(T) + RT \ln P \tag{4}$$

where $G°_{\text{guest}}(T)$ is the gas enthalpy at standard pressure P° and temperature T, and $P = P_{\text{guest}}$. Since variations of G for solids as a function of P are very small, we write (3) as:

$$\Delta G = G°_{\text{clathrate}} - G°_{\text{host}} - G°_{\text{guest}} - RT \ln P$$

$$\Delta G = \Delta G°(T) - RT \ln P \tag{5}$$

At equilibrium, $\Delta G = 0$ and:

$$\Delta G°(T) = RT \ln P° \tag{6}$$

where P° is the 'equilibrium vapour pressure' of the clathrate. Equation (6) is a special case of (2) where K = 1/P. Therefore:

$$\Delta G = RT \ln P°/P \tag{7}$$

The last equation is a clear illustration of the meaning of the sign of ΔG: it becomes positive or negative (thus indicating the direction of the equilibrium reaction) depending on whether $P° > P$ or $P° < P$.

Finally, the host/(guest)$_{gas}$ affinity (defined as $-\Delta G$) is obtained from the value of P° which is accessible to experiment: the smaller the value of equilibrium pressure P° of the gaseous guest above the clathrate, the higher the host/guest affinity or, alternatively, the higher the clathrate stability (Table 3) <25,26,27,28,29,30>.

REMARK When one considers the opposite reaction, clathrate → host crystal + guest, P is referred to as the decomposition pressure. Note the opposite signs of ΔG, ΔH and ΔS.

Table 3

Equilibrium vapour pressures P° (in atm)

gas	hydrate I T = 273.15 K	β-quinol T = 290 K
Ar	95.5	3.4
Kr	14.5	0.4
Xe	1.1	0.058
N_2	140.0	5.8
O_2	100.0	
CO_2	12.5	
NO_2	10.0	
CH_4	26.0	
CF_4	41.5	
C_2H_6	5.2	
HCl		0.01

The decrease of P° from Ar to Xe is correlated with the polarizability of rare gases which, in turn, is related to the strength of van der Waals bond formations. For other gases, the interpretation is more complex.

In urea clathrates, the following formulae have been proposed for the homologous series:

$$- \Delta G^\circ \ (\mathrm{kJ(mol.guest)^{-1}}) = -9.11+1.59x \qquad \text{n-alkanes}$$
$$= -8.39+1.62x \qquad \text{n-acids}$$
$$= -10.47+1.62x \qquad \text{n-alcohols}$$

where x is the number of carbon atoms in each guest molecule <31>.

ΔG° becomes negative only for guests with more than six non-hydrogen atoms. The following diagram shows a succession of included molecules with 'empty spaces' between non-H atoms belonging to adjacent guest molecules. An intuitive explanation of a minimum chain length is given by considering the number of host/guest van der Waals interactions. In the space between included molecules, these interactions are lost and the overall stability of the clathrate decreases with an increasing ratio of 'empty' channel space/channel space filled by the guest. This view is corroborated by the observation that the interatomic distance between end groups, 3.74 Å, is indeed smaller than the normal van der Waals distance of about 4 Å. This can be interpreted as an attempt to minimize the 'wasted space' despite steric hindrance between end groups.

3.2.2 *Enthalpy: use of empty metastable and semi-clathrates*

The variation of enthalpy accompanying a chemical reaction is equal to the heat of reaction at constant pressure. The host and guest molecules are usually allowed to interact in solution or in the liquid state (hydrates). The enthalpy of formation ΔH_f is thus equal in absolute value and has the opposite sign to the enthalpy of solution of the clathrate:

$$(\text{host})_{\text{liquid}} + (\text{guest})_{\text{liquid or gas}} \rightarrow \text{clathrate} + \Delta H_f$$

This last quantity can be obtained by calorimetry.

A further quantity is relevant to the specific study of inclusion phenomena: the enthalpy of enclathration ΔH_c corresponding to the reaction:

$$\text{'empty clathrate'} + \text{guest} \rightarrow \text{clathrate} + \Delta H_c$$

The value of ΔH_c reflects the strength of the molecular host/guest interactions, and it is important to measure its value experimentally. Attempts to eliminate the guest molecules from a clathrate (by heating and lowering the pressure) usually result in the collapse of the clathrate framework. However, it is still possible to determine ΔH_c by indirect experiments.

Partially-filled and metastable empty clathrates

It is possible to prepare certain classes (β-quinol, β-phenol) of clathrates with a given fraction of cavities y occupied by guest molecules. Let us call ΔH_y the corresponding enthalpy of solution. Using such clathrates in a differential calorimetry technique, with pure α-quinol crystal as reference, it was possible to determine experimentally the quantities ΔH_y linked to ΔH_c by:

$$\Delta H_{y_{obs}} = \Delta H_{\beta-\alpha} + \Delta H_c \cdot y$$

where $\Delta H_{\beta-\alpha}$ is associated with the process α-quinol $\rightarrow$ β-quinol.

The linearity of the relation between ΔH_y and the occupancy factor y denotes that inclusion of a given molecule at any stage of the process involves the same host/guest interactions. These interactions are therefore of a local character, and one can neglect, for this approximation, any cavity/cavity interaction. It is interesting to compare the enclathration of gases with the Langmuir adsorption model on monolayers, where the enthalpy of adsorption is also independent of the fraction of surface already covered.

$\Delta H_{\beta-\alpha}$ and ΔH_C are determined by analyzing the data for several values of y (Table 4) <32>.

Table 4

Enthalpies of enclathration

	$-\Delta H_C$ in $kJ \cdot mol^{-1}$		
	hydrate I	β-quinol	β-phenol
Ar	26	25	
K_2	28	26	24
Xe	31	41	44
N_2		24	
O_2		23	
CO_2			65
CH_4		25 to 30	27
C_2H_6			47
HCl		38	

Similarly it can be shown that ΔG_y increases, i.e. the stability of the clathrate decreases with decreasing y. Clathrates with y less than a certain value y_0 are not thermodynamically stable. They can still exist in a metastable form, the decomposition kinetics (with transformation to the stable pure host form) being slow. For β-quinol, $y_0 = 0.34$. It is possible, by careful heating, to obtain the empty β form in a metastable state and to perform direct differential calorimetric measurement of $\Delta H_{\beta-\alpha}$.

In conclusion, clathrate formation is characterized by establishing a number of van der Waals bonds which link a guest molecule to the surrounding host atoms. The bonds involve induced-dipole/induced-dipole (or London) forces as well as permanent-dipole/induced-dipole forces. Theoretical treatment involving semi-empirical constants has led to a measure of agreement between calculated and observed values of ΔH. In β-quinol, the values of ΔH_C in Table 4 are reproduced with an error of 1 to 4 $kJ \cdot mol^{-1}$.

Semi-clathrates

In hydrates, the occurrence of two types of cavity with different dimensions is favourable for experimental measurements pertaining to enclathration reactions. Indeed, it is possible to fill the large cavities with chloroform, whilst leaving the small ones empty. Crystals like this are called semi-clathrates and can be used to study the enclathration of rare gases into the smaller cages <33>:

$$\text{hydrate II.CHCl}_3 + \text{rare gas} \rightarrow \text{hydrate II.CHCl}_3\text{.rare gas}$$

In this way, both equilibrium pressures P° and enclathration enthalpies can be measured or evaluated by indirect methods. Results are given in Table 4.

3.2.3 *Temperature dependence: the PHTP clathrates*

The thermodynamic properties of a recently discovered (1963) perhydrotriphenylene (PHTP) clathrate were studied in depth <34>.

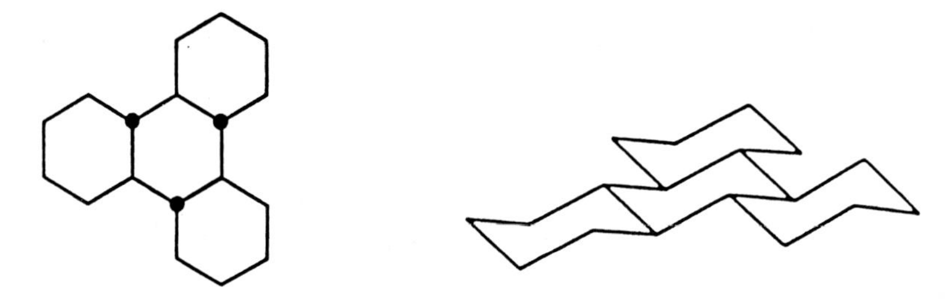

PHTP

PHTP, like TOT, forms clathrates with a variety of space groups, depending on guest and crystallization conditions. The PHTP/n-heptane clathrate belongs to the space group $P6_3/m$ (Figure 9). The host structure is composed of stacks of superimposed PHTP molecules. At the vertices of the cell are channel-type cavities. The guest molecules are disordered.

The phase diagram of the binary system PHTP/n-heptane is given in Figure 10 as an example to illustrate the entire class of these clathrates. This is a typical case of a binary system involving a compound C (clathrate) of definite composition. We focus our attention on the temperature dependence of the decomposition reaction:

$$\text{clathrate} \rightarrow (\text{PHTP})_{\text{crystal}} + (\text{guest})_{\text{gas}}$$

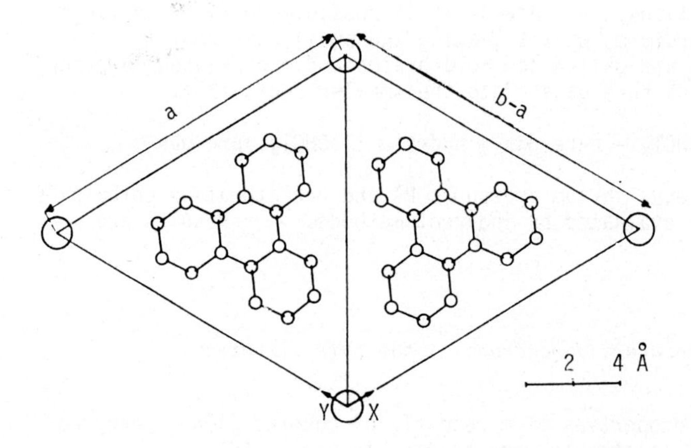

Figure 9 x-y projection of the crystal cell of the inclusion compound PHTP/n-heptane

(Reproduced from Inclusion Compounds, Vol.2, with permission from Academic Press, London, 1984)

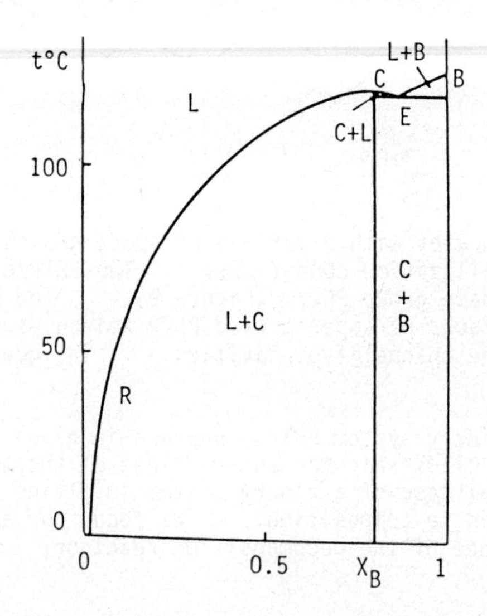

Figure 10 t-x phase diagram of the binary system PHTP/n-heptane

(Reproduced from Inclusion Compounds, Vol.2, with permission from Academic Press, London, 1984)

given by the van't Hoff formula:

$$\frac{d \ln K}{dT} = \frac{\Delta H}{RT^2}$$

As often happens, the variation of ΔH with temperature can be neglected. This leads to:

$$\ln P^\circ = -\frac{\Delta H_{vap}}{RT} + D$$

where P° is the equilibrium vapour pressure or decomposition pressure, ΔH_{vap} is the corresponding enclathration enthalpy variation, and D is a constant characteristic of the clathrate. Figure 11 shows the experimental results: the slope of the straight lines is proportional to ΔH_{vap}.

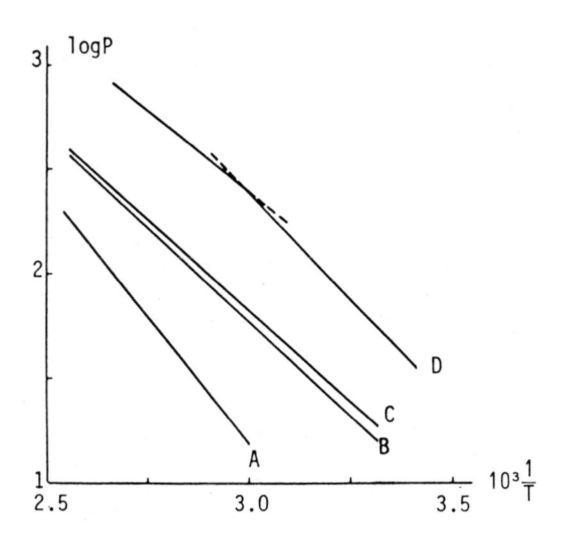

Figure 11 Decomposition pressure of some inclusion compounds versus 1/T,

The curves A, B, C and D refer respectively to the inclusion compounds with n-heptane, dioxan, cyclodexane and chloroform

(Reproduced from Inclusion Compounds, Vol.2, with permission from Academic Press, London, 1984)

The value of ΔH_{vap} is 45.2 kJ for n-heptane. Let us compare it with that of the system urea/n-heptane which is 67 $kJ \cdot mol^{-1}$. The smaller value for PHTP could be ascribed to the smaller number of van der Waals contacts between host and guest atoms. It is interesting to guess a rough estimate of the ΔH increment per CH_2. From the homologous series C_7 to C_{16} in urea, the increment has been estimated as being 11 $kJ \cdot mol^{-1}$. It would be considerably smaller for PHTP, 6.5 $kJ \cdot mol^{-1}$ (although caution should be required in such comparisons because of the unknown effect of end groups). It is interesting to compare these values with those observed in cage clathrates (Table 4), which are much larger per carbon atom (of the order of 20 $kJ \cdot mol^{-1}$). This fact is ascribed, again, to a larger number of host/guest contacts in cage clathrates.

Up to now, we have studied phenomena when the guest/guest interactions are not taken into consideration, although we are aware from the 'hilfgas' experiments that such interactions exist (Section 3.1). In the next section we shall consider phenomena where these interactions cannot be neglected.

3.2.4 *Phase transitions: the entropy factor*

So far we have not described the orientation and precise position of guest molecules. Between the two extreme situations, isotropic disorder in cages and total order as in TOT/trans stilbene, there exists an infinite number of intermediate situations. They are often described as hindered rotation and discrete re-orientation. It may happen that the state of motion undergoes changes as a function of temperature. This phase transition <35> is accompanied by sharp changes in thermodynamic quantities such as the specific heat at constant pressure (Figure 12):

$$C_p = T\frac{\partial S}{\partial T}$$

The entropy S is a measure of the disorder or uncertainty pertaining to the position, rotation or orientation, vibration etc of molecules or groups of atoms. Its variation in certain cases takes on a simple meaning, as in the re-orientation of linear alkanes within the urea channels: we recall that, as a first approximation, the alkanes are in the extended form and lie on a plane parallel to the axis of the channel. If m is the number of equally-probable orientations of the molecular plane, the associated entropy is given by:

$$S = R \ln m \qquad \text{with } R = 8 \text{ JK}^{-1} \text{ mol}^{-1}$$

Assuming that, below a certain transition temperature T_t, the molecular plane is fixed in one orientation, then:

$$S = R \ln 1 = 0$$

and the transition entropy is:

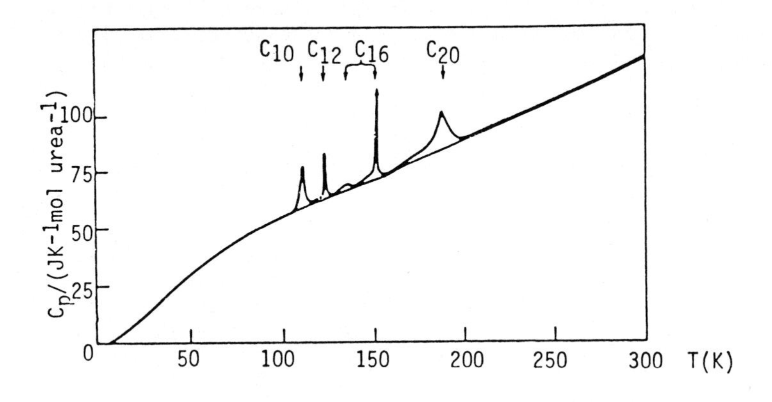

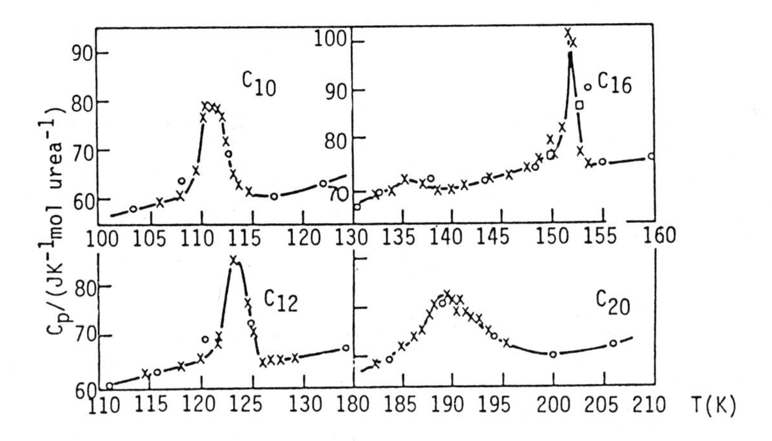

Figure 12 Heat capacity versus temperature for four urea/n-alkane inclusion compounds

(Reproduced from R.C. Pemberton and N.G. Parsonage, Trans.Faraday Soc., 61 2112 (1965)

$$\Delta S_t = R \ln m$$

At temperature T_t:

$$\Delta G_t = 0 = \Delta H_t - T_t \Delta S_t$$

In alkanes C_nH_{2n+2}, for n between 10 and 15, the experimental values $\Delta S_t = \Delta H_t/T_t$ range between 8 and 11 JK^{-1} mol^{-1}, clustering around R ln 3 = 9.14 JK^{-1} mol^{-1}. For $16 \leqq n \leqq 20$, the values of ΔS_t range between R ln 4 and R ln 6, denoting additional orientational disorder, for instance, more than three orientations, or internal rotations. It is interesting to note that, for molecules with much larger internal conformational freedom, the values of ΔS can increase substantially as in the thiourea/cyclooctane clathrate, where $\Delta S_t = R \ln 45$ <36,89>.

In general, when guest molecules leave the gaseous or liquid state and are included in a clathrate cavity, there is partial loss of freedom and decrease in entropy. Thus, the entropy term in:

$$\Delta G = \Delta H - T\Delta S$$

is positive. The enthalpy term must be negative and large in absolute value, so that ΔG is negative. Indeed, the values of the enclathration enthalpy per mole of included compound from the gaseous state is of the order of 20 to 60 $kJ \cdot mol^{-1}$ for small molecules with up to three non-hydrogen atoms. These values are certainly smaller than those relative to a chemical reaction involving covalent bonds, yet of the same order of magnitude. Let us illustrate this point by comparing it with the dissociation enthalpy of the gas/solid reaction where covalent bonds are broken and formed:

$$CaO + CO_2 \rightleftarrows CaCO_3 + 150 \ kJ \cdot mol^{-1}$$

The values of enthalpies in clathrates are ascribed to the large number of van der Waals bonds between a guest and the surrounding host molecules. It should be stressed that the above experiments lead to direct determination of the van der Waals energy associated with one reference molecule. This is not possible for ordinary mixed crystals where individual lattice energies cannot be separated from the interaction energy between components.

4 SEARCH FOR NEW HOST MATRICES

The discovery of the inclusion properties of host molecules seems to have all the characteristics of a random phenomenon. The proportion of host matrix-forming molecules compared with the total number of organic compounds is extremely small. One then can pose the question: can we endeavour to design molecules expected to form inclusion compounds? Two levels of design are to be considered: (1) relatively small modification of known hosts, and (2) new chemical compounds. We shall describe here some of the achievements along both lines, in particular, the 'hexahost' structure which constitutes a positive answer to the question of new hosts.

4.1 Dianin clathrates

4-p-hydroxylphenyl-2,2,4-triphenylchroman 1 was prepared by Dianin (a Russian chemist, student of the composer and professor of chemistry, Alexander Borodin).

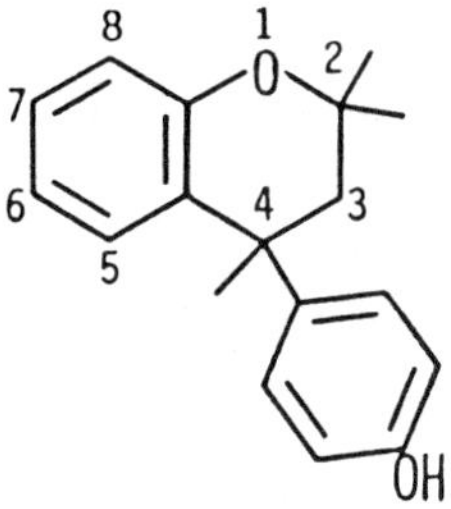

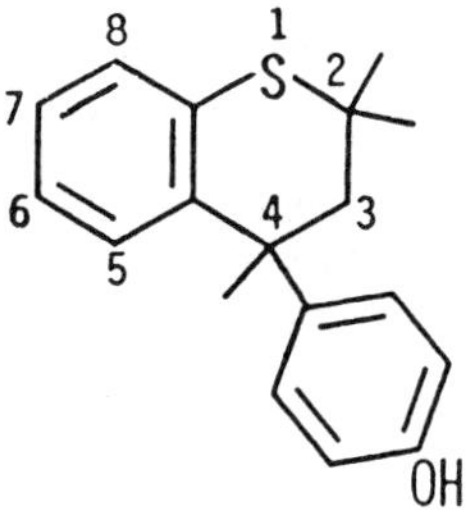

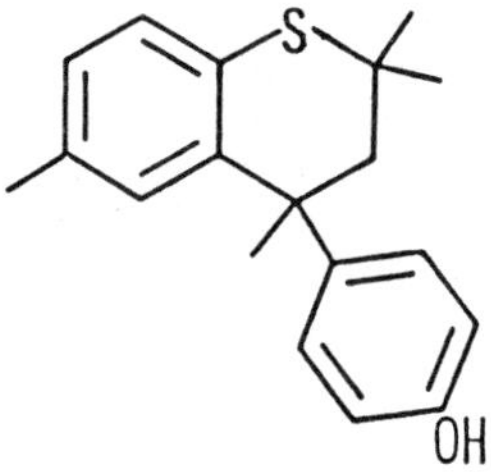

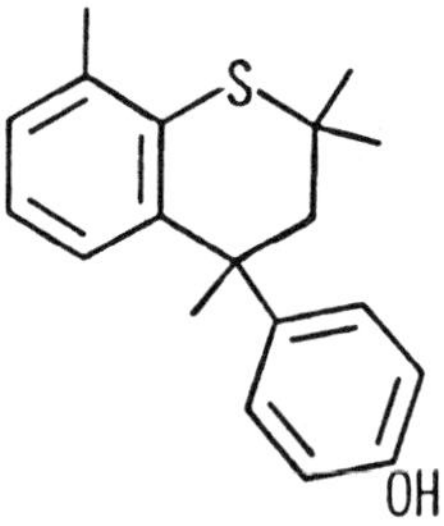

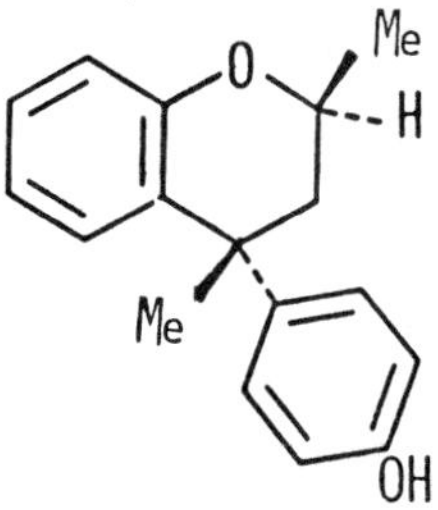

The history of this compound ressembles that of several other important clathrates such as hydrates, cyclodextrins ... etc. In 1914, Dianin discovered that 1 had the remarkable ability to retain certain organic solvents tightly and in fixed amounts but the exact molecular structure of the host and the crystal structure of the clathrate were established in the mid-fifties <10>. This structure is characterized by hexagons formed by the phenolic O-H...O bonds (Figure 17). This is an interesting fact recalling that other phenolic compounds act as hosts, such as hydroquinone and phenol itself. The remainder of the chroman molecule is symbolized by R or R', which represent the R and S enantiomers above and below the hexagon respectively. Two hexagons form a cage, which encloses a variety of guest molecules: Ar, SO_2, I_2, NH_4, glycerol, decalin, SF_6 etc.

This host structure was the starting point for several modifications <46>. The simplest modification is the replacement of the ether oxygen 01 by S. The structure of the thio compound is almost identical to that of the parent one and has similar inclusion properties.

Going from the small ethanol molecule to the large acetylene derivatives, we observe concomitantly (a) an increase in the cell parameters, (b) an increase in cage dimensions, which are constructed from progressively longer and consequently weaker O...H...O hydrogen bonds, and (c) an increase in the infrared frequency ν(OH) reflecting the hydrogen bond strength, as shown in Table 5. The guest 2,5,5-trimethylhexa-3-yn-2-ol provides a good example of host/guest complementarity (Figures 13 and 14).

Table 5

Crystal data for Dianin's compound and related molecules of space group $R\bar{3}$

compound	formula	lattice[1] parameters (Å) a	c	guest	mole ratio host/ guest	O..O	ν_{O-H} (cm^{-1})	Ref.
Dianin	1	26.94	10.94	none	-			<37>
		26.969	10.990	ethanol	3:1[2]			<38>
		27.116	11.023	chloroform	6:1			<38>
		27.12	11.02	n-heptanol	6:1			<39>
	5	26.936	10.796	carbon tetrachloride	6:1			<40> <41>
S-Dianin	2	27.81	10.90	ethanol	3:1	2.96	3345	<42>
		27.91	10.99	2,2,5-trimethyl-hexa-3-yn-2-ol	6:1	3.03	3400	<43>
		28.00	11.08	di-t-butylacetylene	6:1	3.17	3435	<44>
	3	29.22	10.82	cyclopentane	6:1			<45>
	4	33.629	8.239	cyclooctane	4.5:1			<46>

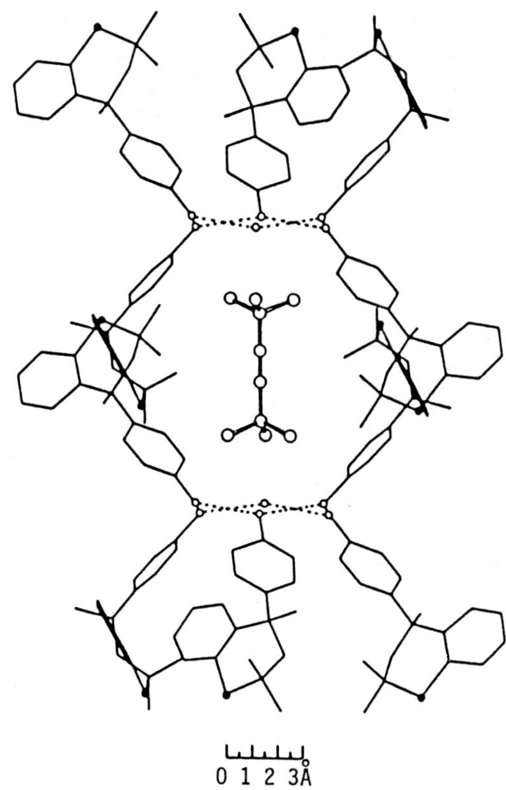

Figure 13. The structure of the Dianin clathrate 2 with 2,5,5-trimethylhexa-3-yn-2-ol as guest

Two host molecules have been excluded (apart from their hydroxyl oxygen atoms) to reveal the guest, which is accurately aligned along the c-axis

(Reprinted from Chem. Comm., 1971, with permission from the Royal Society of Chemistry)

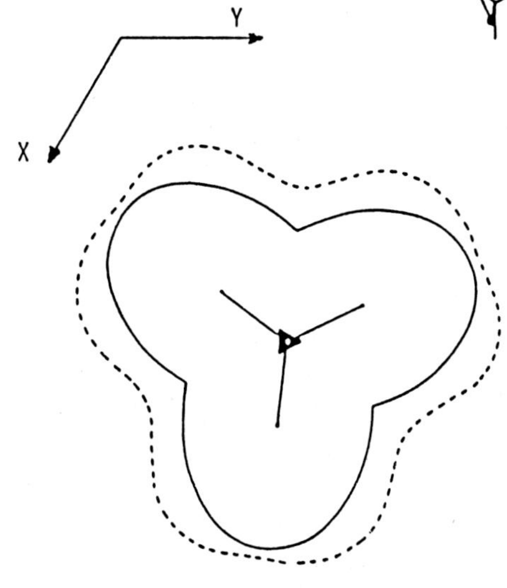

Figure 14. The van der Waals contacts, as viewed along the c-axis, for a section at z = 0.26

The broken lines represent the van der Waals volumes of the atoms comprising the cage, and the full lines the approximate van der Waals volume of the guest

(Reprinted from Chem. Comm., 1971, with permission from the Royal Society of Chemistry)

Ring substitutions lead to clathrates with markedly altered cage geometry. Compounds 3 and 4 are quasi-isomorphous with the parent compound, although a drastic decrease in cavity length occurs in 4, and the 'waist' of 4.3 Å was replaced by a 'bump' of 7.7 Å (Figure 15). The cavity alteration is accompanied by changes in the cell parameters: c decreases and a increases by several Å.

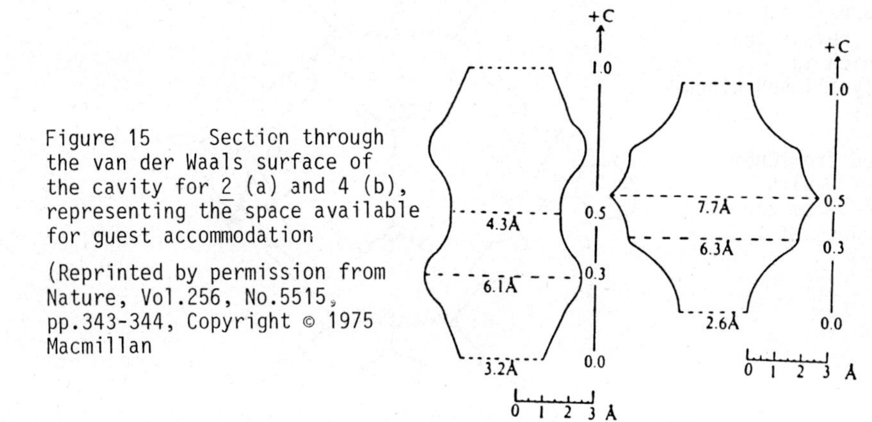

Figure 15 Section through the van der Waals surface of the cavity for 2 (a) and 4 (b), representing the space available for guest accommodation

(Reprinted by permission from Nature, Vol.256, No.5515, pp.343-344, Copyright © 1975 Macmillan

An interesting consequence of the drastic change of cavity shape is the selective enclathration from a mixture of cyclopentane, cyclohexane and cycloheptane: 2 favours cyclopentane (respective percentages 85, 10 and 5), and 4 favours the larger rings (respectively 20, 50 and 30).

Instead of increasing the host size as in 3 and 4, a decrease can be performed as in the demethylated 2-normethyl analogue 5. If, in the computer drawing of the structure, we remove the methyl group in position 2, but without altering the rest of the crystal structure, then the contour in curved broken lines in Figure 16 is obtained. The position of the methyl group corresponds precisely to the 'waist' of the cavity. Direct X-ray analysis shows that the new cavity shape is very close to the predicted one.

Changes in the functional group involved in hydrogen bonding are illuminating with respect to the degree of sensitivity of the inclusion properties to chemical modifications. No clathrates of the amino analogue have been reported. The behaviour of the thiol analogue depends on the solvent: with cyclohexane, the spontaneous resolved pure host crystallizes without inclusion. In the presence of CCl_4, the familiar R3 is obtained (host/guest ratio of 3:1).

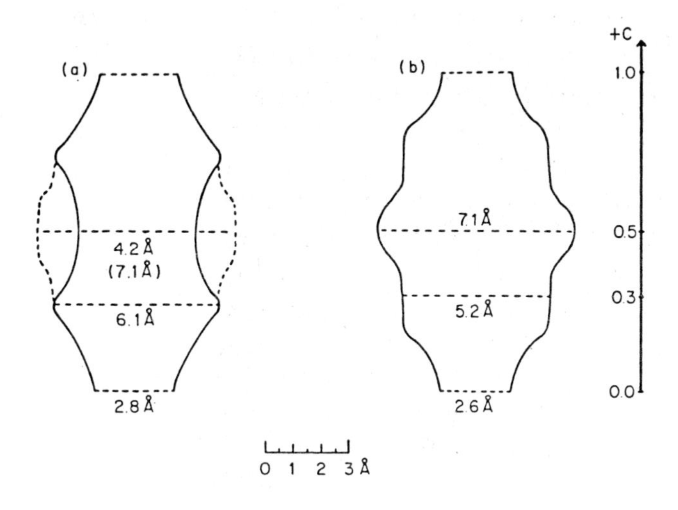

Figure 16 Section through the van der Waals surface for (a) Dianin's compound 1 as chloroform clathrate (curved broken lines represent the effect of formal replacement of the 'waist' methyl groups) and (b) 2-normethyl analogue 5 as its CCl_4 clathrate

(Reprinted from J.Chem.Soc., Chem.Commun. 1976 and Perkin Trans., 2 1979 with permission of the Royal Society of Chemistry)

4.2 The hexahost geometric concept

We have emphasized the occurrence of six-membered O-H..O rings in phenolic clathrates (quinol in Section 2, and Dianin and other clathrates above). What is the essential feature of these building blocks for clathrate formation? Is the system of hydrogen bonds a prerequisite for clathrate formation or should the essential part be sought in the three-dimensional geometry of the hexameric building block?

Indeed, we can pose a more general question: is there a characteristic feature common to clathrate-forming molecules? Most host molecules have an inherent tendency to form open structures, as opposed to the close packing common to many molecular crystal structures. Water is the outstanding case where the density of ice (0.927) is 8% lower than that of water, as a consequence of a markedly open structure. This is ascribed to the strict tetrahedral coordination in ice, whereas in water the tetrahedral surrounding is not strict and holds only on average. Visibly the ice structure presents cavities too small for inclusion, but we have seen in Section 2 that a slight rearrangement of the H_2O molecules into polyhedra, still tetrahedrally surrounded, results in large cavities. In hydrates, these cavities are filled with guest molecules, and the density is again very close to 1.

In the quinol and Dianin structures, the molecular shape alone does not seem to be significant for clathrate formation. It is the hexameric building block which bears the 'clathrate information'. The packing coefficient (i.e. the ratio of the van der Waals molecular volume to that of the crystallographic unit cell) is remarkably low, 0.62, compared with the normal range, 0.65 to 0.77 for most organic molecular crystals, thus achieving an open-type structure. From such indications emerges the strategy of creating molecules which, by their shape, oppose the usual close packing in crystals and creates rigid open structures. Thus by analogy with the quinol and Dianin and other phenol clathrates, the synthesis of molecules with sixfold symmetry was undertaken and a new range of inclusion compounds, named hexahosts, was discovered (Figure 17) <47,48>.

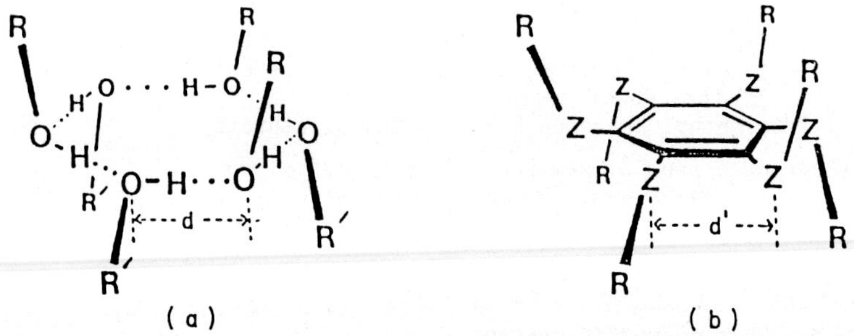

Figure 17 Comparison of (a) hydrogen-bonded hexamer unit with (b) hexasubstituted benzene analogue

(Reproduced from Inclusion Compounds, Vol.2, with permission from Academic Press, London, 1984)

The prototype <u>6</u> (Table 6) crystallizes in the same space group $R\bar{3}$ but the structures are not isomorphous. However, common features appear: the cavity is elongated along c, and the host molecule is again centered at a site of symmetry 3. Figure 18 shows that the cavity contains two CCl_4 molecules.

Table 6

Dianin's clathrates and related compounds

X	m	n	spacer length m+n+1	host	Ar	guest
S	0	0	1	6	Ph	CCl_4,CCl_3CH_3 CCl_3Br,CCl_3SCl CCl_3NO_2
O	1	0	2	7	Ph	toluene,1,4-dioxan tetrahydrothiophen
				7a	ortho-iso-propyl-Ph	toluene,chloroform p-chlorotoluene acetonitrile
				7b	para-iso-propyl-Ph	none
S	1	0	2	8	• Ph	toluene,1,4-dioxan
				8a	para-t-butyl-Ph	cyclohexane,cycloheptane cyclooctane,toluene iodobenzene phenylacetylene 1-methylnaphthalene 2-methylnaphthalene bromoform,squalene hexamethyldisilane
S	1	1	3	9	Ph	cyclohexane,toluene 1,4-dioxan,acetone 1,1,1-trichloroethane ethyl acetate,benzene p-chlorotoluene
S	1	2	4	10	Ph	1,4-dioxan benzene,toluene tetrahydropyran fluorobenzene chlorobenzene bromobenzene iodobenzene 1,1,1-trichloroethane
S	1	3	5	11	Ph	none
NR'	1	1	3	12	Ph R' = $COCF_3$	nitromethane tetramethylurea N,N-dimethylformamide N,N-dimethylacetamide N-n-butyl-N-methylformamide

Replacing a single S-group by a longer one leads to a vast family of molecules endowed with inclusion properties. The general formula can be represented by the following diagram:

with R = $-(CH_2)_m-X-(CH_2)_n-Ar$

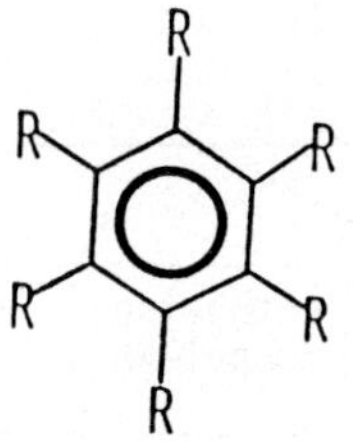

Figure 18 An illustration of the host/guest packing in the crystal of the CCl_4 clathrate 6 as viewed on the a-c plane

Two host molecules that lie above and below the cavity as viewed in this direction have been excluded to show the guest molecules more clearly.

(Reprinted by permission from Nature, Vol.266, No.5603, pp.611-612, Copyright © 1977 Macmillan Journals Limited)

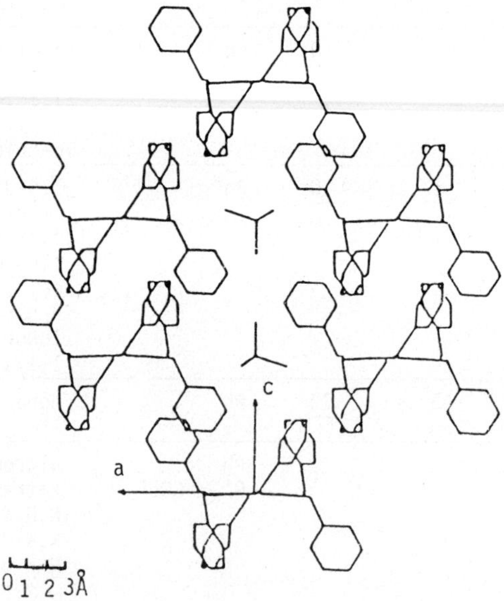

With such a wealth of data, it is possible to investigate the effect of the length of the spacer between the central and outer rings, as well as that of the substitution on the outer rings on the inclusion properties. The substituents investigated in the o-, m- and p-positions include methyl, isopropyl, methoxy, t-butyl, chlorine, -OH, $-NH_2$, and some combinations of these. As a sample of the data, we give, in Table 6, a few examples of inclusion.

We note for instance that 7a is endowed with inclusion properties, whereas the para analogue 7b is not. An interesting feature is the modulation of the inclusion selectivity by different substitution on Ar. Thus, in the 9 family, the compound with Ar = o-methyl-phenyl has a significantly greater selectivity than the parent (Ar = Ph) compound for p-xylene, from a mixture of equimolar ortho- and para-xylene.

A drastic change occurs in the clathrate architecture of 8a when squalene is present: a channel forms with an exact 2:1 host/guest ratio. Incidentally, the conformation of squalene itself is changed compared with that found in the crystal structure of pure squalene.

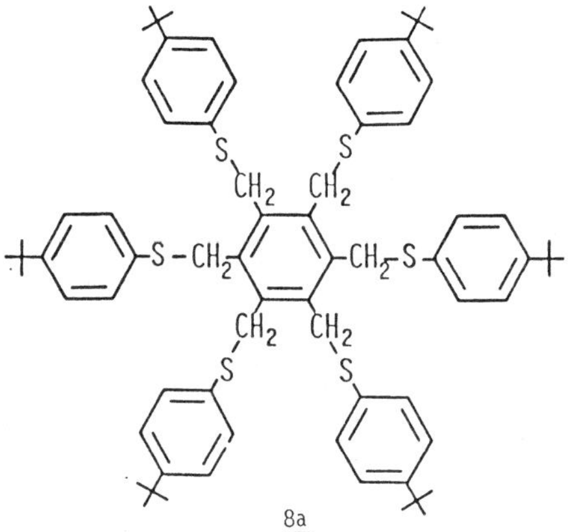

8a

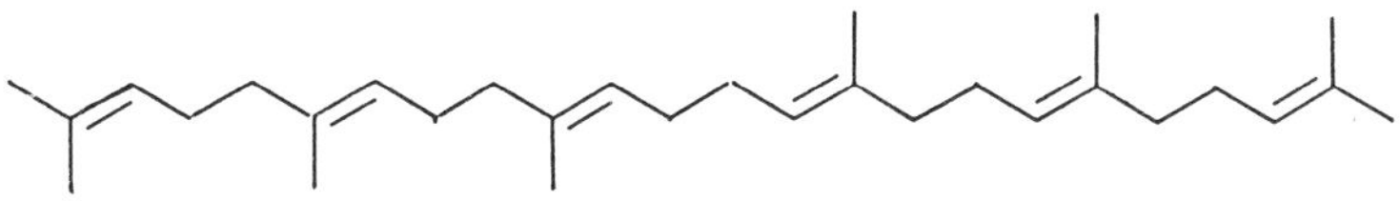

Table 6 shows that compound 11, with a 5-atom spacer, has not been found to form an inclusion compound, whereas 10, with a 4-atom spacer, does.

Change of the heteroatom in the spacer to an amine group 12 maintains and considerably modulates the inclusion properties: an effectively complete configurational selection occurs of the thermodynamically less stable Z form of the following guest:

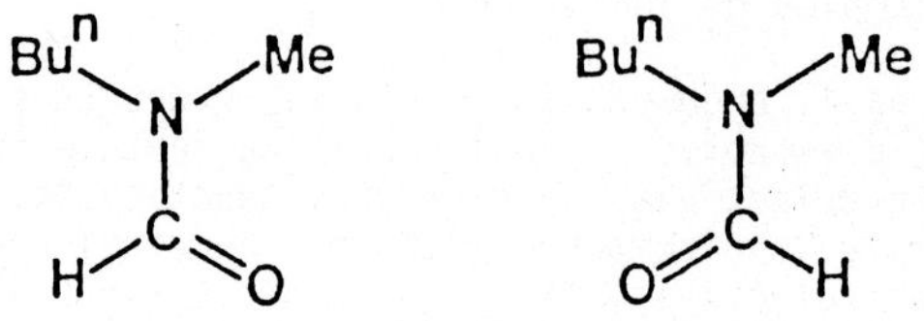

E   Z

This fact led again to a new host family 13 <47>. We make mention also of the extensive studies on tri-o-anthranilides 14: although R_1, R_2 and R_3 are in general different, destroying thus the strict 3 symmetry, 14 still leads to clathrates <49>. Thus it seems that the symmetry argument is far from absolute. Despite the large recent effort and the wealth of results, it is still too early to apply statistical methods to clathrate engineering.

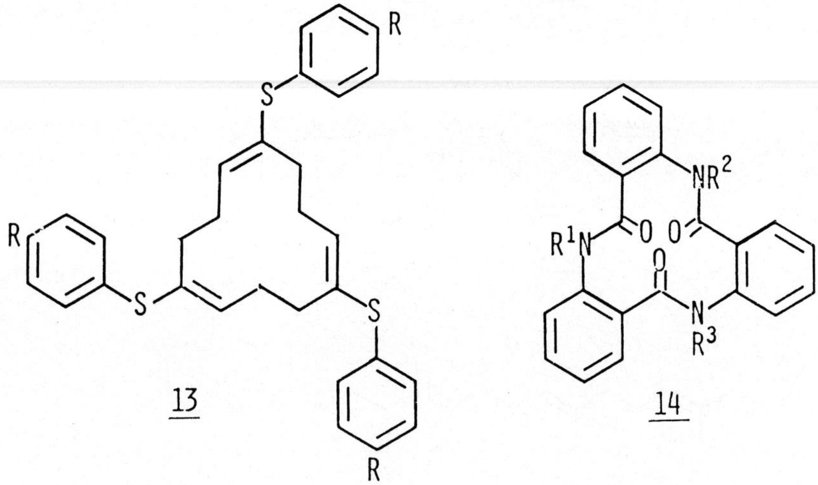

13   14

Figure 19. New hosts with threefold or pseudothreefold symmetry.

5 CHIRAL DISCRIMINATION

Clathrates are prototypes of crystalline templates, and it is natural to enquire about their ability for chiral discriminations <50>. Chirality may appear at two levels, molecular and crystalline.

5.1 Crystalline versus molecular chirality

The first step is to choose a host eventually leading to chiral clathrates. Table 7 indicates the relationships between molecular and crystalline chirality for known clathrates. Clearly, the only certain relationship, e.g. the only implication, is: homochiral host $\Rightarrow$ homochiral clathrate. Crystallization involving a racemic or achiral host may or may not lead to chiral clathrates.

Table 7

Crystalline versus molecular chirality in clathrates

molecular chirality	clathrate chirality	
	chiral	achiral
homochiral	deoxycholic acid cyclodextrins anthroates perhydrophenylene	impossible
racemic	TOT $P3_12$ TOT $P6_1,P6_2$	TOT $P\bar{1}$ Dianin Werner-α-arylalkylanine perhydrophenylene cycloveratrylene
achiral	urea	quinol hexahosts Hoffman Werner-pyridine hydrates

On the other hand, in the course of investigations for new chiral clathrates, the following questions are relevant.

Among host 'candidates' is there a preference for chiral or achiral molecules? The number of known clathrates today is not sufficient to provide a clear indication. We note, however, that the general statistics show that only about 10% of known organic structures belong to chiral space groups. The larger proportion appearing in clathrates may be due to a 'historical bias', and a larger sample of structures is needed to draw such an inference.

- If a clathrate is obtained from a racemic mixture of host molecules, is there also clathrate formation from the optically-pure material? In two cases, the answer is negative: Werner complexes (Section 6) and Dianin clathrates. Highlighting the importance of the packing of both enantiomers to form these two clathrates, no inclusion compound formation has been found for the resolved materials, either for chiral or achiral potential guest components. But, racemic PHTP leads to $P6_3/m$, whereas optically-pure PHTP leads to $P6_3$ clathrates.

5.2 Achiral host: the urea clathrates

Urea is an achiral planar molecule (Section 2.2.1) yielding $P6_122$ clathrates where the host molecules are arranged along a helix (Figure 5). Various techniques have been used in order to perform asymmetric induction: seeding, asymmetric surfaces (silk, hair), optically-active cosolutes (sugars, tartaric acid) etc. A major difficulty in the study of this class of clathrates is the lack of handy methods to determine the absolute configuration of the lattice, compared with clathrates from chiral or racemic hosts <57>.

5.3 Racemic host molecules: the TOT clathrates

Triorthothymotide TOT (Section 2.2.2) is quite a flexible chiral molecule. In solution, TOT exists primarily in a chiral, propeller-like conformation that undergoes rapid interconversion between P (right-handed) and M (left-handed) forms, the activation energy for enantiomerization being 21 kcal/mol (half-life of ca. 30 min at 0 °C). In its guest-free form, TOT crystallizes in the achiral space group $Pna2_1$. But, in the presence of solvent molecules of appropriate size, chiral clathrates are obtained, and each single crystal contains preferentially one of the guest enantiomers. The enantiomeric excess e.e. is given in Table 8. This chiral discrimination relies on precise relationships between the structure of the guest and the geometry of the host cage in the crystal structure. The host/guest interactions will be described at three levels as shown in Table 9.

REMARK The data in table 8 represent the enantiomeric excess in a crystal grown from a racemic solution of guest. But enantiomeric discrimination can be considerably enhanced when optically-active solutions of guests are used: this results in amplification of guest optical purity after repeated crystallizations. For instance, after three cycles, the enantiomeric purity of 5 reaches 80%, despite the very low enantiomeric purity (4%) resulting from the first crystallization.

We shall now comment on these three levels.

Table 8

Enantiomeric excess of guest and correlation of guest and host chirality in P-(+)- TOT clathrate crystals

No.	guest	guest e.e. (%)	guest configuration	Ref.
	CAGE			
1	2-chlorobutane	32	S-(+)	<52>
		45		<53>
2	2-bromobutane	34	S-(+)	<52>
		35		<53>
3	2-iodobutane	<1	-	<54>
9	trans-2,3-dimethyloxirane	47	S,S-(-)	<52>
10	trans-2,3-dimethylthiirane	30	S,S-(-)	<52>
11	trans-2,4-dimethyloxetane	38		<54>
12	trans-2,4-dimethylthietane	9		<54>
13	propylene oxide	5	R-(+)	<52>
14	2-methyltetrahydrofuran	2	S-(+)	<52>
15	methyl methanesulphinate	14	R-(+)	<52>
16	2,3,3-trimethyloxaziridine	7		<54>
17	ethylmethylsulphoxide	83	S-(+)	<55>
18	2-butanol	<5	S-(+)	<53>
19	2-aminobutane	<2	-	<53>
	CHANNEL			
4	2-chlorooctane	4	S-(+)	<52>
5	2-bromooctane	4	S-(+)	<52>
6	3-bromooctane	4	S-(+)	<54>
7	2-bromononane	5	S-(+)	<52>
8	2-bromododecane	5	S-(+)	<52>

5.3.1 *Space group and crystal structure type*

The obtention of cage versus channel clathrates depends essentially on the shape and size of the guest. Molecules containing up to six non-hydrogen atoms tend to adopt cage structures, whilst long-chain molecules lead to channel clathrates. An explanation analogous to that given for urea clathrates can be applied to TOT. We recall however that no cage clathrates have been found with urea.

Channel clathrates display uniformly low, but measurable and significant chiral discrimination whereas discrimination varies widely in cage clathrates, the values ranging from 2 to 46% e.e. (the latter value signifies an enantiomer ratio of 73:27). In channel clathrates the guests are expected to be severely disordered.

Table 9

System: (guest)solution + (TOT)solution → clathrate

The symbol ⇒ usually means an implication
The symbol → means 'main influence on'
The symbol ↔ means interaction

guest		clathrate
(1) Size of the guest	→	Crystal structure type : cage $P3_21$ channel $P6_1, P6_2$ $P\bar{1}$ or others
(2) Absolute configuration of guest	⇒	Absolute configuration of TOT and of clathrate
a) racemic mixture of guest (without seed)	⇒	Spontaneous resolution
		↓
b) preferential incorporation of one enantiomer from a racemic mixture of guest into the growing crystal seed	⇐	Single crystal seed
		↑
c) homochiral guest	⇒	Homochiral TOT and clathrate
(3) Precise molecular shape	↔	Fine structure of the cage
	⇓	
	fit and enantiomeric excess	

5.3.2 *Chiral discrimination and absolute configuration*

(a) In the presence of a racemic solvent, TOT undergoes spontaneous resolution and a conglomerate is obtained. Each single crystal of space group $P3_121$ contains TOT of the same chirality, but it includes both guest enantiomers, the enantiomeric excess (e.e.) varying for different 'host lattice/guest molecule' combinations. Separation and identification of these crystals as (+)- or (-)-TOT can be achieved by taking a chip from each crystal and then measuring its rotation in solution at ca. 0 °C.

The chirality and enantiomeric excess of the guest included in a clathrate crystal were determined by several different methods: direct VPC analysis of the guest or of a suitable derivative on chiral phases, NMR using a chiral shift reagent, or, for large crystals (50 to 100 mg), polarimetric measurements. In some cases, large single crystals (up to 0.5 g) were prepared by repeatedly seeding a hot saturated solution of TOT in the guest with single clathrate crystals, and slowly cooling the solutions.

(b) If a single crystal seed of given chirality is now introduced into the solution of TOT in racemic guest, then a large single crystal of the same chirality can be obtained by incorporating preferentially the same enantiomer of guest and the same enantiomer of TOT.

(c) In the presence of optically-pure solvent, optically-pure TOT clathrates are obtained: thanks to the rapid P-M interconversion of TOT, the guest chirality imposes itself on the whole clathrate, i.e. a given configuration of guest always leads to single crystals or powder containing the same configuration of TOT.

The absolute configuration of TOT has been assigned by determining the configuration of TOT relative to that of the guest for several guest molecules of known absolute configuration: (+)-TOT has a P-propeller-like configuration <51,55>. Conversely, the absolute configuration of new guest molecules can now be determined by X-ray analysis of TOT clathrates. But even without such an analysis, a reasonable inference can be made about the absolute configuration of the preferred guest enantiomer. Indeed it has been noted that homologous atoms or groups of atoms in related molecules occupy similar positions in space (Figure 20). The combined measurement of TOT and guest optical rotation would then lead to the determination of the absolute configuration of the guest.

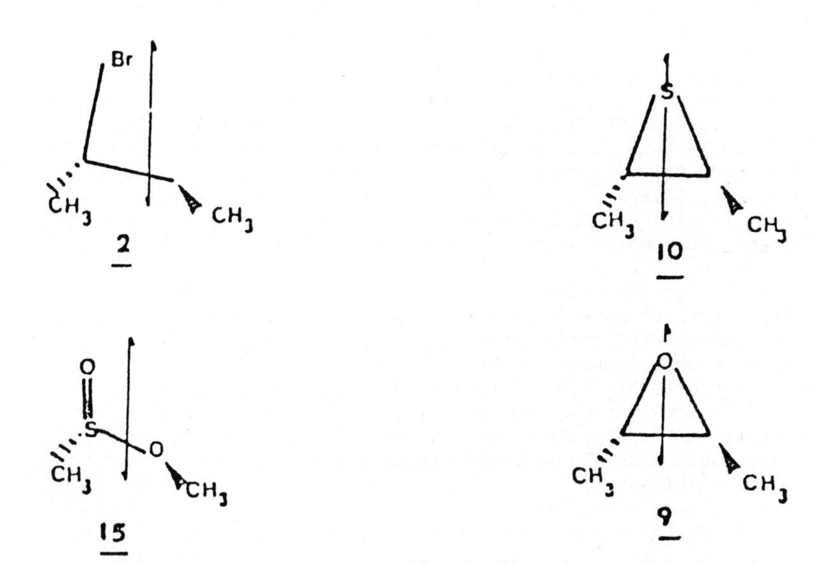

Figure 20 Enantiomers of guest molecules 2, 9, 10 and 15 which are preferentially included in a clathrate crystal built of P-(+)-TOT molecules, The guests are presented in orientations that they adopt relative to the twofold axis of the cage cavity

5.3.3 *Enantiomeric excess and fine structural fit*

The characteristics of the cage are quite constant for several clathrates studied by X-ray analysis (guests 2, 9, 10 and 17). The clathrate cavity is comprised of eight TOT molecules: six form a cylindrical wall that is approximately parallel to the c axis, one molecule is on top, and one on the bottom. The cage is located on a crystallographic twofold axis, and there are three cages in each unit cell. This accounts for the observed 2:1 TOT/guest stoichiometry. Guests lacking molecular symmetry 2 are necessarily disordered. The unexpected fact is that optically-pure guests, like 9 and 10, endowed with 2 symmetry, are also disordered, as a result of lack of spatial coincidence between the molecular and the crystallographic twofold rotation axis. For racemic 9 and 10, the two enantiomers are present in the proportion ca. $^2/_3$:$^1/_3$, as shown in Figure 21. Clearly, the crystallographic 2 axis along the x-axis generates a symmetry equivalent pair of enantiomers (not shown in Figure 21). Despite the presence of four distinct molecular positions within one cage, all guest coordinates have been determined.

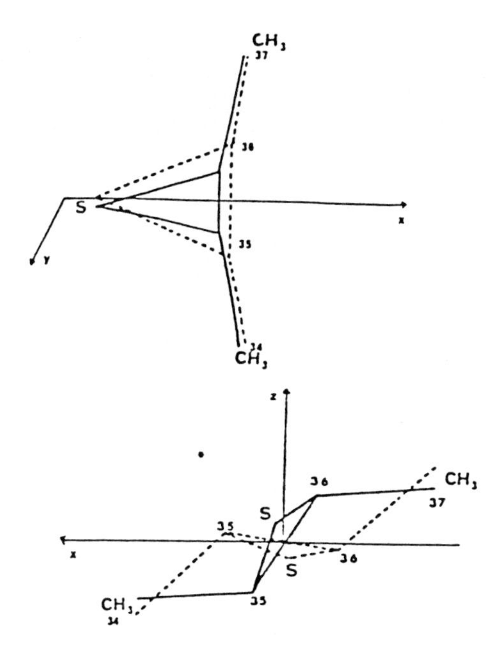

Figure 21 Projections on the xOy and xOz planes of the major (full line) and minor (dotted line) enantiomers of 2,3-dimethylthiirane in TOT.10 grown from racemic material

Enantiomeric selectivity depends on several factors pertaining to the guest molecular properties relative to the host cage.

(a) Molecular symmetry 2 seems a favourable factor, despite the disorder discussed above. Indeed 9 to 12 tend to afford higher e.e. We note however the high discrimination of 1 and 2, although both lack a twofold axis.

(b) The size of the guest is an important factor: the smallest, 13, and largest, 14, molecules of the order 9 to 16 exhibit the least e.e.

(c) The shape of the guest and its orientation in the cage apparently play an important rôle. Indeed, for guests 2, 9 and 10, which exhibit a rather high e.e., it seems that there is a certain complementarity of guest and cavity shape as shown in Figure 22. Moreover, small cage variations depending on the guest may be a way of 'fine tuning' of the clathrate structure in order to achieve the best fit for the guest. Indeed, it seems that there is a correlation between the van der Waals (VDW) volume of the guest, the volume of the VDW envelope of space accessible to the guest, and the volume of the unit cell.

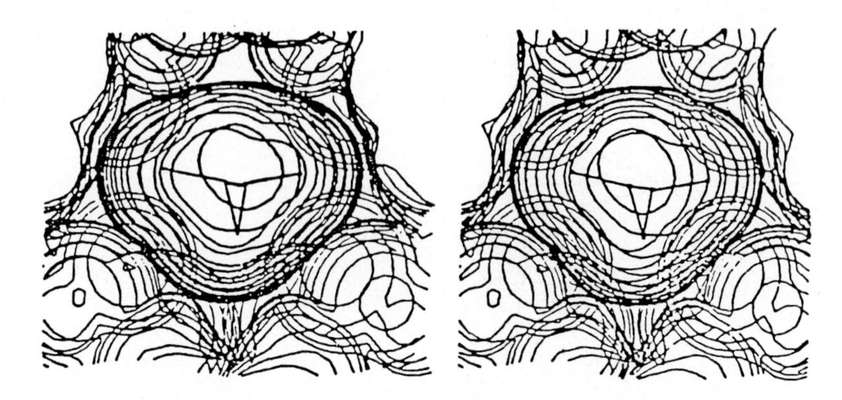

Figure 22 Contours of the van der Waals envelope of TOT (light lines) and of the van der Waals envelope of the volume accessible to guest atoms (heavy lines), viewed along the c axis, Sections are plotted every 0.3 Å parallel to the ab plane in a cube having 9.2 Å sides, The position of (+)-R,R)-2,3-dimethyloxirane is shown

(d) Temperature dependence: this is illustrated by the racemization of methyl methanesulphinate within the TOT cage. Crystals heated to 115 °C, for up to 12 h, showed no change in e.e. (14.1%). However, on heating at 125 °C for 12 h, racemization proceeded and the e.e. fell to zero. In solution, racemization already takes place at 115 °C. Thus the host affords some stabilization of the guest to racemization. Powder photography before and after heating showed that the integrity of the cage clathrate structure had not been destroyed, and it is therefore concluded that enantiomerization takes place within the TOT cage cavities. This clearly shows that increased thermal movements of both guest atoms and TOT atoms at the cage internal surface result in the loss of chiral discrimination ability of the clathrate above a certain temperature. The influence of thermal factors on a different kind of phenomenon, photoisomerization, is described in Section 7.

If X-ray analysis shed light on different aspects of enantiomeric discrimination, there is still no obvious way of predicting either the preferred configuration or the e.e. for new molecules whose shape is markedly different from those of Table 8. The cage is quite featureless, and it seems very difficult to provide a lock/key (or glove/hand) image.

Discrimination seems to result from an overall cage/guest interaction rather than a precise geometrical feature. In addition to the static geometry, dynamic considerations may also be involved. It is remarkable however that, despite the lack of striking features of the host cage, and despite disorder, enantiomeric selectivity is appreciable, and optical purity can reach values close to 100% after repeated crystallizations. This is especially of interest for small, chemically-inert molecules.

5.4 Chiral host

Details on several systems are given in reference <50>. The cyclodextrin family, a recently growing field of research, will be studied in Section 8, and an asymmetric synthesis case is given in Section 7.2.

6 PHYSICOCHEMICAL APPLICATIONS

An important aspect of clathrate structures is the specific micro-environment created by the complete enclosure of the guest by the host, and which leads to a variety of applications. Moreover, we have seen that the fine host molecular structure is reflected in the guest microenvironment, and results in turn in a fine tuning of the desired property. Separation of isomers is perhaps a prototype of this strategy, described in Section 6.2. The clathrate microenvironment can exert an influence on the physical, spectroscopic, chemical and other properties of the guest. Other applications arise from the mere fact of inclusion, such as the storage of rare gases in a solid matrix <56>.

6.1 Stabilization and retrieval of chemicals

Enclathration implies a degree of subtraction of the guest from the environment. As most clathrates dissociate in a solvent, it is easy to recover the guest. The hydroquinone clathrate, for example, can be used as a way of handling radioactive krypton and provides a safe and useful radioactive source. An ingenious application is based on the fact that this clathrate is sensitive to SO_2, O_3, ClO_2 etc, thus liberating ^{85}Kr which is then easy to detect. This clathrate can be used as an air-pollution monitor, sensitive in the ppm range.

In certain applications, the host is recovered rather than the guest. Thus, experimental plants of water desalination have given great hope, but none is in operation. The process relies on the formation of hydrocarbon or other inexpensive gas hydrates (propane was used in particular) at a temperature of a few degrees Celsius and then dissociation at a higher temperature and/or low pressure.

Chemical stabilization in atmospheric conditions is another tempting application. For instance, inclusion in urea or thiourea clathrates of fatty acids has been a method of protecting them from oxidation: linoleic acid is unaltered in such clathrates after several months.

Similarly, inclusion of CCl_4 was suggested as a method for the recovery of certain volatile solvents.

6.2 Separation techniques

Such techniques are being widely applied, and here we can give only some examples from the vast scientific and industrial literature.

6.2.1 *Mixtures and isomers: the Werner complexes*

Separation of rare gases via hydrates can be accomplished by adjustment of the pressure conditions (see Section 3.2).

Linear hydrocarbons can be separated from branched ones, and this has been a great incentive to study urea and thiourea clathrates. The initial discovery by Bengen in Germany <13> was kept secret until the end of World War Two. Since then, research and pilot plant construction have been pursued. Fatty acids and other long-chain molecules can be separated with urea clathrates. The method can be used also for identification and analytical purposes.

Extensive studies on the separation of xylenes and derivatives from a mixture of hydrocarbons were carried out over decades on the Werner complexes <58,59> of general formula MX_2A_4 where:

- M = Ni, Fe, Co, Cu, Zn, Cd, Mn, Hg, Cr,
- X = NCS, NCO, CN, NO_3, NO_2, Cl, Br, I,
- A = substituted pyridines,
  α-arylalkylamines.

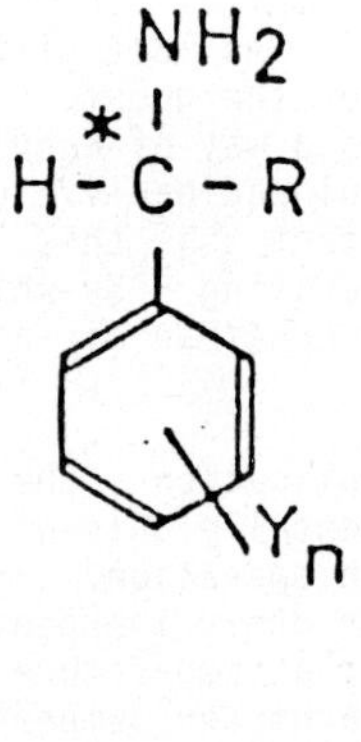

The prototypes are $Ni(NCS)_2.(4\text{-methylpyridine})_4$ (Figure 20) and $Ni(NCS)_2.(Y\text{-}C_6H_4\text{-}CHR\text{-}NH_2)_4$. The o-, m- and p-selectivity of xylenes was extensively studied by varying R from CH_3 to C_5H_{11}, Y = H, halogen, CH_3 to t-butyl. Typically, from an equimolar mixture of the three isomers, an enrichment of one of them is obtained up to about 90% in a single pass. Also, ethyltoluenes, cymenes, diethylbenzenes and other chemicals were investigated and yielded similar results.

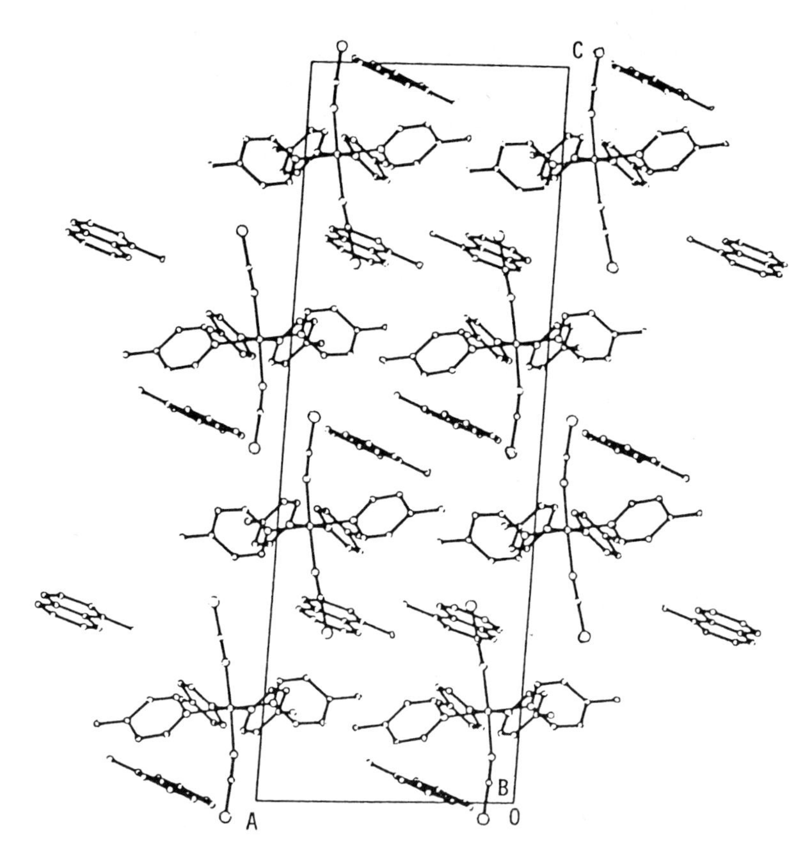

Figure 23 Molecular packing in the $Ni(NCS)_2$ $(4\text{-}MePy)_4\cdot$ 2(1-methylnaphthalene) inclusion compound viewed along (010) (hydrogen atoms are omitted)

(Reprinted from Acta Crystallogr., 1982, B38, with permission from Munksgaard, Noerre Soegade, Denmark)

In addition to steric factors, aromaticity appears to be a prerequisite for the inclusion of a guest. All attempts to include paraffins, cycloparaffins, olefins and cycloolefins have so far failed. Cyclohexyltoluene, on the other hand, can be included. Similarly, the amine coordinated to the central metal atom has to be aromatic in order to act as a host. From an analysis of the Hammet constants and the associated electronic behaviour, it emerges that interactions of a charge/transfer nature occur between the host lattice and the guest molecule. Depending on the nature of the substituents, both the guest and host aromatic rings may act as donors or acceptors, suggesting that Π-charge/transfer might be the primary driving force for the formation of these clathrates. More precisely, the inclusion phenomenon for a given host can be shown to have the transitivity property: if we consider the aromatic compounds A, B, C, it has been shown that if A is selectively included over B, and B over C, then A will be selectively included over C. Thus, for the host ($Ni(NCS)_2$(α-phenylethylamine)$_4$), the electro-releasing substituents of benzene derivatives increase the selective inclusion, whereas a decrease is observed with electro-attracting substituents: nitrobenzene and benzotrifluoride in pure form are not included at all. But a stabilization phenomenon - somewhat similar to the hilfgas effect in hydrates - occurs: each of these two molecules is co-included in the presence of suitable aromatic compounds.

It should be noted, however, that the scale based on electronic factors can be applied only to molecules which fit comfortably into the cavities. Although the host lattice presents some extensibility, for larger molecules steric factors intervene again.

If optically-pure amine is used in the host, no inclusion occurs at all: for a racemic mixture, the inclusion capacity is maximal, and for intermediate cases, the capacity is intermediate. No clear explanation has been offered.

Despite the detailed studies and the effort for setting up pilot plants, no commercial application has been developed to this day.

6.2.2 *Chromatography*

We have seen that, handled with precaution, some inclusion compounds can lose partly or even totally the guest without collapse of the matrix. Conversely, the partly-empty matrices can absorb the same or another guest, opening up possibilities in chromatography. Clearly, partial dissociation of the inclusion compound at the working temperature is a necessary condition for selective sorption. For these reasons, molecular sieves, not described in this chapter, are the most widely-used. However, clathrates may also open up new possibilities <60,61>.

In Werner complex liquid chromatography (LC), the basic solvent of the stationary phase is the partially-filled clathrate β-$Ni(NCS)_2$(4-MePy)$_4$. 0.7(4-MePy). The mobile phase is composed of an aqueous solution of methyl alcohol (or acetone or formamide) 4-MePy and NCS ions. Examples of separations include o-, m- and p-nitrotoluenes, and cis and trans azobenzenes. It is worth emphasizing that the Werner complexes exhibit zeolitic properties, i.e. they allow great absorption/desorption cycles without collapse of the host lattice.

6.2.3 *Isotopic fractionation*

Discrimination of isotopes by the inclusion phenomenon may be unexpected at first sight, as the main factor involved in discrimination in general is the steric factor. However, it has been shown that such discrimination occurs.

Experiments have been performed with Werner complexes: $Ni(4\text{-}MePy)_4 \cdot (NCS)_2$ as hosts, and mixtures of p-xylene and totally deuteriated p-xylene as guests <62>. It was found that the distribution of the two isotopically-differing molecules over the solid clathrate and the solution in equilibrium is different by a factor of 1.10.

The explanation of such selectivity resides probably in the different dynamic behaviour of the isotopes, reflected by the fugacities of the two guests in the liquid phase. Separation in the H_2/D_2 system has been reported with zeolites <63>. Much smaller separation factors are found for the $^{41}K/^{39}K$ system (1.001) in the presence of crown ethers <64>. The isotopic selectivity is again ascribed to dynamic factors: zero point energy differences.

6.3 Miscellaneous clathrates

We quote here several other clathrates whose potential applications have been demonstrated: Hoffman-type clathrates <78>, cyclophosphazene <79>, deoxycholic acid <80>, crown ethers <81>, cryptates <82>, cyclotriveratrylene <83>, and 'liquid clathrates' <84>.

7 CHEMICAL REACTIONS

The microenvironment of the guest molecule entrapped in the clathrate cavity is certainly different from that of the same molecule in solution <65,66>. The effect on the pathway of a chemical reaction would be particularly neat in the case where the guest molecule is located at a single position and orientation, i.e. without any kind of disorder. Indeed, in such cases we shall see that the guest's crystalline environment dramatically modifies the chemical behaviour compared with that in solution.

7.1 Photochemical reactions of the guest: cis/trans isomerization

We have seen (Section 2.2.2) that cis and trans stilbene, as well as methyl cinnamates, form 'bottleneck' type clathrates. Under UV irradiation, they undergo photoisomerization <16,17>.

The essential facts can be summarized by the following scheme:

	solution	TOT clathrate	pure guest crystal
. stilbenes	cis $\rightleftarrows$ trans 80% 20%	cis $\rightarrow$ trans cis $\nleftarrow$ trans	cis $\nrightarrow$ trans cis $\nleftarrow$ trans
. cinnamates	cis $\rightleftarrows$ trans 50% 50%	cis $\rightleftarrows$ trans 50% 50%	cis $\leftarrow$ trans cis $\nrightarrow$ trans

In both types of guest molecule, stilbenes and cinnamates, the cis/trans isomerization pattern in the clathrate is different from that observed in pure guest crystals. This is generally ascribed to specific host/guest interactions, totally different from the guest/guest interaction occurring in the pure guest crystal. Furthermore, cinnamates, but not stilbenes, exhibit the same behaviour in clathrates as is observed in solution. An interpretation of the facts required combined crystallographic and chemical studies.

All these clathrates crystallize in space group $P\bar{1}$ with two (2 TOT. guest) units per unit cell. The TOT molecules are in the general position, but the two guest molecules occupy two crystallographically-independent special positions, i.e. they lie on centres of symmetry (0,1/2,0) and (1/2, 1/2, 1/2), respectively, and are located in two perpendicular, non-interconnected channels parallel to the a and b axes respectively (Figures 6 and 24).

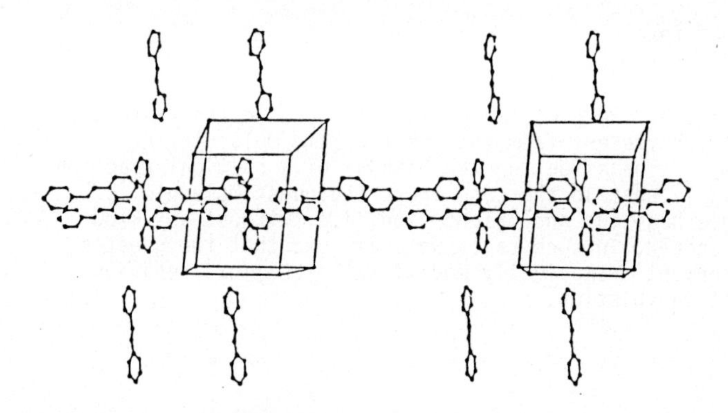

Figure 24 Stereoscopic drawing of the trans-stilbene molecules in both channels of the TOT.trans-stilbene clathrate, The axial directions are a→, b↑, and c up out of the plane of the paper

The crystal structures provide a simple and important clue for understanding the unidirectional cis/trans conversion: the coincidence between the molecular symmetry of trans-stilbene and the symmetry of the crystallographic site at which the guest molecule is located. Cis-stilbene, lacking a $\bar{1}$ symmetry, is necessarily disordered in P$\bar{1}$. It follows that it is impossible for this clathrate to 'optimize' the host/guest contacts simultaneously for both moeties, and we can reasonably infer a considerably larger thermodynamic stability of the trans- over the cis-stilbene clathrate. This higher stability is also reflected in the following experiment: when 1:1 mixtures of cis- and trans-stilbene were subjected to TOT enclathration, the resulting crystals contained a trans/cis isomer ratio of 19:1.

By contrast, both cis- and trans-methyl cinnamate lack $\bar{1}$ symmetry and, in addition, they are smaller than the stilbenes. These facts may account for the photoequilibrium observed, identical to that observed in solution (Figure 25), whether one starts from the cis- or from the trans-cinnamate clathrate.

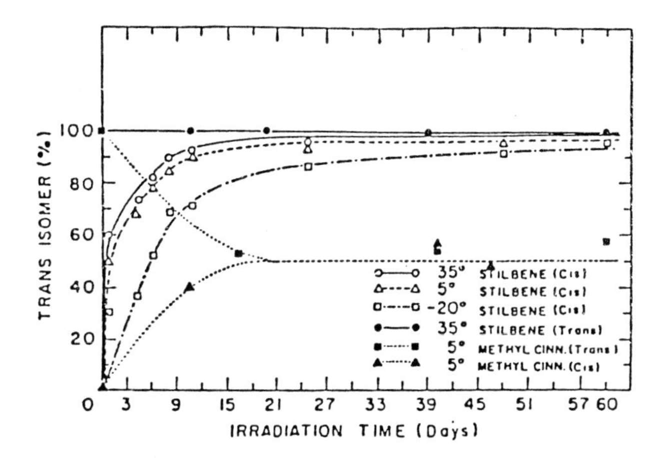

Figure 25 Irradiation of TOT clathrates of cis- (▲) and trans- (■) methyl cinnamate and cis- (○) and trans- (●) stilbene,
For the irradiation of powdered samples of cis-stilbene.TOT clathrate as a function of temperature, the starting samples were taken from the same batch of clathrate

Next we consider the mechanism of the cis→trans isomerization within the cis-stilbene clathrate. Whatever the location of the guest, inspection of the crystal structure shows that there is no room to allow intramolecular rotation to take place inside the channel. A possible explanation resides in the 'loosening' of the surrounding TOT structure, coinciding with the absorption of a photon by stilbene. The TOT molecules would then undergo a concerted movement so as to allow both the intramolecular cis/trans rotation and the necessary displacement of the newly-formed trans-stilbene molecules towards the final ordered position in the corresponding clathrate. This amounts to a local destruction, or at least severe deformation, of the cis-stilbene clathrate structure and subsequent 'recrystallization' in the trans-stilbene clathrate structure. This mechanism accounts for the fact that the initial single crystal is destroyed to yield an aggregate of crystallites in the product phase.

The necessary activation energy being of thermal origin, the kinetic study of the phenomenon is of interest. Indeed, the temperature dependence of the isomerization kinetics most probably reflects the thermal barrier for the above rearrangement of the TOT structure (Figure 25). We note that no such temperature dependence occurs for the isomerization of free cis-stilbene. Furthermore, it has been shown that, in the dark, iodine catalyses the cis/trans conversion in the clathrate. But the iodine-promoted conversion is completely inhibited at -20 °C (whereas for free cis-stilbene it is only retarded at -20 °C). These facts reflect the thermal barrier for the penetration of iodine through the TOT lattice. They favour an *in situ* isomerization within the cavities, and not in interstitial pockets such as dislocations or other defects.

In conclusion, the cis/trans photoisomerization within TOT clathrates is primarily under the control of 'crystal/molecular symmetry', whereas considerations about the 'crystal packing/stereochemical configuration' may play a rôle in a detailed discussion of the corresponding mechanism. Again, as in chiral discrimination within TOT $P3_12$ clathrates, dynamic (thermal) considerations are also highly significant.

7.2 Thermal reaction of the guest: asymmetric decarboxylation

Cyclodextrin (see Section 8) forms a clathrate with phenylethylmalonic acid . By heating this clathrate at 100 °C, decarboxylation occurs and, in the resulting 2-phenylbutyric acid , there is a 7% excess of the S-(+) enantiomer. Figure 28 shows the crystal structure of the complex and its discussion in Section 8 may provide a clue for the mechanism <67>.

7.3 Singlet oxygen reaction on the guest

Asymmetry can also be induced in an organic reaction involving an external agent penetrating the clathrate structure without damaging the framework (cf. I_2 penetration in TOT $P\bar{1}$ clathrates). This is the case of photooxygenation of the prochiral molecule 2-methoxybut-2-ene <68>:

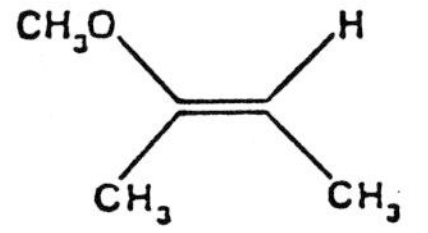

Incidentally, the inclusion phenomenon provides a nice example of the fine influence of a guest candidate on the 'choice' of a particular clathrate framework among several known polymorphic forms. The E isomer 1 leads to achiral P$\bar{1}$ crystals (bottleneck type), whilst the Z isomer 2 gives the chiral $P3_121$ crystals. Irradiation by UV in the presence of a sensitizer leads to 3 (30% yield) with a significant optical enrichment (the optical purity is unknown).

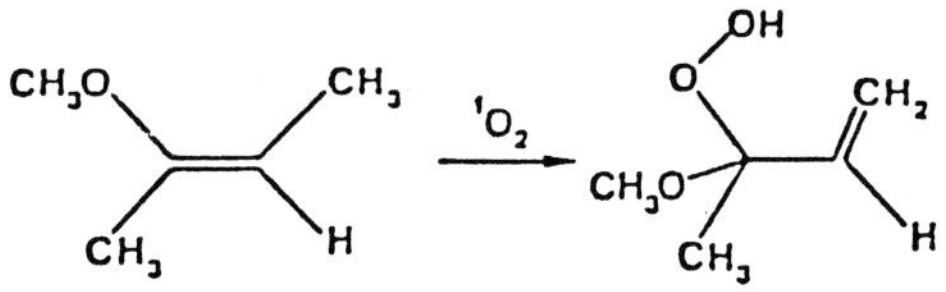

7.4 Host/guest reactions: deoxycholic acid clathrates

These reactions involve hydrogen abstraction from the host deoxycholic acid (DCA) molecules by the photoexcited guest ketones which then add to form a new carbon/carbon bond <69,70>. A new chiral centre is also formed at the carbonyl carbon atom. The ratio of diastereomers varies with the guest.

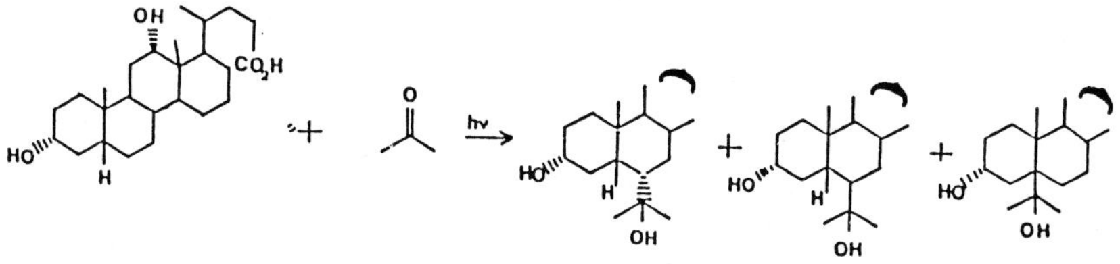

Selectivity can generally be understood in terms of the orientations and distances of the carbonyl chromophore and the potentially reactive H-C groups on the steroid. Perhaps the most intriguing aspect of these studies was the discovery of a topochemical strictly lattice-controlled reaction in the DCA/acetophenone complex wherein the stereochemical outcome is the opposite to that anticipated on the basis of bond formation along the path of least motion. Thus, on irradiation under argon, the 10:4 DCA/acetophenone complex gave only one photoproduct of addition to position 5 (Figure 26).

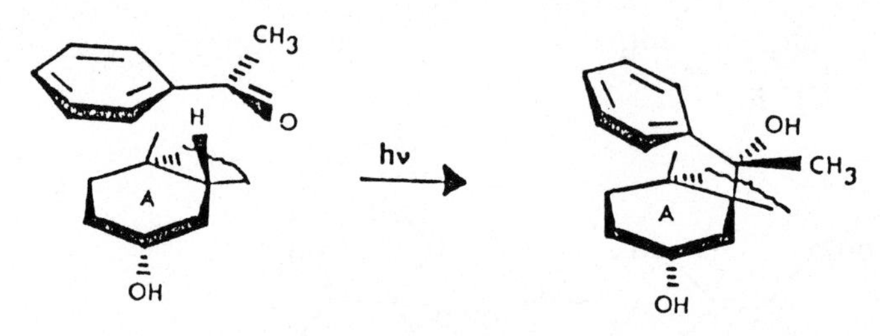

Figure 26 Model describing abstraction of an H atom from the steroid and formation of a new chiral centre in the DCA/acetophenone inclusion compound

The absolute configuration of the new chiral carbon was determined by crystal structure analysis at -170 °C. Despite the disorder of the guest and the complexity of the problem, a model has been proposed which reasonably fits most chemical and crystallographic data. The unexpected fact is that the ketone adds from the face of the acetophenone which is the most distant from the steroid in the starting structure. This implies the need for unusual motion of the guest acetyl group on reaction, as suggested by a detailed X-ray analysis on a crystal which had undergone a partial (40%) reaction.

On the other hand, it is to be noted that precise determination of the crystal structure of the DCA/diethyl ketone inclusion compound has clearly shown the necessity for a correct distance and orientation prior to a host/guest reaction (Figure 27).

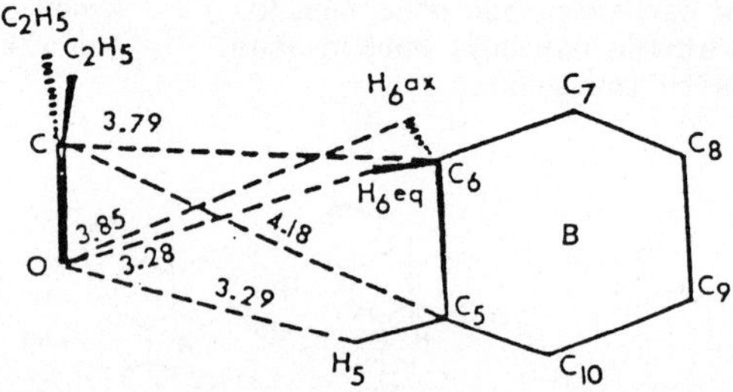

Figure 27 Orientation and distances between the atoms of the carbonyl group of diethylketone and sites C_5-H, C_6-H_{eq} and C_6-H_{ax} in the DCA inclusion compound

The ketone oxygen O is 3.3 Å from $H_{6_{eq}}$ and H_5 but 3.9 Å from $H_{6_{ax}}$. The distance from the ketone C to C_6 is 3.8 Å whilst the distance from the ketone C to C_5 is 4.2 Å. Indeed, the experiment (irradiation under argon) confirms that the reaction gives only the addition product at 6_{eq}.

This work illustrates the interest of clathrates in elucidating reaction mechanisms arising from the exact knowledge of the spatial dispositions of reactants before reaction, and of the stereochemical configuration of the product revealed *in situ*.

7.5 Inclusion polymerization

Polymerization of monomers (butadiene, dimethylbutadiene, 1,3-trans pentadiene etc) within clathrate channels may lead to highly-ordered polymers, as in urea, deoxycholic acid and PHTP <71>. The reaction proceeds:

(a) by irradiating (α, β and X-rays) a previously-formed clathrate,

(b) by irradiating the host (PHTP) alone in the crystal state, whilst the successive addition of monomer in the liquid or gaseous phase, over a suitable pressure and temperature range, makes it possible to obtain the polymer.

The feasibility of this process is linked with the stability of the active species in the crystal state and with the ease with which inclusion compounds are formed with the monomer. By this technique, it was possible to ascertain that inclusion polymerization in PHTP is a living polymerization, in which termination and chain transfer reactions are suppressed or considerably reduced.

An asymmetric, lattice-controlled polymerization has also been achieved within the chiral $P6_3$ PHTP clathrates.

8 CLATHRATES AND INCLUSION COMPOUNDS: CYCLODEXTRINS

These macrocycles possess the outstanding property of forming both clathrates in the solid state and complexes in solution. The formation of cyclodextrin (cyd) clathrates is most probably preceded by the formation in solution of a cyd.guest inclusion compound.

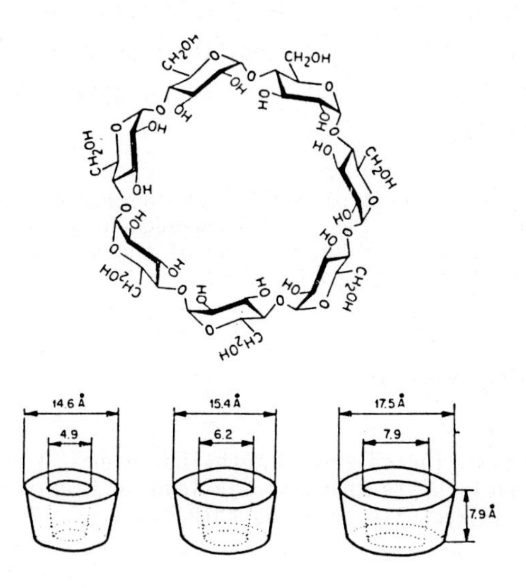

8.1 Inclusion in solution

The interior of the cyclodextrin cavity has the aptitude of accommodating molecules held by van der Waals forces. It follows that in principle most of the molecules of appropriate size could be included in cyclodextrins, regardless of the presence of a hydrophilic group. We quote a few out of hundreds of known inclusion compounds:

- α-cyclodextrin: benzene, phenol, p-nitrophenol, p-nitrophenolate, iodides, benzaldehyde etc,
- β-cyclodextrin: benzophenone, diphenyl, naphthalene and substituted derivatives etc,
- γ-cyclodextrin: anthracene and analogues, vitamin K_3 (menadione, digitalis glycosides etc.

Inclusion has wide-range applications: solubilization in water of hydrophobic and lipophilic molecules, especially in the pharmacological and pharmaceutical fields (pesticides for instance), handling of volatile substances, and chemical stabilization against oxidation and light. Several new fields are being developed in this rapidly-growing area <72,73,74,91>.

8.2 Host/guest interaction

The guest molecule is surrounded by the cyclodextrin atoms which constitute a specific microenvironment. It follows that several physical and chemical properties of the guest are modified accordingly. In particular, optical properties are sensitive to a spatially precise environment: UV, fluorescence, photochromism, electronically-induced circular dichroism etc. The geometry of conformationally flexible molecules may also be dependent on inclusion. Thus, achiral molecules, such as bilirubin, acquire new chiroptical properties which are easy to detect and record in solution <75,77>.

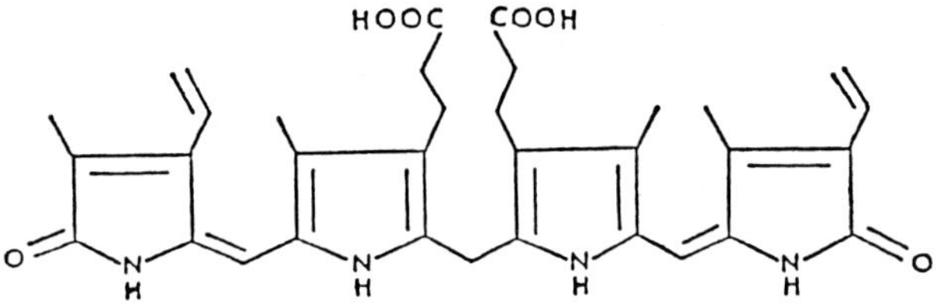

bilirubin IXα

Cyclodextrins present a particular interest in fundamental research: for the same included molecule, they allow the determination of both the spectroscopic properties in solution, and the three-dimentionsal structure in the crystal state, i.e. in the clathrate.

8.3 Crystallization and clathrates

A β-cyclodextrin clathrate crystal structure is given in Figure 28. It is interesting to consider in general whether, in the crystal state, a guest is located in intramolecular or intermolecular cavities or in some intermediate situation. First, we note that solvent molecules, H_2O and C_2H_5OH, are necessary parts of the crystal construction. They are linked to the cyclodextrin hydroxyls and, eventually, to the guest molecule. Thus, one of the carboxyls of phenylethylmalonic acid is linked only to H_2O, whilst the other is linked to two hydroxyls (one belonging to the cyclodextrin molecule which includes the phenyl ring, and the other to a neighbouring cyclodextrin). Thus, the inclusion cavity is both intramolecular and intermolecular, the former being predominant. In most cases of β-cyclodextrin clathrates, the building blocks are cyclodextrin dimers, where the two molecules are linked by 14 H-bonds formed by the 14 secondary hydroxyls. In all known α- and γ-cyclodextrin clathrates, monomers are involved.

In an attempt at classification, two main patterns can be distinguished: channel and cage clathrates. Figure 29 shows model structures for α-cyclodextrin. In β-cyclodextrin clathrates (Figure 30), a third, intermediate, irregular channel type is distinguished <76,77>. Cyclodextrin clathrates also have a wide range of applications as stabilizers and slow release agents.

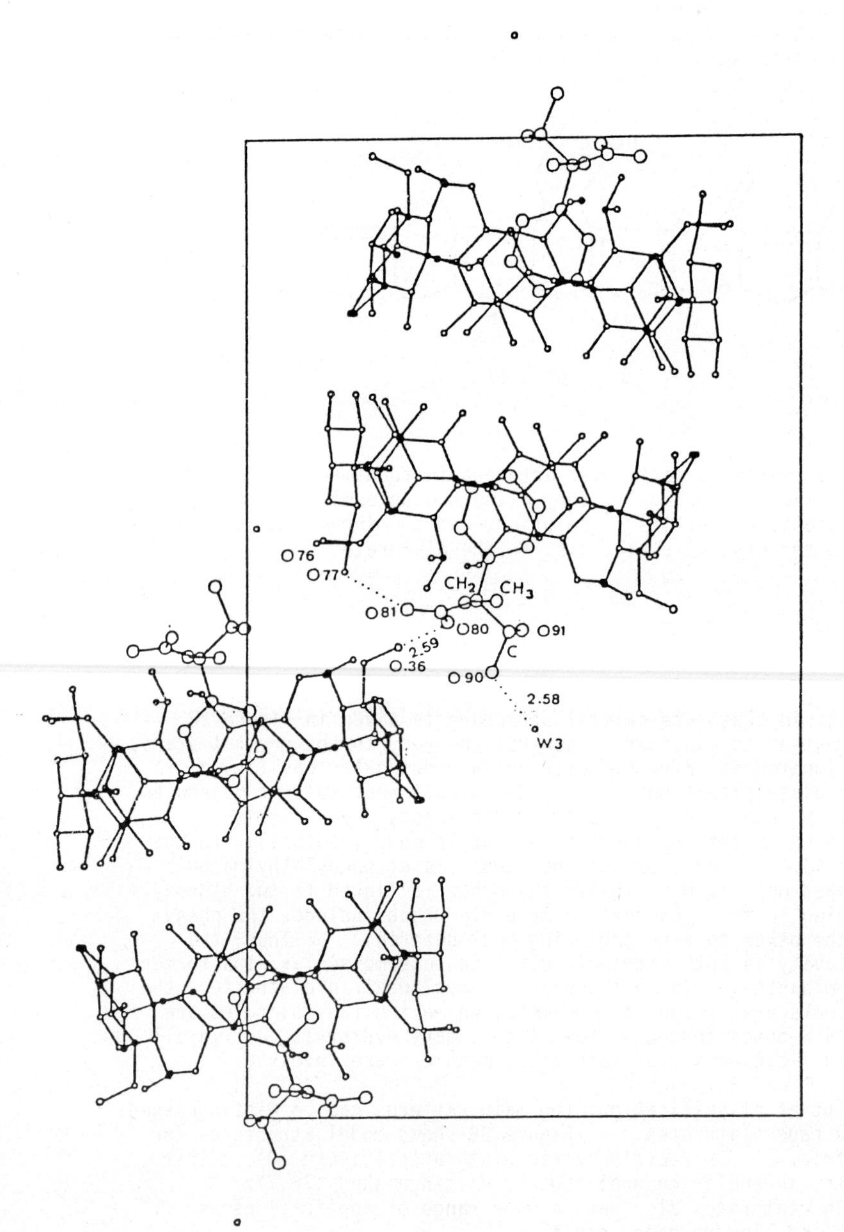

Figure 28 Projection on the ac plane of the structure of cyclodextrin.phenylethylmalonic acid clathrate

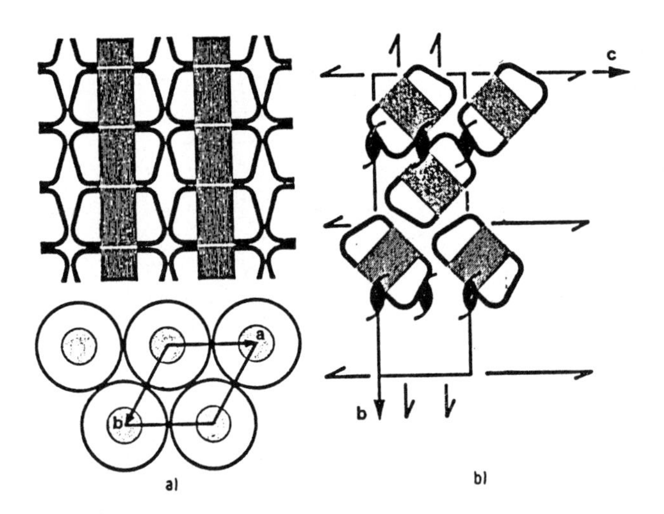

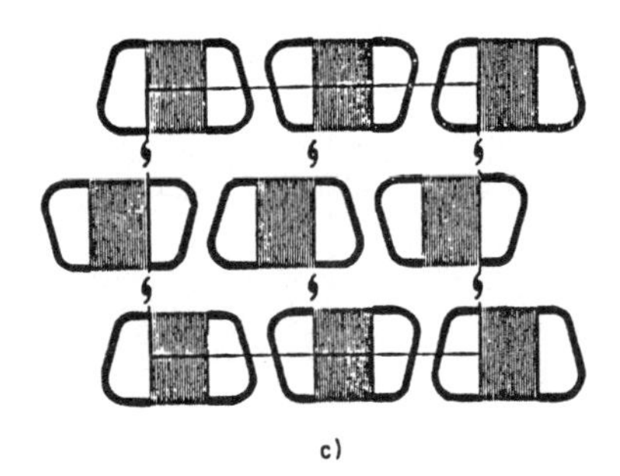

Figure 29 Schematic description of
(a) channel type,
(b) cage herringbone type,
(c) brick type
crystal structures formed by crystalline
cyclodextrin inclusion complexes

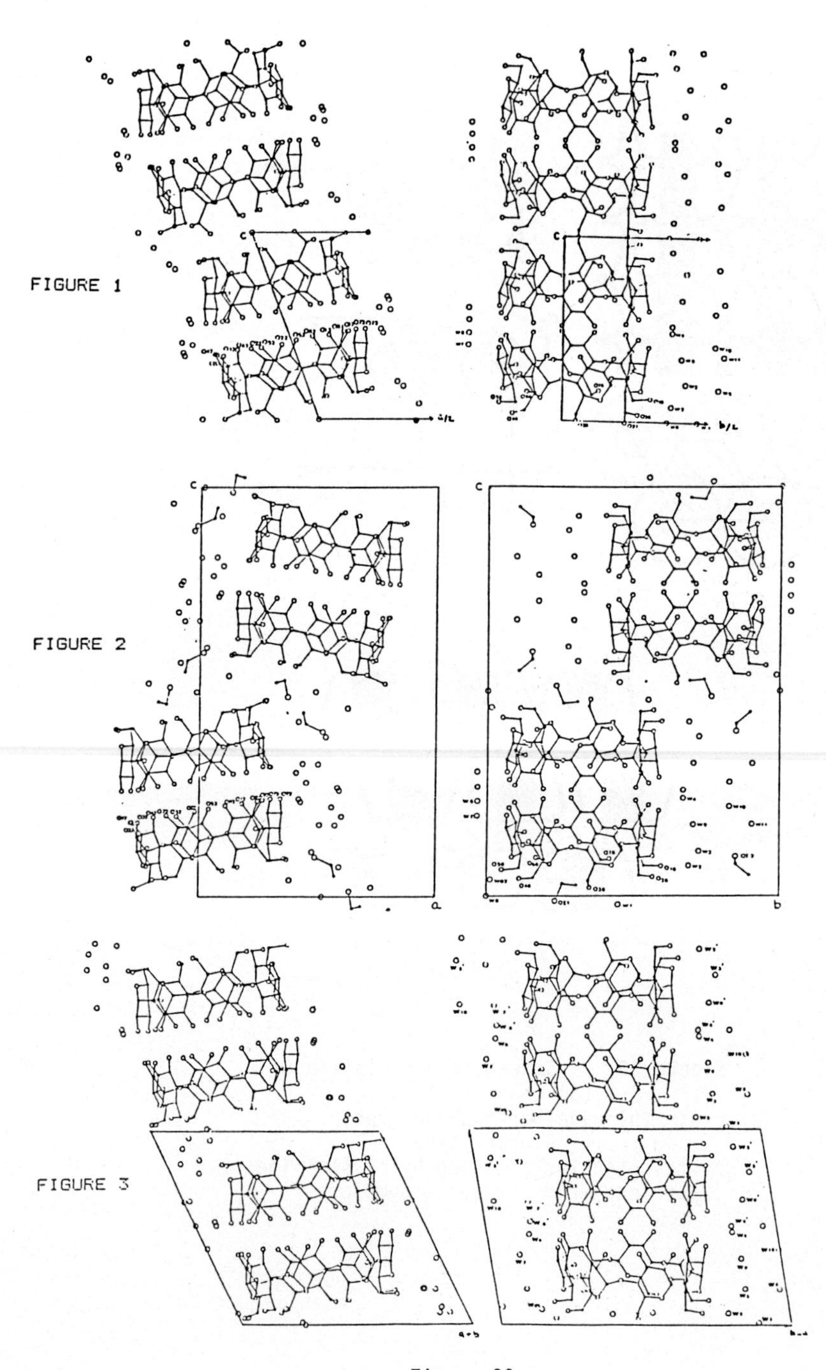

Figure 30

9 CONCLUSION AND PERSPECTIVES

Crystallization experiments in the presence of solvents are a routine practice in chemistry, and therefore new matrices are discovered. The systematic study of the most promising ones requires the combined effort of several disciplines, such as thermodynamics, spectroscopy, crystallography etc.

New developments are expected in the light of recent achievements in 'host design'. Moreover, the perspectives for the creation of new host lattices are considerably enlarged by the quick development of inclusion compounds in solution.

In general, when a clathrate is put in solution, the host/guest specific interactions vanish, and the compound is dissociated in its elements, with subsequent loss of the associated properties. Chemists have focused their attention on inclusion compounds whose specific interactions are maintained in solution: this is the case for cyclodextrins <90>, cryptates, cavitands, calixarenes etc.

Thus the building blocks of clathrates are already present in solution although not necessarily in the same geometry as that of the solid state. Moreover, chemists endeavour to make modifications to the above host molecules so as to achieve a specific function in solution. Such modifications can also be designed bearing in mind the desired specific properties of the corresponding clathrates.

The combined efforts concerned with the investigation of clathrates both in the solid state and in solution will considerably enlarge the domain of research and their field of application.

ACKNOWLEDGEMENT

The author wishes to thank Dr. Vassiliki Tsomi for her assistance and pertinent remarks.

REFERENCES

1 H. Davy, Philos. Trans. R. Soc. (London), 101, 155-162 (1811)
2 M. Faraday, Quant. J. Sci. Lit. Arts, 15, 71-74 (1823)
3 F. Wöhler, Justus Liebigs Ann. Chem., 69, 297-300 (1849)
4 F. Mylius, Ber. Bunsenges. Phys. Chem., 19, 999-1009 (1886)
5 D.E. Palin and H.M. Powell, J. Chem. Soc., 208-224 (1947)
6 M. von Stackelberg, Naturwiss, 36, 327-333 and 359-362 (1949)
7 W.F. Claussen, J. Chem. Phys., 19, 259, 662, 1425 (1951)
8 L. Pauling and R.E. Marsh, Proc. Nat. Acad. Sci. USA, 38, 112-118 (1952)
9 Inclusion Compounds, Edited by J.L. Atwood, J.E.D. Davies and D.D. MacNicol, Academic Press, London, 1984
10 H.M. Powell, ibid. ref. 9, 1, 1-28
11 G.A. Jeffrey, ibid. ref. 9, 1, 135-190
12 G.A. Jeffrey and R.K. McMullan, Prog.Inorg. Chem., 8, 43-108 (1967)
13 M.F. Bengen and W. Schlenk, Experientia, 5, 200 (1949)
14 K. Takemoto and N. Sonoda, ibid. ref. 9, 2, 47-67
15 J. Allemand and R. Gerdil, Acta Cryst., C39, 260-265 (1983)
16 R. Arad-Yellin, S. Brunie, B.S. Green, M. Knossow and G. Tsoucaris, J. Am. Chem. Soc., 101, 7529-7537 (1979)
17 R. Arad-Yellin, B.S. Green, M. Knossow, N. Rysanek and G. Tsoucaris, J. Incl. Phenom., 3, 317-333 (1985)
18 D.D. MacNicol, ibid. ref. 9, 2, 1-45
19 S.V. Lindeman, V.E. Shklover and T.Yu. Struchkov, Cryst. Struct. Comm., 10, 1173-1179 (1981)
20 W.C. Ho and T.C.W. Mak, Z. Kristallogr., 161, 87-90 (1982)
21 T.M. Polyanskaya, V.I. Alekseev, V.V. Bakakin and G.N. Chekhova, J. Struct. Chem. (Engl. transl.), 23, 101-104 (1982)
22 T.C.W. Mak, J. Chem. Soc. Perkin Trans., 2, 1435-1437 (1982)
23 J.C.A. Boeyens and J.A. Pretorius, Acta Cryst., B33, 2120-2124 (1977)
24 T.L. Chan and T.C.W. Mak, J. Chem. Soc. Perkin Trans., 2, 777-781 (1983)
25 J.N. Helle, D. Kok, J.C. Platteeuw and J.H. van der Waals, Recl. Trav. Chim. Pays Bas, 81, 1068-1074 (1962)
26 V. McKoy and O. Sinanoglu, J. Chem. Phys., 38, 2946-2956 (1963)
27 S.L. Miller, E.J. Eger and C. Lundgren, Nature, 221, 468-469 (1969)
28 N.G. Parsonage and L.A.K. Staveley, ibid. ref. 9, 3, 1-36
29 J.C. Platteeuw and J.H. van der Waals, Mol. Phys., 1, 91-96 (1958)
30 J.H. van der Waals and J.C. Platteeuw, Adv. Chem. Phys., 2, 1-57 (1959)
31 O. Redlich, C.M. Gable, A.K. Dunlop and R.W. Millar, J. Am. Chem. Soc., 72, 4153-4160 (1950)
32 D.F. Evans and R.E. Richards, J. Chem. Soc. 3932-3936 (1952)
33 R.M. Barrer and A.V.J. Edge, Proc. R. Soc. London, Ser.A, 300, 1-24 (1967)

34 M. Farina, ibid. ref. 9, 2, 69-95

35 R.C. Pemberton and N.G. Parsonage, Trans. Faraday Soc., 61, 2112-2121 (1965)

36 A.F.G. Cope, D.J. Gannon and N.G. Parsonage, J. Chem. Thermodyn., 4 829-842 (1972)

37 H.H. Mills, D.D. MacNicol and F.B. Wilson, unpublished results, cited in : ibid. ref. 9, 2, 1-45

38 J.L. Flippen, J. Karle and I.L. Karle, J. Am. Chem. Soc., 92, 3749-3755 (1970)

39 J.L. Flippen and J. Karle, J. Phys. Chem., 75, 3566-3575 (1971)

40 J.H. Gall, A.D.U. Hardy, J.J. McKendrick and D.D. MacNicol, J. Chem. Soc. Perkin Trans., 2, 376-380 (1979)

41 A.D.U. Hardy, J.J. McKendrick and D.D. MacNicol J. Chem. Soc. Chem. Comm., 355-356 (1976)

42 D.D. MacNicol, H.H. Mills and F.B. Wilson, Chem. Comm., 1332-1333 (1969)

43 D.D. MacNicol and F.B. Wilson, Chem. Comm., 786-787 (1971)

44 A.D.U. Hardy and D.D. MacNicol, unpublished results, cited in : J. Chem. Soc. Perkin Trans., 2, 729-734 (1979)

45 A.D.U. Hardy, J.J. McKendrick and D.D. MacNicol, J. Chem Soc. Perkin Trans., 2, 1072-1077 (1979)

46 D.D. MacNicol, A.D.U. Hardy and J.J. McKendrick, Nature (London), 256, 343-344 (1975)

47 D.D. MacNicol, ibid. ref. 9, 2, 123-168

48 D.D. MacNicol and D.R. Wilson, J. Chem. Soc. Chem. Comm., 494-495 (1976)

49 W.D. Ollis and J.F. Stoddart, ibid. ref. 9, 2, 169-205

50 R. Arad-Yellin, B.S. Green, M. Knossow and G. Tsoucaris, ibid. ref. 9, 3, 263-295

51 R. Arad-Yellin, S. Brunie, B.S. Green, M. Knossow and G. Tsoucaris, Tetrahedron Lett., 21, 387-390 (1980)

52 R. Arad-Yellin, B.S. Green and M. Knossow, J. Am. Chem. Soc., 102, 1157-1158 (1980)

53 R. Gerdil and J. Allemand, Helv. Chim. Acta, 63, 1750-1753 (1980)

54 R. Arad-Yellin, B.S. Green, M. Knossow and G. Tsoucaris, J. Am. Chem. Soc., 105, 4561-4571 (1983)

55 R. Gerdil and J. Allemand, Tetrahedron Lett., 3499-3502 (1979)

56 M. Hagan (Sister), Clathrate Inclusion Compounds, Rheinhold Publishing Corporation, Chapman and Hall Limited, New York and London, 1962

57 W. Schlenk Jr., Justus Liebigs Ann. Chem., 1145, 1156, 1179, 1195 (1973)

58 J. Hanotier and P. de Radzitzky, ibid. ref. 9, 1, 105-134

59 J. Lipkowski, ibid. ref. 9, 1, 59-104

60 A. Sopkova and M. Singliar, ibid. ref. 9, 3, 245-256

61 D. Sybilska and E. Smolkova-Keulemansova, ibid. ref. 9, 3, 173-243

62 N.O. Smith, ibid. ref. 9, 3, 257-261

63 I.D. Basmadjian, Can. J. Chem., 38, 141-148 (1960)

64 B. Schmidhalter and E. Schumacher, Helv. Chim. Acta, 65, 1687-1693 (1982)

65 J.E.D. Davies, ibid. ref. 9, 3, 37-68

66 D.W. Davidson and J.A. Ripmeester, ibid. ref. 9, 3, 69-128

67 M. Knossow, R. Arad-Yellin and B.S. Green, unpublished work (1980)

68 R. Gerdil, G. Barchietto and C.W. Jefford, J. Am. Chem. Soc., 106 8004-8005 (1984)

69 R. Popovitz-Biro, C.P. Tang, H.C. Chang, N.R. Shochet, M. Lahav and L. Leiserowitz, Nouv. J. Chim., 6, 75-77 (1982)

70 R. Popovitz-Biro, H.C. Chang, C.P. Tang, N.R. Shochet, M. Lahav and L. Leizerowitz, Pure Appl. Chem., 52, 2693 (1980)

71 M. Farina, ibid. ref. 9, 3, 297-329

72 J. Szejtli, ibid. ref. 9, 3, 331-390

73 R.J. Bergeron, ibid. ref. 9, 3, 391-443

74 I. Tabushi, ibid. ref. 9, 3, 445-471

75 E. Hadjoudis, I. Moustakali-Mavridis, G. Tsoucaris and F. Villain Mol. Cryst. Liq. Cryst., 134, 255-264 (1986)

76 W. Saenger, Angew. Chem. Int. Ed. Engl., 19, 344-362 (1980)

77 G. Le Bas and G. Tsoucaris, Mol. Cryst. Liq. Cryst., 137, 287-301 (1986)

78 T. Iwamoto, ibid. ref. 9, 1, 29-57

79 H.R. Allcock, ibid. ref. 9, 1, 351-374

80 E. Giglio, ibid. ref. 9, 2, 207-229

81 I. Goldberg, ibid. ref. 9, 2, 261-335

82 B. Dietrich, ibid. ref. 9, 2, 337-405

83 A. Collet, ibid. ref. 9, 2, 97-121

84 J.L. Atwood, ibid. ref. 9, 1, 375-405

85 H.M. Powell, Nature (London), 170, 155 (1952)

86 H. Gies, Zeitschrift für Kristal., 164, 247-257 (1983)

87 H. Gerke and H. Gies, Zeitschrift für Kristal., 166, 11-22 (1984)

88 H. Gies, J. of Inclusion Phenomena, 4, 85-91 (1986)

89 R. Clément, J. Jegoudez and C. Mazieres, J. Solid State Chem., 10, 46 (1974)

90 Cyclodextrins and their industrial uses, to be published by Editions de Santé, Paris.

91 R. Breslow, ibid. ref. 9, 3, 473-508

Chapter 8

SOLID STATE CHEMISTRY OF PHENOLS AND POSSIBLE INDUSTRIAL APPLICATIONS

R. PERRIN, R. LAMARTINE, M. PERRIN and A. THOZET

1 INTRODUCTION

Phenol is the name of the class of compounds containing one or more hydroxyl groups attached to an aromatic ring. These compounds are very important for industry and too for biology. To be convinced, you just have to notice the example of the 2.000.000 tons world yearly non-captive production of phenolic resins and to remember that hormones such as thyroxine, adrenalin and estrone are phenolic compounds.

The arrangements of phenol molecules in solid state are various, as it is shown by the study of cristalline structures of these substances presented in the first part. In these arrangements, hydrogen bonds mostly appear. Their influence is important, like in biological materials, and we have been led to classify the observed arrangements according to the number and to the type of the hydrogen bonds.

Studies about solid state phenol chemistry, presented in the second part, are rather numerous and illustrate the various fields of organic solid state chemistry. Thus the presented results practically constitute a full statement of organic solid state chemistry.

The last part is devoted to the possibilities of industrial developments of organic solid state chemistry. A particular attention is paid to the preparation of chemical substances with solid organic raw materials.

2. CONFORMATIONS AND CRYSTAL STRUCTURES OF PHENOLS

2.1. CONFORMATIONS OF PHENOL MOLECULES

First we examine the phenol ring ; nomenclature of carbon atoms and angles is shown on Fig. 1.

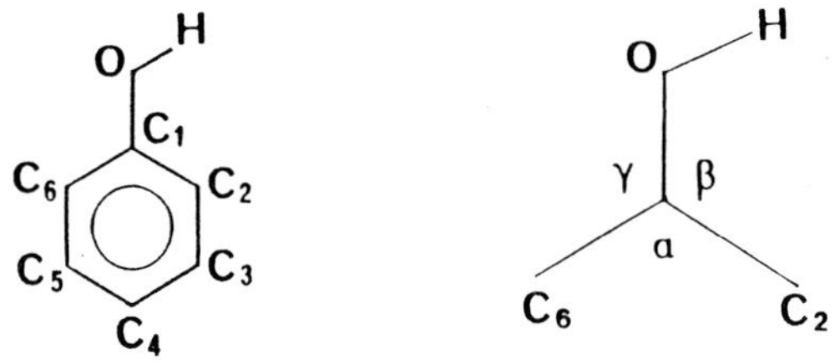

Fig. 1. Nomenclature used for phenol ring.

As reported by different authors (ref. 1, 2, 3, 4, 5, 6, 7) the internal angles of the benzene ring differ from 120°, the values depending on the electronic properties of the substituents ; so angle α, at carbon C_1 bonded to OH group, is generally greater than 120° ; nethertheless we find some values smaller than 120° when there are many substituents on the ring particularly at ortho position (118.4° for tetrachloro-hydroquinone (ref. 8)) and in the case of nitrophenols ; Iwasaki and Kawano (ref. 9) give a comparison of some bond lenghts and angles for these compounds and note for this particular angle the following values : 118.6(4)°, 116.0(5)°, 116.7(4)°, 117.2(5)°, 117.2(5)°, 120.5(4)° respectively for 2,4-dinitrophenol (ref. 9), 2,6-dinitrophenol (ref. 10), 2-chloro-4,6-dinitrophenol (ref. 11), 2-bromo-4,6-dinitrophenol (ref.12), 2-nitro-4-chlorophenol (ref. 13) and β p-nitrophenol, (ref. 14) ; the mean value is 117.1°.

The benzene ring is planar with atomic deviations from the mean plane within experimental errors (between 0.09 Å and 0.005 Å). Only two cases were published showing deviations larger than usual : the maximum and r.m.s. values are 0.069 Å and 0.052 Å respectively for 3,5-diamino-2,4,6-trinitrophenol (ref. 15) ; the second case, 2-methyl-3-bromophenol (ref. 16) shows an hexagon bended along the line $O\text{-}C_1\text{-}C_4$ giving two planes $C_1C_2C_3C_4$ and $C_1C_6C_5C_4$ with an angle of 3.6°.

The inequality of the two C-C-O angles (β and γ) is well known : β on the same side as hydrogen atom is larger than γ ; the explanation is given by Hirshfeld (ref. 17) : the increase of angle is due to repulsions between OH group and H_2 and C_2 atoms. The values for β and γ can be very different as found by Andersen and Andersen (ref. 11) for 2-chloro-4,6-dinitrophenol : 124.8° and 118.5° ; the adjacent nitro group probably increases the repulsion. Table 1 gives some values of β and γ angles for six phenols.

Table 1. Values of β and γ angles for six substituted phenols.

	(1)	(2)	(3)	(4)	(5)	(6)
β	121.8	124.1	123.2	120.7	123.3	121.4
	122.9				121.0	120.5
γ	117.8	116.1	117.4	117.8	117.0	117.5
	116.9				118.9	118.6

(1) : p-cyanophenol (ref. 18) ; (2) : m-nitrophenol (ref. 19) ; (3) : 3,5-diamino-2,4,6-trinitrophenol (ref. 15) ; (4) : 2,5-dimethylphenol (ref. 20) ; (5) : 1,2-dihydroxybenzene (Catechol) (ref. 21) ; (6) : γ hydroquinone (ref. 22).

An other important feature is the lenght of the C-O bond. The normal lenght is 1.368 Å according to Brown (ref. 23) but it is shorter for o-substituted phenols with intramolecular hydrogen bonding ; Kagawa (ref. 24), then Iwasaki (ref. 25) plot the phenolic $C(sp^2)$-O bond lenghts for phenols vs the pK_a values ; the contraction of the C-O bond is possibly related to high conjugation, in o-nitrophenols.

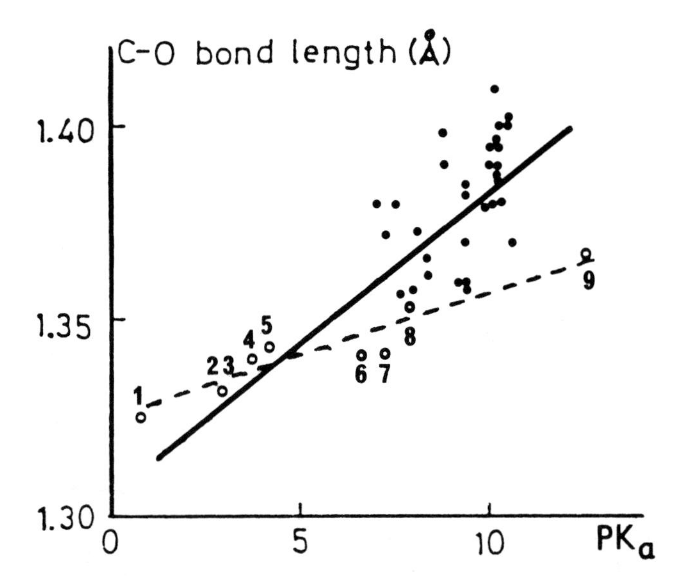

Fig. 2. Diagram showing relationship between C-O bond lenghts vs pK_a values. The solid line and tilled circles are based on the diagram by Kagawa (ref. 24) ; the open circles and broken line represent o-substituted phenols with intramolecular hydrogen bonding from Iwasaki (ref. 25).

Now, we look at positions of different substituents vs phenol ring plane. Distances from mean plane about 10^{-2} Å are found for chlorine atoms in many phenols such as : 0.035 Å in 2-amino-4-chlorophenol (ref. 26), 0.015 Å in 2-chloro-4,6-dinitrophenol (ref. 11), as well as in bisphenols : 0.013 Å and 0.011 Å in 2,2'-methylenebis (4-chloro-3-methyl-6-isopropylphenol) (ref. 27). Pentachlorophenol was studied by Sakurai (ref. 28) who notes the very small deviations of chlorine atoms from the mean ring plane as given in the Table 2.

Table 2. Deviations (Å) of the atoms from the mean plane of the benzene ring in pentachlorophenol (from Sakurai ref. 28).

C1	-0.003	Cl1	-0.043
C2	-0.002	Cl2	-0.037
C3	0.001	Cl3	0.056
C4	0.004	Cl4	0.022
C5	-0.008	Cl5	0.004
C6	0.008	0	-0.020

Carbon of methyl groups, secondary carbon of isopropyl groups and tertiary carbon of tert-butyl groups are also in the benzene ring plane : CH_3 groups of isopropyl and tert-butyl are not symetrically positionned vs the ring as show in many substituted phenols. (ref. 16, 27, 29, 30, 31, 32).

Cyano group in p-cyanophenol deviates significantly from the mean plane as shown on Fig. 3. (ref. 18).

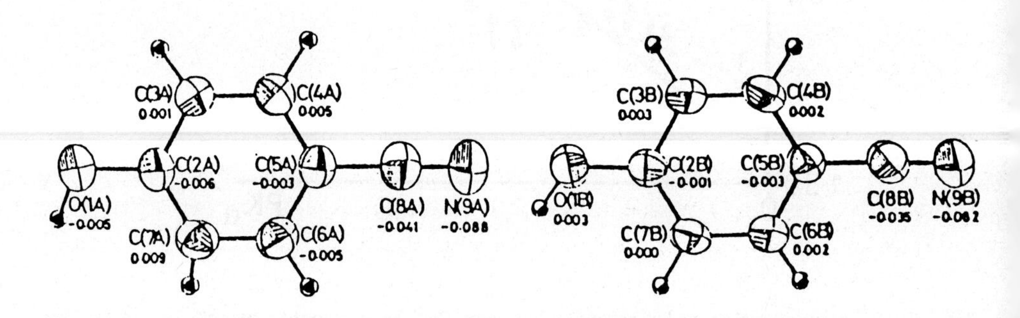

Fig. 3. ORTEP drawings of the two independent molecules of p-cyanophenol showing deviations (Å) from the least squares plane of benzene ring.

Finally a lot of structure determinations were made on phenol with nitro groups. Table 3 gives the dihedral angles (°) between the benzene ring plane and nitro group plane.

Table 3. Dihedral angles (°) between benzene ring plane and nitro group planes.

(1)	(2)	(3)	(4)	(5)	(6)	(7)	(8)
1.58	A 0.65(18)	α1.53	2.58(o)	13.10(o)	3.5	5.06(o)	9.4(o)
	B 0.52(15)	β7.15	4.84(p)	2.7(o')		4.00(p)	5.0(p)
			2.2(o)				
			4.9(p)				

(1) o-nitrophenol (ref. 25) ; (2) m-nitrophenol (A : crystal from melt ; B : crystal from benzene) (ref. 19) ; (3) p-nitrophenol (α and β modifications) (ref. 33, 14) ; (4) 2,4-dinitrophenol (two authors) (ref. 9, 24) ; (5) 2,6-dinitrophenol (ref. 10) ; (6) 2-amino-5-nitrophenol (ref. 34) ; 7) 2-chloro-4,6-dinitrophenol (ref. 11) ; (8) 2-bromo-4,6-dinitrophenol (ref. 12).

The polysubstituted 3,5-diamino-2,4,6-trinitrophenol (ref. 15) is an interesting case. As seen before, ring is not quite planar ; more, with the exception of the phenolic OH, the most in-plane of the six ring substituents, the out of plane deviation of each substituent atom directly linked to the ring is of the same sign but greater magnitude than the associated ring atom. The three nitro groups make angles of 8.30°, 15.50° and 54.50°respectively with the benzene ring. Other structural features of amino and nitro groups are given by Bhattacharjee and Ammon (ref. 15) and by Haisa et al. (ref. 34) and Ashfaquzzaman et al. (ref. 26).

When the substituent is hydroxymethyl at ortho position, OH group of the substituent is out of the phenol ring plane ; however when the two ortho positions are occupied by hydroxymethyl group, one OH is very near the plane and the other out of it as seen on Fig. 4 (ref. 35, 36).

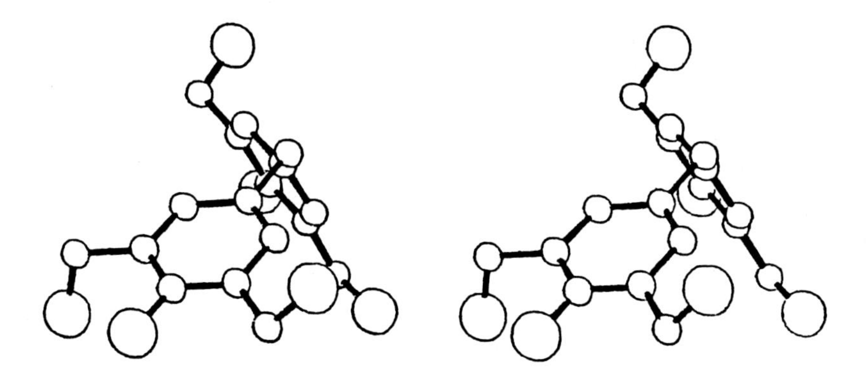

Fig. 4. Stereoscopic view of 3,5,3',5'-tetrahydroxymethyl-4-4'-dihydroxydiphenylmethane.

When substituents are phenyls the question is : are the rings in the same plane or not ? Some compounds were studied : p-phenylphenol (two crystalline modifications) (ref. 37), o-phenylphenol (ref. 38) and 2,6-diphenylphenol (ref. 39). For the first the two rings make a very small angle (about 2°) whereas for the two others angles are 57.0° (o-phenylphenol) and 52° and 44° (2,6-diphenylphenol).

We find phenols with rings linked by N = CH or CH_2 groups ; so p-[(p-methoxybenzylidene) amino] phenol (ref. 40) exhibits the geometry shown on Fig. 5.

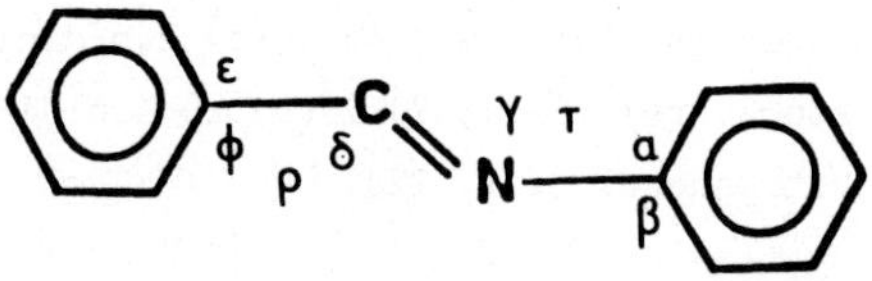

α = 123.7° β = 117.6° γ = 118.4°
δ = 123.7° ε = 118.7° ϕ = 126.0°
Torsion angles : ρ = 0.3°
τ = 41.4°

Fig. 5. Torsional geometry around C_{Ar}-N = C-C_{Ar} in p-[(p-methoxybenzylidene) amino] phenol.

With CH_2 group as linkage, the angle C_{Ar}-CH_2-C_{Ar} is greater than the usual tetraedral angle (≈ 109°) ; Whittaker (ref. 41) in 3,3'-dichloro-4,4'-dihydroxydiphenylmethane and Rantsordas (ref. 27, 31, 32) in a series of 2,2'-methylenebis phenols give values near 119°.

Recently compounds obtained from condensation of para-substituted phenols and formaldehyde have received renewed attention (Fig. 6).

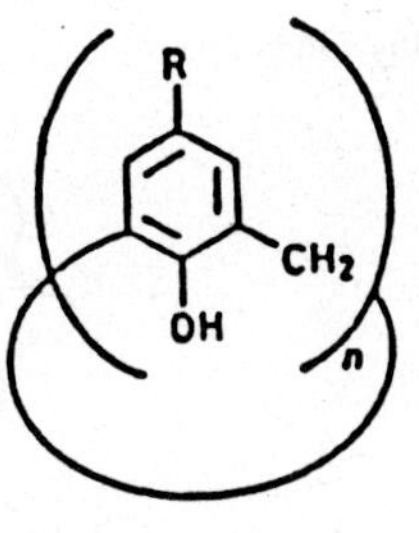

n = 4,8

Fig. 6. General "calixarene" formula

The structure of these compounds called "calixarenes" are now investigated Fig. 7 shows two of them (ref. 42, 43). Angles at the bridging methylene carbon atoms have values of 112.5(5)° (ref. 42) and from 106.2 (5)° 108.2(8)° and 107.6(5)° in the orthorhombic phase to 112.7(3)° and 113.0(3)° in the hexagonal one (ref. 43).

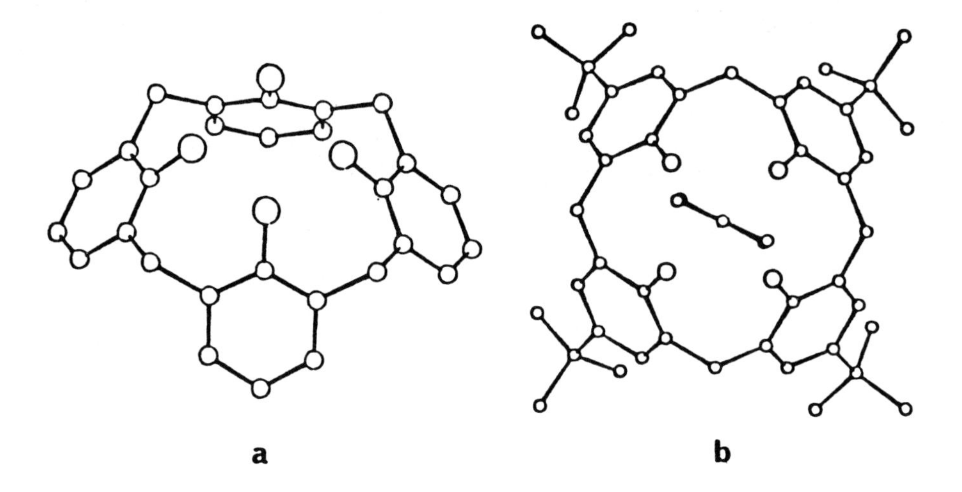

Fig. 7. Calix[4]arene molecule : a) perspective view ; b) view along the fourfold axis with toluene in the cage.

Others features about conformation of such molecules are described in many papers of Andreetti et al. and C.D. Gutsche, A.E. Gutsche and A.I. Karaulov (ref. 44).

In naphtol, the naphtalene ring is planar and C-C bonds have values between 1.36 and 1.45 Å (ref. 45) but in a more complicate compound, 8-benzoyl-5-ethoxy-1-naphtol (ref. 46), the angle of tilt between the least square planes defined by the rings of the naphtol group is 4.0° ; C-C bonds distances range from 1.353 Å to 1.437 Å ; the dihedral angle between plane of phenyl ring and naphtol ring is 86.8°.

The case of a diphenol, α resorcinol, was studied by J. M. Robertson in 1936 (ref. 47) and 1938 (ref. 48). It is difficult to take in account these old results. However G.E. Bacon and R.J. Jude in 1972 (ref. 49) study this molecule by neutron diffraction. The two oxygen atoms lie at 0.11 Å and 0.060 Å. Finally, phloroglucinol (ref. 50) has its benzene ring planar but oxygen atoms deviates significantly from it (+ 0.024 Å, + 0.039 Å, + 0.045 Å).

Now we look at intramolecular hydrogen bonds if they exist. Only phenols with substituents at ortho positions can exhibit such H-bonds.

An interesting case is the intramolecular O-H----π bond between O-H group and π electrons of a phenyl substituent. Spectroscopic studies in CCl_4 solutions showed intramolecular interaction between the hydroxyl group and π electrons of the o-phenyl group.

Two structures were known : 2,6-diphenylphenol (ref. 39) and o-phenylphenol (ref. 38). The geometry of such a bond is given on Fig. 8.

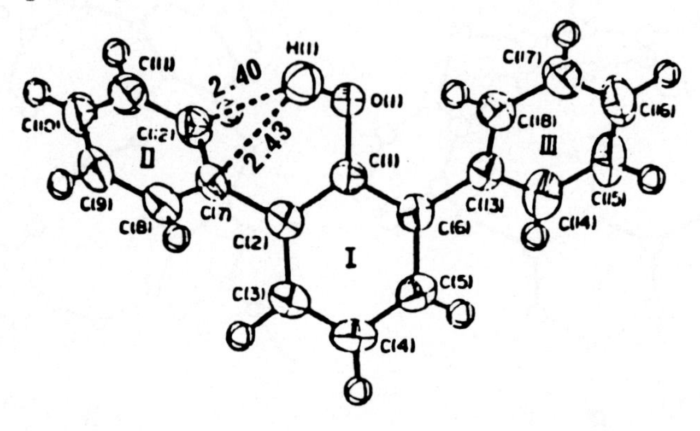

Fig. 8. Structure of 2,6-diphenylphenol viewed perpendicular to ring I.

O_1 is coplanar with ring I, while H_1 slightly protudes from plane I by 0.09 Å and approaches C_7 and C_{12}. The O-H direction is almost parallel to ring II. The distance between H_1 and the midpoint of the bond C_7 C_{12} is 2.31 Å and the angle (O_1 H_1....midpoint) is 122°; thus the O-H group is favorably located to interact equally with the π-electrons on both C_7 and C_{12}.

Intramolecular hydrogen bonds are described in 2,5-diacetylohydroquinone (ref. 51) in agreement with spectroscopic investigations.

An other series of phenols have intramolecular hydrogen bonds : o-nitrophenols. The O....O distances are short : between 2.54 and 2.60 Å. The characteristics around the O....O contacts have been given by Kagawa et al. (ref. 24) for 6 phenols. The numbering scheme is given on Fig. 9 and results on Table 4.

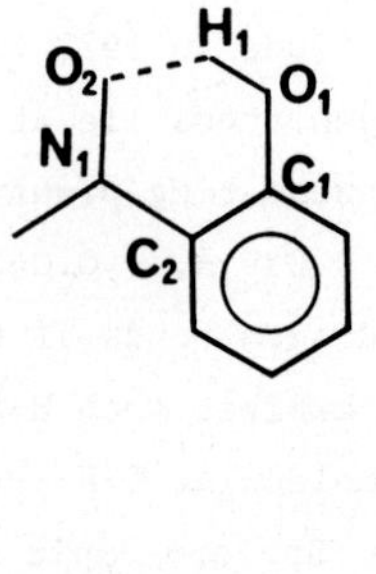

Fig. 9. Geometry around O----O contact for o-nitrophenols.

TABLE 4. The O----O contact (Å), torsion angles τ (°) and bond angles (°) characteristic of the intramolecular hydrogen bonding of o-nitrophenols

		(a)	(b)	(c)	(d)	(e)	(f)
O(1)...O(2)		2.546	2.557	2.558	2.587	2.593	2.599
τ[H(1)-O(1)-C(1)-C(2)]	1.9	-4.2	13.6	-5.3	-13.5	-5.7	
τ[O(1)-C(1)-C(2)-N(1)]	-2.3	0.9	-1.7	-0.1	-0.4	0.2	
τ[C(1)-C(2)-N(1)-O(2)]	3.6	0.8	-4.7	-0.2	3.2	8.6	
C(2)-C(1)-O(1)		125.0	123.8	124.8	125.5	124.9	125.5
N(1)-C(2)-C(1)		119.7	120.9	120.1	120.6	121.8	120.7

(a) π-Molecular compound anthracene-picric acid (ref. 52). (b) 2,6-dinitrophenol(ref. 10). (c) 2-chloro-4,6-dinitrophenol (ref. 11) (d) 2-nitro-4-chlorophenol (ref. 13). (e) 2,4-dinitrophenol (ref. 24). (f) 2-bromo-4,6-dinitrophenol (ref. 12).

H atom is displaced far from the line joining the two O atoms forming a six membered chelate ring. The angle O_1-H_1....O_2 is 143°, 138° and 152° for o-nitrophenol, 2,4-dinitrophenol and 2,6-dinitrophenol respectively.

For 3,5-diamino-2,4,6-trinitrophenol (ref. 15), the O----O distance between hydroxyl group and the adjacent nitro group is 2.48 Å indicating a very strong hydrogen bond ; further the four amino H atoms form strong intramolecular contacts to the O atoms of adjacent nitro groups. 2-hydroxy-3-methoxybenzaldehyde (o-vanillin) (ref. 53) is a molecule with 2 ortho substituents OCH_3 and CHO ; a strong intramolecular hydrogen bond O----O, 2.603 Å, is found between hydroxyl and aldehydic groups.

2.2 CRYSTAL STRUCTURES OF PHENOLS

Phenols crystallise in different systems ; we note for fifty of them : 1 triclinic, 29 monoclinic, 14 orthorhombic and 6 of higher symmetry (trigonal or tetragonal or hexagonal). Twenty of them are non-centrosymmetric, fourty one have a P mode. If we look at the symmetry groups, fourtheen are $P2_1/c$, heigh are $P2_12_12_1$ and 5 are C2/c.

In general phenol molecules are held together by hydrogen bonds of different types :

O-H----O
O-H----N
O-H----Cl

A classification is suggested by the number of bonded molecules : none, two (dimers) three (trimers)... infinite (chains or sheets or tridimensional nets). By analogy with silicates structures, A. Thozet (ref. 54) proposes the next classification :

- neophenols for non H-bonded molecules
- sorophenols for a finite number of H-bonded molecules
- inophenols for infinite chains of H-bonded molecules
- phylophenols for sheets
- tectophenols for a tridimensional net

2.2.1 Neophenols

A few number of phenols do not exhibit intermolecular H-bonds ; bulky substituents, such as tert-butyl or phenyl close to hydroxyl groups, seem to prevent the formation of hydrogen bonds. In 1959, Aihara (ref. 55) estimated the energy of hydrogen bonds for phenols and found for o-phenylphenol, 2-tert-butyl-4-methylphenol, energies near 0 $Kcal.mole^{-1}$. The structure determinations are in good agreement with these results : 2-tert-butyl-4-methyl phenol shows molecules along axis parallel to "c" but the O....O distances are greater than 4 Å as in o-phenylphenol (ref. 38). o-vanillin (ref. 53) does not exhibit intermolecular H-bond ; the shortest contact between molecules is 3.14 Å between oxygen of aldehydic group and carbon of methoxy group.

2.2.2. Sorophenols

These structures have a finite number of hydrogen-bonded molecules giving dimers, tetramers and hexamers. These "supra molecules" are held together in the crystal by Van der Waals forces.

Among different bisphenols molecules, we note that 2,2'-methylenebis (4-chloro-5-isopropyl-3-methylphenol) (ref. 56) shows dimer between two independant molecules.

4-methylphenol (p-cresol) and 4-chlorophenol are isomorphous if we look at stable form of the first and metastable of the second (ref. 57, 58) ; the phase diagram of these two compounds showing the existence of three cristalline modifications called α, β, γ as well as parameters of the cells are given later (Table 6, Fig. 29). In the case of β form,two independent molecules are held together by intermolecular hydrogen bonds ; these two and their symmetric by a center give a tetramer ; along the short "a" axis tetramers form piles held together by Van der Waals forces as shown on Fig. 10. The same situation is found for 5-methyl-4-chloro-2 isopropylphenol (ref. 59).

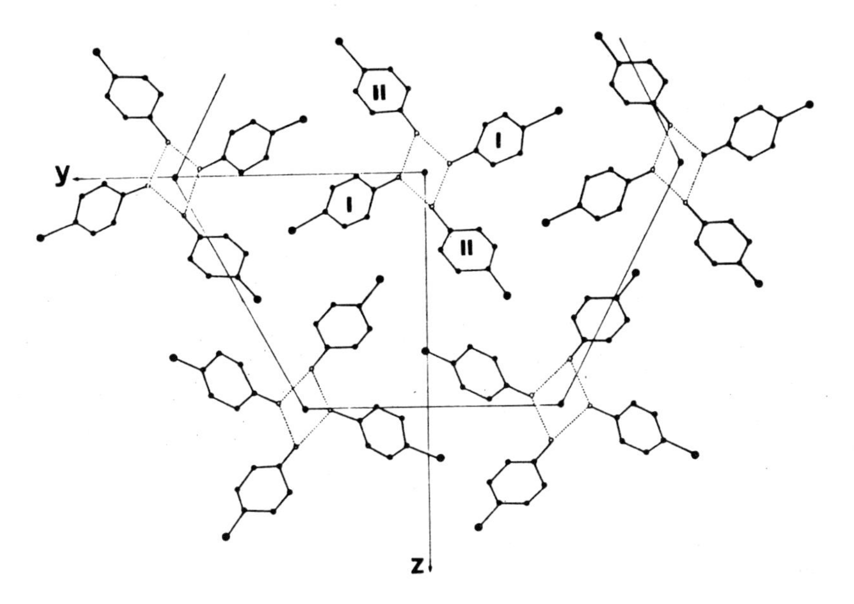

Fig. 10. Projection of the crystal structure of the β-form of p-chlorophenol on the bc plane.

For 2-isopropyl-3-methylphenol (ref. 60) and 2-isopropyl-3-methyl-4-chlorophenol (ref. 59), four independant molecules give a tetramer.

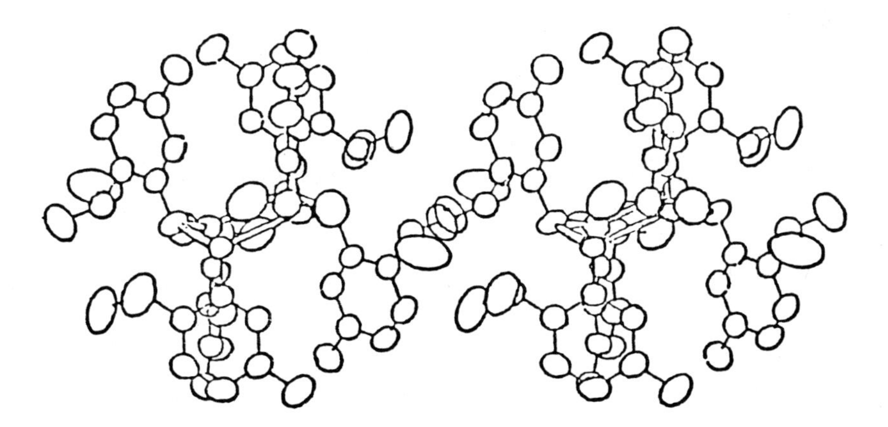

Fig. 11. Stereoscopic view of a hexamer in the crystal structure of 3-methyl-6-isopropylphenol or thymol.

The case of 3-methyl-6-isopropylphenol (ref. 30) is illustrated on Fig. 11. Molecules are linked by six hydrogen bonds related by a $\bar{3}$ axis, giving an hexamer ; the six O atoms are at vertices of a ring in the chair conformation ; the centers of the hexamers are repeated at the vertices of a rhomboedra. The arrangement with hexamers has also been found for 3,4-dimethylphenol (ref. 61).

2.2.3. Inophenols

In general, hydrogen-bonded molecules form infinite chains. Interaction can be between OH groups or depending of the substituents, between OH and NO_2 or NH_2 groups showing frequently head-to-tail associations. Illustration of the last is given by α p-nitrophenol (ref. 33) ; hydrogen bonds are established between oxygen atom of phenol group and one oxygen atom of nitro group : OH----O distance is 2.818 Å ; infinite chains of glide-plane related molecules are held together by Van der Waals contacts. The β form (ref. 14) shows the same sort of infinite chains (OH----O = 2.84 Å) ; the angle between the benzene ring planes in adjacent molecules is 74°2' in the α form and 28°0' in the β form ; the shortest intermolecular contacts are very similar in the two modifications and discussed in the paper of Coppens for explanation of the thermodynamic relationship and photochemical behaviour of the two forms (Fig. 12).

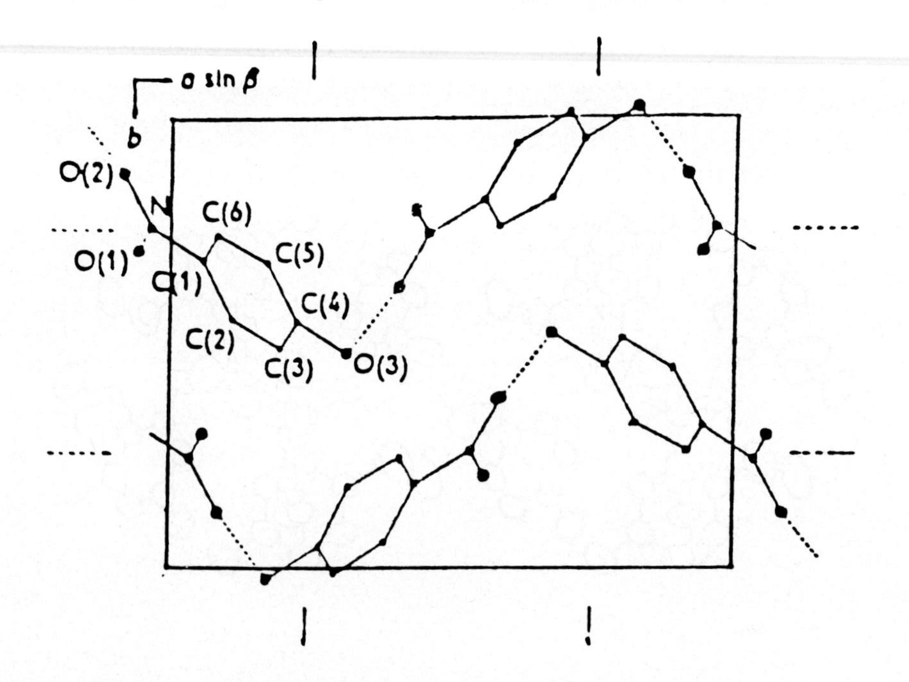

Fig. 12. Chains of H-bonded molecules in p-nitrophenol (α form).

An other example of head-to-tail alternatively joined molecules is the structure of p-cyanophenol (ref. 18) ; two non equivalent molecules are hydrogen-bonded and form infinite helical chains along the 2_1 axis parallel to "c" ; the two O-H----N bonds with distances 2.82 Å and 2.84 Å have similar geometrical environments.

4-isopropylidene-aminophenol (ref. 62) as shown on Fig. 13 shows a bond between the nitrogen atom of one molecule and oxygen atom of a molecule related to the first by a 2_1 axis parallel to "b" ; physical properties, namely good cleavage perpendicular to the "a" axis and positive birefringence with the vibration direction of greatest refractive index along "b", are in accordance with the structure.

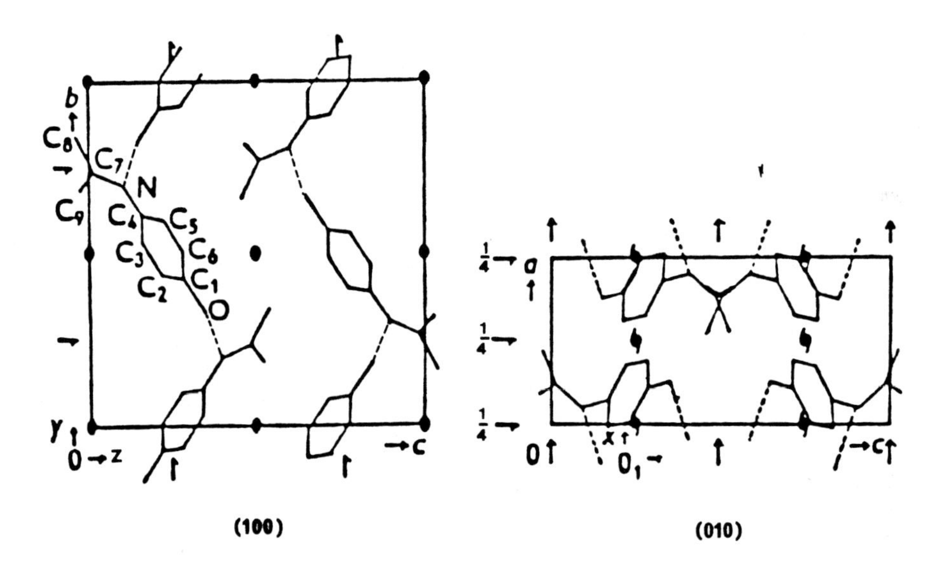

Fig. 13. H-bond on two projections of the structure of 4-isopropylidene-aminophenol.

Bonds are often between OH groups, giving chains running along helicoïdal axis. Among phenol compounds, we note the cases of phloroglucinol (ref. 50), 3,5-dichlorophenol (ref. 63), 2,3 dimethylphenol, (ref. 16, 20) 2,5 dimethylphenol (ref. 20) 2,6-dimethylphenol (ref. 69), 2,5 dichlorophenol (ref. 65), 8-benzoyl-5-ethoxy-1-naphtol (ref. 64) for which helicoïdal axis is a 2_1 axis ; for 2-3 dichlorophenol (ref. 66) three independant molecules form chains along the 3_1 axis (see Fig. 14) ; for 3,4-dichlorophenol (ref. 67), 4-bromophenol (ref. 68), 4-isopropylphenol (ref. 29) and 3-methyl-4 isopropylphenol (ref. 29) helicoïdal axis is 4_1 (see Fig.15).

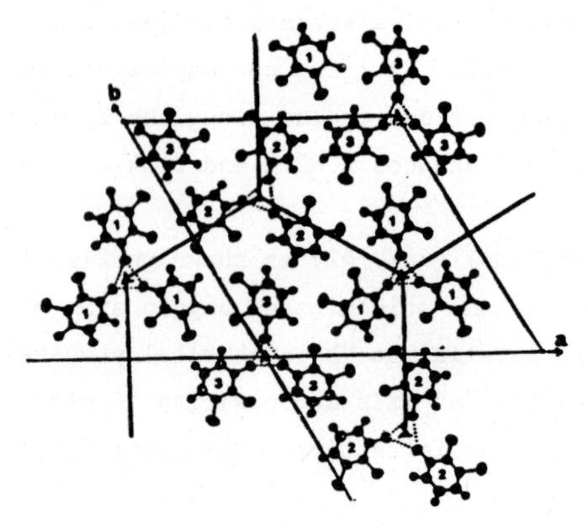

Fig. 14. Projection down "c" axis of the structure of 2,3-dichlorophenol.

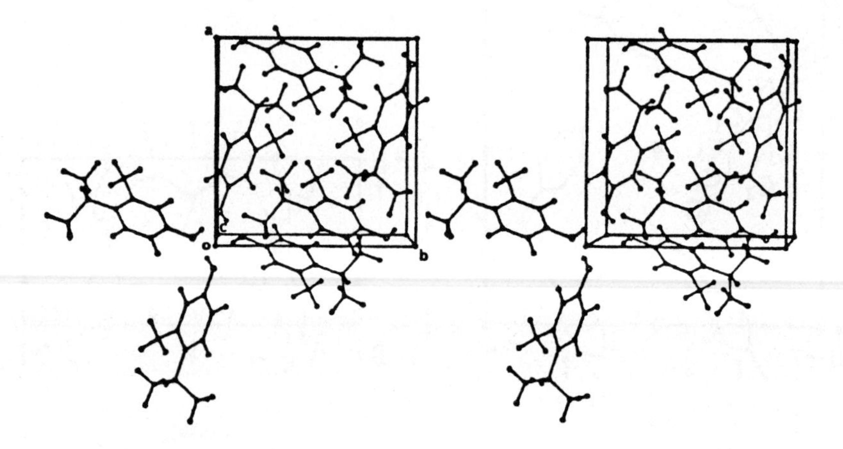

Fig. 15. Stereoscopic view of the structure of 3-methyl-4-isopropylphenol.

The two last compounds crystallise in $P4_1$ space group, consequently with a polar axis ; this fact will be use later for reactivity studies on single crystals (see 3.6.2). We note too that these two compounds have isomorphous structures as well as 2,6-dimethylphenol and 3,5-dichlorophenol.

Chains can be parallel to a non-helicoïdal axis ; it is the case of 4-methylphenol (p-cresol) (metastable form γ) (ref. 69)) : the three independant molecules are linked by 2.63 Å, 2.65 Å, 2.67 Å bonds into infinite chains parallel to the "c" axis (group C2/c) ; two chains related by a center of symmetry are linked by Van der Waals forces, as shown on the projection (see Fig. 16). The situation is very similar for 2,4-dichlorophenol (ref. 70).

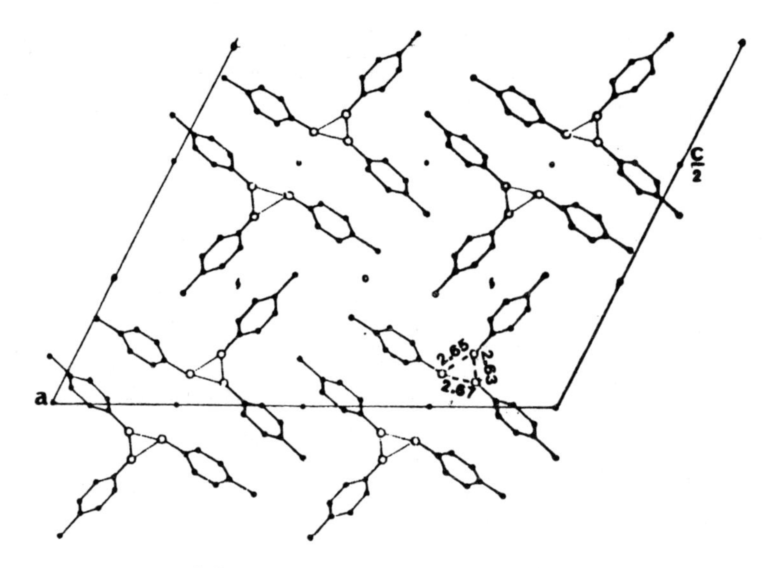

Fig. 16. Projection down "b" axis of γ form of p-cresol.

The case of phenol molecule is particular (ref. 71, 72) : three independant molecules are linked by hydrogen bonds around a pseudo 3 axis ; this compound was studied in orthorhombic system ($P2_122_1$) and in monoclinic one ($P2_1$) and the authors thought that crystals are twinned.

Sometimes a glide plane permits existence of hydrogen-bonded chains as seen in α p-nitrophenol previously. We give here an other example : o-nitrophenol (ref. 25) ; the molecules related by the a-glide symmetry are stacked to form a column along the "a" axis ; the overlapping of the aromatic ring is considerable (see Fig. 17).

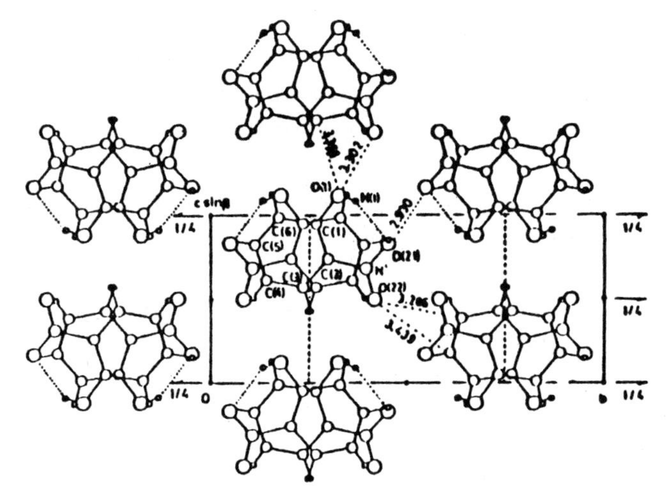

Fig. 17. Projection of the structure of o-nitrophenol along the "a" axis.

Infinite interlocking zig-zag chains are found for p-[(p-Methoxybenzylidene)amino] phenol (ref. 40) ; the terminal hydroxy groups act as donors and the central nitrogen atoms as acceptors with O-H----N = 2.756 Å (see Fig. 18).

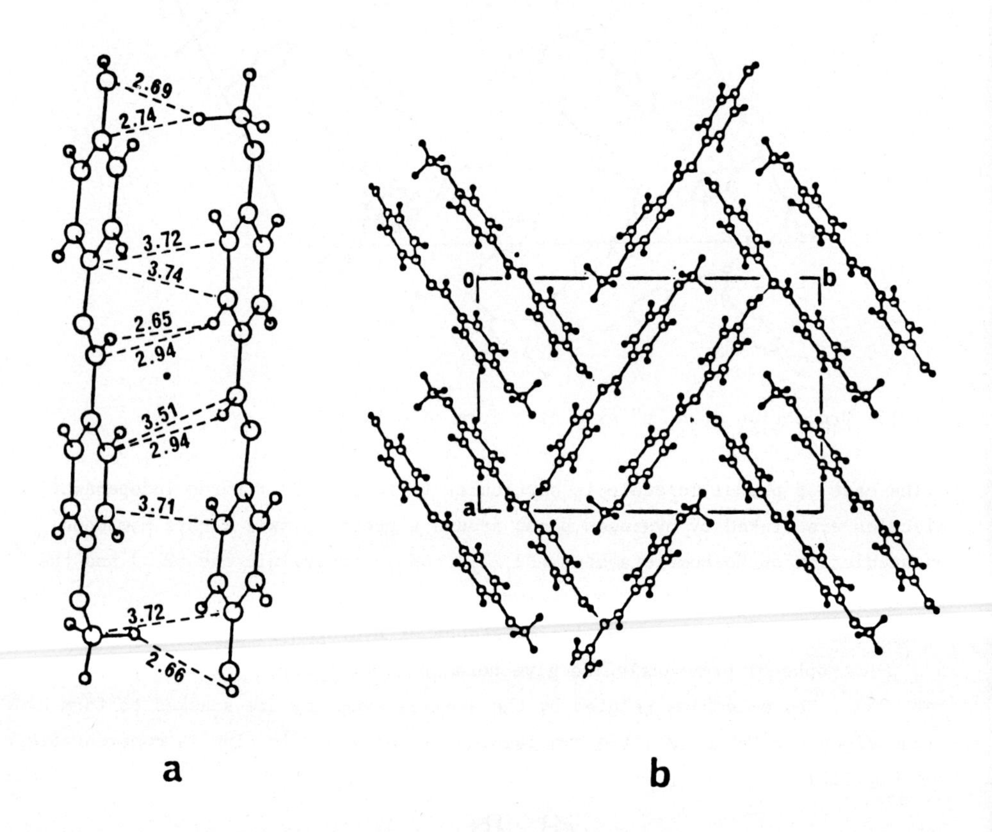

Fig.18 a). Intermolecular contacts (Å) between centrosymmetrically related pair of p-[(p-Methoxybenzylidene)amino] phenol molecules ; b) view of the molecular packing in the crystal seen in projection down "c" axis.

The authors show for this structure the unsuitability of the packing to act as a precursor for mesophase formation.

Several authors speak about "bifurcated hydrogen bond" ; G. Albrecht (ref. 73), then Sakurai, (ref. 8, 28) study examples of such bonds. The geometry of these bonds is described by C. Bavoux (ref. 74). Among phenol molecules we give three examples ; tetrachloro-hydroquinone (ref. 8), pentachlorophenol (ref. 28), 2,4-dinitrophenol (ref. 9). The first example shows two intermolecular bonds : O-H----O = 2.92 Å and O....Cl = 3.92 Å and the position of the hydrogen atom suggests a bifurcated hydrogen bond system (see Fig. 19). The same situation exists for the other examples.

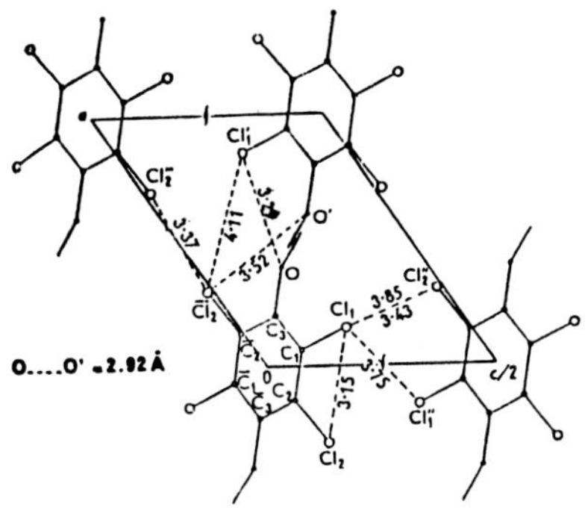

Fig. 19. Bifurcated intermolecular bond in tetrachloro-hydroquinone.

2.2.4. Phylophenols

Two studies were made on 2,4-dinitrophenol. Iwasaki et al. (ref. 9) speak about the bifurcated intra and inter hydrogen-bonding system whereas Kagawa et al. (ref. 24) describe the crystal structure as sheets parallel to (010) and show that the interaction cannot be a bifurcated hydrogen-bond.

2,6-dinitrophenol (ref. 10) gives intermolecular contacts as shown on Fig. 20. The shorter bonds N(1)....O(4) : 2.853 Å and O(3)....O(5) : 2.900 Å give a sheet of molecules ; in the "c" direction, intermolecular contacts are longer.

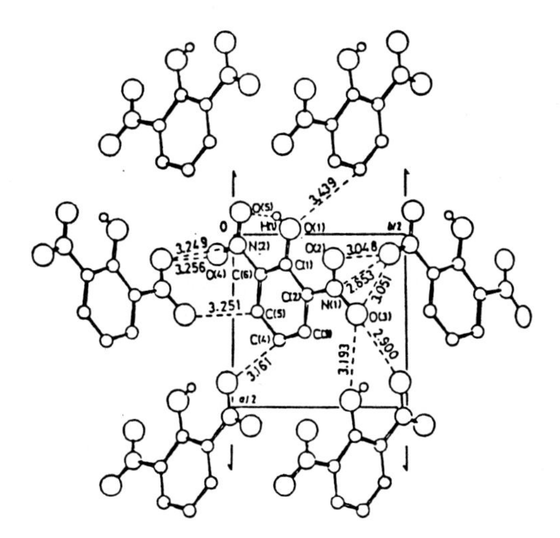

Fig. 20. Projection along the "c" axis of the structure of 2,6-dinitrophenol.

The same situation is found for 3,5-diamino-2,4,6-trinitrophenol (ref. 15) : a lot of intra and intermolecular contacts form layers in which the normal to the benzene ring is inclined by 44° from "a". Along "a" molecules are separated by the cell repeat (a = 4.9653(5)) and have no C....C contacts less than 3.6 Å.

2-amino and 2-amino-4-chlorophenol (ref. 26) have OH....N bonds forming chains along "a" crystallographic axis. Then other contacts form sheets explaining the cleavage of the crystals.

p-amino-salicylic acid (ref. 75) shows dimeric molecules held together by hydrogen bridges between carboxyl groups as in salicylic acid (ref. 76, 77). The other intermolecular distances are of Van der Waals type, except the distances between the phenolic hydroxyl and the amino group and between two amino groups which seem to indicate the presence of very weak O----N and N----N hydrogen bonds.

Recently Perrin et al. (ref. 35) describe the structure of two precursors of phenolic resins. One of them, 2,4,6-trihydroxymethylphenol, has a lamellar structure : hydrogen bonds form a double plane of molecules almost parallel to (301) ; between double plane there are only Van der Waals contacts (see Fig. 21).

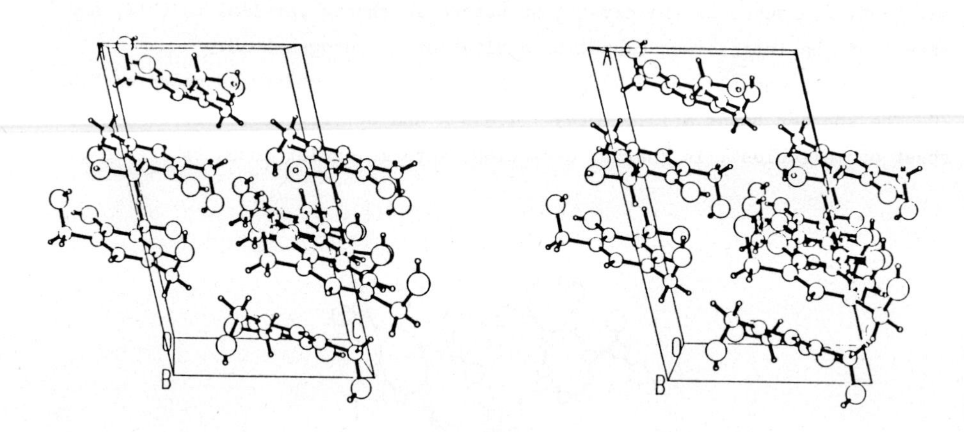

Fig. 21. Stereoscopic view of 2,4,6-trihydroxymethylphenol.

2.2.5. Tectophenols

These phenols show a tridimensional net of hydrogen bonds. First we speak about 3,4-dichlorophenol (ref. 67) which forms a tridimensional net because of the presence of Cl....Cl interactions (Cl....Cl = 3.27 Å, Cl....Cl = 3.48 Å) ; the structure consists of infinite chains of H-bonded molecules along a 4_1 axis ; then these chains are bonded by Cl....Cl bond (see Fig.22) ; two neighbouring chains are antiparallel.

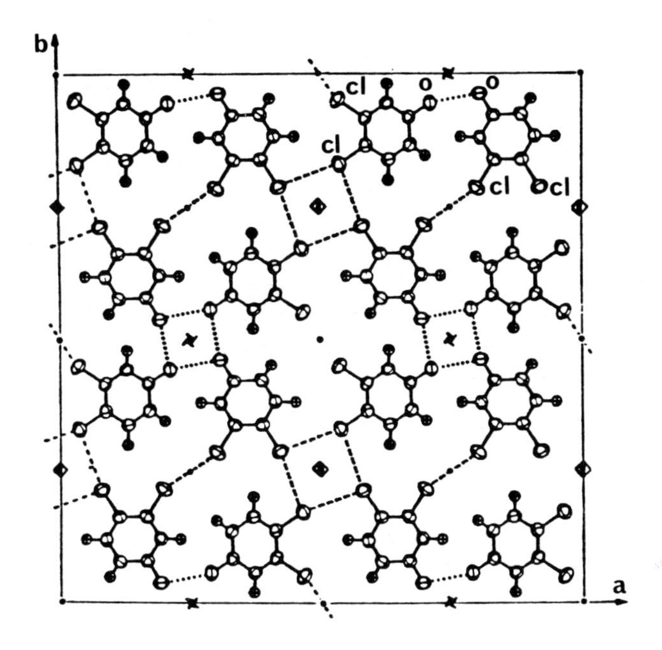

Fig. 22. Projection on ab plane of the structure of 3,4-dichlorophenol.

When there are many OH groups on the molecule, the number of H-bonds is increased ; so a net of H-bonded molecules is built. Two cases are shown here : 3,5, 3',5'-tetrahydroxymethyl-4,4'-dihydroxydiphenylmethane (ref. 35) and phloroglucinol (ref. 50). For the first case, authors give a table of the twelve H-bonds between molecules in the crystal. The great number of bonds explains the high density (1.46 $Mg.m^{-3}$) and the stability of such a phenol-alcohol compound. The second case is a triphenol molecule ; the authors find infinite helices around screw axis parallel to "b" ; however the projection of the structure (see Fig. 23) and the list of bonds given in the paper suggestes a tridimensional net of molecules.

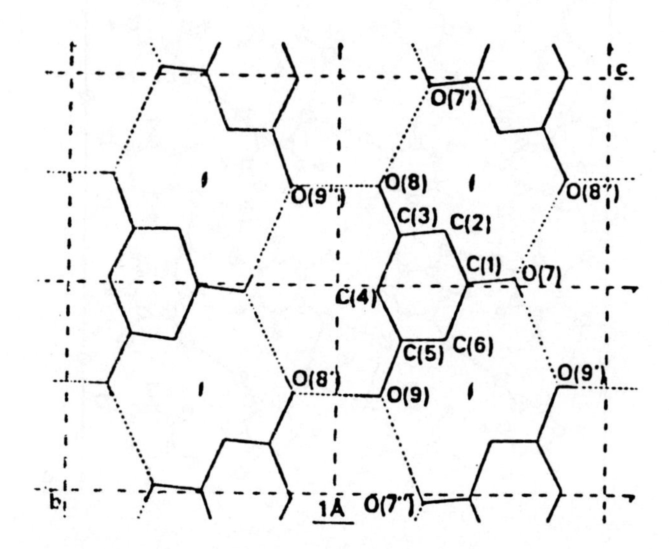

Fig. 23. Projection of the structure of phloroglucinol down the "a" axis.

2.2.6. Complexes with phenol molecules

In suitable circonstances two or more different compounds may crystallize from melt or solution giving a single new phase. In general there is some action between the different partners : hydrogen bonds, donor-acceptor interactions.... A complete study of this last kind of molecular compounds was made by Herbstein as a chapter in the book of Dunitz (ref. 78). So here we shall give only some examples of hydrogen bonded complexes.

Bavoux et al. (ref. 79) study 1 : 1 molecular compound between 3,5-dichlorophenol and 2,6-dimethylphenol. The two pure compounds crystallize in $P2_1/c$ symmetry group whereas the molecular compound in $P2_1$ acentric group. Arrangements are very similar for pure substances and complex : molecules are linked by hydrogen bonds giving infinite chains of alternate molecules along the 2_1 axis (see Fig. 24).

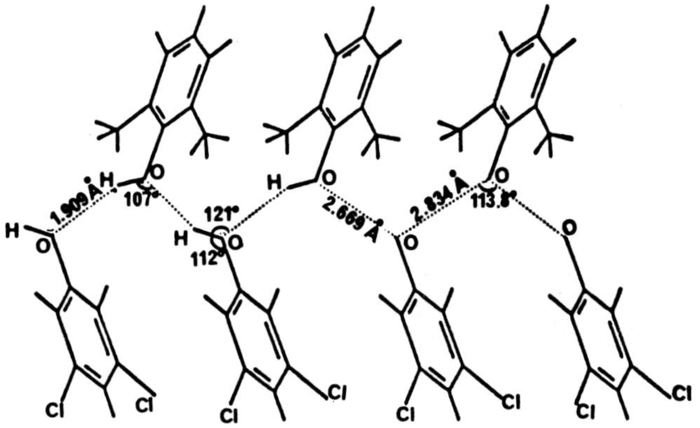

Fig. 24. Chains of molecules for 1 : 1 molecular complex between 3,5-dichlorophenol and 2,6-dimethylphenol.

Hexamethylenetetramine (HMT) forms a variety of crystalline complexes with phenol molecules. The X-ray analysis of one of them is given by Mak et al. (ref. 80). The complex consists of 1 molecule of HMT on a site of symmetry 2 and 2 molecules of m-cresol. N----H-O bonds link the different molecules as seen on Fig. 25 giving a tridimensional network.

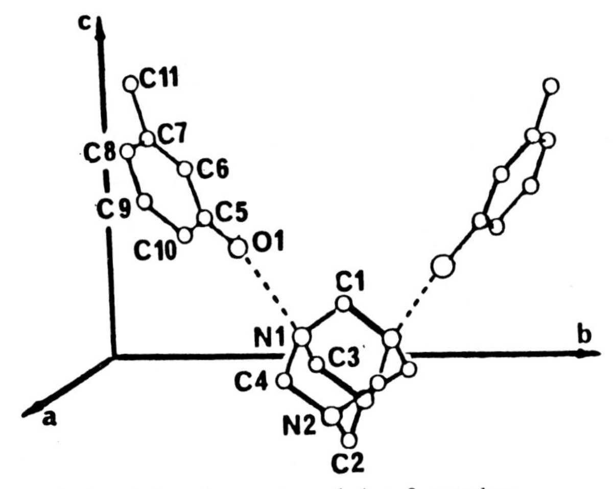

Fig. 25. Hexamethylenetetramine-m-cresol 1 : 2 complex.

Other tridimensional networks are found for complexes with p-nitrophenol. So it is the case for 1 : 1 complex of 2-methylthio-6-benzamidopurine and p-nitrophenol (ref. 81). For 1 : 1 complex of theophylline and p-nitrophenol (ref. 82), the theophylline molecule is linked to p-nitrophenol molecule through a hydrogen bond between carbonyl oxygen and OH group of p-nitrophenol as shown on Fig. 26.

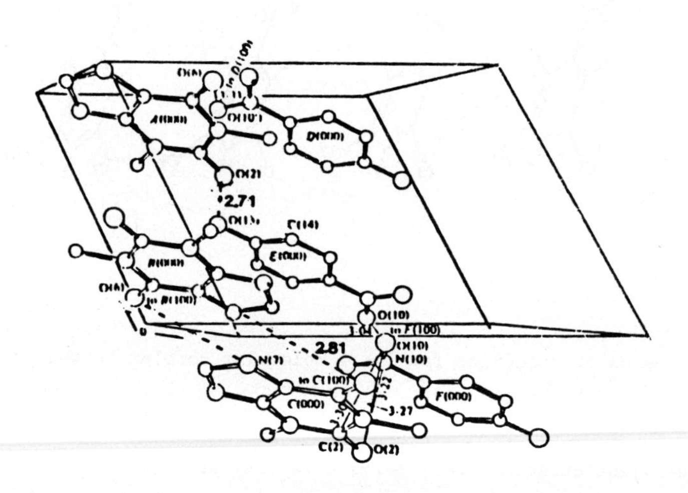

Fig. 26. Crystal packing viewed down "a" axis of 1 : 1 complex of theophylline and p-nitrophenol.

Gramstad et al. study complexes with pentafluorophenol and proton acceptors such as ethers, ketones, and phosphoryl compounds. The problem is to ascertain whether the association takes place by formation of a H-bonded complex or proton-transfer complex ; by example for pentafluorophenol and triphenylphosphine oxyde (ref. 83) is the bond can be written P = O----H-O or P = OH----O ?. The conclusion of the authors is the existence of strong hydrogen bond with the geometry given on Fig. 27.

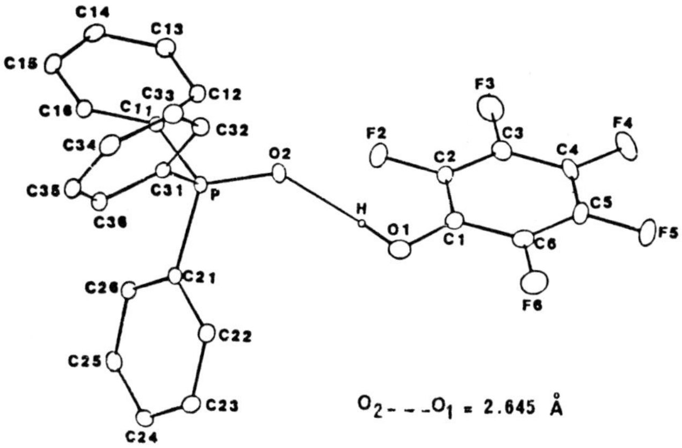

Fig. 27. The hydrogen-bonded complex between pentafluorophenol and triphenylphosphine oxyde. P, O_2 and O_1 are in the plane of the paper.

Diphenols as resorcinol can be a partner in complexes . So 1 : 1 progesterone-resorcinol complex (ref. 84) shows hydrogen bonds giving chains of alternate molecules along the "b" axis.

An other interesting example is the 1 : 1 complex of 2,9-dimethyl-1,10-phenanthroline and resorcinol (ref. 85) where parallel sheets of 2,9-dimethyl-1,10-phenanthroline molecules are almost perpendicular to parallel sheets of resorcinol molecules. The hydroxyl H of one resorcinol binds in a bifurcated manner to the two N atoms from above while the second hydroxyl H binds from below to the two N atoms of another molecule.

To close the chapter about complexes, it is necessary to speak of complexes with molecules such as 1,4,7,10,13,16 hexaoxocyclooctadecane (18-crown-6) and calixarenes.

C. Bavoux et al. (ref. 74, 86) study supramolecules between 18-crown-6 and on one hand 3,5-dichlorophenol and on other hand 2-isopropyl-3-methylphenol. Other studies were made with 3-nitrophenol (ref. 87) and with 4,4'-bis(phenol) (ref. 88). In each case the crystal is built with molecules of water and there are several H-bonds between water, hydroxyl group of phenol and oxygen atom of the crown.Fig. 28 gives one example of such a compound.

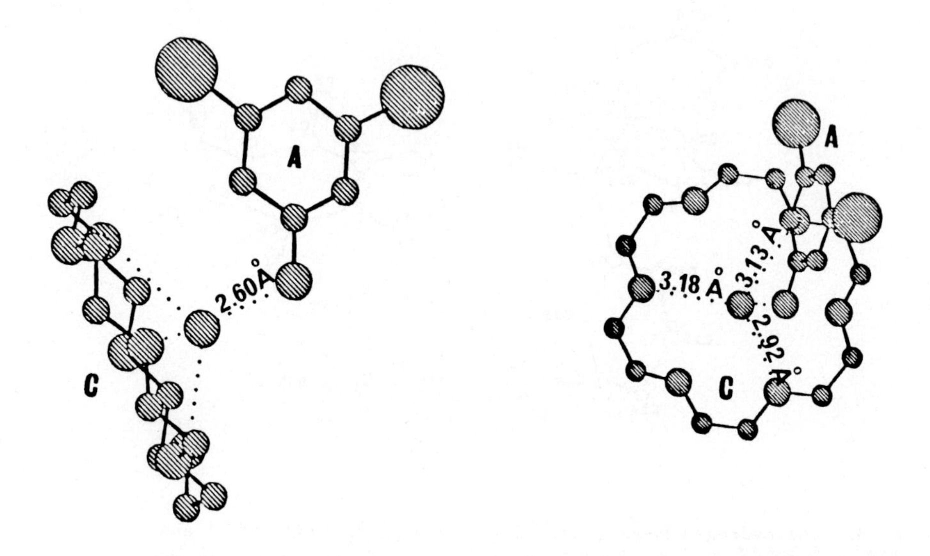

Fig. 28. Complexe between 18-crown-6, 3,5-dichlorophenol and water.

The case of calixarenes is different ; many of them retain the solvent from which they are crystallized. Many exemples are given in the book of C.D. Gutsche (ref. 89). Andreetti et al. give the structure of p-tert-butyl [4]arene with toluene (ref. 42) ; in our laboratory we crystallize the p-isopropyl [4]arene with chloroform ; the two structures are isomorphous (P4/n symmetry group) and the solvent molecule is in disorder position because of the symmetry 4 ; tert-butyl is also disordered but not the isopropyl group.

The CH_3---π interactions between methyl groups and aromatic ring of the guest molecule seem to built the crystal of tert-butyl derivative ; probably chlorine atoms interact in the isopropyl derivative. Actually a lot a such compounds are studied.

3. THE IMPLICATIONS OF THESE STRUCTURES DURING REACTIONS CARRIED OUT IN THE SOLID STATE

3.1. INTRODUCTION

In this part are described the solid state reaction of phenols. These included solid state polymorphic transformations, thermal reactions, polycondensation, solid-solid, solid-liquid, and solid-gas reactions.

The solid state reactions may be classified as follows (ref. 90) :

a) Solid transformations, e.g.

$A(s) \rightarrow B(s)$

$A(s) \rightarrow B(g) + C(s)$

$A(s) \rightarrow B(s) + C(s)$

b) Solid-solid reactions, e.g.

$A(s) + B(s) \rightarrow C(s)$

$A(s) + BC(s) \rightarrow AB(s) + C(s)$

c) Solid-gas reactions, e.g.

$A(s) + B(g) \rightarrow C(s)$

$A(s) + B(g) \rightarrow C(l)$

$A(s) + B(g) \rightarrow C(s) + D(g)$

Relatively few reactions of organic solid with liquids have been studied. For these reactions and for solid gas reactions some authors argue that the reaction occured in a fluid phase existing at the surface. Some convincing criteria are, therefore, necessary for proving that any observed reaction takes place in the solid rather than the liquid or gaseous state.

Morawetz (ref. 91) has suggested criteria to establish the existence of true solid state reaction :

1. A reaction occurs in the solid state, when the liquid reagent is either totally unreactive or reacts at a much slower rate.

2. Pronounced differences, which cannot be accounted for by known chemical principles, are found in the reactivity of crystals of closely related compounds.

3. Different reaction products are obtained when the reaction is carried out in the solid and the liquid state.

4. The same reagents in different crystal modifications has a different reactivity or leads to different reaction products.

5. Reaction product are formed whose molecules have a preferred orientation in a crystallographic direction of the crystal of the parent reagent (topotaxy) (ref. 92, 93).

Paul and Curtin (ref. 94) have added an important criterion : a reaction occurs in the solid if it occurs at a temperature below the eutectic point of a mixture of the starting material and products.

When it has been established that the reactions are occuring in the solid state, the solid state reactions of phenols can be viewed in terms of a four stages process. Paul and Curtin (ref.94) have proposed in the case of solid state reactions induced thermally the following four stages :

1. Loosening of molecules at the reaction sites

2. Molecular change

3. Solid-solution formation
4. Separation of product

Molecular loosening and molecular change are the most important steps of this process ; they are the rate determining steps for most solid-state reactions like solid-gas reactions and thermal reactions.

Polymorphic transformation converts one crystal form to another, during the transformation the molecular structure is unacted. Polymorphic transformations thus do not involve a molecular change but involve three steps : molecular loosening, solid solution formation and separation of product.

3.2. POLYMORPHISM OF PHENOLS

Many phenols exist in more than one crystalline forms (Table 5). The existence of polymorphism is best established by X-ray crystallographic examination. Polymorphs can differ in their melting point, density, hardness, crystal shape and reactivity. Two types of polymorphism are encountered in nature, reversible called enantiomorphism and irreversible called monotropism (ref.95). Enantiotropic polymorphs can be interconverted below the melting point of either polymorph, while monotropic polymorphs cannot.

3.2.1. Polymorphism of 4-chlorophenol

Polymorphism of 4-halogenophenols has been well studied (ref. 96, 97). Perrin and Michel (ref. 98) have determined the crystal structure of both the stable (α) and unstable (β) forms of 4-chlorophenol and studied the behaviour of these polymorphs. When 4-chlorophenol is melting it easily remains in the liquid state for long period even if the temperature is lower than the melting point. If the liquid is in a glass sealed tube the crystallization can be induced at a low temperature. Such a cristallization gives transparent needles. After a variable time the crystals become opaque, each needle becomes a collection of micro-crystals. A transformation occurs giving more stable form. The β form is less stable than the α form. The transformation from metastable form to stable one can be observed at all temperature below the melting point. It is easy to obtain directly the stable form as transparent crystals. On the other hand it is not possible to transform the stable form into the metastable one.

Molecules of 4-chlorophenol and 4-methylphenol have a similar shape and volume. Metastable 4-chlorophenol and stable 4-methylphenol are similar and their space groups are the same $P2_1/c$ (see Table 6). A very interesting study of the phase diagram has been done by Perrin and al (ref. 97). Mixtures of the two products were made and for each mixture the solidus and liquidus values were determined, Fig.29. Three crystals forms, labeled α, β and γ are possible for the two compounds. The crystal structures of the three forms were determined by X-ray diffraction ; they correspond to three arrangements of molecules in

Table 5. Phenols with more than one crystalline form.

Compound	Structure	References
4-Chlorophenol	OH, Cl	(Ref. 96, 97, 99)
4-Bromophenol	OH, Br	(Ref. 96, 97)
4-Iodophenol	OH, I	(Ref. 96, 97)
4-Fluorophenol	OH, F	(Ref. 101, 102)
4-Methylphenol	OH, CH_3	(Ref. 96, 97, 98, 99)
4-Isopropylphenol	OH, $CH(CH_3)_2$	(Ref. 96)
4-Nitrophenol	OH, NO_2	(Ref. 14, 33, 100)
Resorcinol	OH, OH	(Ref. 103)
4-Hydroxybiphenol	OH	(Ref. 105)

Table 6. Cell parameters.

	metastable form	stable form
OH	a = 4.14 (1) Å	a = 8.841 (3) Å
	b = 12.85 (2) Å	b = 15.726 (2) Å
	c = 23.20 (3) Å	c = 8.790 (2) Å
	β = 93°.0 (5)	β = 92°.61 (2)
Cl	Z = 8	Z = 8
	$P2_1/c$	$P2_1/c$
4-chlorophenol (1)	F = 34.85°C	F = 43.45°C
4-methylphenol	a = 26.412 (5) Å	a = 5.72
OH	b = 8.001 (2) Å	b = 11.74
	c = 27.842 (5) Å	c = 18.68
	β = 117°.61 (2) Å	β = 98°.0
	Z = 24	Z = 8
	C2/c	$P2_1/c$
(2) CH_3	F = 31.9°C	F = 34.7°C

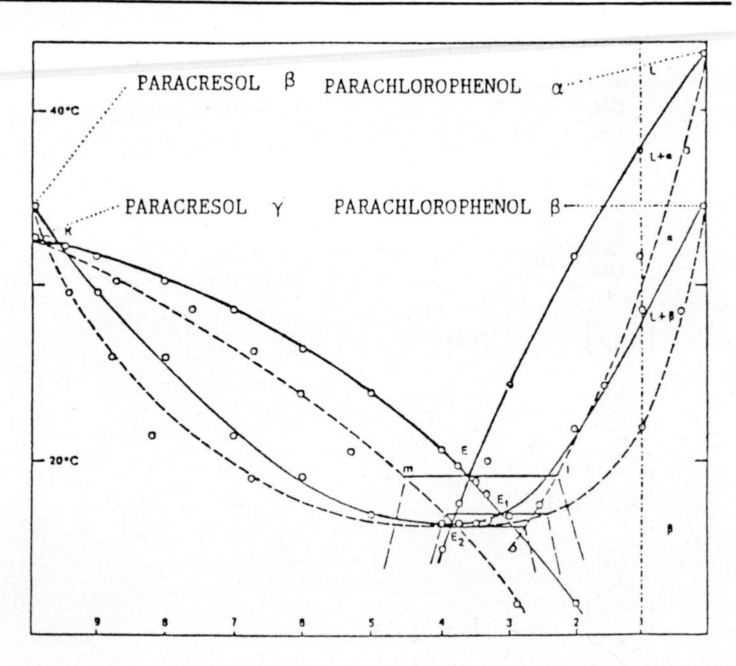

FIGURE 29. Phase diagram of 4-methylphenol and 4-chlorophenol.

crystals. In α and γ forms molecules are hydrogen bonded to form chains while in the β form the molecules are linked by hydrogen bonds to give tetramers. It can be noted that the γ form is an intermediate stable form. It is not stable for the two pure compounds but only for the compositions between the peritectic point K and the eutectic point E. This implies that the γ form of 4-chlorophenol is not stable.

The stabilities of organic crystals of monotropic polymorphic substances has been investigated in the case of 4-chlorophenol and 4-methylphenol (ref. 99). The calculated order of stability for the different polymorphic varieties is the same as the experimental one. In particular the β form which is the more stable for 4-methylphenol becomes unstable for 4-chlorophenol. In order to compare calculated energies with crystal stability a detailed analysis of the total energy was undertaken. A partition of the total energy in two parts was carried out ; a stabilization energy, E intrapile, due to interaction with other molecules in the same pile of tetramers or chains and a stabilization energy, E interpile, due to interaction with molecules from other piles. The results obtained show that the interaction energy between piles is an important factor of the stability of these crystals.

3.2.2. Polymorphism of 4-nitrophenol

The phase transformation of β to α 4-nitrophenol has been carefully studied (ref. 14, 33, 100). The β to α phase transformation occurs upon heating a crystal of the β form. In this reaction, needle shaped single crystals rearrange with the phase boundary moving approximately perpendicular to the needles axis. The structural change from β to α form is subtle and involves a twist of rings in adjacent links in the chains. During the transformation, the crystals retain their original shape and the product crystal consists of one or a small number of single crystals. The orientation of the product crystals was not related to the starting crystal.

3.2.3. Polymorphism of 1,3-dihydroxybenzene

The phase transformation of resorcinol (ref. 103) is very interesting. Resorcinol is one of the few substances known for which the denser polymorph is the metastable polymorph at room temperature. Careful studies show that upon transformation to the high temperature there is a contraction in volume, no change in crystal symmetry and a transition heat not exceeding 5 % of the heat of fusion (ref. 104). In addition, the temperature factors show that the thermal motion in α and β resorcinol is nearly the same.

3.2.4. Polymorphism of 4-hydroxybiphenyl

Recently the two crystal modifications of 4-hydroxybiphenyl have been determined (ref. 37). The crystals grown from solution belong to the orthorombic system while the crystals grown by sublimation belong to the monoclinic system. Unfortunately the stabilities of the different forms have not been studied.

3.2.5. Polymorphism of drugs

Some substances bearing a phenolic ring, like estrone, exist in different polymorph forms.

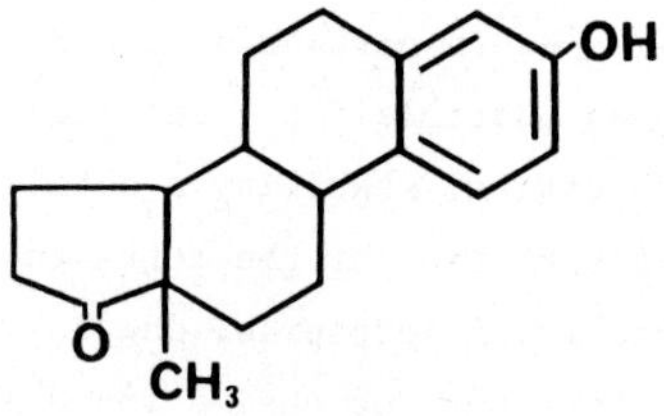

OH

O

CH_3

(3)

The polymorphism of these compounds can have an important effect on their pharmaceutical and biological properties (ref. 104). Estrone exists in three polymorphs (ref. 106). The conformation of the estrone molecule is similar in all three polymorphs. Form I contains layers of estrone molecules but no obvious stacks of molecules. Form II as a herringbone arrangement of molecules and Form III contains both layers and stacks of molecules. No transformations or interconversions of these forms have been reported ; however and as Byrn concludes (ref. 104) "it is likely that the densest form, form II is the most stable".

3.3. THERMAL REACTIONS

Heat induces in organic crystals various reactions like rearrangements, polymorphic transformations, dehydrations, desolvations, decarboxylations, polycondensations... Polymorphic transformations of phenols have been considered precedently. In this chapter are discussed rearrangements, polycondensations and reactions that involve gas as products.

3.3.1. Rearrangement reaction

The rearrangement in the solid state of the yellow form (y-1) to the white form (w-1) of dimethyl 3,6-dichloro-2,5-dihydroxyterephtalate has been investigated by Byrn, Curtin and Paul (ref. 107).

YELLOW FORM Y-1 WHITE FORM W-1

The yellow form y-1 transforms to the white isomer w-1 upon heating. Photomicrographic analysis of this reaction shows that the transformation begins at one ore more nucleation sites and spreads through the crystal in a front. The spread is highly anisotropic ; it is rapid in the plane normal to the long morphological crystal axis and slow in the direction of this axis. Reaction must be rapid through the two-dimensional layers of molecules but slow in moving from one layer to the next.

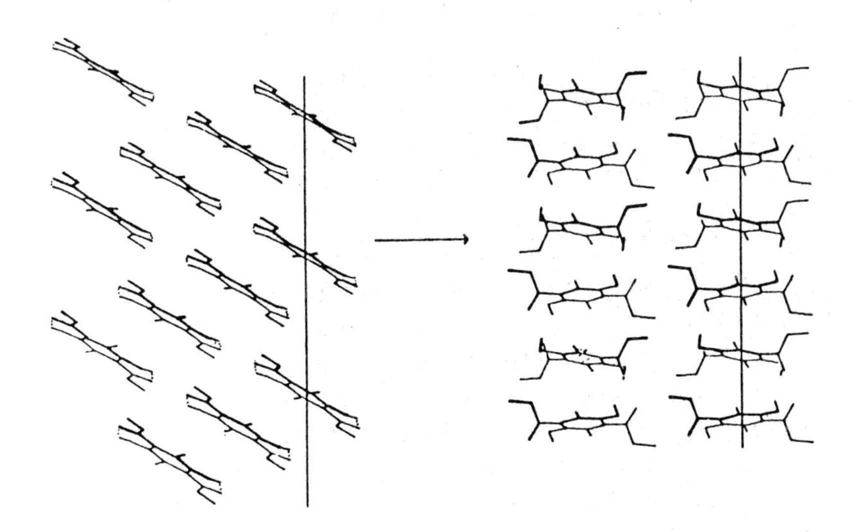

Fig. 30. Change in the molecular structure that occurs in the conversion of the yellow form y-1 to the white form w-1. From I.C. Paul and D.Y Curtin (ref. 94).

This reaction involves a change in the hydrogen bonds formed by the two protons and a conformational change. During the conversion of y-1 to w-1 (see Fig. 30), the ester groups rotate out of the plane of the aromatic ring. The ester groups which are essentially planar in y-1 become nearly perpendicular to the benzene ring plane in w-1 and form hydrogen bonds to the hydroxyl group of adjacent molecules in the stack. Moreover, every other molecule in a stack in y-1 form rotates 180° about the Cl-Cl axis or an equivalent motion to change to the stable packing of the w-1 form.

The behaviour of yellow crystals is explained by the authors in terms of rapid reaction in a given stack and transmission to adjacent stacks in one direction but not to stacks in the other direction.

3.3.2. Dehydration reaction

Thermal, photochemical and photonucleated thermal dehydration of p-hydroxytriarylmethanols have been investigated by Lewis et al (ref. 108, 109). These triarylmethanols 4a, 4b, 4c dehydrate in the solid state to form the corresponding yellow or orange fushsones 5a, 5b, 5c.

$$\text{HO-C}_6\text{H}_2\text{R}_2\text{-C(OH)(C}_6\text{H}_5)_2 \xrightarrow[-\,OH^-]{-\,H^+} \text{O=C}_6\text{H}_2\text{R}_2\text{=C(C}_6\text{H}_5)_2$$

4a R = H
4b R = Br
4c R = CH_3

5a R = H
5b R = Br
5c R = CH_3

The crystal structure of 4a, 4b, 4c, which are isostructural have been determined by Lewis (ref. 108) and Stora (ref. 110). The molecules of 4a, 4b and 4c are aligned in the solid state with the phenolic OH group hydrogen bonded to an alcoholic OH group of an adjacent molecule. These compounds constitute a series in which crystal packing aligns the molecule in a geometry favorable to reaction.

Microscopic observation of the thermal dehydration of single crystals shows that the reaction begins by development of solid solution of the product crystal in the parent in well-demarcated triangular regions at each end of the crystal.

Cracks develope in the reactive regions, and finally the product phase separates out and reaction spreads through the remainder of the crystal.

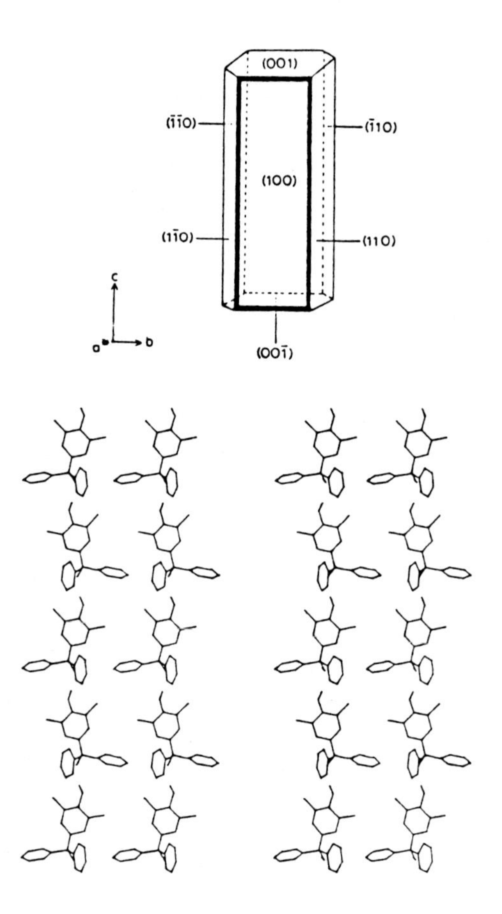

Fig. 31. The upper drawing shows a crystal habit of 4c with the major faces indexed. The lower stereopair drawing shows the hydrogen bonded chains of a crystal of 4c in the same orientation as the crystal above. From T.W. Lewis, D.Y. Curtin and I.C. Paul (ref. 108).

Examination of the behaviour of large numbers of single crystals shows that dibromo crystals of 4b reacted much faster than the dimethyl crystals of 4c. Time for 90% reaction at 110°C was about 10 h for 4b and approximately 100 h for 4c. In solution, 4b is about 10 000 times more acidic than 4c, perhaps explaining the increased solid state reactivity of 4b.

Crystal reaction of 4a, 4b and 4c can be made to occur photochemically as well as thermally (ref. 111). When 4b or 4c are irradiated for 48 hours with 254 nm light the yellow or orange color of fuchsone 5b or 5c appears. This reaction is probably due to an acidity effect, since solution photochemistry of phenols have established that the lowest singlet excited state of phenol has an acidity which is some 10^6 greater than the acidity of phenol in the ground state. This experiment has interesting implications with respect to attemps which have been made to carry out excited-state-acid-catalyzed reaction.

The most interesting aspect of this work, has been the demonstration that the photochemical reaction of 5b and 5c could be used to nucleate the thermal reactions. It was shown that exposure of a part of a crystal to ultraviolet light photonucleated the thermal dehydration process ; when heated a crystal to 110°C for a short time underwent thermal reaction on the side which had been irradiated and then the reaction moved slowly to the unexposed parts of the crystal.

The controlling feature is the nucleation process. Although thermal dehydration of phenols such as 4b or 4c show both heterogeneous and homogeneous stages, it is of particular interest that it is the first reaction of this type where such complete control of nucleation is attainable.

3.3.3. Decarboxylation reaction

Solid state decarboxylation reactions have been well studied by Lin, Siew and Byrn (ref. 112). These reactions are of the type A (solid) B (solid) + CO_2. These authors have investigated the decomposition of salicylic acids 6a, 6b, 6c, 6d, 6e, 6f.

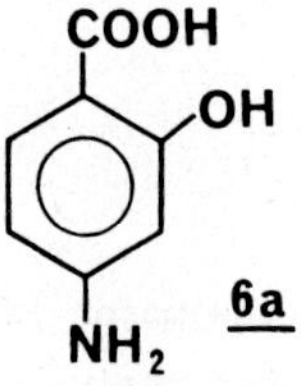

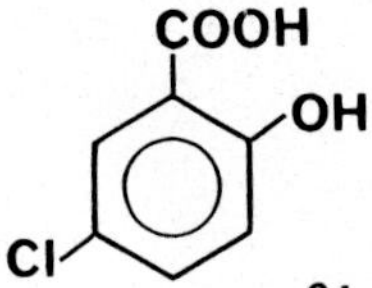

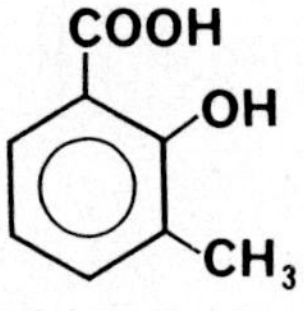

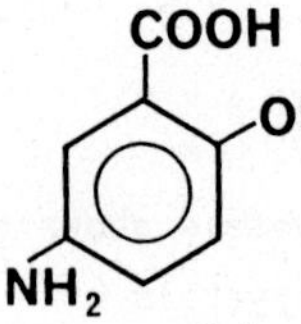

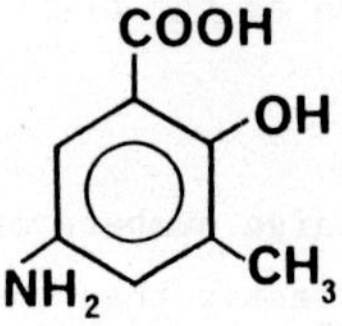

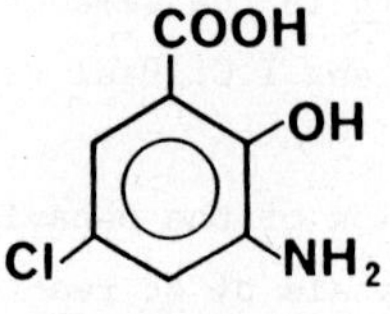

At 100°C only p-aminosalicylic acid 6a decarboxylates in 10 hours, while the other salicylic acids do not decarboxylate even after one week. This is in good agreement with the decomposition rates of salicylic acids based on the mechanism of the solution reaction and studies of the solid state decomposition of benzoic acids. The rate constants for decarboxylation of p-salicylic acids obeyed a Hammett ρ^+ plot. The predicted relative rates of decarboxylation are 1 for 6a and around 10^{-5} for the others.

The solid state decarboxylation follows a path similar to the solution reaction. The proton requires for the electrophilic addition may come from an adjacent ArCOOH or ArOH group. An analysis of the crystal packing shows that there are at least four carboxylic protons and two phenolic protons at 5 Å of the aromatic carbon atom.

3.3.4. Polycondensations

Phenol-alcohols, precursors of phenolic resins constitue a very important class of products. These substances are relatively stable and highly reactive. They give starting material able to produce in the solid state phenolic resins and also they are intermediate leading to products which are in great demand in chemistry.

Thermal decomposition of precursors of phenolic resins has been the subject of a detailed study reviewed in the book of Megson (ref. 113). This decomposition is illustrated by 2-hydroxymethylphenol as model :

2 OH CH_2–OH → OH CH_2–O–CH_2 OH + H_2O

→ OH CH_2 OH + H_2O + CH_2O

The formation of ether bridges is preferentially made at low temperature. At higher temperatures only methylen bridges are formed. The rate of polycondensation depends on the chosen precursor.

Very recently Perrin et al (ref. 35 and 114) have isolated characterized and studied the solid state thermal behaviour (ref. 115) of 2,4,6-trihydroxymethylphenol (7) and 3,5,3',5'-tetrahydroxymethyl-4,4'-dihydroxydiphenylmethane (8).

Their crystalline structures have been determined and their density calculated. It is noticed that the value of densities $d_c = 1.365$ Mg.m^{-3} for 7 and $d_c = 1.46$ Mg.m^{-3} for 8 are high for substances containing only C, H and O atoms. The high degree of packing of 7 and 8 is explained by the great number of intramolecular H-bond arrays which constitute either a two or three dimensional network in the crystal. No intramolecular H-bonds have been found.

By heating it found that phenol-alcohols 7 and 8 are stable up to 80°C. Above this temperature a loss of weight is registered (see Fig. 32). Compound 8 starts to decompose in the solid state between 80 and 145°C. It is concluded that the polycondensation reactions of 7 and 8 as solid pure compound are occuring up to 80°C with chiefly formation of water and formation of ether bonds ; up to 230°C formaldehyde is evolved and methylen bridges are formed.

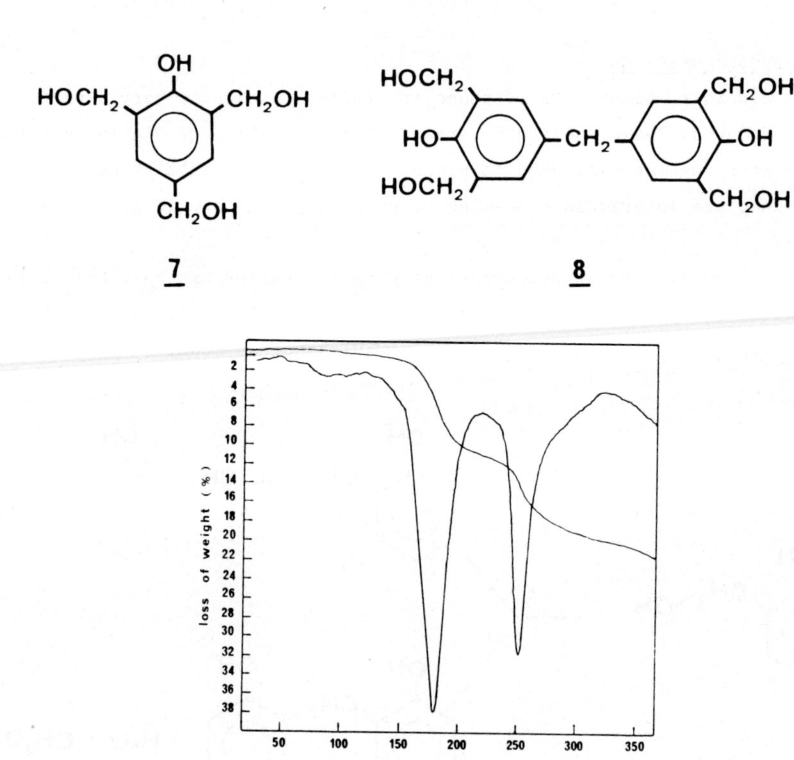

Fig. 32. Loss of weight versus temperature for compound 8.

3.4. SOLID-SOLID REACTIONS

Organic solid-solid reactions are relatively rare. However and in spite of complexity, many aspects of solid-solid reactions, have been resolved.

Solid-solid reaction have been reviewed by Rastogi (ref. 90, 116), with special emphasis on addition reactions of the type A(s) + B(s) $\rightarrow$ C(s).

Rastogi (ref. 117, 118, 119)) and Singh (ref. 120) have studied the addition reactions between solid picric acid and several solid aromatic compounds, (see Table 7).

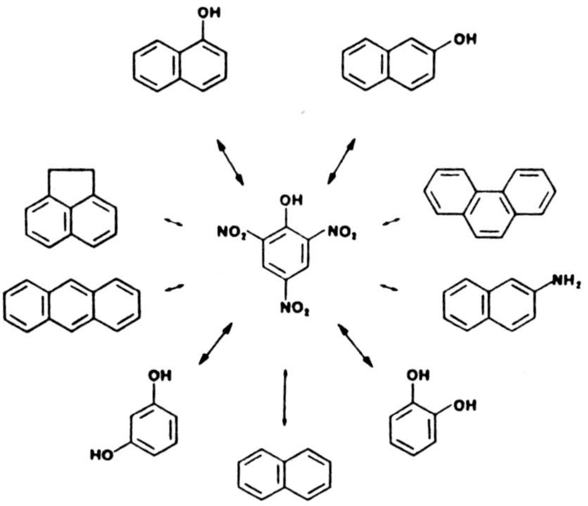

OH
OH
OH
NO_2
NO_2
NO_2
NH_2
OH
OH
OH
HO

Table 7. Addition reaction of solid picric acid with various solid aromatic compounds.

The kinetics have been studied at several temperatures and for particles of various sizes. On the basis of the available evidence, it was concluded that the reaction is diffusion controlled. If ξ is the thickness of the diffusion layer the kinetic results were found to obey the empirical equation :

$$\xi^2 = 2k_i t \exp(-P\xi)$$

where :

k_i = C exp $(-E/RT_{max})$

C = certain constant

E = energy of activation

P = proportionality constant

T_{max} = maximum temperature attained instantaneously in the mixture

The course of the addition reaction may be represented schematically in the following manner, where the encircled A, P and AP represent aromatic compound, picric acid, and additive product, respectively.

a) In the preliminary stages, the reaction takes place at the interface of the reactants :

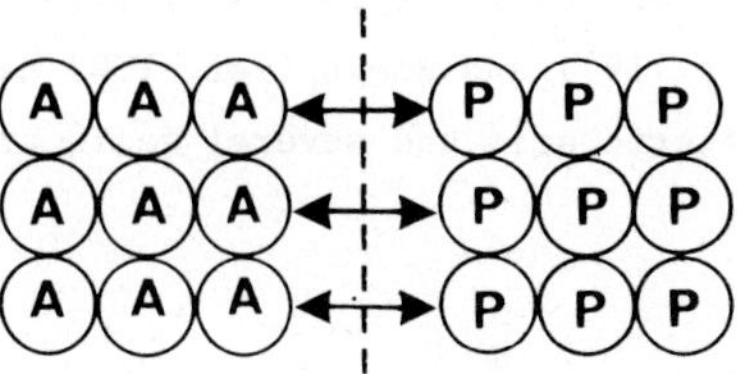

b) Subsequently, diffusion of aromatic compound across the product layer occurs either by vapor phase or surface migration or grain boundary diffusion :

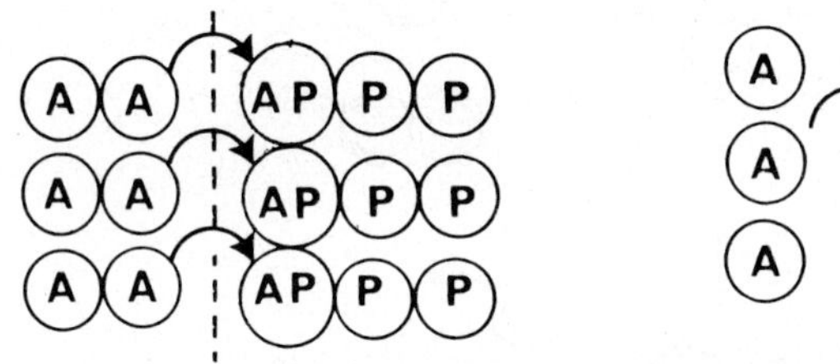

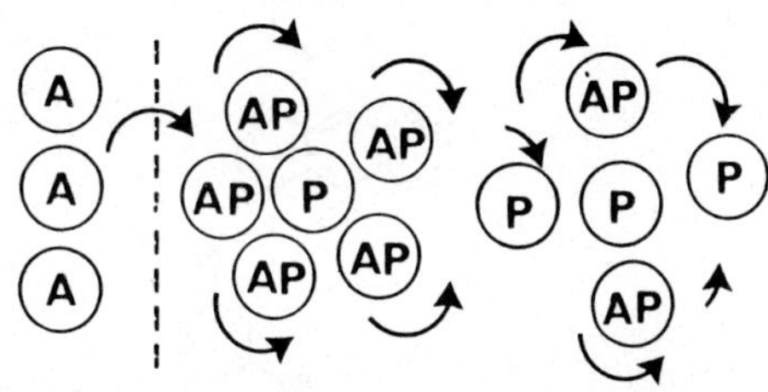

c) Finally, propagation of the reaction in the individual picric acid grain occurs.

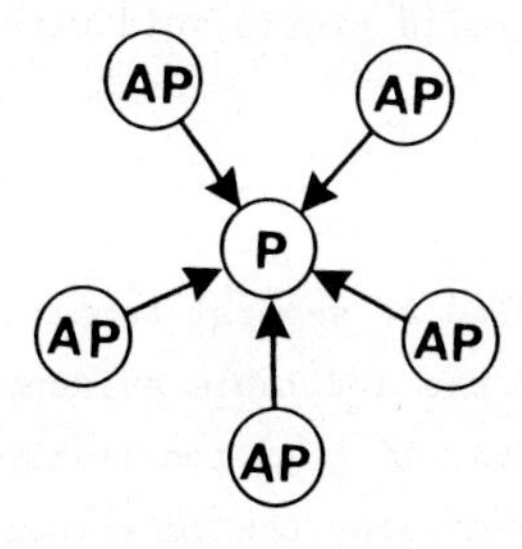

The diffusion in the solid state may occur by any one of the following mechanisms : bulk diffusion ; lattice diffusion ; grain boundary diffusion ; surface diffusion and vapor phase diffusion. Rastogi and Singh (ref. 121) have shown that diffusion in the solid state is controlled by surface diffusion. It was found that molecules having a smaller size and molecules having greater symmetry are more favorable for surface migration. The migration is easier when the migrating molecules and the substrate molecules have the same size and similar symmetry.

Lin et al (ref. 122) have investigated the reaction of single crystals of p-aminosalicyclic acid hydrochloride with powdered sodium carbonate.

$$2\ \mathrm{HOOC(OH)C_6H_3NH_3^+Cl^-} + \mathrm{Na_2CO_3} \longrightarrow 2\ \mathrm{HOOC(OH)C_6H_3NH_2} + \mathrm{H_2O} + \mathrm{CO_2} + 2\,\mathrm{NaCl}$$

The solid-solid reactions were run by placing crystals of Na_2CO_3 in contact with crystals of p-aminosalicylic acide hydrochloride. The p-aminosalicylic acid exists in three crystal habits (see Fig. 33).

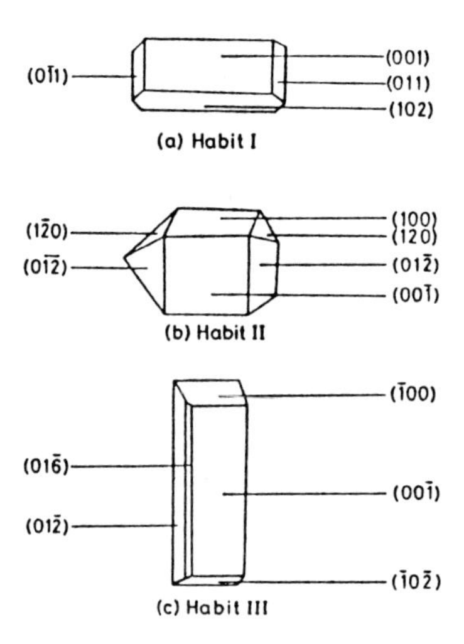

Fig. 33. Schematic diagram of the three crystal habits of p-aminosalicylic acid.

The three habits undergo solid-solid reaction with sodium carbonate. These reactions always start at the point of contact of the base and proceed through the crystal anisotropically. They are quite slow and take more than 40 days to reach completion at 60°C. The anisotropic behaviour of two of the crystal habits I and II is consistent with crystal packing.

3.5. SOLID-LIQUID REACTIONS

Relatively few reactions of organic solids with liquids have been studied. Kornblum and Lurie (ref. 123) investigated the alkylation of sodium and potassium salts of phenol and p-tert-octylphenol.

These authors compared alkylation of phenoxide ions in homogeneous and heteregenous reactions. In the case of potassium p-tert-octylphenoxyde and benzyl chorid the set of experiments can be summarized as follows.

O^-K^+ / C_8H_{17} + $C_6H_5CH_2Br$ —(MeOCH$_2$OMe, Homog.)→ $OCH_2C_6H_5$ / C_8H_{17} (97%)

—(Et_2O, Heterog.)→ $OCH_2C_6H_5$ / C_8H_{17} (74%) + OH / $CH_2C_6H_5$ / C_8H_{17} (21%)

The same experiments carried out in a single solvent, toluene, show that under heterogeneous conditions the reaction of the potassium salt of p-tert-octylphenol with benzyl chloride give 79 % O-alkylation and 13 % C-alkylation. In contrast when an homogeneous solution of potassium p-tert-octylphenoxide is treated with benzyl chloride, the product is the benzyl p-tert-octylphenyl ether (97-99 % yields) and none of the C-benzylated phenol is detected.

During the heterogeneous process the reaction mixture becomes completely homogeneous after the reaction has gone 50-60 % to completion. This suggested that the 13 % yields of C-alkylation obtained from the heterogeneous reaction actually derives from the early stage of the reaction when a minimal amount of the phenoxide salts is in solution. The results (see Table 8) obtained by Kornblum and Lurie for reactions allowed to go only 46, 16 and 4 % completion confirm the conclusion that heterogeneity is a necessary condition for carbon alkylation and constitute a demonstration that the truly heteregeneous reaction gives exclusively carbon alkylation.

% to completion	O-benzylation	C-benzylation
100	86	14
46	58	42
16	5	95
4	0	100

Table 8. Heteregeneous reaction of benzyl chloride with potassium p-tert-octylphenoxide in toluene at 25°C.

Kornblum and Lurie explained alkylation on carbon by postulation of a reaction of alkyl or benzyl chloride at the surface of solid potassium p-tert-octylphenoxide.

Curtin and al (ref. 124) have also investigated the importance of heterogeneity in the alkylation of sodium phenoxides with benzyl chloride. The alkylation of sodium 4-tert-butyl-2,6-dimethylphenoxide was examined in toluene and in toluene containing amounts of tetrahydrofuran.

O^-Na^+; CH_3; CH_3; $C(CH_3)_3$ + $C_6H_5CH_2Br$ $\xrightarrow[-NaBr]{Heterog.}$ x [O; CH_3; $CH_2C_6H_5$; CH_3; $C(CH_3)_3$] + y [$OCH_2C_6H_5$; CH_3; CH_3; $C(CH_3)_3$]

The ratio of carbon to oxygen alkylation varies from a value to 1.4 in an heterogeneous experiment in toluene to about 0.17 in an homogeneous solution containing 16 % of tetrahydrofuran. By kinetic studies these authors showed that the effect to added tetrahydrofuran is to accelerate the oxygen alkylation much more than the carbon alkylation. A change in the amount of undissolved sodium salt by a factor of about 10 fails to change appreciably the carbon to oxygen attack by benzyl chloride in toluene. It is concluded that in these reactions, at least, heterogeneity has no significant effect on the ratio of carbon to oxygen alkylation.

In these two studies a solid (s) reacts with a liquid (1_1) to give a liquid (1_2). But it has not been firmly established that the structure of the solid phase imposes specificity on the reaction. The existence of true solid-liquid reactions is problematic. It is always possible to imagine that a slight part of the solid is dissolved and the reaction occurs in a liquid phase. Bittner and Idelsohn (ref. 125) have studied the solid state bromination of different phenols. In most cases the substrate remained in the solid state throughout the reaction. In all cases the same product was obtained as was produced by bromination in solution. In was concluded by Bittner and Idelsohn : "These results are in accord with Buckles' (ref. 126) assumption that the reaction of aromatic solids with bromine occurs in an adsorbed phase on the surface of the crystal or in a film of solution formed by the aromatic compound dissolved in liquid bromine".

3.6. SOLID-GAS REACTIONS

There are a large number of literature references to solid-gas reactions, many of them deal with phenols. The list includes such diverse examples as reaction of dry ammonia gas with dried and pulverized phenols (ref. 127), carbonation of solid sodium phenoxide (ref. 128, 129), reaction of aromatic solids with halogen vapors (ref. 130), alkylation of solid sodium and lithium 2,6-dimethylphenoxide (ref. 131), aspirin hydrolysis (ref. 132, 133), reaction of solid aromatic carboxylic acids and related compounds with ammonia and amines (ref. 134), akylation of paraphenylphenol crystals (ref. 135), carbonation and chlorination of solid metallic phenoxides (ref. 136, 137), polycondensation of solid phenol-alcohols with hydrochloric acid vapours (ref. 115). Very recently the solid state reactivity of crystalline hydroquinones with quinone vapour has been studied (ref. 138). The reaction was carried out with thin single crystals of hydroquinone surrounded by powdered 1,4-benzoquinone (not in direct contact). The reaction showed evidence of selective attack at nucleation sites at the reactive surface.

3.6.1. Chlorination of phenols

Lamartine and Perrin (ref. 139) have investigated the solid state chlorination of a large number of alkylphenols (see Table 9). The calibrated powder of phenols are exposed to chlorine under anhydrous conditions. The products formed are essentially chlorophenols, chlorocyclohexadienones and chlorocyclohexenones. The main characteristic of these gas-solid reactions is that they are very rapid and nearly always lead to the total transformation of the initial product. In view of the conditions, the chlorination reaction can be represented as follows :

Table 9. Chlorination of alkylphenols in the solid state by gaseous chlorine

Phenol	Product (yield, %)
Phenol	2,4,6-Trichlorophenol (97) 2,4-Dichlorophenol (3)
4-Methyl-	2-Chloro-4-methylphenol (20) 2,6-Dichloro-4-methylphenol (80)
4-Ethyl-	2-Cloro-4-ethylphenol (30) 2,6-Dichloro-4-ethylphenol (70)
4-Isopropyl-	2-Chloro-4-isopropylphenol (30) 2,6-Dichloro-4-isopropyphenol (65) Chlorocyclohexadienone (5)
4-tert-Butyl-	2,6-Dichloro-4-tert-butylphenol (68) 2,4,6-Trichloro-4-ter-butyl-2,5-cyclohexadienone (32)
2,6-Dimethyl-	Chlorophenols (~55) o-Chorocyclohexadienones (~25) Chlorocyclohexadienones (~20)
3,5-Dimethyl-	2,4,6-Trichloro-3,5-dimethylphenol (20) 2,4-Dichloro-3,5-dimethylphenol (78) 4-Chloro-3,5-dimethylphenol (2)
3-Methyl-5-ethyl-	2,4,6-Trichloro-3-methyl-5-ethylphenol (98) 2,4-Dichloro-3-methyl-5-ethylphenol (2)
3,5-Diethyl-	2,4,6-Trichloro-3,5-diethylphenol (80) 2,4-Dichloro-3,5-diethylphenol (20)
3-Methyl-5-isopropyl-	2,4,6-Trichloro-3-methyl-5-isopropylphenol (94) 2,4,4,6-Tetrachloro-3-methyl-5-isopropyl-2,5-cyclohexadienone (6)
3-Methyl-2-isopropyl-	6-Chloro-3-methyl-2-isopropylphenol (5) 4,6-Dichloro-3-methyl-2-isopropylphenol (95)
2-Methyl-3-isopropyl-	6-Chloro-2-methyl-3-isopropylphenol (8) 4,6-Dichloro-2-methyl-3-isopropylphenol (92)
3,5-Diisopropyl-	2,4,6-Trichloro-3,5-diisopropylphenol (73) 2,4-Dichloro-3,5-diisopropylphenol (4) 2,4,4,6-Tetrachloro-3,5-diisopropyl-2,5-cyclohexadienone (23)
2,6-Di-tert-butyl-	4-Chloro-2,6-di-tert-butylphenol (45) 4-Chloro-2,6-di-tert-butyl-2,5-cyclohexadienone (20) 4,6-Dichloro-2,6-di-tert-butyl-2,4-cyclohexadienone (35)
2,5-Di-tert-butyl-	Chlorophenols (40) 4,4,6-Trichloro-2,5-di-tert-butyl-2,5-cyclohexadienone 4,6,6-Trichloro-2,5-di-tert-butyl-2,4-cyclohexadienone (60) 2,4,6-Trichloro-2,5-di-tert-butyl-3,5-cyclohexadienone
2,4-Di-tert-butyl-	Chlorophenols (37) 4,6-Dichloro-2,4-di-tert-butyl-2,5-cyclohexadienone (63)
3,5-Di-tert-butyl-	Chlorophenols (37) 2,4,4-Trichloro-3,5-di-tert-butyl-2,5-cyclohexadienone 2,4,4,6-Tetrachloro-3,5-di-tert-butyl-2,5-cyclohexadienone (63)
2,4,6-Trimethyl-	Chlorocyclohexenones (~30)
3,4,5-Trimethyl-	Chlorocyclohexenones (~40)
4-Methyl-3,5-diisopropyl-	Chlorophenols (~90) Chlorocyclohexadienones (~10)
4-Methyl-2,5-diisopropyl-	Chlorophenols (~85) Chlorocyclohexadienones (~15)
2,6-Dimethyl-4-tert-butyl	2,5,6-Trichloro-2,6-dimethyl-4-tert-butyl-3-cyclohexenone (95)
4-Methyl-2,6-di-tert-butyl	4-Chloro-4-methyl-2,6-di-tert-butyl-2,5-cyclohexadienone (47)
2,4,6-Triisopropyl-	Chlorocyclohexenone (~40) Chlorocyclohexadienone (~60)
2,4,6-Tri-tert-butyl-	4-Chloro-2,4,6-tri-tert-butyl-2,5-cyclohexadienone (92).

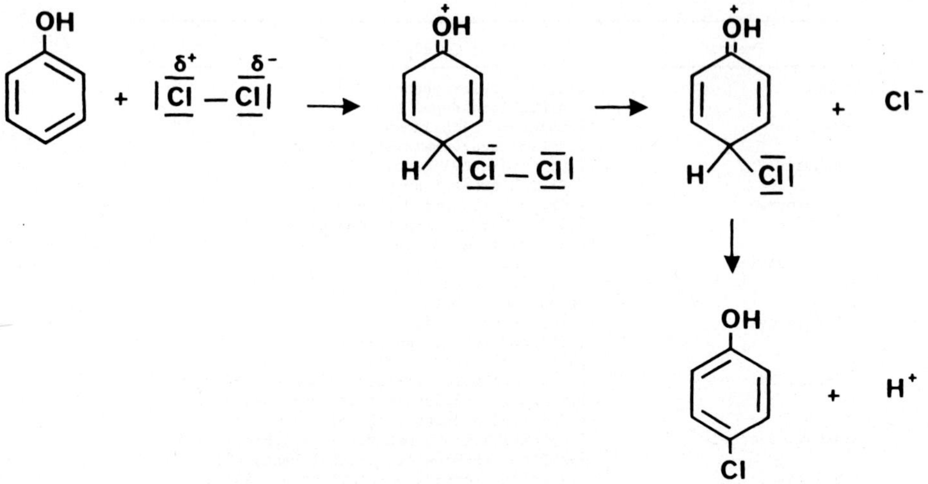

Fig. 34. Clorination of phenols. It is not obvious that this representation usually accepted is exact. During the chlorination by molecular chlorine the donor effect of the aromatic ring gives a negative charge on the halogen molecule (ref. 148).

The gaseous chlorine molecules can only be polarized by the corresponding nucleus of the initial phenol. It is proposed (see Fig. 34) that the reaction proceeds via an addition elimination mechanism, where chlorine is added and then both a chloride ion and a proton are eliminated. Consideration of the steric and electronic effects of the groups attached to the phenolic nucleus leads to a consistent interpretation. An increase in the number of donor effect substituents increase the rate of formation of chlorophenols. If the phenols with a great number of donor effect substituents have some positions for which steric hindrance is reduced, chlorocyclohexenones are obtained preferentially.

The reaction of single crystals of 2-methylphenol with chlorine gas has been studied in detail (ref. 140, 141). Lamartine et al (ref. 142) have shown the anisotropic character of this reaction. The products of reaction are mainly the 4-chloro- and 6-chloro-2-methylphenol.

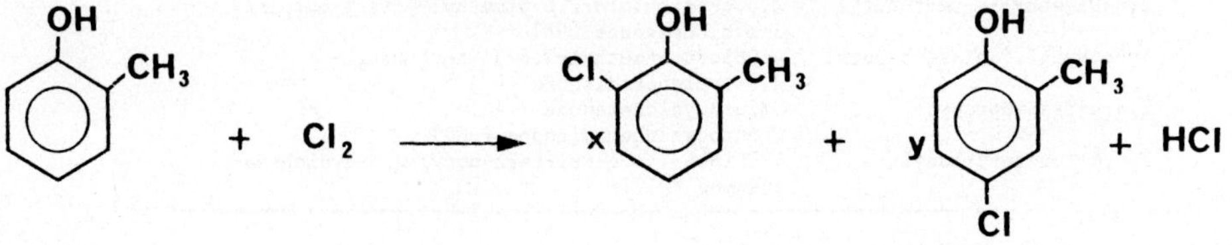

The ratio of the reaction products x/y depended on which face of the crystal is exposed to gaseous chlorine. It is observed (see Fig. 35) that in all cases,

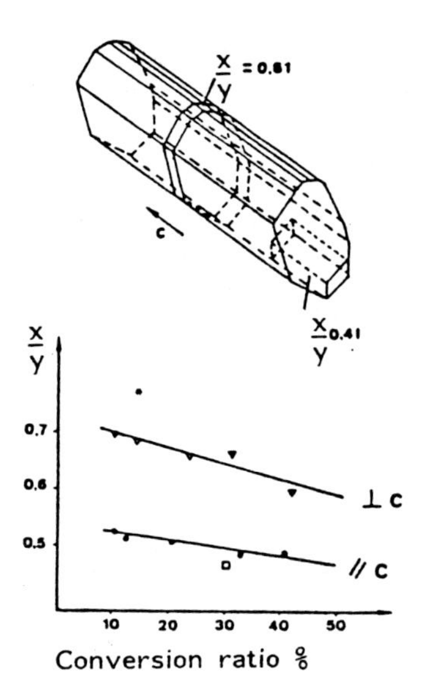

Fig. 35. A typical crystal of 2-methylphenol and x/y ratio plotted against conversion ratio.

the x/y ratio for a plate perpendicular to the "c" axis is higher than the x/y ratio obtained with a plate cut parallel to the "c" axis. The results are correlated with the crystal structure of the 2-methylphenol and with the accessibility of the reactive sites : with a plate cut perpendicular to the "c" axis the accessibility to the ortho or para positions is the same if the attack is allowed on the two opposite faces of the plates. For plates cut parallel to the "c" axis, some of the molecules have the same orientation as for a plate cut perpendicular to "c" but the others expose on one face, the non reactive groups OH or CH_3 and on the other face, the para position. These explanations do not take into account the possibility of chlorine molecules diffusing inside the bulk crystal ; in fact the minimum dimension of a chlorine molecule is about 3.6 Å and the largest available channels in the crystal are about 3.1 Å. So, the orientation of the 2-methylphenol molecules within the crystal and the accessibility to the reactive molecule position explain the anisotropic behaviour of this true solid state reaction.

Similar conclusions have been drawn by Lamartine and Perrin (ref. 143) in the case of reaction of gaseous chlorine with stable and metastable varieties of powders and single crystals of 4-chlorophenol. Kinetic studies have shown that the two crystalline varieties of 4-chlorophenol reduced to fine powder react appreciably in the same way whereas single crystals of the metastable variety react more rapidly than those of the stable form. In the case of powders the rate is proportional to the remaining quantity of reagent (first order reaction) while in the case of single crystals the rate is proportional to the remaining amount of reactant and also to the quantity of product formed.

The indentical behaviour of the powders is explained in the following manner. The powders obtained by grinding show faces which correspond to many crystallographic directions. Under these conditions the approach of the gaseous reagent cannot be differenciated.

The difference in the rates observed between single crystal of the stable and metastable varieties is explained by the difference in accessibility to the reactive centers. As shown on figure 36, for the different faces on the stable variety most of the molecules have their C-Cl and C-OH bonds approximately perpendicular to the face leading to an inaccessible zone for the chlorine molecules whereas for the metastable variety a great number of molecules have their axes OH-C---C-Cl approximately parallel to the faces.

For the chlorination of single crystals of 2-methylphenol and 4-chlorophenol the knowledge of the crystal structure and the crystal morphology allows to explain the different results. In these reactions quantitative relationships exist between crystal structure and reactivity. Thus the topochemical principle stated by G.M.J. Schmidt for photochemical reaction taking place within a solid can be extended to gas-solid reactions which take place at the surface of organic solids.

The pecular effects of organic solid state can be used to produce and isolate new substances. In this way the solid state chlorination of 4-alkylphenols (ref. 144) permits the obtention of ipso-compounds : 4-chloro 4-alkylcyclohexa-2,5-diene-1-ones. These substances can not be formed in solution, the solid state favours ipso-chlorination.

3.6.2. Hydrogenation of phenols

Hydrogenation of thymol is a very interesting reaction because, in this way, menthols can be obtained. In the case of hydrogen it is necessary to use a catalyst in order to obtain reactive species because hydrogen is not naturally reactive as chlorine, bromine or carbon dioxyde are. Lamartine and Perrin (ref. 145, 146) have shown that it is possible to hydrogenate solid phenols and phenoxides with molecular hydrogen in presence of various catalysts under mild

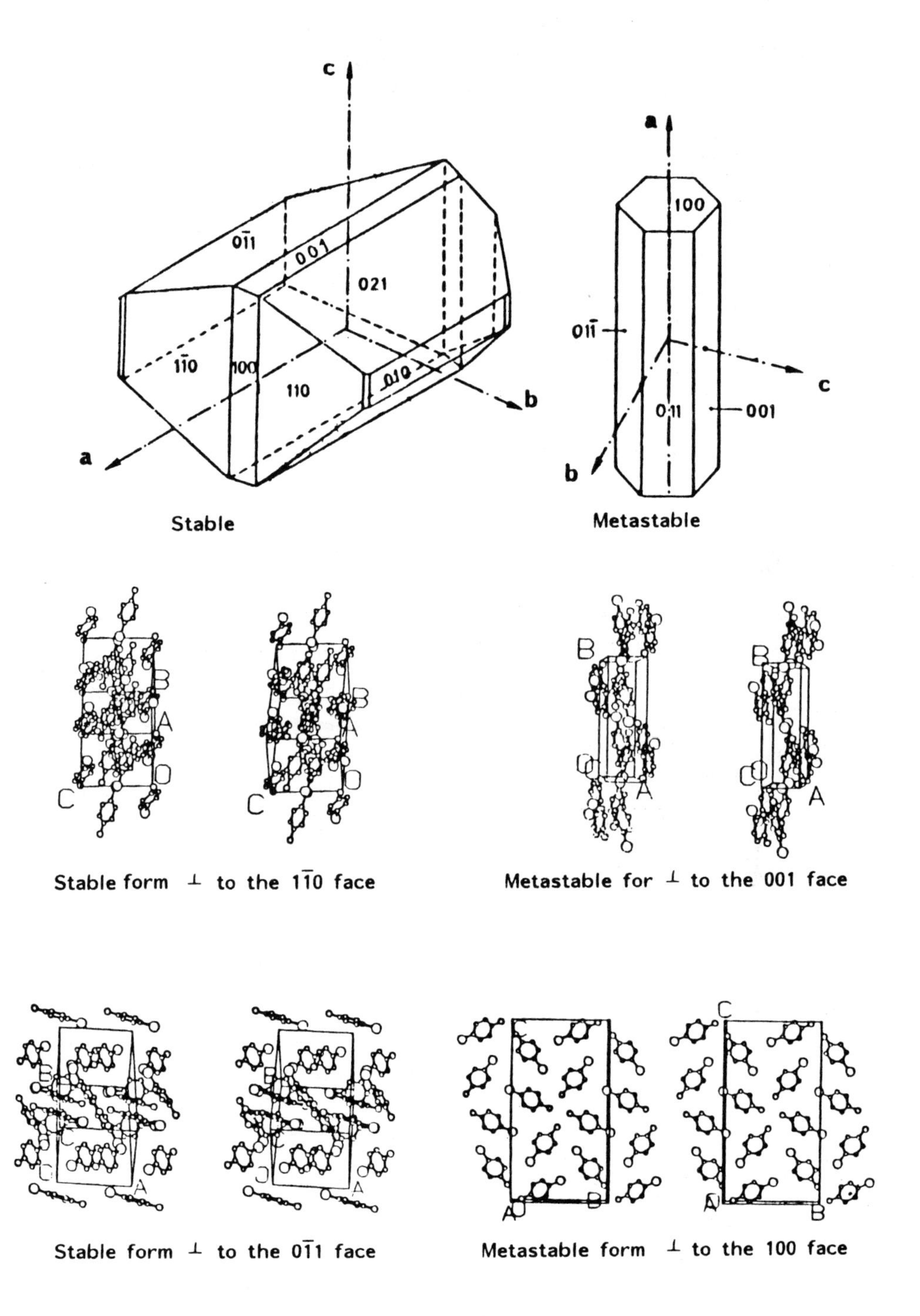

Fig. 36 Single crystals of 4-chlorophenol and stereoscopic views.

conditions. The different catalysts used allow to hydrogenate phenols. In the case of thymol, platinum and rhodium are the best actives ; the reaction can be carried out with a metal without support (Raney nickel) and with an oxyde (Adams platinum). The reactions performed in the same conditions but in solutions show that thymol is not hydrogenated in protic solvents like methanol or ethanol while in non polar or non protic polar solvents hydrogenation occured.

The catalyst support effects determined with 4-tert-butylphenol hydrogenation show the stereoselectivity is highly dependent on the support of the catalyst. The ratio cis/trans goes from 6.43 to 0.82 according to the support of rhodium is carbon or alumina.

Hydrogenation of a chiral crystal of 3-methyl-4 isopropylphenol (space group $P4_1$ or $P4_3$) gives a mixture of menthols and menthones with rotatory power. This reactions of an achiral compound with an achiral catalyst leading to rotatory power is an absolute asymmetric synthesis using only the chirality of the crystalline arrangement. Thus, it is well established that the results obtained depend solely on the molecular arrangement in the organic solid and rule out the possibility that the reaction proceeds via liquid or gaseous phases (ref. 147). Because of the lack of fluid phase, the organic reagent cannot be adsorbed on the catalyst as suggested the classical catalysis mechanism. On figure 37 the solid state hydrogenation of phenols is described. The molecular hydrogen on contact with metal gives activated hydrogen. This hydrogen is transferred on the surface or in the bulk. This is the first spillover effect. By means of contact between support and organic solid, activated hydrogen moves across the organic reagent and reacts. This is the second spillover effect. The support and the organic reagent make up solid phases which allow the migration of hydrogen in its activated state.

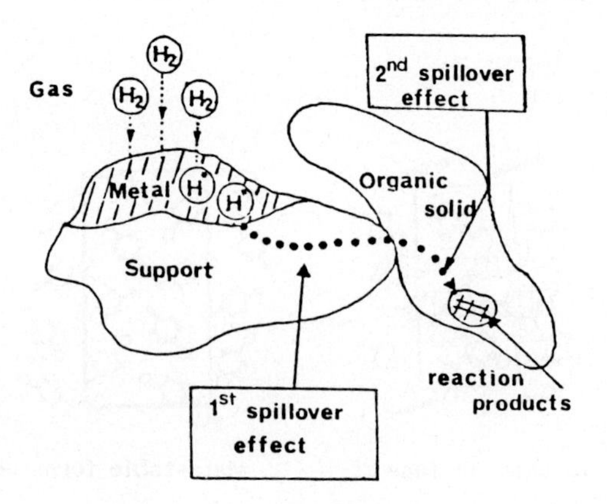

Fig. 37. Solid state hydrogenation of phenols.

Organic solid-gas reactions are the most interesting for industrial applications. In peculiar hydrogenation of solid phenols which gives menthols and hydrogenated phenols used in perfumery could be performed in large scale (for practical treatments see the last part).

4. THE POSSIBLE INDUSTRIAL APPLICATIONS

4.1 Principal industrial fields of the organic solid state chemistry

The chemistry of organic solid state, and of solid phenols particularly, is not a purely academic chemistry as you could think when reading some basic papers. The principal fields of this chemistry are :

- Materials : molecular and macromolecular materials with particular physical properties : non linear optic, electrical conduction, superconductivity, ferroelectricity, luminescence,...
  Materials for information recording and particularly materials for reprographic process.
- Drugs
- Agrochemical products (pesticides)
- Dyes
- Explosives.

We will not detail the part of the organic solid state chemistry in these different fields but present some references and examples about the phenolic substances.

We shall mention the use of phenolic polymeric materials as photoresists. Photoresists are light sensitive resins that are used for either positive or negative imaging. Commercial positive photoresists use a phenolic host resin such as cresol-formaldehyde novolak resin. This novolak 1 is manufactured with a mixture of three isomeric cresols (ref. 148, 149).

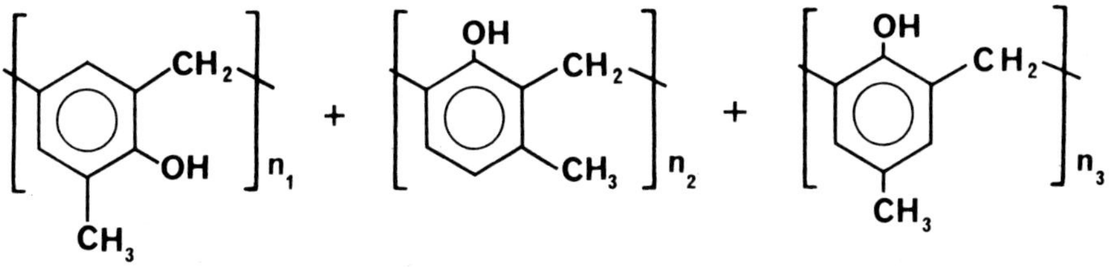

Other phenolic materials are synthesized by using raw materials like orthochlorometacresol 2 and benzaldehyde 3.

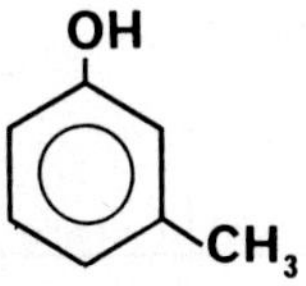

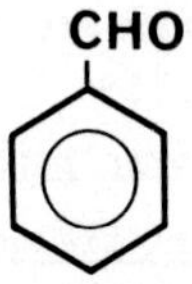

As concerns drugs, you will have to read Byrn's book (ref. 104). Interesting stability studies about biologically active compounds will be found, such as p-aminosalicylic acid 4 which is decarboxylated from 100°C.

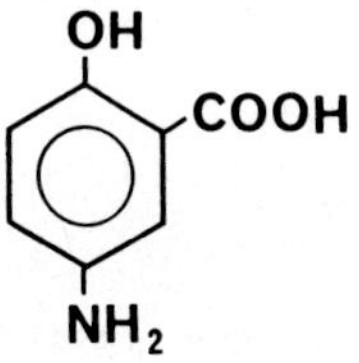

As for the pesticides, "Pesticide manual" (ref. 150) could be consulted. Substances like Dinoseb or 2-(1-methyl-n-propyl)-4,6-dinitrophenol 5 and pentachlorophenol 6 are studied.

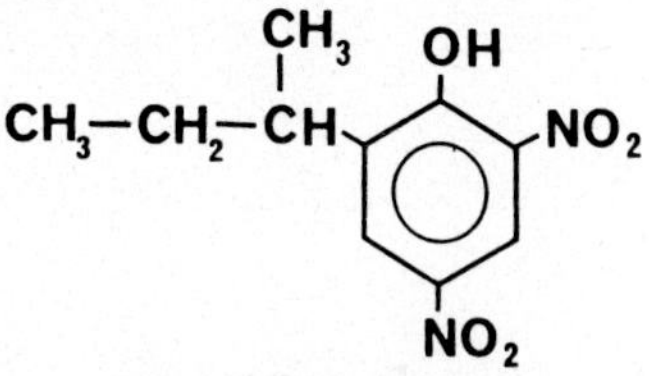

The influence of water, heat and light on solid substances has almost not been studied. We have only found some works about their absorbed phase transformation. We think important to develop works about these solid products.

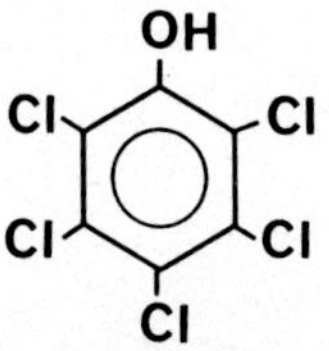

Two books seem interesting about dyes : Venkataraman's 7 volumes book (ref. 151) and Abahart's book (ref. 152). For example, the dinitroanthrarufin 7 and the dinitrochrysazin 8 have been used for many years.

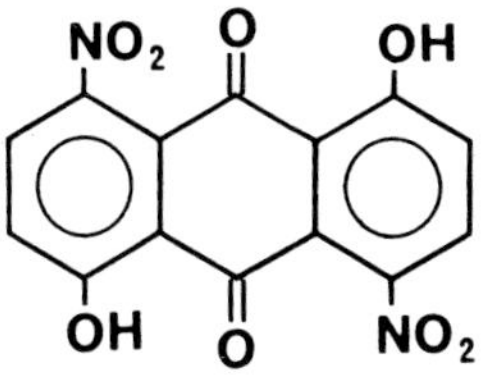

7

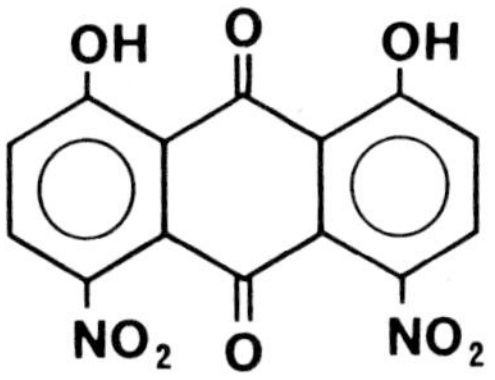

8

Stability towards light and different gas of such dyes is an important property.

Finally, the organic solid state chemistry must play an essential part in the field of explosives, because of the importance of studying the decomposition of these substances (see Morawetz (ref. 153) and Urbanski (ref. 154)).

For example, 2,4,6-trinitrophenol or picric acid 9 is an explosive called melinite. Diazodinitrophenol 10 is a primary explosive or initiator. It is the most sensitive explosive to heat, friction and shock.

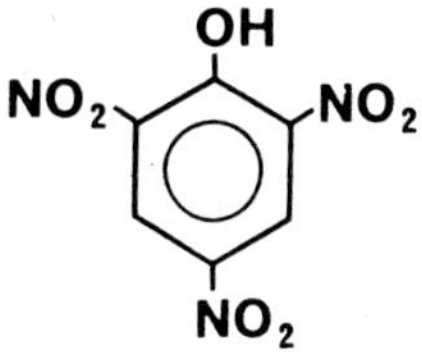

9

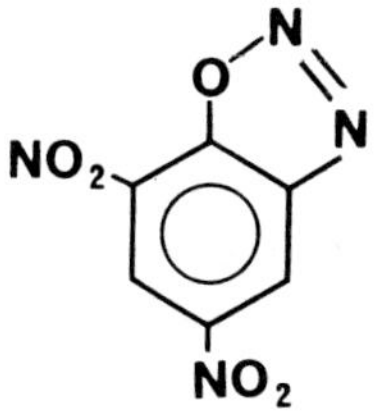

10

4.2 Reaction with organic solid or organic-inorganic solid at industrial scale

The Kolbe-Schmitt reaction which is a true organic solid state reaction is a good example of industrial reaction.

In this reaction, the solid alkaline phenates, submitted to the action of carbone dioxide, yield ortho and parahydroxybenzoic acids. The orthohydroxybenzoic or salicylic acid is by far the most important, as it yields aspirin by means of acetylating phenolic function. Aspirin is very often used as an analgesic, antipyretic and antiinflammatory agent.

The Kolbe-Schmitt reaction can be written as follows for the formation of the salicylic acid.

$$C_6H_5ONa + CO_2 \rightarrow \underset{\underline{11}}{C_6H_5O{-}Na{\cdots}O{=}C{=}O} \rightarrow o\text{-}HOC_6H_4COONa$$

However, the mecanism is not clearly understood (ref. 155). The first laboratory synthesis of salicylic acid seems due to Piria (ref. 156, 157). The first commercial means of manufacturing salicylic acid was the saponification of natural methyl salicylate. In 1874, a suitable commercial process giving synthetic salicylic acid was introduced and involved the reaction of dry sodium phenate with carbon dioxide under pressure at a high temperature. You will note that the dry sodium phenate melts at 384°C (ref. 158).

At lower temperature (120-130°C) and under pressure (7 bars), the carbon dioxide forms the intermediate 11 almost quantitatively (ref. 128). The intermediate rearranges to give primarily the sodium salicylate. The Schmitt conditions , which involved relatively low reaction temperatures and reaction times, significantly increases the yields. Studies showed that the transformation rate of the alkaline phenates into ortho and parahydroxybenzoic acid increases in the series lithium, sodium, potassium, rubidium, cesium (ref. 159). Moreover, for a given transformation rate, at 200°C, the ratio orthohydroxybenzoic acid/parahydroxybenzoic acid considerably increases in the series sodium, potassium, rubidium. It is observed that traces of water can widely lower reactivity without modifying the order of selectivities (ref. 136).

Modern methods of commercial manufacture of salicylic acid still employ the basic Kolbe-Schmitt reaction. The phenate is prepared by mixing phenol and soda in water then by drying under atmospheric pressure and then under vacuum. The dry sodium phenate is placed in a "carbonator" with dry carbon dioxide at 5-6 bars. Air is excluded to minimize oxydation and the formation of coloured compounds. For more details, you can read the excellent review by Lindsey and Jeskey (ref. 160).

Industrial production of parahydroxybenzoic acid is made in the same way as for orthohydroxybenzoic acid, but the reaction is oriented into para to OH group by replacing sodium phenate by potassium phenate according to the results given in a previously quoted paper (ref. 159). Of course, the rubidium and cesium phenates yield a still better selectivity, though the transformation rate is lower at the same temperature.

However, the prices of rubidium and cesium derivatives are too high to be used at the industrial stage.

P-hydroxybenzoic acid is of significant commercial importance. The must familiar application is in the preparation of several of its esters, which are used as preservatives.

4.3 Uses of organic solids

Generally speaking, an organic solid is used in a reaction in its powder form. Naturally, it is not possible to quote all the devices which could be used to make a solid-solid, solid-liquid or solid-gas reaction with an organic solid. Examples will follow ; they correspond either to tests carried out in semi pilot-plant scale, or to coming tests.

The first test corresponds to the production of the tert-butylphenylether according to the Stevens and Bowman patent (ref. 161). Doped phenol in liquid state with 5.10^{-4} sulfuric acid mole per phenol mole is solidified and reduced into thin powder. This powder is placed and stirred, at 8°C temperature, in a semi-batch reactor shown Fig. 38.

Addition of isobuten under 1 bar pressure yields fluid phase mainly constituted by tert-butylphenylether 12 according to the following reaction.

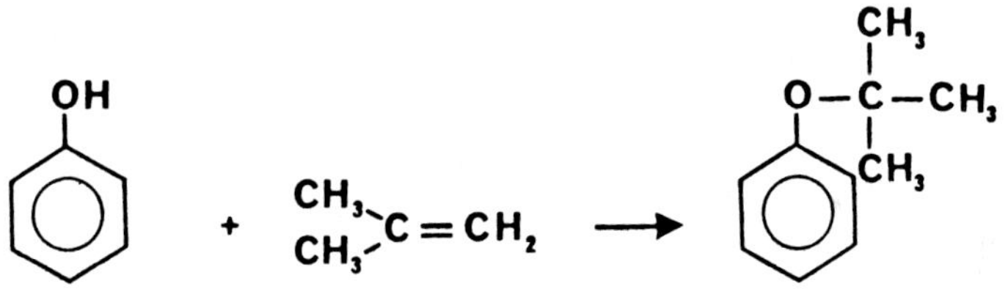

We use solid phenol at low temperature to make ether rather than alkylphenols.

The second test corresponds to the formation of quinhydrone by reacting solid hydroquinone with solid quinone. This reaction can be performed from mixing two finely divided solids in a batch reactor.

Phenol chlorination in solid state by gaseous chlorine has been seriously studied in the laboratory. Several studies about this theme are presented in the second part. Reaction rate is generally very high and polychlorinated derivatives are usually made. Considering these results, it could be possible to use a fluid-solid reactor containing chlorine under atmospheric pressure ; in the reactor, phenol powder is dispersed in chlorine according to the device presented Fig. 39.

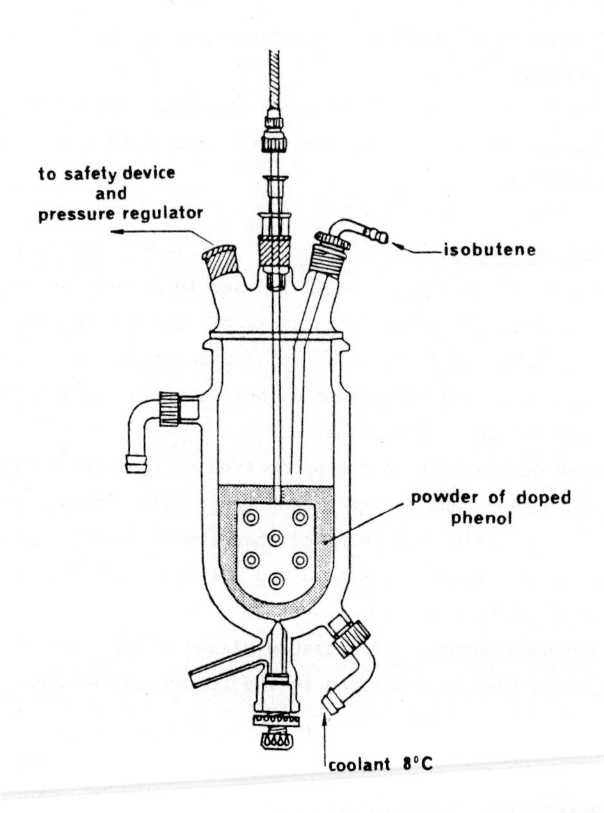

Fig. 38. Tert-butylation of solid phenol.

The relatively dense chlorine is introduced in the upper part of the reactor then falls by gravity at the same time as the phenol powder. The important factor, besides temperature and pressure, is the mean residence time of particles of size R_1 in seconds.

The final example is that of catalytic hydrogenation of solid phenols. Though such reactions cannot be possible according to the classical catalysis mecanisms, various studies appearing in the second part of this paper show that it is quite possible and that yields can be very important in relatively soft conditions. It is possible, for example, to use the reactors of Fig. 40 and Fig. 41.

Products from phenol hydrogenation generally have vapor pressures higher than those of the corresponding phenols. Therefore, they can be carried away by the hydrogen flow possibly diluted by an inert gas. The temperature must be adapted, particularly to avoid clogging for a fixed bed reactor.

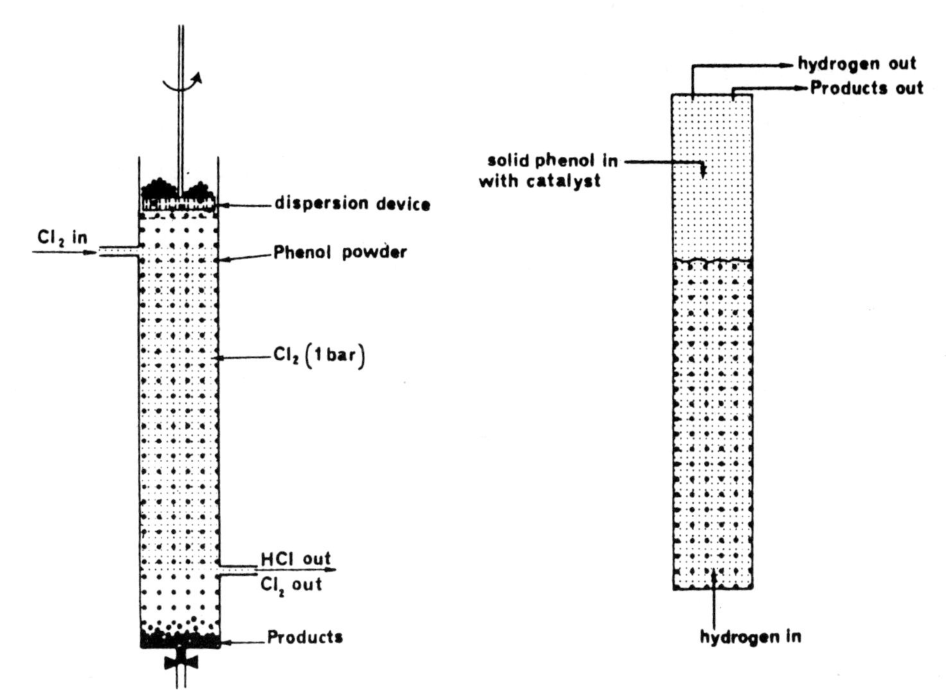

Fig. 39. Cocurrent fluid-solid reactor.

Fig. 40. Fluidized-bed reactor.

The different examples illustrate reaction possibilities with organic solids. In these conditions, costly solvents are avoided, at least during the reaction . Of course, there are other possibilities and Levenspiel's book can give other ideas (ref. 162).

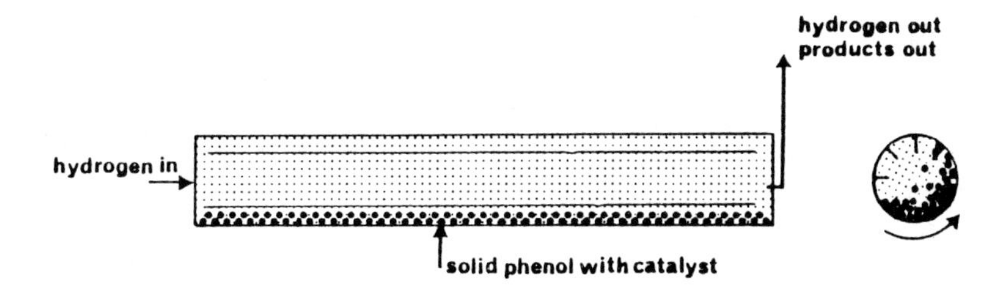

Fig. 41. Rotary reactor.

The conclusion is that the industrial development of the organic solid state chemistry must be carried out as well in the production of new higher value materials as in the improvement of processes. Moreover, we must keep in mind the fact that there are fields peculiar to the organic solid state chemistry which are very well developped at the industrial scale, even if the chemist is not always aware of these fields.

REFERENCES

1 A. Domenicano, A. Vaciago and C.A. Coulson, Acta Cryst., B 31, 1975, pp. 221-234.
2 A. Domenicano, A. Vaciago and C.A. Coulson, Acta Cryst., B 31, 1975, pp. 1630-1641.
3 A. Domenicano and A. Vaciago, Acta Cryst., B 35, 1979, pp. 1382-1388.
4 A. Domenicano, P. Mazzeo and A. Vaciago, Tetrahedron Lett., 1976, pp. 1029-1032.
5 A. Domenicano and P. Murray-Rust, Tetrahedron Lett., 1979, pp. 2283-2286.
6 R. Norrestam and L. Schepper, Acta Chem. Scand., A 35, 1981, pp. 91-103.
7 A. Domenicano, P. Murray-Rust and A. Vaciago, Acta Cryst., 1983, pp. 457-468.
8 T. Sakurai, Acta Cryst., 15, 1962, pp. 443-447.
9 F. Iwasaki and Y. Kawano, Acta Cryst., B 33, 1977, pp. 2455-2459.
10 F. Iwasaki, M. Sato and A. Aihara, Acta Cryst., B 32, 1976, pp. 102-106.
11 E.K. Andersen and I.G.K. Andersen, Acta Cryst., B 31, 1975, pp. 387-390.
12 R.J. Neustadt and F.W. Cagle Jr, Acta Cryst., B 31, 1975, pp. 2727-2729.
13 R. Kawai, S. Kashino and M. Haisa, Acta Cryst., B 32, 1976, pp. 1972-1975.
14 P. Coppens and G.M.J. Schmidt, Acta Cryst., 18, 1965, pp. 654-663.
15 S.K. Bhattacharjee and H.L. Ammon, Acta Cryst. B 37, 1981, pp. 2082-2085.
16 M. Maze-Baudet, Acta Cryst., B 29, 1973, pp. 602-614.
17 F.L. Hirshfeld, Israël J. Chem., 2, 1964, pp. 87-90.
18 T. Higashi and K. Osaki, Acta Cryst., B 33, 1977, pp. 607-609.
19 F. Pandarese, L. Ungaretti and A. Coda, Acta Cryst., B 31, 1975, pp. 2671-2675.
20 A. Neuman and H. Gillier-Pandraud, Acta Cryst., B 29, 1973, pp. 1017-1023.
21 C.J. Brown, Acta Cryst., 21, 1966, pp. 170-174.
22 K. Maartmann-Moe, Acta Cryst., 21, 1966, pp. 979-982.
23 H.C. Brown, D.H. Mc Daniel and O. Häfliger, Determination of Organic Structures by Physical Methods, edited by E.A. Braude and F.C. Nachod, New York Academie Press, pp. 567-662.
24 T. Kagawa, R. Kawai, S. Kashino and M. Haisa, Acta Cryst., B 32, 1976, pp. 3171-3175.
25 F. Iwasaki and Y. Kawano, Acta Cryst., B 34, 1978, pp. 1286-1290.
26 S. Ashfaquzzaman and A..K. Pant, Acta Cryst., B 35, 1979, pp. 1394-1399.
27 S. Rantsordas, M. Perrin, A. Thozet, Acta Cryst., B 34, 1978, pp. 1198-1203.
28 T. Sakurai, Acta Cryst., 15, 1962, pp. 1164-1173.
29 M. Perrin, C. Bavoux and A. Thozet, Acta Cryst., B 33, 1977, pp. 3516-3520.
30 A. Thozet and M. Perrin, Acta Cryst., B 36, 1980, pp. 1444-1447.
31 S. Rantsordas, M. Perrin, A. Thozet and S. Lecocq, Acta cryst., B 37, 1981, pp. 1253-1257.
32 S. Rantsordas and M. Perrin, Acta Cryst., B 38, 1982, pp. 1871-1873.
33 P. Coppens and G.M.J. Schmidt, Acta Cryst., 18, 1965, pp. 62-67.
34 M. Haisa, S. Kashino and T. Kawashima, Acta Cryst., B 36, 1980, pp. 1598-1601.
35 R. Perrin, R. Lamartine, J. Vicens, M. Perrin, A. Thozet, D. Hanton and R. Fugier, Nouveau Journal de Chimie, 10, 1986, pp. 179-190.
36 M. Perrin and M. Cherared, Acta Cryst., in press.
37 C.P. Brock and K.L. Haller, J. Phys. Chem., 88, 1984, pp. 3570-3574.
38 M. Perrin, K. Bekkouch and A. Thozet, Acta Cryst., to be published.
39 K. Nakatsu, H. Yoshioka, K. Kunimoto, T. Kinugasa and S. Ueji, Acta Cryst., B 34, 1978, pp. 2357-2359.
40 R.F. Bryan, P. Forcier and R.W. Miller, J.C.S. Perkin II, 1977, pp. 368-372.
41 E.J.W. Whittaker, Acta Cryst. 6, 1953, pp. 714-720.
42 G.D. Andreetti, R. Ungaro and A. Pochini, J.C.S. Chem. Comm., 1979, pp. 1005-1007.
43 R. Ungaro, A. Pochini, G.D. Andreetti and V. Sangermano, J. Chem. Soc. Perkin Trans II, 1984, pp. 1979-1985.
44 C.D. Gutsche, A.E. Gutsche, A.I. Karaulov, Journal of Inclusion Phenomena, 3, 1985, pp. 447-451.

45 H.C. Watson and A. Hargreaves, Acta Cryst. 11, 1985, pp. 556-562.
46 B. Deppisch, G.D. Nigam, E. Bernhard and R. Neidlein, Acta Cryst. B 34, 1978, pp. 3840-3842.
47 J.M. Robertson, Proc. Roy. Soc (London) A 157, 1936, pp. 77-99.
48 J.M. Robertson and A.R. Ubbelohde, Proc. Roy. Soc (London) A 167, 1938, pp. 122-135.
49 G.E. Bacon and R.J. Jude, Z. Krystallogr. 138, 1973, pp. 19-40.
50 K. Maartmann-Moc, Acta Cryst. 19, 1965, pp. 155-157.
51 E. Wajsman, M.J. Grabowski, A. Stepien and M. Cygler, Cryst. Struct. Comm, 7, 1978, pp. 233-236.
52 F.H. Herbstein and M. Kaftory, Act Cryst., B 32, 1976, pp. 387-396.
53 F. Iwasaki, I. Tanaka and A. Aihora, Acta Cryst., B 32, 1976, pp. 1264-1266.
54 A. Thozet, Thesis, LYON (1981)
55 A. Aihara, Bull. Chem. Soc. Japan, 33, n° 2, 1960, pp. 194-200.
56 S. Randtsordas, M. Perrin, A. Thozet and S. Lecocq, Acta Cryst., B 37, 1981, pp. 1253-1257.
57 C. Bois, Bull. Soc. Chim. Fr., 12, 1966, pp. 4016-4023.
58 M. Perrin and P. Michel, Acta Cryst., B 29, 1973, pp. 258-263.
59 S. Rantsordas et al., to be published
60 A. Thozet and M. Perrin, to be published.
61 M.T. Vanderborre, H. Gillier-Pandraud, D. Antona and P. Becker, Acta Cryst., B 29, 1973, pp. 2488-2492.
62 D.R. Holmes and H.M. Powell, Acta Cryst. 6, 1953, pp. 256-259.
63 C. Bavoux and A. Thozet, Acta Cryst., B 29, 1973, pp. 2603-2605.
64 D. Antona, F. Longchambon, H.J. Vandenbone and P. Becker, Acta Cryst., B 29, 1973, pp.1372-1376.
65 C. Bavoux and M. Perrin, Acta Cryst. B 29, 1973, pp. 666-668.
66 C. Bavoux and A. Thozet, Cryst. Struct. Comm. 5, 1976, pp. 259-263.
67 C. Bavoux, M. Perrin and A. Thozet, Acta Cryst. B 36, 1980, pp. 741-744.
68 M. Perrin, S. Rantsordas and A. Thozet, Cryst. Struct. Comm. 7, 1978, pp. 59-62.
69 M. Perrin and A. Thozet, Cryst. Struct. Comm. 3, 1974, pp. 661-664.
70 C. Bavoux and M. Perrin, Cryst. Struct. Comm. 8, 1979, pp. 847-850.
71 Von C. Scheringer, Z. fur Elektrochimie Kristallographie, 119, 1963, pp. 273-283.
72 H. Gillier-Pandraud, Bull. Soc. Chim. Fr., 6, 1967, pp. 1988-1995.
73 G. Albrecht and R.B. Corey, J. Am. Chem. Soc. 61, 1939, pp.1087-1103.
74 C. Bavoux, thesis, Lyon 1986.
75 F. Bertinotti, G. Giacomello and A. M. Liquori, Acta Cryst., 7, 1954, pp. 808-812.
76 W. Cochran, Acta Cryst., 6, 1953, pp. 260-268.
77 M. Sundaralingam and L. H. Jensen, Acta Cryst., 18, 1965, pp. 1053-1058.
78 J.D. Dunitz and J.A. IBERS, perpectives in structural chemistry, vol. 4, John Wiley and Sons, 1971, pp. 166-391.
79 C. Bavoux, A. Thozet, Cryst. Struct. Comm. 9, 1980, pp. 1115-1120.
80 T.C.W. Mak, W.H. Yu and Y.S. Lam, Acta Cryst. B34, 1978, pp. 2061-2063.
81 T. Ichikawa , K. Aoki and Y. Iitaka, Acta Cryst. B 34, 1978, pp. 2336-2338.
82 K. Aoki, T. Ichikawa, Y. Koinuma and Y. Iitaka, Acta Cryst. B 34, 1978, pp. 2333-2336.
83 T. Gramstad, S. Husebye and K. Maartmaan Moe. Acta. Chemica Scand., B 40, 1986, pp. 26-30.
84 O. Dideberg, L. Dupont and H. Compsteyn, Acta Cryst., B 31, 1975, pp. 637-640.
85 W.H. Watson, J. Galloy, F. Vögtle and W.M. Müller, Acta Cryst., C 40, 1984, pp. 200-202.
86 C. Bavoux, B. Belamri and M. Perrin, to be published.
87 W.H. Watson, J. Galloy and D.A. Grossie, J. Org. Chem., 49, 1984, pp. 347-353.
88 D.A. Grossie and W.H. Watson, Acta Cryst. B 38, 1982, pp. 3157-3159.
89 C.D. Gutsche, The calixarenes in L. Boschke Ed. in current chemistry, Springer Verlag, 122, 1984, pp. 1-47.

90 R.P. Rastogi, J. Sci. Ind. Res., 29, 1970, pp. 177-189.
91 H. Morawetz, Science, 152, 1966, pp. 705-711.
92 G.M.J. Schmidt, J. Chem. Soc., 1964, pp. 2014-2046.
93 M.D. Cohen, Angew. Chem. Int. Ed., 14, 1975, pp. 386-392.
94 I.C. Paul and D.Y. Curtin, Acc. Chem. Res., 6, 1973, pp. 217-225.
95 M. Kuhnert-Brandstatter. Thermomicroscopy in the Analysis of Pharmaceuticals, Pergamon Press, New York, 1971.
96 M. Perrin, Thesis, Lyon, 1974.
97 M. Perrin, P. Michel and R. Perrin, J. Chim. Phys., 7-8, 1975, pp. 851-854.
98 M. Perrin and P. Michel, Acta Cryst., B29, 1973, pp. 253-255.
99 R. Lamartine, C. Décoret, J. Royer and J. Vicens, Mol. Cryst. Liq. Cryst., 134, 1986, pp. 197-228.
100 M.D. Cohen, P. Coppens and G.M.J. Schmidt, J. Phys. Chem. Solids, 25, 1964, pp. 258-263.
101 B. Jones, J. Chem. Soc., 1938, pp. 1414-1421.
102 K.W.F. Kohlrausch and G.P. Ypsilanti, Monatsh. Chem., 66, 1935, pp. 285-292.
103 J.M. Robertson and A.R. Ubbelohde, Proc. R. Soc., London, Ser. A., 167, 1936, pp. 136-139.
104 S.R. Byrn, Solid State Chemistry of Drugs, Academic Press, New York, 1982.
105 P. Busetta, C. Courseille and M. Hospital, Acta. Cryst., B29, 1973, pp. 288-290.
106 S.R. Byrn, D.Y. Curtin and I.C. Paul, J. Amer. Chem. Soc., 94, 1972,pp. 891-898.
107 T.W. Lewis, D.Y. Curtin and I.C. Paul, J. Amer. Chem. Soc., 101, 1979, pp. 5717-5726.
108 T.W. Lewis, E.N. Duesler, R.B. Kress, D.Y. Curtin and I.C. Paul, J. Amer. Chem. Soc., 102, 1980, pp. 4659-4664.
109 C. Stora, Bull. Soc. Chim., 1971, pp. 2153-2161.
110 D.Y. Curtin, I.C. Paul, E.N. Duesler, T.W. Lewis, B.J. Mann and W. Shiau, Mol. Cryst. Liq. Cryst., 50, 1979, pp. 25-41.
111 C.T. Lin, P.Y. Siew and S.R. Byrn, J. Chem. Soc., Perkin Trans., 2, 1978, pp. 957-962.
112 N.J.L. Megson, Phenolic Resin Chemistry, Butterworths Scientific Publications, London, 1958.
113 M. Perrin, R. Perrin, A. Thozet and D. Hanton, Polymer Preprint, 24, American Chemical Society, Meeting, Washington, USA, 1983, pp. 163-164.
114 R. Lamartine, R. Perrin, J. Vicens, D. Gamet, M. Perrin, D. Oehler and A. Thozet, Mol. Cryst. Liq. Cryst., 134, 1986, pp. 219-237.
115 R.P. Rastogi, N.B. Singh and R.P. Singh, J. Solid State Chem., 20, 1977, pp. 191-200.
116 R.P. Rastogi, S. Bassi and S.L. Chadha, J. Phys. Chem., 66, 1962, pp. 2707-2708.
117 R.P. Rastogi, S. Bassi and S.L. Chadha, J. Phys. Chem., 67, 1963, pp. 2569-2573.
118 R.P. Rastogi and M.B. Singh, J. Phys. Chem., 70, 1966, pp. 3315-3324.
119 N.B. Singh, Indian J. Chem., 8, 1970, pp. 916-918.
120 R.P. Rastogi, and N.B. Singh, J. Phys. Chem., 72, 1968, pp. 4446-4449.
121 C.T. Lin, P.Y. Siew and S.R. Byrn, J. Chem. Soc., Perkin Trans., 2, 1978, pp. 963-968.
122 N. Kornblum and A.P. Lurie, J. Amer. Chem. Soc., 81, 1959, pp. 2705-2715.
123 D.Y. Curtin and D.H. Dybvig, J. Amer. Chem. Soc., 84, 1962, pp. 225-233.
124 S. Bittner, M. Idelsohn, Chemistry and Industry, 1975, pp. 838.
125 R.E. Buckle, E.A. Hausman, N.G. Wheeler, J. Amer. Chem. Soc., 1950, pp. 2494-2499.
126 G. Pellizzari, Gazz. Chim. Ital., 14, 1884, pp. 362-368.
127 H. Kolbe, J. Pr. Chem., (2), 10, 1874, pp. 89-95.
128 R. Schmitt, J. Pr. Chem., (2), 31, 1885, pp. 397-402.
129 M.M. Labes and H.W. Blakeslee, J. Org. Chem., 32, 1967, pp. 1277-1278.
130 D.Y. Curtin and A.R. Stein, Canad. J. Chem., 47, 1969, pp. 3637-3639.
131 L.J. Leeson and A.M. Mattocks, J. Am. Pharm. Assoc., 47, 1958, pp. 330-336.

132 A.Y. Gore, K.B. Naik, D.O. Kildsig, G.E. Peck, V.F. Smolen and G.S. Banker, J. Pharm. Sci., 57, 1968, pp. 1850-1857.
133 R.S. Miller, D.Y. Curtin and I.C. Paul, J. Amer. Chem. Soc., 96, 1974, pp. 6329-6349.
134 R. Lamartine and R. Perrin, Canad. J. Chem., 50, 1972, pp. 2882-2886.
135 R. Lamartine, M.F. Vincent-Falquet and R. Perrin, 3rd International Symposium Chemistry of the Organic Solid State, Glasgow, 18-22 September 1972.
136 M.F. Vincent-Falquet and R. Lamartine, Bull. Soc. Chim., 1-2, 1975, pp. 47-52.
137 W.T. Pennington, A.O. Patil, I.C. Paul and D.Y. Curtin, J. Chem. Soc., Perkin II, 1986, pp. 557-563.
138 R. Lamartine and R. Perrin, J. Org. Chem., 39, 1974, pp. 1744-1748.
139 R. Lamartine, C.R. Acad. Sci., Paris, 279, 1974, pp. 429-431.
140 R. Lamartine, R. Perrin, G. Bertholon and M.F. Vincent-Falquet, Mol. Cryst. Liq. Cryst., 32, 1976, pp. 131-136.
141 R. Lamartine, R. Perrin, G. Bertholon and M.F. Vincent-Falquet, J. Amer. Chem. Soc., 99, 1977, pp. 5436-5438.
142 R. Lamartine, R. Perrin and M. Perrin, Nouveau Journal de Chimie, 7, 1983, pp. 185-189.
143 J. Vicens, A. Thozet, M. Perrin and R. Perrin, C.R. Acad. Sci., Paris, 294, 1982, pp. 717-720.
144 R. Lamartine, G. Bertholon and R. Perrin, Mol. Cryst. Liq. Cryst., 52, 1979, pp. 293-302.
145 R. Lamartine, R. Perrin, Spillover of Adsorbed Species, Elsevier Science Publishers B.V., Amsterdam, 1983, pp. 251.
146 R. Lamartine, J. Vicens and R. Perrin, Reactivity of Solids, Elsevier Science Publishers, B.V., Amsterdam, 1985, pp. 721.
147 P.B.D. De La Mare. Electrophilic Halogenation, Cambridge University Press, London, 1976, p. 25.
148 A. Knop, L.A. Pilato, Phenolic Resins, Springer-Verlag, Berlin, 1985, pp. 166-168.
149 H. Hiraoka, Fonctionally Substitued Novolak Resins, in Materials for Microlithography, L.F. Thompson, C.G. Willson, J.M.F. Frechet, Eds, ACS Symposium series 266, Washington, 1984, pp. 339-360.
150 H. Martin, Pesticide Manual, British Crop. Protection Council.
151 K. Venkataraman, The Chemistry of Synthetic Dyes, vol. I-VII, Academic Press, New-York, 1952-1974.
152 E.N. Abahart, Dyes and their Intermediates, Chemical Publishing, New York, 1977.
153 H. Morawetz, Organic Explosives, in Physics and Chemistry of the organic Solid State, D. Fox, M.M. Labes, A. Weissberger, Eds, Interscience, 1963, pp. 316-318.
154 T. Urbanski, Chemistry and Technology of Explosives, vol. I, II and III, MacMillan, New York, 1964-1967.
155 J. March, Advanced Organic Chemistry, third Edition, Wiley, New York, 1985, pp. 491.
156 R. Piria, Ann. Chim. Phys., 69, 1838, pp. 281.
157 R. Piria, Prakt. Liebigs, Ann., 30, 1839, pp. 151.
158 M.F. Berny and R. Perrin, Bull. Soc. Chim. Fr., 1967, pp. 1014.
159 M.F. Berny and R. Perrin, C.R. Acad. Sc. Paris, 265, 1967, pp. 492-493.
160 A.S. Lindsey and H. Jeskey, Chem. Rev., 57, 1957, pp. 583-620.
161 D.R. Stevens and R.S. Bowman, U.S. Patent, 2, 655, 546, 1953.
162 O. Levenspiel, Chemical Reaction Engineering, Second Edition, John Wiley and Sons, New York, 1972.

Chapter 9

GAS-SOLID REACTIONS AND POLAR CRYSTALS

IAIN C. PAUL AND DAVID Y. CURTIN

1 REACTIONS OF SINGLE CRYSTALS WITH GASES

1.1 Introduction

Reactions of organic crystals with gases such as ammonia or bromine have been known for many years (ref. 1) and the literature on this subject is much too voluminous to be reviewed here (ref. 2). The present discussion will be confined to gas-solid reactions of single crystals of certain carboxylic acids and anhydrides. Such reactions typically occur anisotropically and it has been possible to correlate qualitatively the rates of reaction in various crystal directions with the internal structure of the crystal. One of the simplest such reactions is that of ammonia gas with crystals of benzoic acid and its ring-

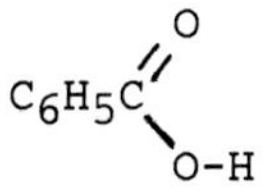

$$C_6H_5C(=O)-O-H$$

substituted derivatives. Compounds in this group generally crystallize as cyclic hydrogen-bonded dimers (Fig. 1).

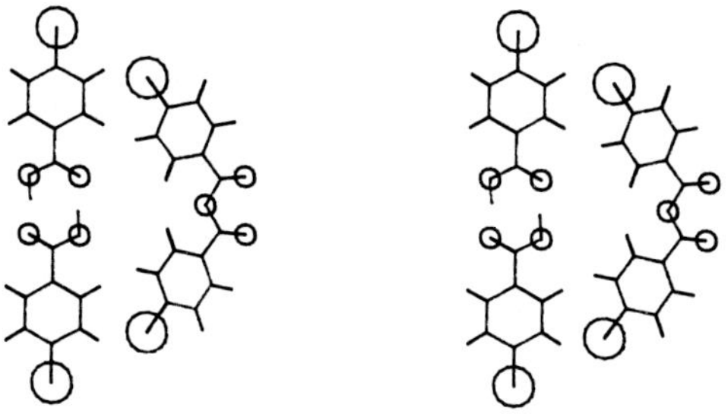

Fig. 1. Stereopair drawings. Left: the hydrogen-bonded dimer of p-chlorobenzoic acid (ref. 3). Chlorine atoms are represented by the larger circles, oxygen atoms by the smaller. Right: a molecule of p-chlorobenzoic anhydride (ref. 6) in the conformation found in the crystal. There is a two-fold rotation axis through the central oxygen atom. The hydrogen atoms of the anhydride were not located in the X-ray structure determination and have been estimated by the present authors.

In Fig. 2 are shown crystals of p-chlorobenzoic acid undergoing reaction with ammonia gas (ref. 3a) and in Fig. 3, a stereopair drawing of the internal structure of such a crystal. Ammonia molecules attack preferentially from those crystal faces where carboxyl groups are accessible and, as reaction occurs, the ammonium p-chlorobenzoate formed separates from the reactive face of the parent crystal and permits passage of the ammonia to the

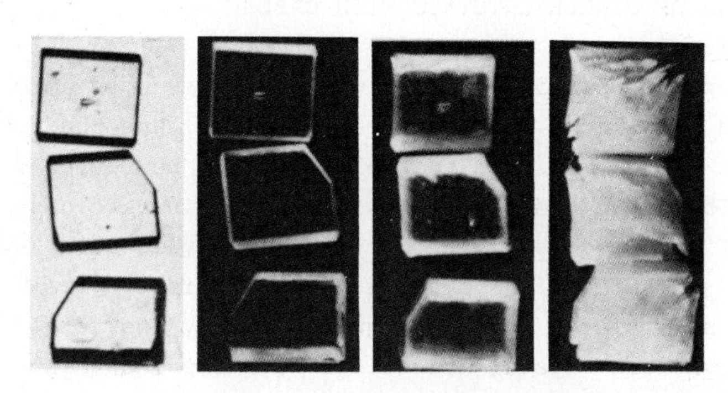

Fig. 2 Crystals of p-chlorobenzoic acid reacting with ammonia gas. Left picture (with illumination from behind the crystal): before reaction. The two middle pictures illuminated from above show the regions where reaction has occurred as white due to scattering of light by the microcrystallites formed. Right picture: the product after essentially complete reaction. (Reprinted with permission from ref. 3a.

next layer of acid molecules. Reaction thus continues into the interior of the acid crystal until it is complete. The ammonium salt produced reassembles as a microcrystalline aggregate which retains approximately the original shape and is held together so firmly that it might appear on casual inspection to be a single crystal. However, observation through a light microscope shows that crystalline order is not retained since the reacted regions consist of microcrystalline aggregates of ammonium salt of a particle size which scatters visible light. Furthermore, such a "crystal" after reaction, when mounted unground in a powder camera, gives a powder photograph essentially identical with that of the ground ammonium salt.

1.2 Relationship of Anisotropy to Crystal Packing

Crystals of benzoic acids, typified by p-chlorobenzoic acid just discussed, are generally composed (Fig. 3, ref. 4, 5) of layers of hydrogen-bonded dimers with alternating slices of nonpolar aromatic rings and polar carboxyl dimer groups. Such structures provide relatively sharp differentiation between the

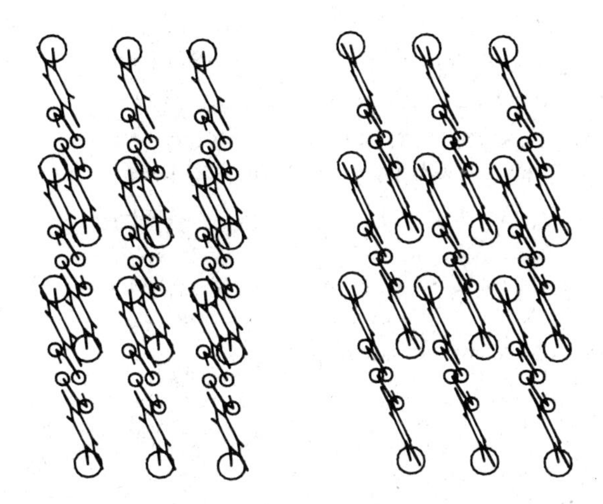

Fig. 3. A stereo pair drawing of the structure of p-chlorobenzoic acid (ref. 3) in nearly the same orientation as the crystals in Fig. 2. The b axis is vertical in the plane of the page. The view is onto the (100) face (but rotated 10° around b in order to reduce overlap of the upper molecules with the lower). The larger circles represent chlorine atoms and the smaller circles oxygen. Note the tight structure of the (100) face which might be expected (as is found experimentally) to be relatively impervious to passage of ammonia gas through gaps or channels.

reactive polar regions and the inert non-polar regions of the crystal. Thus the contrast between the facility of reaction of ammonia gas at the side faces made up of carboxyl groups with the lack of reaction at the major non-polar faces is readily apparent by visual microscopic observation. The rates of migration of the reaction fronts from the various side faces are too similar, however, to permit a correlation of relative rate with structure.

The anhydrides of benzoic acids have structures of alternating polar and non-polar layers similar to the acids but with the reactive anhydride groups instead of the carboxylic acid groups making up the reactive polar regions (Fig. 1,4). The behavior of centrosymmetric crystals of p-chlorobenzoic anhydride (ref. 6) (Fig. 5) shows anisotropic reaction with ammonia gas (ref. 7) similar to that observed with p-chlorobenzoic acid.

Benzoic anhydrides however, unlike the benzoic acids, often

crystallize in non-centrosymmetric space groups. Thus, all of

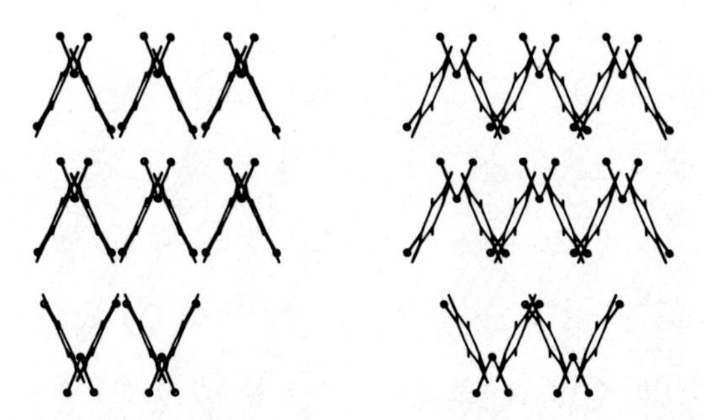

Fig. 4. A stereopair drawing of the crystal structure of p-chlorobenzoic anhydride (ref. 6) in approximately the same orientation as the crystals in Fig. 5. The long morphological axis is a and the major faces {001}. The two rows toward the top of the drawing are made up of molecules of the same chirality with the carbonyl groups pointing in the same direction. The next layer into the plane of the page, of which just two molecules are shown for clarity, is related to the first layer by the operation of a center of symmetry which changes the chirality of the molecules and also reverses their directions, thus making the crystal centrosymmetric.

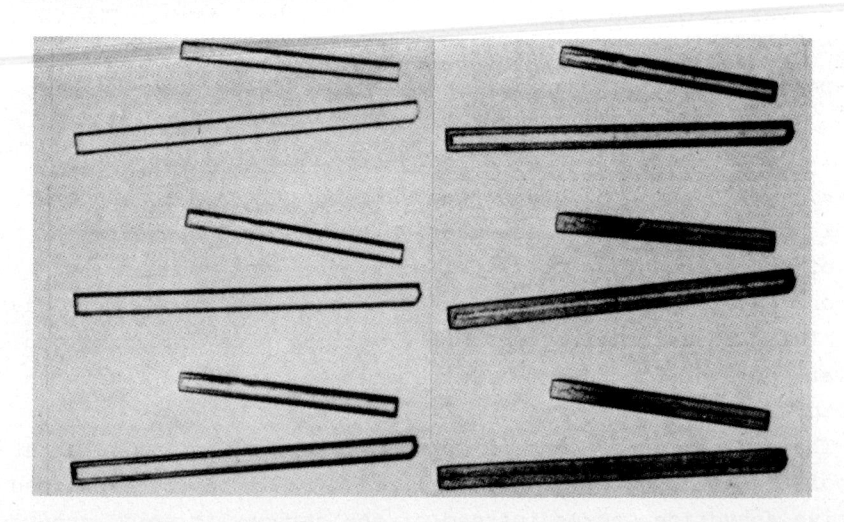

Fig. 5. Crystals of p-chlorobenzoic anhydride (space group P2/c) undergoing reaction with ammonia gas (reprinted with permission from ref. 7). The crystals before reaction are shown in the upper left photograph. The progression of the reaction fronts is shown in the sequence of photographs of the same crystals after exposure to ammonia gas.

the 16 benzoic acids with a single methyl group or halogen atom substituent in an ortho, meta, or para position on the ring (including the parent unsubstituted benzoic acid) crystallize in centrosymmetric space groups (ref. 5, 8)--- in each case the crystal is made up of centrosymmetric and nearly planar hydrogen-bonded carboxylic acid dimers lying at crystallographic centers of symmetry (see Fig. 1). However, of the eight corresponding anhydrides whose space groups are known (ref. 9) six crystallize in non-centrosymmetric space groups. The anhydride molecules show conformational similarity in each of these structures. Because of the C-O-C bond angle of the central -COOCO- group a centrosymmetric molecular geometry is not attainable. Instead the anhydride molecules adopt a structure with two-fold axial symmetry--- either lying on a crystallographic two-fold axis or approximating such a conformation (see Fig. 1). This difference in molecular symmetry seems to be at least partially responsible for the greater tendency of anhydrides to crystallize in non-centrosymmetric space groups. Consequences of this lack of a

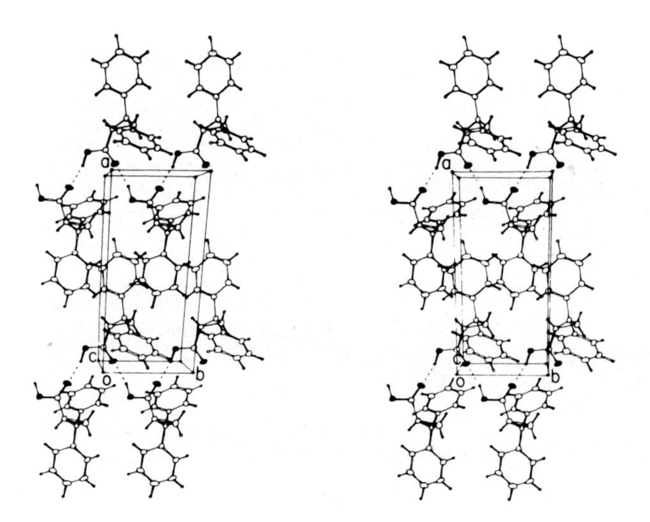

Fig. 6. A stereopair drawing of the crystal structure of (*S*)-(+)-2,2-diphenylcyclopropanecarboxylic acid (space group $P2_1$) (reprinted with permission from ref. 11). The structure has chains formed by hydrogen-bonds between the carboxylic O-H of one molecule and the carboxyl carbonyl group of the next. Note that this structure has all hydroxyl groups to the left and carbonyl C=O bonds pointing toward the right.

center of symmetry will be discussed later in this chapter.

Certain carboxylic acids, particularly single enantiomers of chiral acids, cannot crystallize in pairs related by centers of symmetry. Such compounds can crystallize instead with a different hydrogen bonding arrangement in the form of chains of acid molecules connected by intermolecular hydrogen bonds (Fig. 6). This change in the nature of the hydrogen-bonding was anticipated by Walborsky and his associates (ref. 10) on the basis of differences in the infrared spectra of the two classes of acids and confirmed by X-ray structure analysis (ref. 11). Again the anisotropy observed can be correlated with the crystal structure; single crystals of such acids show a marked preference for reaction from the two ends of the crystal at which the hydrogen-bonded carboxyl chains are accessible (ref. 12). This type of reaction along the two directions of a single crystal axis has been designated as "unitropic".

1.3 Use of the Reaction of Chiral Gases with Solids to Distinguish Between Enantiomeric Crystals.

Pasteur's first method of resolution, hand separation of enantiomeric crystals, requires not only that the compound crys-

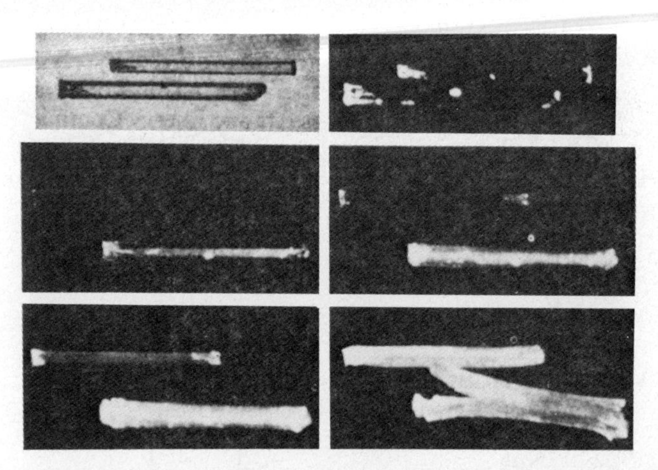

Fig. 7. A (+) (upper) and (-) (lower) crystal of 2,2-diphenylcyclopropanecarboxylic acid competing for (+)-1-phenylethylamine vapor: top left photograph, the crystals before reaction viewed with transmitted light; top right, before reaction viewed between crossed polarizing filters; middle left, after exposure to amine vapor for 5 min; middle right, after 20 min; lower left, 1 hr; and lower right, 24 hr. The selective reaction from the two ends of the crystal along the hydrogen-bonded chain which had been observed with ammonia is not found with the larger phenylethyl amine vapor. (Reprinted with permission from ref. 13.)

tallize as a conglomerate but also that each crystal develop faces which identify it as (+) or (-). If it were not for this second requirement the method would be more generally applicable. An alternative method of marking the (+) and (-) crystals to permit their separation involves their reaction with a chiral gas. It has been found (ref. 13) that a single enantiomer of 1-phenylethylamine, when it is allowed to react with a pair of crystals of enantiomeric acids, reacts preferentially with one of the two enantiomers (Fig. 7). The acids employed included not only the enantiomeric 2,2-diphenylcyclopropanecarboxylic acids discussed above and but also the enantiomeric tartaric acids and mandelic acids to be discussed further in this chapter.

The use of a chiral gas for labeling the (+) and (-) enantiomer thus supplements Pasteur's classic use of visual observation of hemihedral faces to be discussed below.

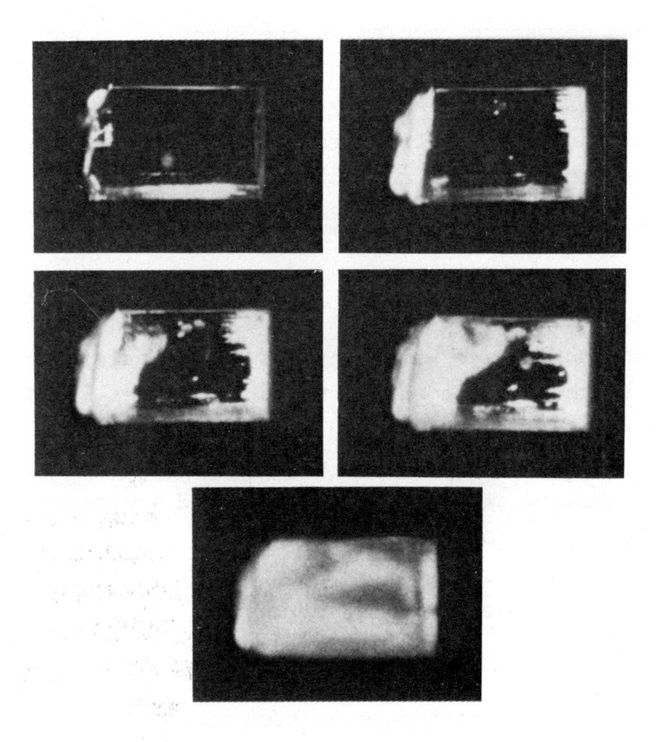

Fig. 8. A crystal of (R)-(+)-2,2-diphenyl-1-methylcyclopropanecarboxylic acid undergoing reaction with ammonia vapor showing unitropic reaction along the long morphological axis of the crystal. The crystal was illuminated between crossed polarizing filters and was set in the extinction position. (Reprinted with permission from ref. 12.)

1.4 Effect of Crystal Defects on Reactions of Carboxylic Acids and Anhydrides with Ammonia Gas. Solid state reactions are often profoundly influenced by the presence of nucleation sites which may be required for initiation of reaction (ref. 14). Reactive anhydrides, on microscopic examination at early stages of reaction appear to have been attacked by ammonia gas in a completely uniform manner. Similarly, frontal migration during such a reaction often seems to occur quite uniformly. However, the unitropic reaction of crystals of (R)-(+)-2,2-diphenyl-1-methylcyclopropanecarboxylic acid has shown clear evidence of non-uniform advance of the reaction front--- tongues of reaction appeared to precede the advancing front (Fig. 8). A similar phenomenon has been observed in the reaction of crystalline acenaphthylene-1-carboxylic acid with ammonia gas (ref. 15). It seems probable that crystal defects may have a more general role in reactions of this type.

A major and pervasive problem in the study of anisotropy of solid-gas reactions of crystals with an unreactive major face is that if the faces are not relatively free from defects the reaction may "break through" the top surface and, by forming a new reactive center, obscure the inherent anisotropy being observed.

1.5 Factors Influencing Rates of Reaction of Crystalline Solids with Gases

Rates of such reactions may be influenced by a number of factors. Among them are the initial rates of attack of the gas molecules at the outer layer of molecules on the various crystal faces, the relative rates at which the various faces are disrupted after reaction occurs, the relative rates at which the reagent gas can pass through to the next molecular layer in the crystal either through gaps in the newly formed microcrystallites or by some other mechanism such as diffusion along surfaces of the microcrystallites formed during reaction. Of course, crystal defects and the presence of impurities are likely to influence many of these rates. Because of our almost total ignorance concerning even the structures of the faces of real crystals which may well have transient species such as water vapor adsorbed on them we must be cautious in drawing conclusions about detailed mechanism from simplistic observations of the type described. Nevertheless observations of this sort seem to provide some insight into how such reactions occur and, in certain cases equally importantly, about the internal orientation of molecules in the crystal.

Beginning with observations of the type described above, there is a natural progression to a consideration of crystal symmetry (chiral and polar crystals) and, finally, to the subject of electrical properties of crystals. In the remaining sections of this review we will follow this progression.

2 THE POLAR AXIS IN SOLID STATE CHEMISTRY

2.1 Structural and Electrical Polarity

The chemical significance of chirality resulting from the absence from a molecule (or crystal) of a plane or center of symmetry is well-known to solution chemists as well as to those interested in the solid state. The implications of the absence of a center of symmetry to produce a "polar" molecule or crystal is less generally appreciated.

As has been pointed out (ref. 16) the word "polar" is a source of potential confusion (ref. 17). It has been used in other areas of chemistry as well as in crystal physics to refer to the presence of a permanent electric dipole moment. On the other hand it has long been used by crystallographers to refer to structural polarity (ref. 18, 19,). In this usage, a crystal is said to be polar if there exist vectors through the crystal such that on progression in a positive sense along the vector the arrangement of atoms is different from that found on progression in the negative sense along the same vector (ref. 18). Although such crystals often have a permanent dipole moment this may not necessarily be allowed by crystal symmetry. For example, in certain point groups there are structurally polar directions without a resultant electric dipole moment because the symmetry elements operating on a molecular electric dipole produce additional dipoles whose vector sum is equal to the original in magnitude and opposite in direction; thus the effect of the original dipole is cancelled. Such a crystal can, however, be piezoelectric--- that is generate a change in separation of centers of positive and negative charge when pressure is applied in an appropriate direction. An example of a common point group whose crystals are chiral and piezoelectric but not pyroelectric is the point group 222. The International Tables for Crystallography (ref. 16) tabulates the occurrence of specific physical properties in non-centrosymmetric crystals and also the polar and non-polar directions in each class. In this review we will avoid confusion by referring to "structural polarity" and "electrical polarity". A compound whose crystals are both structurally and electrically polar is the carboxylic acid in Fig. 6. Note that in an ideal crystal of that acid the components of the all of the carboxylic C=O vectors along the polar b-axis point in the same direction. The crystal is therefore structurally polar but, in addition, because of the dipole moment known from solution studies (ref. 20) to be associated with the carboxyl group, should be expected to have an electric dipole as well.

The presence of a polar axis often confers on a crystal properties which are not only unique but which have been and will continue to be of great technological importance. Some of these have been discussed in previous reviews (ref. 18, 19). It should be emphasized that a crystal can be chiral without having an electric dipole and can be structurally (and electrically) polar without being chiral. Furthermore a crystal can have structural polarity without an electric dipole moment.

p-Bromobenzoic anhydride, unlike p-chlorobenzoic anhydride discussed earlier, crystallizes in space group C2 in a structure which is chiral and also both structurally and electrically polar

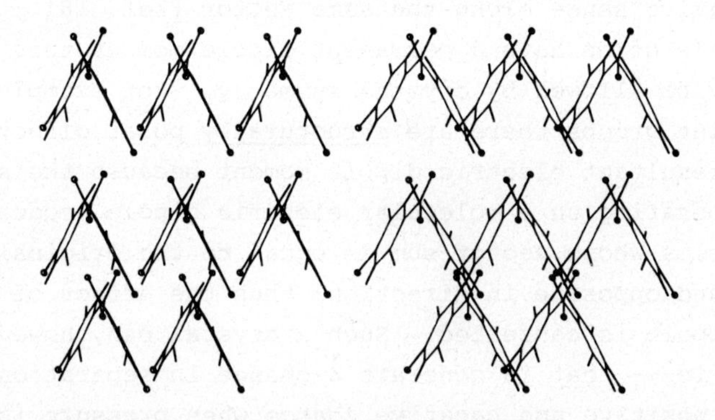

Fig. 9. The crystal structure of p-bromobenzoic anhydride (space group C2) in the same orientation as that of the p-chloro compound in Fig. 3. The structures of molecular layers of the two compounds are similar but in this case the next layer into the page (indicated by the two molecules in the row at the lower part of the drawing) is related to the first by a C-translation so that its molecules have the same chirality and molecular orientation as those in the original layer. Thus this crystal is chiral and both structurally and electrically polar.

(ref. 21) (Fig. 9). Since organic crystals are generally made up of molecules which are electrically only weakly polar compared to salts or other ionic compounds, it might appear unlikely that they could have interesting electrical properties. However, the pyroelectric effect, the change in the separation of centers of positive and negative charge when an electrically polar crystal is heated or cooled, has been recognized since the 19th century as a property of many organic crystals. Pyroelectricity has been given a comprehensive review by Lang (ref. 22). A compact statement of the criterion for pyroelectricity (ref. 23) is that "a crystal is pyroelectric if the primitive cell possess a dipole moment".

Piezoelectricity, the separation of centers of positive and negative charge on expansion or compression of a crystal, has been technically important since it has been employed as the basis of a variety of devices ranging from the microphone to the ultrasonic cleaner. Piezoelectricity, like pyroelectricity, is found only with crystals lacking a center of symmetry; there are, however, differences in the symmetry requirements of the two (ref. 16). With some effort it has been possible to separate experimentally the measurement of the pyroelectric and the piezoelectric effect (ref. 22) but this is not easy and, in later discussions of the pyroelectric effect in this chapter there will be a tacit assumption that there may be a piezoelectric component in the crystals being discussed.

Two other striking properties of electrically polar molecular crystals have received recent attention. Triboluminescence, the emission of light on application of a mechanical force to a solid, is shown by many common substances such as sucrose and tartaric acid. This phenomenon has been investigated most extensively by Zink and his associates (ref. 24). At the beginning of their work it was postulated that in order for a crystal to be triboluminescent it had to be electrically polar. There were, however, at that time a number of seeming exceptions to this postulate; many of these exceptions have been shown on reinvestigation to be based on misassignments of the crystal symmetry. The detailed mechanism of triboluminescence is still under active investigation.

A related phenomenon is pyroelectric luminescence (ref. 25). Here an electrically polar crystal emits light when heated or cooled. This property, too, has been demonstrated with common organic structures, such as resorcinol, phthalic anhydride, and m-bromonitroaniline. A survey has shown that only pyroelectric (electrically polar) crystals show this property, and substances which are piezoelectric but not pyroelectric show no activity. Careful examination has failed to reveal cracking of the crystal during pyroelectric luminescence which differs from triboluminescence in this respect.

Quantitative study of piezoluminescence has led to the development of piezomodulation spectroscopy which has been applied by Eckhardt (ref. 26) and his associates to a study of molecular motion in crystals. These and other studies currently being pursued should lead to a more detailed understanding of triboluminescence and related phenomena.

2.2 Chirality, Structural Polarity and Crystal Morphology. Consequences of Polar Directions in a Crystal

We will examine some of the properties of chemical interest of chiral and polar crystals and the relationship between the absolute direction of orientation of molecules along the polar axis and the absolute configuration of molecules in a crystal. The relationship of structural polarity to crystal growth and reactivity and chemical applications of the pyroelectric effect are of particular interest.

2.3 Use of a Polar Axis to Investigate Mechanism of a Gas-Solid Reaction

p-Bromobenzoic anhydride (Fig. 9) as well as the m-bromo- and m- and p-iodoanhydrides form crystals which are both structurally and electrically polar. A study of the reaction of p-

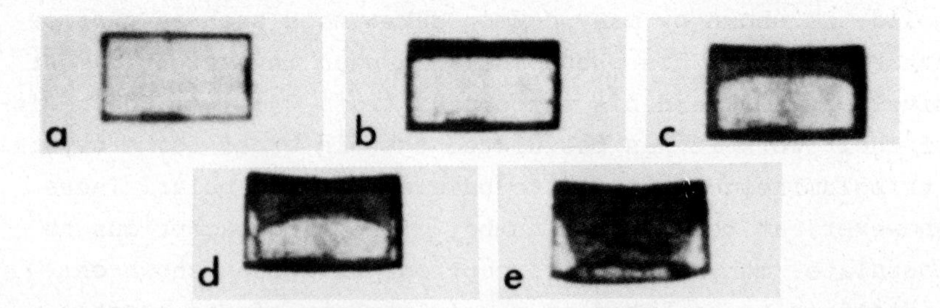

Fig. 10. Crystals of p-bromobenzoic anhydride (in the same orientation as the structural drawing in Fig. 9) undergoing reaction with ammonia gas (reprinted with permission from ref. 2a). In this habit the major face is (100) with the b axis vertical in and the c axis horizontal in the plane of the page.

bromobenzoic anhydride with ammonia gas demonstrated that there is preferred reaction from one end of the polar axis (Fig. 10).

Determination of the absolute orientation of the crystal structure along the polar axis (ref. 21b) by means of anomalous scattering (ref. 27) showed that the preferred attack occurs from the carbonyl-oxygen end of the polar axis. This has suggested that reaction involves initial hydrogen bonding of the ammonia molecules to carbonyl oxygen atoms followed by attack of the nucleophilic nitrogen atom at the sp^2 face of the other carbonyl group in the same molecule. When an ammonia molecule is plotted on a stereopair drawing of a molecule of the anhydride with the coordinates found in the crystal structure determination the geometry seems to be reasonable for such a transition state (Fig. 11).

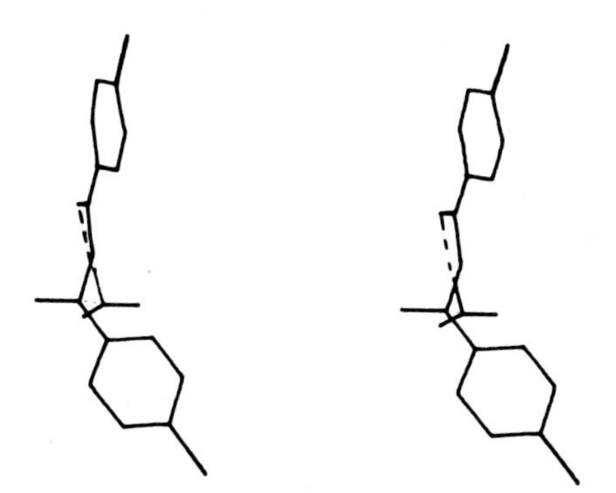

Fig. 11. Possible attachment by hydrogen bonding of an ammonia molecule to *p*-bromobenzoic anhydride in the conformation found in the crystal (reprinted from ref. 21b with permission).

It seems likely that in other cases, also, the study of polar crystals may give information difficult to obtain in any other way about the nature of such a reaction.

2.4 Effect of Polar Axes on Crystal Morphology

If an idealized crystal is centrosymmetric each crystal face is required by symmetry to be accompanied by a parallel face at the opposite side of the crystal. Even when the crystal lacks a center of symmetry, unless a crystal face cuts a polar direction of the crystal, it is still required by symmetry to be accompanied by a parallel opposite face. It has long been recognized that hemihedral crystal faces (faces not accompanied by parallel counterparts at the opposite end of the crystal) can serve to differentiate the two ends of a polar axis. (The parallel opposite face may, of course, develop even though not required by symmetry.) Although simple differentiation of the ends of a polar axis shows the existence of the polar axis and is valuable in its own right, it is less so than it would be were it possible to correlate the differentiating features with the absolute orientation of the internal structure in a particular crystal. Thus, in the classic set of volumes (ref. 28) produced by P. Groth in the years 1906-1919, there are many drawings of crystals of organic compounds including a number whose lack of a center of symmetry gives rise to hemihedral crystal faces indicating the presence of a polar axis. However in very few of these examples, even though many of their crystal structures have since been determined, has the absolute direction of the polar axis been correlated with the internal structure of the crystal.

The relationship of the absolute direction of the polar axis to the absolute configuration and morphology of a crystal can be illustrated with a substance which is both polar and chiral, (+)-tartaric acid, whose X-ray crystal structure, absolute configuration, and morphology are now known. Crystals of the enantiomers of tartaric acid belong in space group $P2_1$ (point group 2). The

(2*R*)(3*R*)-(+) or *L*-(+)   (2*S*)(3*S*)-(-) or *D*-(-)

symmetry of point group 2, with only a two-fold rotation axis parallel to the polar axis, is particularly suited to this kind of argument and several examples of crystals in this point group will be discussed in some detail.

We will show, first, how the crystal morphology (ref. 29) and X-ray structure (ref. 30) of an individual enantiomer, (+)-tartaric acid, of known absolute configuration (ref. 31) can be used to determine the absolute orientation of molecules along the polar axis. The crystal morphology of this compound and its enantiomer were investigated (ref. 29) as early as 1850 but the X-ray structure was first reported 100 years later (ref. 30). In Fig. 12 is a stereopair drawing of a crystal of (2*R*)(3*R*)-tartaric acid looking down on the (001) face with the polar *b* axis oriented toward the top of the page. The two-fold symmetry can be

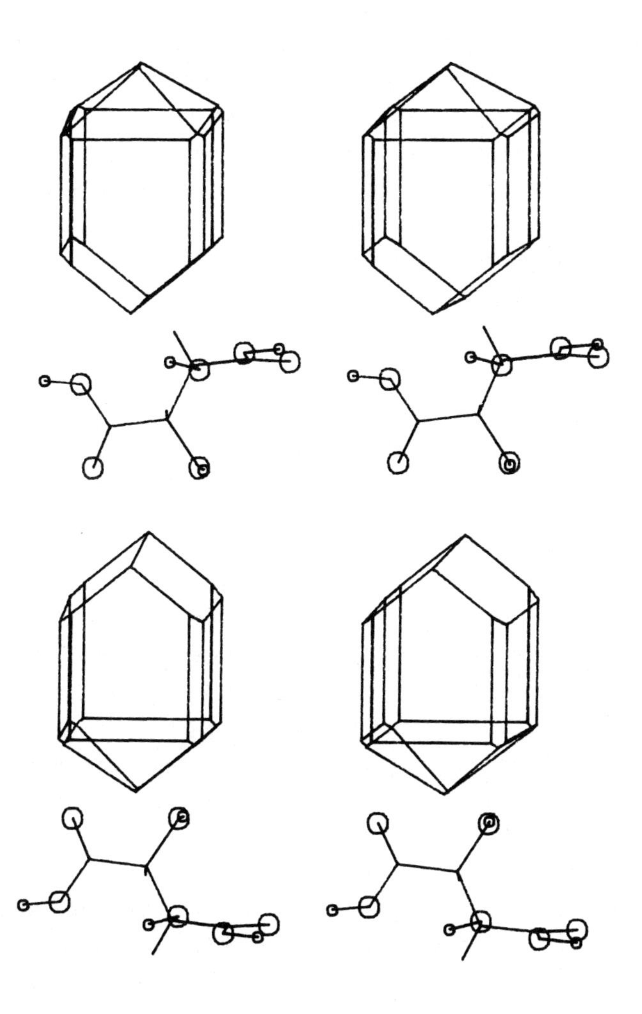

Fig. 12. Upper pair of drawings, a stereopair of an L-(+)-tartaric acid crystal and a molecule in the same orientation. The view is onto the (001) face with the polar b axis vertical in the page. The major faces bounding the crystal on the right and left sides are the {100} faces and slope outward on the right side and inward on the left. Lower pair of drawings, a stereopair of a D-(-)-tartaric acid. Again the (100) face bounding the crystal on the right side slopes outward and that on the left side of the crystal inward.

recognized from the shape of the crystal and the polar nature of the b axis is evident since faces intersecting the b axis are hemihedral. Below the drawing of the crystal is the reference molecule in the same orientation. The structure is made up of two groups of symmetry-related molecules: **I**, the reference molecule, and **II**, related to **I** by operation of a two-fold screw axis involving a 180°-rotation around the polar b axis and a minuscule 3-Å translation along b. The molecule **I** is chiral and, as seen in Fig. 12, can be expected to be electrically polar. The sum of the projections on the electrically polar axis, b, of all interatomic vectors of molecule **I** is identical to the sum of those of molecule **II** related by the screw axis; therefore both molecules make the same contribution to the crystal dipole moment and we need only consider the dipole of molecule **I**. It was long ago reported (ref. 29b) that crystals of the enantiomers of tartaric acid are strongly pyroelectric.

Identification of the +b direction of such a crystal can be made by visual inspection; in the right-handed axial system (used by the X-ray crystallographers who determined the crystal structure), an L-(+) crystal oriented on (001) with the sloping (100) face tilted upward on the right must have +b toward the top of the page as shown in Fig. 12. The crystal structure with the signs of atomic coordinates chosen so as to give the correct absolute configuration of the molecules making up the crystal is that shown in the upper part of Fig. 12. Thus the molecules of L-(+)-tartaric acid are oriented with one carboxylic hydroxyl group pointing along +b and the other almost normal to the polar b-axis. It is important to note that if the right-handed axial system is maintained a crystal of the enantiomer of opposite configuration in the same orientation (with the side face on the right sloping upward) has an external shape with the bullet nose of the crystal pointed in the opposite direction and the molecular orientation reversed as shown in the lower part of Fig. 12. This crystal is related to the first by reflection in a horizontal plane normal to the page. Again one carboxylic C-O-H group points toward the bullet nose of the crystal--- in this case along -b. Thus, the absolute configuration together with the crystal morphology permits the unequivocal deduction of the orientation of molecules along the polar b axis by visual inspection. Note that the hemihedral crystal faces served to mark the ends of the polar axis but it is only with the knowledge of the absolute configuration derived from other sources that the orientation of molecules in the crystal along the polar axis can be assigned.

Attempts to understand the nature of facial development during crystal growth have been made for many years (ref. 32). Unfortunately, as was pointed out by Wells (ref. 33), crystal shape can be drastically altered by changes in the conditions employed for the crystallization and, in particular, by the presence of impurities. (The successful use of deliberately added impurities will be mentioned below.) A further possible difficulty inherent in drawing conclusions from crystal morphology should be recognized. An individual crystal is normally one of a multitude of widely varying sizes and shapes. If observations are too limited there is the possibility that attention may be focussed on unrepresentative individuals whose shapes, having been influenced by some accident of growth, fail to show the true morphology of an ideal crystal. Nevertheless the work of the crystallographers on which Groth's compilation (ref. 28) is based has been remarkably reliable as a source of information about the common crystalline forms of the many organic crystals discussed there.

Other methods of determining experimentally the absolute orientation of the molecular structure along the polar axis (in this example, determining whether the carboxylic hydroxyl group points toward the top or toward the bottom of the crystal in Fig. 12) will be discussed below. Reasons for our interest in making this correlation should be apparent. If the absolute configuration of a crystal were unknown, it could be derived from an experimental determination of the orientation of molecules along the polar axis (the direction of the crystal dipole) together with the crystal morphology.

2.5 Use of Selective Crystal Growth or Dissolution to Mark the Ends of the Polar Axis

The selective growth of a structurally polar crystal in one direction along the polar axis was observed (ref. 33) many years ago to occur when crystals of resorcinol are allowed to grow in water. It was proposed that since the crystal faces at one end of the polar axis are hydroxylic and at the other hand hydrophobic, water blocks crystallization selectively by adsorption at the hydroxylic faces. Under ideal conditions crystals grow so that the macroscopic habit has the point group symmetry corresponding to the symmetry of the molecular arrangement in the unit cell. The reverse process, the dissolution of the crystal, will also be determined by this symmetry and is less susceptible to influence by impurities. Treatment of crystal faces by small amounts of solvent leads to etching and the symmetry relationships of the etch patterns along the different directions of the crystal can be used to determine the presence and orientation of a polar axis (Fig. 13) (ref. 34a). Thus, different rates of

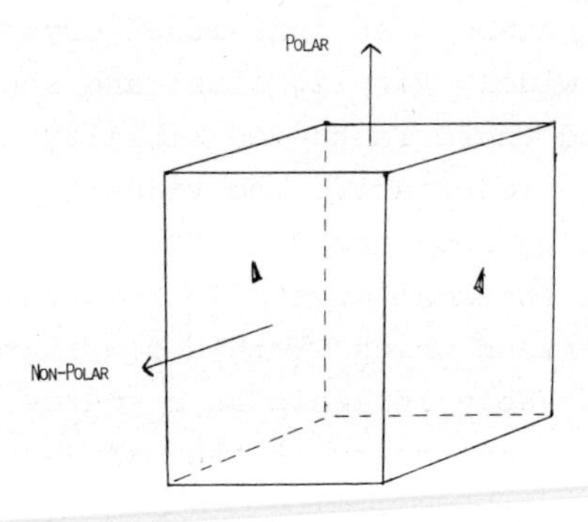

Fig. 13. Idealized view of a crystal showing asymmetric etch pits on opposite faces in the presence of a polar axis. The crystal shown belongs to point group 2.

dissolution in water have been observed at the opposite ends of the polar axis of lithium iodate crystals (ref. 34b).

2.6 Use of the Effect of "Impurities" on Crystal Morphology to Deduce the Absolute Direction of a Polar Axis

In the earlier discussion we have found the orientation of molecules along the polar axis from the known morphology of tartaric acid crystals of a known absolute configuration. As we have seen, the difference in development of crystal faces served as an empirical marker of the ends of the polar axis of (+)- or (-)-tartaric acid. It was pointed out by Waser (ref. 35) that with a sufficient knowledge of principles of crystal growth it should be possible to go further. A knowledge of the morphology of a crystal of known sign of optical rotation could, in principle, be used, together with the X-ray structure (without anomalous scattering), if sufficient information about crystal growth were available, to deduce the absolute configuration of tartaric acid. In other words, the orientation of molecules along the polar axis might be predicted from the crystal morphology and relative rates of growth in directions normal to crystal faces cutting the polar axis. With the advent of X-ray crystallography as a routine tool, substantial progress has been made, largely by a group (ref. 36) at the Weizmann Institute, in gaining a more fundamental understanding of the relationship of crystal growth to the internal structure of the crystal. An extended and systematic study of the relation of crystal structure to the effect of impurities on morphology has been successfully applied to the determination of absolute configuration and also the resolution of racemates.

We will present here a brief discussion of the use of the effect of "impurities" to determine the absolute orientation of molecules along a polar axis.

(+)- [or (-)]-Lysine hydrochloride dihydrate whose structure

$^{+}NH_3(CH_2)_5CH(NH_3^{+})COO^{-}\ Cl^{-}.(H_2O)_2$

had been determined (ref. 37) by X-ray and neutron diffraction, crystallizes in space group $\underline{P}2_1$ as triangular bars shown in Fig. 14 on the following page (ref. 38). The triangular facial development shows the presence of a polar axis and the facial development further makes it possible to distinguish by visual inspection between crystals of the (+) and (-) isomer but even with the availability of the X-ray crystal structure more information is required in order to assign the absolute configuration from crystal morphology or to solve the equivalent problem of deducing the orientation of molecules along the polar $\underline{b}$ axis of a crystal of $\underline{L}$-lysine of known absolute configuration. In this structure the $\underline{L}$-lysine molecules adopt an extended zig-zag conformation with the axis aligned approximately along the polar $\underline{b}$ axis. The carboxylate anions thus emerge at faces at one end of the $\underline{b}$ axis and the terminal NH_3^+ groups at the other--- the question is which faces at the ends of the polar axis are made up of carboxylate ions and which of NH_3^+ groups. The answer was provided experimentally (ref. 38) by growing crystals in a solution containing carefully selected agents which could be counted on to be absorbed selectively on certain faces and so retard growth normal to those faces. Then by examining the change in crystal shape to see which faces had been enlarged (by retardation of their further growth by the adsorbed impurities) the direction of alignment of molecules along the polar axis could be deduced. A schematic representation of a crystal of the hydrochloride dihydrate showing eight molecules of L-lysine is shown in Fig. 15. The molecules are aligned along the polar b axis in approx-

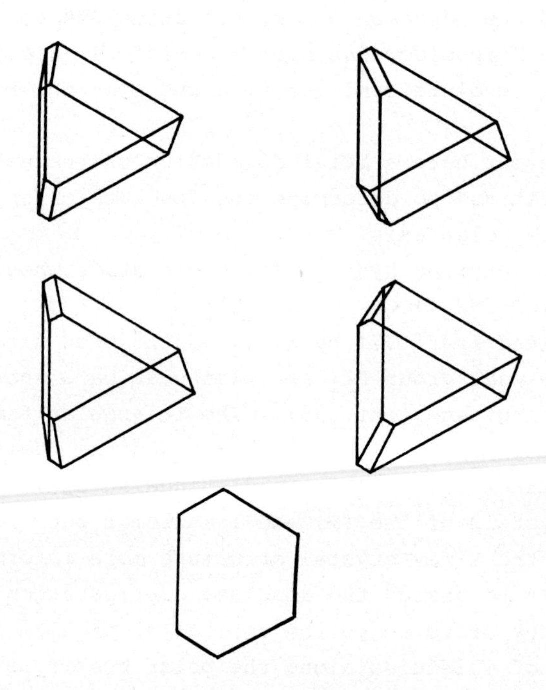

Fig. 14. Stereopair drawings of idealized crystals of (upper left) L-and (lower left) D-lysine hydrochloride. In each case the crystals are shown with the triangular (001) face up and in the plane of the page. The b axis goes to the right in the upper drawing and to the left in the lower. Note that the chirality of the crystal can be identified by inspection because with the triangle pointing along +b to the right and the (001) face in the plane of the page the long morphological axis of the crystal (along +c) tilts toward the top of the page in the L-crystal and toward the bottom of the page in the D-crystal. At the bottom is a sketch of a crystal in the same orientation grown in the presence of methyl lysinate showing the induced development of the (010) face.

imately the orientation of the L crystal above, with carboxylate ions at one end of the crystal and ammonium ions at the other. In normal growth as a (010) face (normal to the b axis on the right) begins to develop molecules of L-lysine add to it so rapidly that the crystals grow along b extinguishing the (010) face and leaving the more slowly developing oblique (110) and

$$
\begin{array}{l}
\quad\quad\quad\quad\quad\quad\quad\; NH_3^+ \quad\quad\quad\quad\quad\quad\quad\quad\quad\quad\; NH_3^+ \\
{}^+HN_3CH_2CH_2CH_2CH_2CHCOO^- \;\; {}^+HN_3CH_2CH_2CH_2CH_2CHCOO^- \\
\quad\quad\quad\quad\quad\quad\quad\quad\quad\quad\quad\quad\quad\; NH_3^+ \quad\quad\quad\quad\quad\quad\quad\quad\quad\quad\; NH_3^+ \\
\quad\quad\quad {}^+HN_3CH_2CH_2CH_2CH_2CHCOO^- \;\; {}^+HN_3CH_2CH_2CH_2CH_2CHCOO^- \\
\quad\quad\quad\quad\quad\quad\quad\quad\quad\quad\quad\quad\quad\; NH_3^+ \quad\quad\quad\quad\quad\quad\quad\quad\quad\quad\; NH_3^+ \\
\quad\quad\quad {}^+HN_3CH_2CH_2CH_2CH_2CHCOO^- \;\; {}^+HN_3CH_2CH_2CH_2CH_2CHCOO^- \\
\quad\quad\quad\quad\quad\quad\quad\; NH_3^+ \quad\quad\quad\quad\quad\quad\quad\quad\quad\quad\; NH_3^+ \\
{}^+HN_3CH_2CH_2CH_2CH_2CHCOO^- \;\; {}^+HN_3CH_2CH_2CH_2CH_2CHCOO^-
\end{array}
$$

Fig. 15. A schematic representation of a crystal of L-lysine hydrochloride dihydrate in the orientation of the crystal in the upper part of Fig. 14. The orientation of the molecules along the horizontal b axis is known because the absolute configuration of the substance used for the crystal structure determination was known.

($\bar{1}$10) faces. The methyl ester of L-lysine serves as an additive which has an unmodified NH_3+ end which should continue to add to a face terminating in carboxylate ions. However the hydrogen bonding ability of the anionic carboxylate groups at the other end of the molecule has been effectively lost. When crystal growth is carried out in the presence of the methyl ester of L-lysine, it can be adsorbed on the right hand (010) face of a growing crystal since the NH_3^+ ends are little altered by ester formation and can fit into the surface structure. Growth normal to an embryonic (010) face at the right side of the crystal which has adsorbed a layer of methyl lysinate molecules, is thus blocked and the face allowed to develop. On the other hand methyl lysinate molecules should not be expected to compete with lysine molecules and be absorbed on the (0$\bar{1}$0) face made up of NH_3^+ groups since the ester groups haven't the same hydrogen-bonding capability as the carboxylate ions of lysine molecules. Development of faces at the -b end of the polar axis should thus be relatively unaffected. The result is shown in Fig. 13.

Turning the argument around, the face along whose normal crystal growth was inhibited is therefore the face containing carboxylate anions. In this case, because the orientation of L-lysine molecules is known from the crystal structure together with the absolute configuration of the lysine crystals, the interpretation of the effect of methyl lysinate on crystal growth is confirmed by experiment. However, had the absolute configuration not been known it could have been found from the crystal structure (without anomalous scattering) and the effect on the morphology of added impurities whose structures were selected to show specific adsorption at faces at one end or the other of the polar axis. Because of the uncertainty inherent in an argument based on crystal morphology the Weizmann group has been careful to confirm the method by doing experiments involving crystallization with a number of different additives. Their paper (ref. 38) gives photographs of crystals, stereopair drawings showing the modification of facial development in a number of cases, and also histograms showing the relative areas of faces.

3 SOME SPECIAL PROPERTIES OF SYMMETRY-RELATED CRYSTAL FACES IN CENTROSYMMETRIC CRYSTALS

The opposite faces of an idealized centrosymmetric crystal are not identical but are related by a center of symmetry. They normally occur in a crystal as a pair of approximately equally developed faces so that it might seem to be futile to attempt to take chemical advantage of their difference. However, if one of them is coated to render it unreactive, reaction can be made to occur selectively at the remaining face in order to take advantage of its special stereochemistry. An ingenious series of experiments applying this principle was carried out by Holland and Richardson (ref. 39) who studied the oxidation of tiglic acid

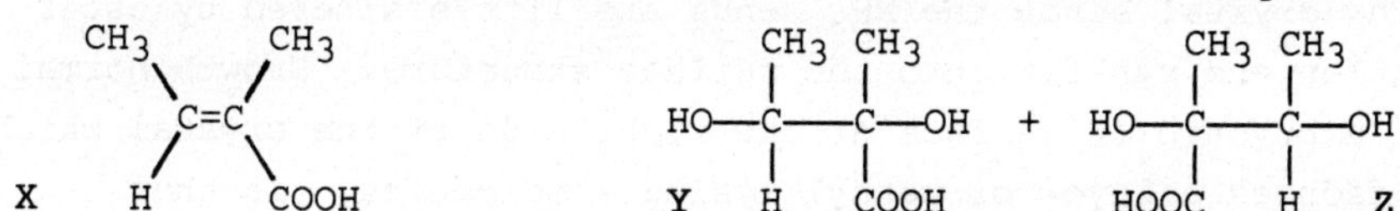

(ref. 40) (space group $P\bar{1}$) to <u>threo</u>-2,3-dihydroxy-2-methylbutanoic acid. In solution stereospecific <u>syn</u> addition to the double bond occurs equally rapidly from each side of a molecule so that there is obtained a mixture of the two enantiomeric <u>threo</u> diols **Y** and **Z** shown. If attack is blocked from one side, the product is either **Y** or **Z** alone depending on the structure of the face. When a large single crystal was cut in two and the two halves partially coated with epoxy resin so that only one face of each was exposed [the (210) face of one and the ($\bar{2}\bar{1}0$) face of the other half]. As is seen in Fig. 16, each face is made up of pairs of molecules related by a two-fold rotation axis normal to the plane. (The plane group is p2). Note that each face has a center of symmetry and therefore no polar directions but the faces are not superimposable by rotation in the plane of the surface. Addition of the two hydroxyl groups from the open side of each face should give a single enantiomer; addition to either of the two differently oriented molecules in face 1 should

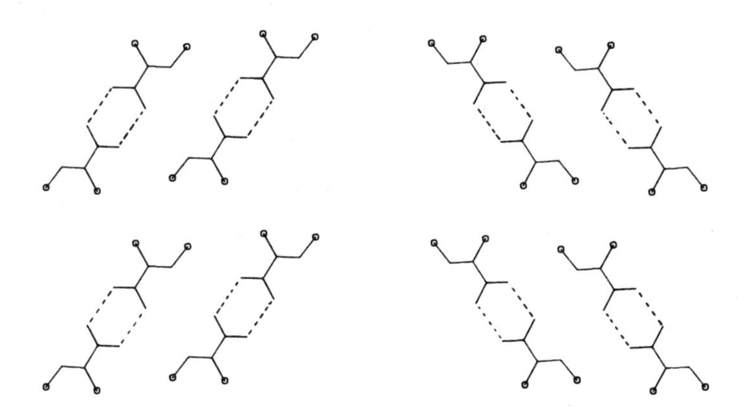

Fig. 16. Left, a view onto face 1 [the (210) face of tiglic acid]. Circles represent methyl groups and dotted lines are the hydrogen bonds of the carboxyl groups. Right, the corresponding view onto face 2 [the ($\bar{2}\bar{1}0$) face].

give the product **Y** of one chirality and each of the molecules in face 2, should give **Z**, the enantiomer of **Y**. When each half-crystal was suspended in its own cold solution of barium chlorate and osmium tetroxide, reaction occurred at the crystal surface and the reaction product dissolved until only the coating material remained. Analysis of the products showed that reaction at the C=C double bonds had occurred from the exterior of the exposed face in each case; the solutions from the two half-crystals gave complimentary o.r.d. curves showing that each half-crystal had produced a single enantiomer as suggested above. Optical purities as high as 95% were obtained. The authors pointed out that the requirement for enantiomeric differentiation of this type requires that the crystal structure at the exposed faces belong to a diptychiral plane group (with non-superimposable two-dimensional faces) with an appropriate molecular orientation at the crystal face.

The solid-state group at the Weizmann Institute (ref. 41) has shown that pairs of faces of centrosymmetric crystals may show chiral differences which can be used to assign absolute configuration to external molecules based on differences in their adsorption during crystallization. For example, the stable α-form of glycine crystallizing in the centrosymmetric space group $\underline{P}2_1/\underline{c}$ contains an equal number of (+)- and (-)-molecules. (Although glycine is not optically active in solution, in the crystal the molecules are held rigid in a chiral conformations.) Consideration of the crystal structure (ref. 42) shows that a major face (010) and its opposite (0$\bar{1}$0) are enantiomeric--- that is, one is composed of glycine molecules in a (+) conformation and the other in a (-) conformation as shown in Fig. 17. A

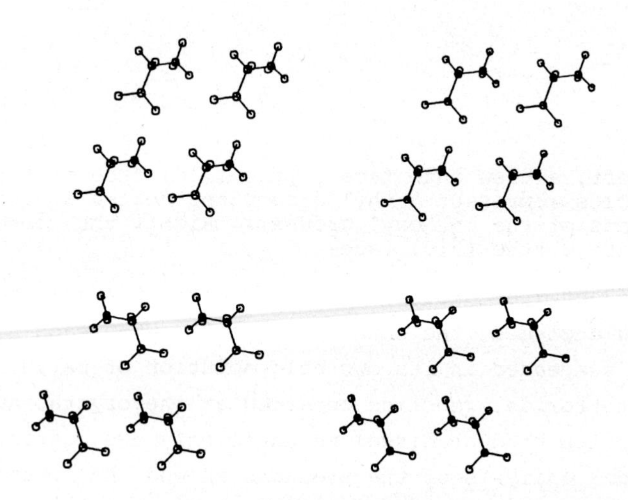

Fig. 17. _Upper, (010) face composed of chiral glycine molecules. Lower, (0$\bar{1}$0) face related to the upper face by a center of symmetry.

remarkable property of such a centrosymmetric crystal is that the absolute direction of the b axis of the X-ray structure can be assigned by visual inspection of the crystal. To clarify this point, in Fig. 18 is a drawing of a monoclinic crystal with the cell constants of α-glycine but showing only the simplest possible faces {100}, {010}, and (001}. Suppose that orientation of the crystal with an X-ray diffractometer has shown that a goes horizontally in the plane of the page and c also lies in the plane of the page. There are two ways to choose a. If a goes to the right then, since the angle between +a and +c is greater than 90° by convention, +c must point down in the page and since the system used in the X-ray structure determination is right-handed b must be out of the page. If on the other hand a is chosen as pointing left then c must point up in the plane of the page and in a right-handed system b must again point out of the page. By a similar argument +b of the second crystal can be seen to be pointing into the page. There is, then, the remarkable result that the absolute direction of the b axis of such a centrosymmetric structure determined by X-ray crystallography can be assigned by visual inspection of the crystal. It follows that the chiralities of the molecules making up the (010) and $(0\bar{1}0)$ faces can be unambiguously assigned. In this case the structure looking onto the (010) and $(0\bar{1}0)$ faces is shown in Fig. 18. A

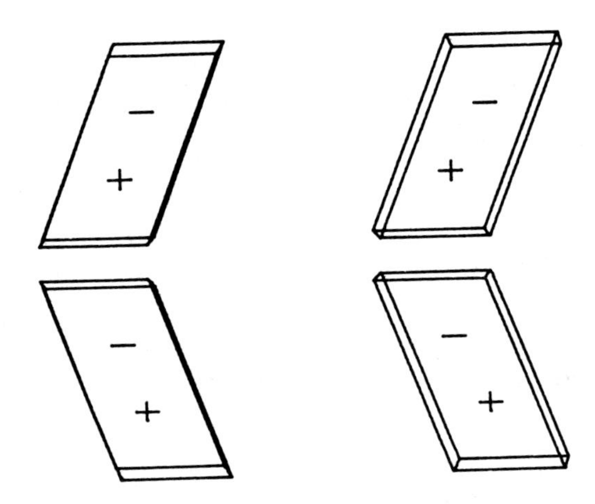

Fig. 18. Two centrosymmetric crystals showing that if the axial system is right-handed there is only one way to choose the absolute direction of the b axis.

molecule of (R)-alanine can readily replace a molecule of glycine

$^{+}NH_3CHCOO^{-}$ with H on the CH — Glycine

$^{+}NH_2CHCOO^{-}$ with CH_3 on the CH — Alanine

on the (010) face because, if the alanine molecule in the conformation similar to that of the glycine molecules making up the face is to be adsorbed, the methyl group of the alanine molecule points out of the surface and has minimum interference with the remainder of the surface structure. On the other hand an (S)-alanine molecule cannot fit since the methyl group in the corresponding glycine-like conformation would point in a direction parallel to the surface and have an unfavorable fit with the surroundings. In the same way an (S)-alanine molecule can fit onto an $(0\bar{1}0)$ but not onto an (010) surface. Growth of a glycine crystal in a direction to normal to (010) should therefore be unaffected by the presence of (S)-alanine during crystallization but should be inhibited by (R)-alanine. The converse should be true for growth at a $(0\bar{1}0)$ surface. In Fig. 19 are shown draw-

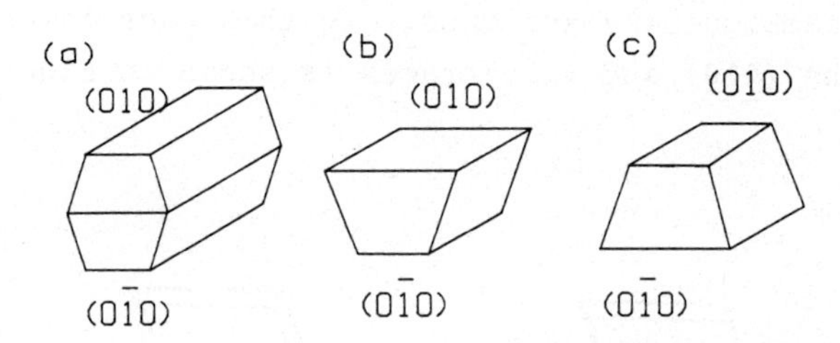

Fig. 19. Crystals of glycine grown (a) without added impurities. (b) In the presence of (R)-alanine which is adsorbed on (010) growth normal to (010) is blocked. (c) In the presence of (S)-alanine which is adsorbed on $(0\bar{1}0)$ growth normal to $(0\bar{1}0)$ is blocked.

ings indicating morphology of crystals of glycine grown without added impurities and in the presence of (R)- and (S)-alanine. In this case the method is being tested by use of "impurities" of known absolute configuration. The argument can, of course, be employed in reverse if an amino acid of unknown configuration is employed, once the relation of structure to the effect of the "impurities" is understood. Then, if the growing crystal has morphology (b) the amino acid has the (R) configuration and if morphology (c) the (S), providing that the compound being tested has no functional groups which would otherwise affect the result.

4 USE OF THE PYROELECTRIC EFFECT TO DETERMINE THE ABSOLUTE DIRECTION OF THE POLAR AXIS

The most direct way of determining the absolute direction of orientation of electrically polar molecules along a polar crystal axis would seem to be to detect the direction of the electric dipole directly. This process is complicated by the fact that polar crystals mask their electric dipoles on standing under ambient conditions by a process which has been described as "acquisition of free electric charges from the surrounding media by its [the crystal's] surface and ... internal conduction of free charges" (ref. 43). Nevertheless when such a crystal is heated the polarity of the crystal is changed and a voltage develops between ends of the crystal. Conversely, if the crystal is cooled, polarization develops but in the opposite sense. In the latter part of the 19th century such pyroelectric effects were demonstrated for a number of organic crystals. The results are incorporated in Groth's Chemische Kristallographie (ref. 28) together with drawings of many polar crystals whose morphology shows the presence of the polar axis. That end of the crystal which becomes positive on heating was called the "analogous pole" and the other end the "antilogous pole". There was no attempt to correlate the observed polarization with crystal structure since no crystal structures were known at this time. Later, when crystal structural data became available, interest in the pyroelectric effect had diminished and there seem to have been few if any attempts to correlate the direction of the observed pyroelectric effect with the molecular dipole moment as deduced from the structure. While accurate direct measurement of the electrical properties of small crystals is experimentally demanding, a method developed by Kundt and modified by Burker (ref. 44) consists of heating or cooling the crystal and spraying it with a mixture of powdered lycopodium powder, carmine, and sulfur ground together. When particles of the powder separate, the lycopodium powder develops a positive electrostatic charge while the sulfur and carmine become negative. The carmine and sulfur are thus attracted to the positive pole of the crystal and the lycopodium powder to the negative. This test has been extensively used by crystallographers as a test for a center of symmetry in crystals but there seems to have been little if any attempt to correlate the polarity of the crystal with its structure until the report of the application of the method to p-bromobenzoic anhydride (ref. 45). In this case the end of the crystal toward which the carbonyl oxygen atoms were directed was found, when the crystal was heated and allowed to cool rapidly to room temperature (ref. 46), to attract the particles of lycopodium powder while the other end of the crystal along the polar axis attracted the carmine and sulfur. Thus, the powder method gave the answer predicted on the basis of the dipole moment estimated from the structure together with the dipole moment of benzoic anhydride measured in solution.

An example of crystal morphology of a chiral crystal somewhat more typical than that of tartaric acid discussed earlier in this review is provided by the crystallization of (R)- and (S)-mandelic acid.

COOH | H—C—OH | C_6H_5 COOH | HO—C—H | C_6H_5

(R)-(-)- and (S)-(+)-Mandelic Acid

Although these enantiomers crystallize in the same space group as do those of tartaric acid the crystal morphology has been found (ref. 47, 48) to give no indication of the presence of a polar axis. As shown in Fig. 20, the faces in the zone with

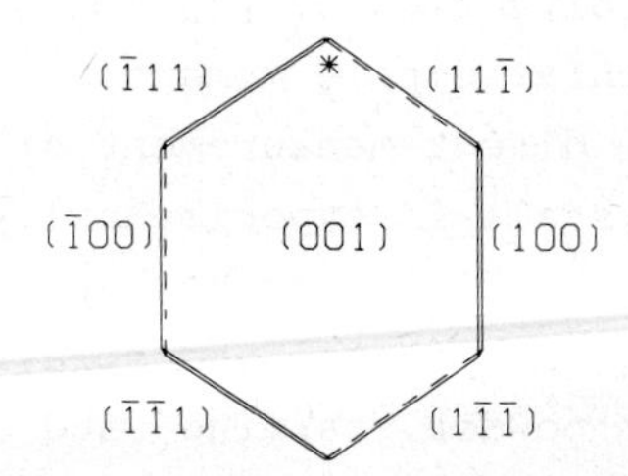

Fig. 20. A crystal of (S)-mandelic acid. Faces occur in parallel pairs so that the (R)-enantiomer would be identical in appearance.

axis [010] are {001} and {100} as was the case with the tartaric acids but the faces at the ends of the polar axis in this case occur in parallel pairs even though there is no crystallographic center of symmetry. There is thus no simple way to determine by visual inspection the orientation of molecules along the polar axis even if the morphology of a crystal of known absolute configuration is available since the directions along the polar axis are not marked by hemihedral facial development in contrast to the situation with tartaric acid above. To apply the pyroelectric effect it is necessary to know the direction of the electric dipoles of molecules in the conformation in which they appear in the crystal. For this purpose *ab initio* calculations were employed to calculate the dipole moments of the structure of each molecule in the asymmetric unit of mandelic acid. From the crystal structure together with the absolute configuration, it is known that an (S)-(+) crystal oriented with its sloping side (100) face tilted upward on the right of the crystal has the carboxylic O-H vector pointing toward the top of the drawing as shown in Fig. 21. The dipole moment calculation then leads to the conclusion that this face is at the (+) end of the crystal's electric dipole. Use of the pyroelectric effect to determine the

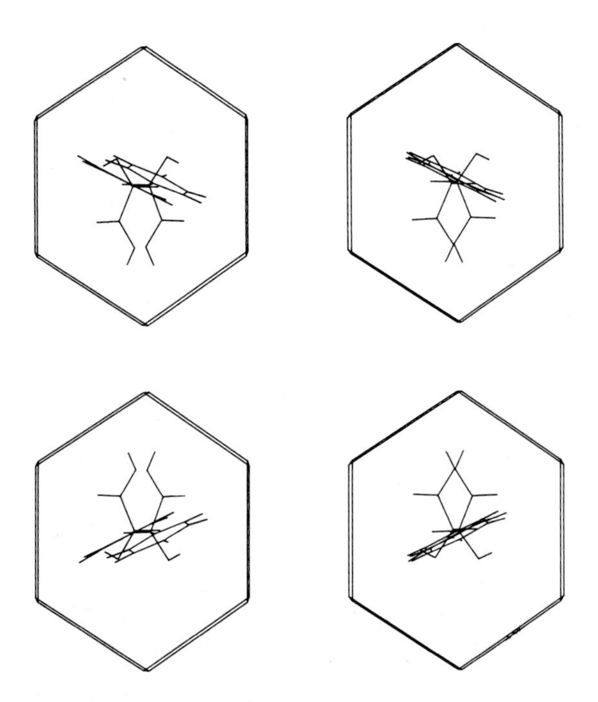

Fig. 21. Stereopair drawings of crystals of (upper drawing) (R)- and (lower drawing) (S)-mandelic with the two independent molecules superimposed in their correct orientation. The positive end ("analogous pole") of the (R) crystal is toward the bottom of the drawing and that of the (S) crystal toward the top. Note that in Fig. 19 the positive end indicated by the Kundt-Burker powder method is toward the top in agreement with the results in Fig. 20.

absolute orientation gives the same conclusion.

The Kundt-Burker powder method offers promise of being a

valuable aid in determining the absolute orientation of molecules along the polar axis in pyroelectric crystals. It is easy to apply, sensitive (ref. 49), and reproducible. However, before it can be accepted implicitly it will be necessary to test it with a wider range of types of structures.

5 USE OF SECOND HARMONIC GENERATION AS A STRUCTURAL TOOL IN SOLID STATE CHEMISTRY

Non-linear optics and, in particular, second harmonic generation (SHG) have been discussed elsewhere in this volume and a book reviewing the nonlinear optical properties of organic materials has appeared (ref. 50).

The method has been applied in studies of the structures of the 1:1 complexes of phenylquinone with phenylhydroquinone (phenylquinhydrone) and the structurally similar p-chloro derivative, which have presented an unusually subtle problem (ref. 51). X-ray structure determination indicated that the complexes are centrosymmetric whereas the crystal morphology suggested that they belong in polar space groups Pc and $P2_1$. Although such centrosymmetry could be explained by disorder of the hydroxylic protons between the two oxygen atoms with which they are associated, substantial chemical grounds existed for rejecting such

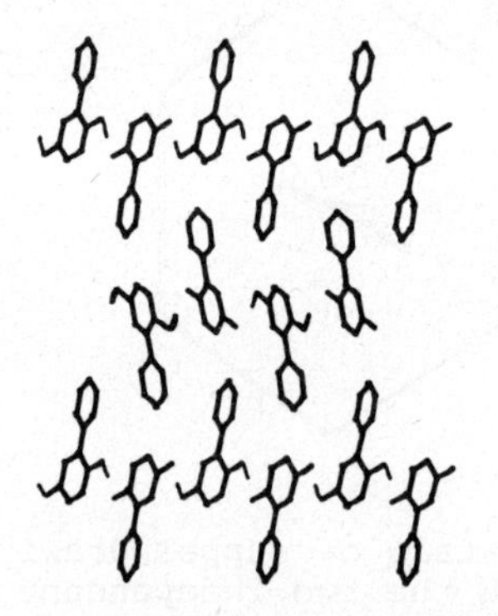
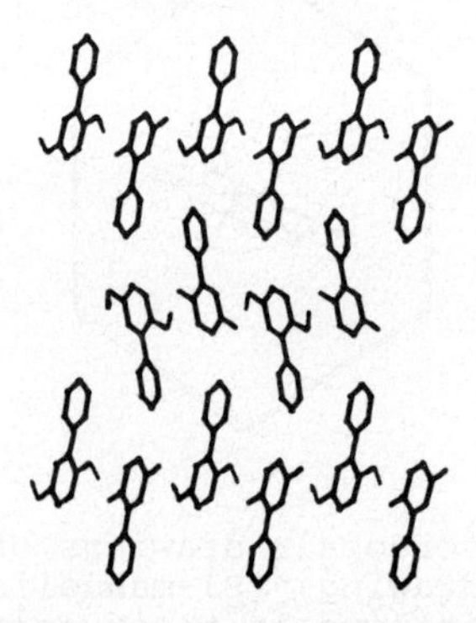

Fig. 22. Structure of phenylquinhydrone. The b axis goes to the left in the page and a is normal to the plane of the page. The hydroxylic protons found by X-ray structure determination to be disordered between the oxygen atoms of adjacent molecules are shown here in positions which would give the crystal $P2_1$ symmetry. Note that the b axis is polar since all phenyl rings attached to hydroquinone molecules lean to the right while those attached to quinone molecules lean to the left. (Reprinted with permission from ref. 51.)

an explanation. Each of these complexes on an SHG test showed strong frequency doubling which would seem to require that the crystal have polar regions of substantial size which when averaged give the centrosymmetric structure determined by X-ray crystallography. The structure shown in Fig. 22 has relatively strong hydrogen-bonding interactions in one direction and relatively strong pi-complexing in a second. In the third direction the crystal is held together with van der Waals forces and the pendant phenyl groups serve as spacers which weaken the transmission of information about the positions of the hydroxylic protons in adjacent layers.

A possible reconciliation of the crystallographic order required by the X-ray results with the disorder required by SHG has been suggested by work of Green and Knossow (ref. 52) in their study of the structure of hexahelicene. Hexahelicene molecules are chiral and configurationally stable at room temperature yet form a racemate crystallizing in space group $\underline{P}2_12_12_1$ and therefore would appear to be a conglomerate--- a mechanical mixture of (+) and (-) crystals. A single crystal from this mixture gave a structure refining to an R of 4%; however, crystals of the mixture were shown to have an optical rotation of only 2% of the value expected for an optically pure sample. The

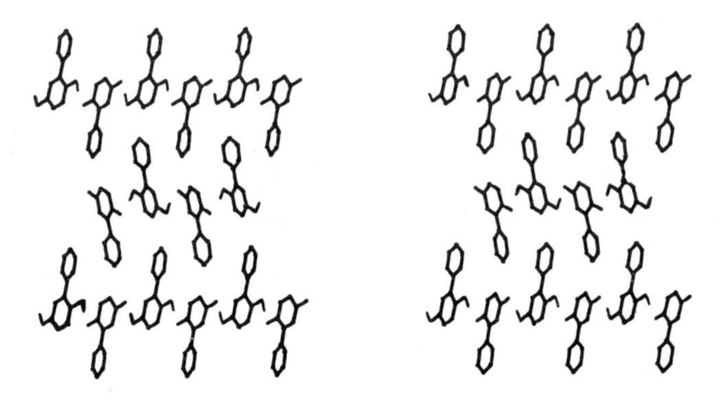

Fig. 23. Hypothetical crystal structure of a layer of phenylquinhydrone in space group Pc. Notice that the heavy atom positions are still those determined by X-ray crystallography as shown in Fig. 19. The protons of every other molecule have been switched to their alternate positions. The a axis is now polar since all phenyl groups attached to hydroquinone rings point up in the page and all phenyl groups attached to quinone rings point down. (Reprinted with permission from ref. 51.)

investigators were able to show that the explanation for a structure which appeared optically pure to X-ray crystallography but largely racemic to the optical polarimeter is that the structure consists of alternating (+) and (-) layers ordered with respect to each other (lamellar twinning). The crystal layers could be located by etching and then sliced apart; individual slices were shown to have optical purities approaching 100%, adjacent layers giving opposite signs of rotation. The nature of the twinning has been investigated in more detail by Ramdas, Thomas, Jordan, and Eckhardt (ref. 52b) whose calculations of interfacial energies make such twinning seem reasonable.

Such lamilar twinning would appear to provide a possible explanation for the phenyl- and p-chlorophenylquinhydrone complexes discussed above (Figs. 22 and 23). Since in this case the molecules are achiral there is not the possibility of using enantiomeric purity to probe the structures of individual layers. It is possible, however, that examination of oriented slices with SHG could provide interesting additional structural

TABLE 1

Solid State Transformations Involving Major Symmetry Changes

Starting Compound (or Type) and Its Point Group		Point Group of Product	Note
Phenanthrene	2	2/m?	a
2,3-Dichloro-quinizarin	2/m	m	b
Naphthazarin C	2/m	m	c
Azo Enol Benzoates	1	mm2	d
1,1'-Binaphthyl	2/m	422	e
A Fuchsone	2/m	222	f
E and Z Fulgides	2	2/m	g
Isopropylcarbazole	mm2	mm2	h
9-Hydroxyphenalenone	2/m	2	i

[a]Ref. 53, 54. [b]R. C. Hall, Ph. D. Thesis, University of Illinois, 1986. [c]Ref. 55. [d]Ref. 56. [e]Ref. 57. [f]Ref. 58. [g]Ref. 59. [h]Ref. 60. [i]Ref. 61. This is a particularly interesting transformation because the transition is piezoluminescent.

information about such crystals.

5.1 Use of Second Harmonic Generation to Follow Chemical Transformations

Among the more interesting solid-state transformations are those which involve a change between structures with centric and non-centric symmetry. In the Table are shown a few reactions involving a change of crystal symmetry from polar to non-polar or the reverse. The use of SHG measurements to study the phase transition of phenanthrene (ref. 53, 54) has provided a striking

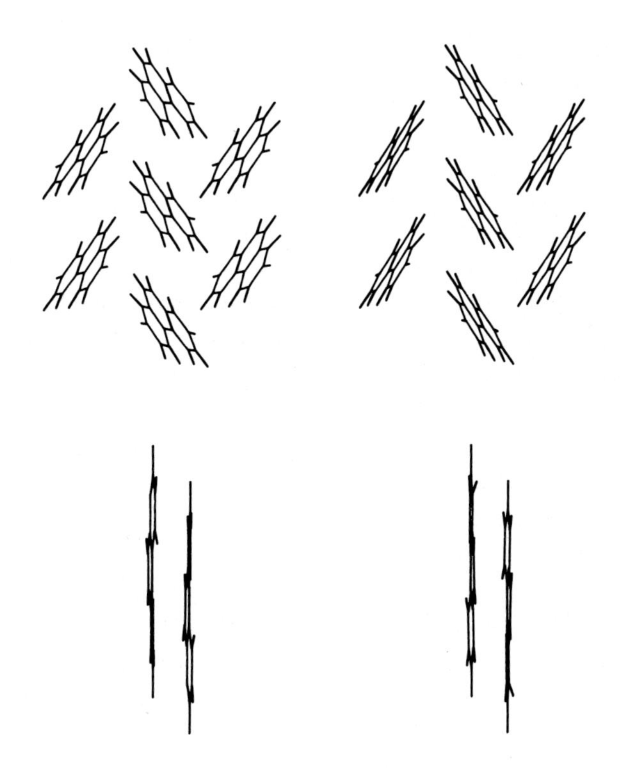

Fig. 24. Upper: crystal structure of phenanthrene (ref. 53c) showing the polar b axis vertical in the page. Lower: enantiomeric molecules of phenanthrene viewed edge-on showing the chirality due to the interference of the 9- and 10- hydrogen atoms.

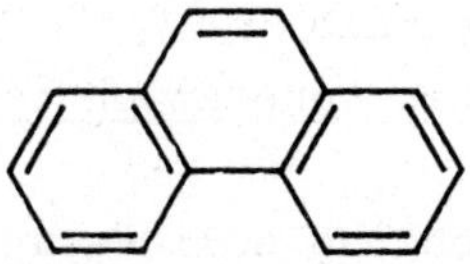

demonstration of the utility of this technique. Phenanthrene in its room temperature form (space group $\underline{P}2_1$) is both chiral and polar (Fig. 24). Molecular chirality results from non-coplanarity due to interference of the hydrogen atoms in the 9- and 10-positions. Phenanthrene had been found when heated to about 70° to undergo a single-crystal to single-crystal transition which seemed to involve no change in space group. However, when the reaction of a powdered sample was examined with second harmonic generation the SHG signal originally present gradually disappeared as the sample was heated and was completely gone at 72°. The transition was reversible; when the sample was cooled the signal reappeared. There is clearly a thermal transition to a centrosymmetric structure (point group probably $2/\underline{m}$). The structure of the high-temperature-stable form of phenanthrene is unknown but since the phenanthrene molecule has no center of symmetry, the centrosymmetric crystal must contain pairs of molecules related by centers. This requires that during the transition one-half of the molecules rotate 180° around some axis, probably the axis normal to the plane of the molecule and, if the centrosymmetric structure is ordered, that one-half of the chiral phenanthrene molecules undergo racemization to the enantiomer by a process in which the 9- and 10-hydrogen atoms pass each other in an approximately planar transition state. A determination of the structure of the high-temperature form would, thus, be of considerable interest.

In the return from a centrosymmetric to a polar crystal structure in a single-crystal-single-crystal transition the question of whether the polar axis assumes the same absolute direction that it had originally or the opposite is very closely related to the well-known problem of which enantiomer would be produced in the spontaneous resolution of chiral crystals (ref. 62) and again points out the close relationship of chirality and polarity of crystals.

The rearrangement on heating of powders of the low temperature centrosymmetric form (space group $\underline{P}2_1/\underline{c}$) of 2,3-dichloro-

1,4-dihydroxyanthraquinone to a polar structure (space group *Pc*) (see the preceding Table) has been followed with SHG. Investigation of the behavior of single crystals may give insight into the factors which influence the orientation of the polar axis being introduced.

Extrapolation from the examples in the Table suggests a more general utility of such SHG studies to obtain information about solid-solid transformations. For example, rearrangement of a centrosymmetric crystal to a centrosymmetric product might proceed through an intermediate polar crystalline state; the presence of the intermediate might be detected by the use of SHG measurements.

6 HOW CAN ORGANIC COMPOUNDS BE INDUCED TO CRYSTALLIZE IN NON-CENTROSYMMETRIC SPACE GROUPS?

Because of the potential technological importance of non-centrosymmetric crystals (ref. 19) methods of inducing molecules to crystallize in polar space groups have been of considerable interest. It has long been recognized that certain types of structures may show a preference for certain crystal symmetries. For example there is a pronounced tendency for *m*-disubstituted benzenes to crystallize in point group *mm*2 (ref. 63). A similar phenomenon was mentioned earlier in this chapter in a comparison of crystal symmetries of benzoic anhydrides with those of the corresponding benzoic acids.

An approach to inducing crystallization in polar space groups has been based on the recognition that any crystal which contains species of only one chirality cannot (unless disordered) crystallize in a centrosymmetric space group and a screening program for polar compounds with suitable non-linear optical properties has been based on this principle (ref. 64).

An ingenious method has been applied to *p*-nitroaniline which, although highly polarizable, crystallizes in a centrosymmetric space group. This substance forms cyclodextrin complexes which are polar and show respectable SHG signals (ref. 65).

Finally there might be mentioned the technique of growing crystals in an electric field to induce crystallization in a polar structure. This technique is well known as applied to inorganic salts and polymers and has had important industrial application (ref. 18,19). Thus the non-polar α form of polyvinylidene fluoride can be induced by application of an electric field of 1 Mv/cm at 170° to undergo a transition to a new polar polymorphic structure which has been designated α' (ref. 66). Although application of the method to crystallization of small organic molecules seems not to have been reported, it might also succeed with such compounds.

Generalizations about symmetry of crystal packing, with our present state of knowledge, cannot be applied to specific cases with reliability. It follows that understanding factors which control crystal symmetry remains an important challenge.

7 A LOOK AHEAD

Predictions of the future direction of science almost always turn out to be wide of the mark. However, it is impossible to resist calling attention to recent developments which permit an optimistic view of the future of solid-state organic chemistry. Attention of the chemical community has recently been most attracted, perhaps, by the dramatic advances made in the discovery and understanding of super-conducting molecular crystals (ref. 67) and in the remarkable properties of diacetylene polymers shown to retain their crystallinity on polymerization. The further understanding of the properties of polar crystals would appear to have much future potential. The increasing importance of optical information processing has been pointed out (ref. 68) and there has been recent demonstration of the superior non-linear properties of organic crystals (ref. 50). There have been proposals of development of a molecular computer which have no doubt been premature (ref. 69). Nevertheless, the variety of structures and properties of polar organic crystals are so extensive and varied (ref. 70) that it seems likely that chemistry of the solid state in general and the subject of crystal symmetry in particular will play an expanding role in science.

8 ACKNOWLEDGEMENT

We are indebted to the National Science Foundation for support of the research on which much of this review is based.

9 REFERENCES

1 C. Graebe and C. Liebermann, Ann. Suppl., 7 (1870) 257-322.
2 For reviews of solid-gas reactions and the related processes, loss of solvent of crystallization, see (a) I. C. Paul, and D. Y. Curtin, Science, 187 (1975) 19-26. (b) S. R. Byrn, Solid State Chemistry of Drugs, Academic Press, 1982. (c) S. R. Byrn, J. Pharm. Sci., 65 (1976) 1-22. (d)

R. Lamartine, G. Bertholon, M.-F. Vincent-Falquet, and R. Perrin, Ann. Chim. (1976) 131-147.
3 (a) R. S. Miller, I. C. Paul, and D. Y. Curtin, J. Am. Chem. Soc., 96 (1974) 6334-6339. (b) M. Colapietro and A. Domenicano, Acta Crystallogr., B38 (1982) 1953-1957.
4 L. Leiserowitz, Acta Crystallogr., B32 (1976) 775-801.
5 A. A. Patil, D. Y. Curtin, and I. C. Paul, Israel J. Chem., 25 (1985) 320-326.
6 M. Calleri, G. Ferraris, and D. Viterbo, Att Acad. Sci. Torino Cl. Sci. Fis. Mat. Nat., 100 (1966) 145-173.
7 R. S. Miller, D. Y. Curtin, and I. C. Paul, J. Am. Chem. Soc., 94 (1972) 5117-5119.
8 T. Taga, N. Yamamoto, and K. Osaki, Acta Crystallogr., C41 (1985) 153-154.
9 E. N. Duesler, R. B. Kress, A. A. Patil, R. B. Wilson, unpublished work.
10 H. M. Walborsky, L. Barash, A. E. Young, and F. J. Impastato,J. Am. Chem. Soc., 83 (1961) 2517-2525; H. M. Walborsky and A. E. Young, ibid., 86 (1964) 3288-3296.
11 C. C. Chang, C-T. Lin, A. H-J. Wang, D. Y. Curtin, and I. C. Paul, J. Am. Chem. Soc., 99 (1977) 6303-6308.
12 C-T. Lin, I. C. Paul, and D. Y. Curtin, J. Am. Chem. Soc., 96 (1974) 3699-3701.
13 C-T. Lin, D. Y. Curtin, and I. C. Paul, J. Am. Chem. Soc., 96 (1974) 6199-6200.
14 See for example, A. I. Kitaigorodsky, Yu. V. Mnyuch, and Yu. G. Asadov, J. Phys. Chem. Solids, 26 (1965) 463-472; J. M. Thomas, Phil. Trans. Royal Soc. London, 277 (1974) 251-286.
15 J. P. Desvergne and J. M. Thomas, Chem. Phys. Lett., 23 (1973) 343-344.
16 See Th. Hahn and H. Klapper "Point Group Symmetry and Physical Properties of Crystals" in Th. Hahn (Ed.), International Tables for Crystallography, Vol. A, D. Reidel Pub. Co., Dordrecht, Holland, 1983, pp. 780 ff.
17 We are indebted to Dr. Carolyn Brock for helpful discussion of these matters.
18 See D. Y. Curtin and I. C. Paul, Chem. Rev. 81 (1981) 526-541 for an earlier discussion of the chemical consequences of the polar axis in organic chemistry.
19 J. C. Burfoot and G. W. Taylor, Polar Dielectrics and Their Applications, Univ. of California Press, Berkeley, 1979.
20 L. G. Wesson, Tables of Electric Dipole Moments, The Technology Press, Massachusetts Institute of Technology Cambridge, 1948.
21 (a) C. S. McCammon and J. Trotter Acta, Crystallogr., 17 (1964) 1333-1334. (b) E. N. Duesler, R. B. Kress, C-T. Lin, W-I. Shiau, I. C. Paul, and D. Y. Curtin, J. Am. Chem. Soc., 103 (1981) 875-879.
22 S. B. Lang, Sourcebook of Pyroelectricity, Gordon and Breach Science Publishers, London, 1974.
23 G. Burns, Solid State Physics, Academic Press, N. Y., 1985, pp. 90 ff.
24 (a) G. E. Hardy, W. C. Kaska, B. P. Chandra, and J. I. Zink, J. Am. Chem. Soc., 103 (1981) 1704-1709. (b) B. P. Chandra and J. I. Zink, Phys. Rev., B21 (1980) 816-826.
25 J. S. Patel and D. M. Hanson, Nature, 293 (1981) 5832-5833.
26 See R. C. Dye and C. J. Eckhardt, Mol. Cryst. Liq. Cryst., 134 (1986) 265-268.
27 J. D. Dunitz, X-ray Crystallography, Cornell University Press, 1983.

28 P. Groth, Chemische Kristallographie, 5 Volumes, Verlag von Wilhelm Engelmann, Leipzig, 1906-1919.
29 (a) L. Pasteur, Ann. chim. phys. [3], (1850) 56-61. (b) See P. Groth, Chemische Kristallographie, Vol. 3, Verlag von Wilhelm Engelmann, Leipzig, 1910, pp 303-304.
30 (a) F. Stern and C. A. Beevers, Acta Crystallogr., 3 (1950) 341-346. (b) Y. Okaya, N. R. Stemple, and M. I. Kay, Acta Crystallogr., 22, (1966) 237-243. (c) J. Albertsson, A. Oskarsson, and K. Stahl, J. Appl. Crystallogr., 12 (1979) 537-541.
31 There is an unfortunate non-uniformity in the convention used in the early literature. Thus (+)-tartaric acid was called D- by some investigators and L- by others. We are indebted to Professor Ernest Eliel for pointing out that the assignments of R- and S- in ref. 2a are incorrect. R- should in every case be (S)(S)- and (S)- should be (R)(R)-. (See E. L. Eliel, Stereochemistry of Carbon Compounds, McGraw-Hill, New York, 1962, pp. 95 ff.
32 See P. Hartman, "Crystal Form and Crystal Structure" in Physics and Chemistry of the Organic Solid State, Vol 1, D. Fox, M. Labes, and A. Weissberger, Interscience Publishers, New York, 1963, Chapter 6.
33 A. F. Wells, Discussions Faraday Soc., 13, (1949) 197-201.
34 (a) M. J. Buerger, Elementary Crystallography, MIT Press, Cambridge, MA, 1978, pp 177-183. (b) A. Rosensweig and B. Morosin, Acta Crystallogr., 20 (1966) 758-761.
35 J. Waser, Angew. Chem., 85 (1973) 498-499.
36 For references see Z. Berkovitch-Yellin, L. Addadi, M. Idelson, M. Lahav, and L. Leiserowitz, Angew. Chem. (Engl. Ed.), 21 (1982) 631-632. Z. Berkovitch-Yellin, L. Addadi, M. Idelson, L. Leiserowitz, M. Lahav, Nature, 296 (1982) 27-34.
37 D. A. Wright and R. E. Marsh, Acta Crystallogr. 15 (1962) 54-64. T. F. Koetzle, M. Lehmann, J. J. Verbist, and W. C. Hamilton, ibid. B28 (1972) 3207-3214.
38 Z. Berkovitch-Yellin, L. Addadi, M. Idelson, L. Leiserowitz, and M. Lahav, Nature, 296 (1982).
39 H. L. Holland and M. F. Richardson, Mol. Cryst. Liq. Cryst., 58 (1980) 311-314. P. Ch. Chenchaiah, H. L. Holland, and M. F. Richardson, JCS Chem. Commun., (1982) 436-437. M. F. Richardson, P. Ch. Chenchaiah, and H. L. Holland, Abstract of talk presented at the International Summer School on Crystallographic Computing, Ottawa, Ontario, August, 1981.
40 A. L. Porte and J. M. Robertson, J. Chem. Soc. (1959) 825-829.
41 L. Addadi, Z. Berkovitch-Yellin, I. Weissbuch, M. Lahav, L. Leiserowitz, and S. Weinstein, J. Am. Chem. Soc., 104 (1982) 2075-2077. I. Weissbuch, I. Addadi, Z. Berkovitch-Yellin, E. Gati, M. Lahav, and L. Leiserowitz, Nature, 310 (1982) 161-164.
42 P. Jonsson and A. Kvick, Acta Crystallogr., B28 (1972) 1827-1833.
43 Ref. 22, p. 1.
44 A. Kundt, Ann. Phys. (Leipzig), 28 (1886) 145-150. K. Burker, Ann. Phys. (Leipzig), 1 (1900) 474-478.
45 A. A. Patil, D. Y. Curtin, and I. C. Paul, J. Am. Chem. Soc., 107 (1985) 726-727.
46 The method of heating and allowing the crystal to cool rapidly has been used in an attempt to minimize piezoelectric effects; it has been found to give the same polarity as when the crystal was simply heated and sprayed

while still hot (unpublished work of Drs. Anjali Patil and Sumita Chakraborty.

47 P. Groth, Chemische Kristallographie, Vol. 4, Verlag von Wilhelm Engelmann, Leipzig, 1917, p. 559. H. Traube, Chem. Ber., 32 (1899) 2385.

48 A. O. Patil, W. T. Pennington, I. C. Paul, D. Y. Curtin, and C. E. Dykstra, manuscript in preparation.

49 The method has been shown capable of detecting twinning in a small crystal where the crystal dipole was reversed at the twin plane [W. T. Pennington, A. O. Patil, I. C. Paul, and D. Y. Curtin, J. Chem. Soc. Perkin Trans. 2 (1986) 557-563.

50 Non-linear Optical Properties of Organic and Polymeric Materials, Edited by D. J. Williams, ACS Symposium Series, American Chemical Society, 1983.

51 G. R. Desiraju, D. Y. Curtin, and I. C. Paul, Mol. Cryst. Liq. Cryst., 52 (1979) 259-266. For a review of other work on solid-state chemistry of quinhydrones see A. O. Patil, W. T. Pennington, G. R. Desiraju, D. Y. Curtin, and I. C. Paul, Mol. Cryst. Liq. Cryst., 134 (1986) 279-304.

52 (a) B. S. Green and M. Knossow, Science, 214 (1981) 795-797. (b) S. Ramdas, J. Thomas, M. E. Jordan, and C. J. Eckhardt, J. Phys. Chem., 85 (1981) 2421-2525.

53 (a) J. P. Dougherty, S. K. Kurtz, J. Appl Cryst.,9 (1976) 145-158. (b) J. Trotter, Acta Crystallogr., 16 (1963) 605-608. (c) M. I. Kay, M. I., Y. Okaya, D. E. Cox, Acta Crystallogr., B27 (1971) 26-31.

54 S. Matsumoto and T. Fukuda, Bull. Chem. Soc. Jpn., 40 (1967) 743-746. D. H. Spielberg, R. A. Arndt, A. C. Damask, and I. Lefkowitz, J. Chem. Phys., 54 (1971) 2597-2601.

55 F. H. Herbstein, M. Kapon, G. M. Reisner, M. Lehman, R. B. Kress, R. B. Wilson, W-I. Shiau, E. N. Duesler, I. C. Paul, and D. Y. Curtin, Proc. Roy. Soc. London, A399 (1985) 295-319.

56 D. B. Pendergrass, I. C. Paul, D. Y. Curtin, J. Am. Chem. Soc., 74 (1972) 8730-8737.

57 R. B. Kress, E. N. Duesler, M. C. Etter, I. C. Paul, D. Y. Curtin, J. Am. Chem. Soc., 102 (1980) 7709-7714.

58 T. W. Lewis, I. C. Paul, D. Y. Curtin, Acta Crystallogr., B36 (1980) 70-77.

59 M. Kaftory, Acta Crystallogr., C40 (1984) 1015-1019.

60 (a) F. Baert, A. Mierzejewski, B. Kucjta, R. Nowak, Acta Cryst., B42 (1986) 187-193. (b) N. Kitamura, O. Saravari, H-B. Kim, and S. Tazuke, Chem. Phys. Lett., 125 (1986) 360-363.

61 C. Svensson and S. C. Abrahams, Mat. Res. Soc. Symp. Proc. Vol. 21 (1984) 149-154.

62 (a) R. E. Pincock, R. R. Perkins, A. S. Ma, K. R. Wilson, Science, 174 (1971) 1018-1020; K. B. Wilson, R. E. Pincock, J. Am. Chem. Soc. 97 (1975) 1474-1478. (b) R. B. Kress, E. Duesler, M. C. Etter, I. C. Paul, and D. Y. Curtin, J. Am. Chem. Soc., 102 (1980) 7709-7714.

63 I. S. Rez, Kristallografiya, 5 (1960) 63-67.

64 See G. Meredith, in ref. 51, pp. 28 ff.

65 (a) S. Tomaru, S. Zembutsu, M. Kawachi, and M. Kbayashi, J. Chem. Soc. Chem. Comm. (1984) 1207-1208. (b) Y. Wang, D. F. Eaton, Chem. Phys. Lett., 120 (1985) 441-444.

66 B. Servet and J. Rault, Journ. Phys., 40 (1979) 1145-1148.

67 See, for example, R. L. Greene and G. B. Street, Science, 226 (1984) 651-656.

68 A. M. Glass, Science, 226 (1984) 657-662.

69 See, for example, the description of a meeting on the subject held at the Crump Institute for Medical Engineering, University of California, Los Angeles, in November, 1983 [C&EN 1983, 33-34.

70 S. C. Abrahams, Abstracts of American Crystallographic Society Winter Meeting, March, 1981, H4, p. 28.

Chapter 10

PHASE TRANSITIONS IN ORGANIC SOLIDS

C. N. R. RAO

1. INTRODUCTION

A variety of solids transform from one crystal structure to another as the temperature or pressure is varied. The subject of phase transitions has grown explosively in recent years and there has been much interest in understanding the unifying principles of the subject (1,2). Many new types of phase transitions have been discovered, besides large classes of solids undergoing transitions. Traditionally, most studies of phase transitions pertain to metallic and inorganic solids, but many organic solids exhibit phase transitions, often involving interesting structural changes. It was believed not too long ago by many workers that there are no structural relations between phases in transitions of organic solids. More recently, however, phase transitions of several organic solids have been well characterized and we shall discuss a few typical examples in this Chapter. An important development in the last five years pertains to the study of phase transitions by computer simulation techniques and we shall refer to some of the recent results from molecular dynamics and Monte Carlo calculations. We shall first briefly review the general features of phase transitions.

2. GENERAL FEATURES OF PHASE TRANSITIONS

The free energy of a solid remains continuous during a phase transition, but thermodynamic quantities such as entropy, volume and heat capacity exhibit discontinuous changes. Depending on which derivative of the Gibbs' free energy, G, shows a discontinuous change at the transition, phase transitions are generally classified as first order or second order. In a first-order transition, the G(P,T) surfaces of the parent and product phases intersect sharply and the entropy as well as the volume show singular behaviour (Fig. 1). In a second-order transition, the heat capacity, compressibility, or thermal expansivity shows singular behaviour.

Landau introduced the concept of an order parameter, ξ , which is a measure of the order resulting from a phase transition. In a first-order transition (e.g., liquid-crystal), the change in ξ is discontinuous, but

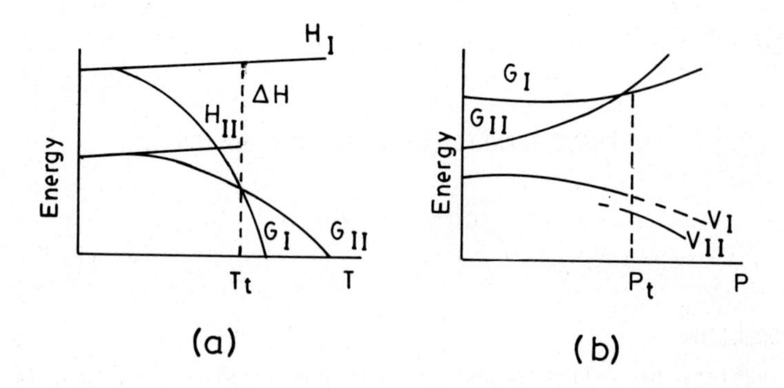

Fig. 1 Changes in Gibbs' free energy and enthalpy (or volume) with temperature (or pressure) in a first-order transition. T_t and P_t are transition temperature and pressure respectively.

in a second-order transition where the change of state is continuous, the change in ξ is also continuous. Landau proposed that G in a second-order or structural phase transition, besides being a function of P and T, is also a function of ξ and expanded G as a series in powers of ξ around the transition point. The order parameter vanishes at the critical temperature, T_c, in such transitions (Fig. 2). Landau also examined the symmetry changes

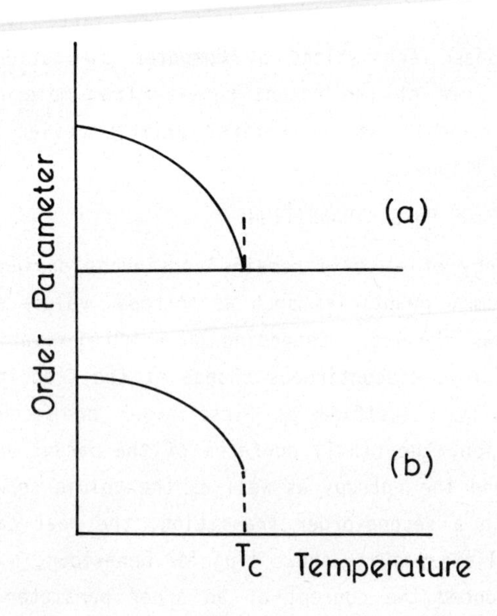

Fig. 2 Variation of order parameter in (a) second order and (b) first-order transitions.

across phase transitions. Thus, a transition from a phase of high symmetry to one of low symmetry is accompanied by an order parameter. In a second-order transition, certain elements of symmetry appear or disappear across the phase transition.

In phase transitions of ordered, low-temperature phases of organic solids (e.g. neopentane, CCl_4) into orientationally disordered phases, we can define an angle between one of the crystal directions and a bond (or a molecular axis as an order parameter). In a ferroelectric-paraelectric transition, polarization is the order parameter while in the transition of a ferromagnet to a paramagnet, magnetization is the order parameter.

Many physical properties diverge near the critical temperature, T_c; that is, they show anomalously large values as T_c is approached from either side. The divergences in different phase transitions are, however, strikingly similar. These divergences can be quantified in terms of critical exponents, λ :

$$\lambda = \lim_{\varepsilon \to 0} \left| \frac{\ln f(\varepsilon)}{\ln |\varepsilon|} \right|$$

where $\varepsilon = (T - T_c)/T_c$ and $f(\varepsilon)$ is the function whose exponent is λ. The most important exponents are those associated with the specific heat, α, the order parameter, β, the susceptibility, γ, and the range over which individual constituents like atoms and atomic moments are correlated, ν. It so happens that the individual exponents for many different transitions are roughly similar (e.g., $\beta \approx 0.33$). More interesting is the fact that $\alpha + 2\beta + \gamma = 2$ in most transitions, independent of the detailed nature of the system. In other words, although individual values of exponents may vary from one transition to another, they all add up to 2. Such a universality in critical exponents is understood in the light of Kadanoff's concept of scale invariance associated with the fluctuations near T_c. The exponents themselves can be calculated by employing the renormalization group method developed by Wilson (3).

Another important aspect of phase transitions in solids is the presence of soft modes. Operationally, a soft mode is a collective excitation whose frequency decreases anomalously as the transition point is reached. In second-order transition, the soft mode frequency goes to zero at T_c, but in first-order transitions, the phase change occurs before the mode frequency goes to zero. Soft modes have been found to accompany a variety

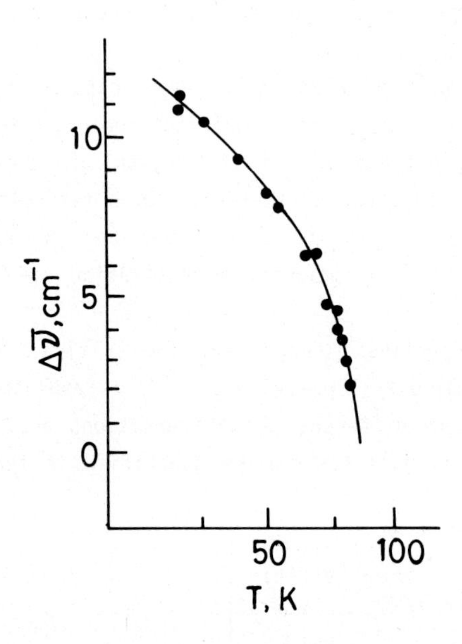

Fig. 3. Raman soft mode in chloranil (from ref. 4).

of solid-state transitions, including those of superconductors and organic solids. Occurrence of soft modes in phase transitions can be inferred from Landau's treatment wherein atomic displacements may themselves be considered to represent an order parameter. In Fig. 3 we show a typical example of a soft mode in the case of chloranil (4).

It is convenient to classify phase transitions in solids on the basis of the mechanism as well. Three important kinds of transitions are of common occurrence: (i) nucleation and growth transitions, a typical example being the anatase-rutile transformation of TiO_2, (ii) positional and orientational order-disorder transitions, and (iii) martensitic transitions. A typical example of a positional order-disorder transition is that of AgI; orientational order-disorder transitions are exhibited by many solids such as ammonium halides and orientationally disordered organic crystals. The entropy change in such order-disorder transitions is given by R ln (ω_2/ω_1) where ω_2 and ω_1 are the number of possible configurations of phases 2 and 1 respectively. A martensitic transition is a structural change caused by atomic displacements (and not by diffusion) corresponding to a homogeneous deformation wherein the parent and product phases are related by a substitutional lattice correspondence, an irrational habit plane and a precise orientational relationship. These transitions occurring with high velocities of the order of sound velocity were originally discovered in steel but are now known to occur in several inorganic as well as organic solids.

3. ORGANIC SOLIDS

The various types of transitions found in organic solids include those involving the liquid crystalline, the plastic crystalline and the glassy states. Crystals ordinarily possess both translational and orientational order and phase transitions occur between one ordered crystalline state to another. Some of the transitions occur from an ordered crystalline state to a disordered state as mentioned earlier. When the disorder is associated with molecular orientation, the state is referred to as the plastic state. The transition from the crystalline to the plastically crystalline state is exhibited by molecular crystals as the molecules acquire orientational degrees of freedom on raising the temperature (5,6). In the liquid crystalline state, there is no translational order, but there is orientational order. A large number of molecules and polymers exhibit the liquid crystalline state. Rod-shaped as well as disc-shaped molecules form liquid crystals, well-known examples being, p-azoxyanisole, p-heptyl-p'- cyanobiphenyl and hexasubstituted benzoic esters (7). Many organic liquids when quenched to low temperatures form glasses, exhibiting the characteristic glass transition (8). The glass transition is found not only in positionally disordered glasses but also in glasses formed by quenching orientationally disordered (plastic) crystals (6,8).

The occurrence of a soft mode in the Raman spectrum of chloranil was referred to earlier. This transition in chloranil at 92K is from the space group $P2_1/a$ to $P2_1/n$ as demonstrated by x-ray and NQR studies. The soft mode behaviour is confirmed by the measurement thermal diffuse scattering of x-rays in the high-temperature phase, at positions in reciprocal space where the extra Bragg reflections of the low-temperature phase appear. The inverse x-ray intensity extrapolates to zero at the transition temperature (9). N-Nitrodimethylamine is reported to show a soft mode in the Raman spectrum at the transition temperature of 107K, but the structure of the low-temperature phase is not known (10).

Mechanistic studies on the phase transitions of organic solids are of relatively recent origin and there have been some interesting investigations giving structural and other details. It is noteworthy that Robertson and Ubbelohde (11) showed nearly five decades ago that there was preferred orientation in the transformation of the β-form of resorcinol to the α-form. The thermal transformation of the β-phase of p-nitrophenol to the stable α-phase occurs with the movement of the phase boundary perpendicular to the [001] axis of the crystal (12). Jones et al (13) have investigated the stress-induced phase transition of 1,8-dichloro-10-methylanthracene and found it to proceed by a diffusionless displacive transition (somewhat similar to a martensitic transition) with definite orientational relationships. The irrational habit plane seems to be composed of close-packed planes and the properties of the interface could be formulated in terms of slip dislocations. The mechanism of this transition is somewhat similar to that proposed for the stress-induced (martensitic) transition of polyethylene. The reversible topotactic phase transition of 5-methyl-1-thia-azoniacyclo-octane-1-oxide perchlorate at 280K which was earlier considered to involve a cooperative inversion and rotation of half the molecular cations (14) was later shown to proceed through the movement of recurrent glissile partial dislocations (15). During this transition, the unit cell $\underline{a}$ parameter doubles while the $\underline{c}$ parameter halves.

Paraterphenyl exhibits a phase transition at 110K when the space group changes from $P2_1/a$ to $P\bar{1}$ accompannied by the doubling of the $\underline{a}$ and $\underline{b}$ parameters. This phase transition involves rotational disorder in which the molecule becomes non-planar. Ramdas and Thomas (16) have elucidated the nature of the transition by evaluating pair-wise interactions between nonbonded atoms. Bis-(p-methoxy)-trans-stilbene exhibits an unusual photo-induced conformational polymorphism involving a single crystal - single crystal transformation (17). The change in conformation has been rationalised in terms of an intermediate excimeric state.

The α - γ - α - β transitions of p-dichlorobenzene have been investigated by employing several techniques including IR and NQR spectroscopy (Fig. 4). The γ phase is characterized by unusually high intramolecular vibration mode frequencies. The α - γ transition shows athermal nucleation behaviour as in martensitic transitions (Fig. 5) while the α - β transition seems to be associated with some disorder (18). 3,5-Dichloropyridine exhibits two phase transitions, the one at 285K being a first-order reconstructive type while the other at 168K is displacive; librational modes seem to play a major role in both the transitions (19). Vibronic bands of TCNQ salts show marked changes in the intensity of some of the infrared bands at the phase transition temperatures where the dimeric units of

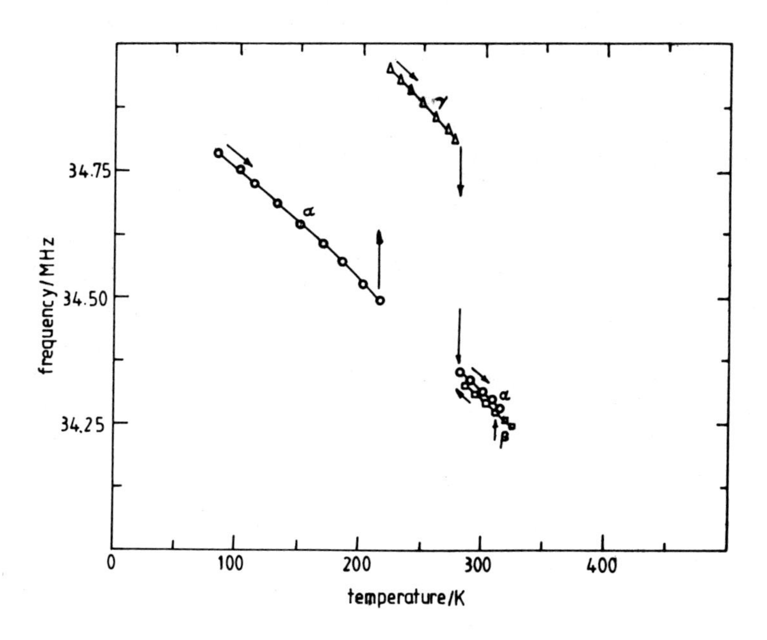

Fig. 4 Variation of ^{35}Cl NQR frequency of p-dichlorobenzene (from ref. 18): α (circles); γ (triangles) and β (squares).

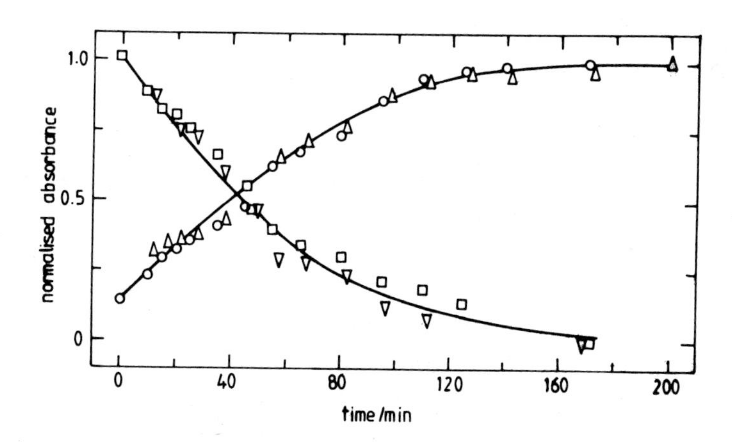

Fig. 5 Time-dependence of infrared bands of p-dichlorobenzene through the α - γ transition (from ref. 18) obtained from IR spectroscopy.

the radical anions readjust themselves to become uniformly stacked monomeric units (20).

When the asymmetry of molecules is not too high, organic crystals often exhibit order-disorder transitions. For example, benzothiophene undergoes a transition with an entropy gain of 2.75 cal deg^{-1} mol^{-1} which is exactly Rln 4 suggesting that the molecule uses four possible configurations in the disordered state. The order-disorder transitions in benzothiophene and other compounds have been investigated by making use of changes in infrared band intensities and half-widths (20). Raman spectroscopy would be specially useful in investigating such order-disorder transitions. The disordered state gets frozen during crystallization in many organic compounds and no transitions occur to the ordered state; azulene and hexa-substituted benzene derivatives are typical examples.

Phase transitions of compounds of the type $(C_nH_{2n+1}NH_3)_2MCl_4$ with M = Mn, Fe, Cd or Cu, provide interesting model systems to investigate magnetic phenomena in two dimensions. Earlier spectroscopic investigations seemed to indicate that in $(CH_3NH_3)_2MCl_4$, the phase transitions are determined essentially by the motions of the methyl-ammonium groups. A detailed investigation of the infrared spectra of several $(C_nH_{2n=1}NH_3)_2MCl_4$ systems through their phase transitions has been carried out to see whether the intramolecular vibration modes show the expected changes (21). In the high-temperature phases of these solids, the $(CH_3NH_3)^+$ ion has C_{3v} symmetry, but the symmetry goes down to C_1 or C_s, in the ordered low-temperature phases. The spectra indeed show the expected site-group as well as factor-group splittings in the low-temperature phases, the degenerate bending modes of NH_3 and CH_3 being particularly sensitive. The phase transitions of the tetrachlorometallates are similar to those of the corresponding alkylammonium chlorides, $C_nH_{2n+1}NH_3Cl$, thereby establishing that the phase transitions in the former are entirely controlled by the motions of the $(C_nH_{2n+1}NH_3)^+$ group. Accordingly, it is found that the $(C_nH_{2n+1}NH_3)_2MBr_4$ system shows transitions similar to those of the chloro compounds. Solids of the type $[N(CH_3)_4]_2$ MX_4 (M = Mn, Co, Cu or Zn, X = Cl or Br) show phase transitions involving incommensurate structures (22). Transitions in such solids are ideally investigated by NMR and NQR spectroscopy (7).

Phase transitions of hydrogen-bonded solids such as ferroelectric hydrogen phosphates and Rochelle salt have been investigated widely in the literature. Thiourea exhibits several transitions with two ferroelectric phases. Many alkanedioic acids exhibit two polymorphic forms and the phase transitions of the lower alkanedioic acids have been investigated by employing vibrational spectroscopy (20). Oxalic acid itself can exist in two crystalline forms with unique hydrogen bonded structures which have been characterized by x-ray crystallography and vibrational spectroscopy (23). The phase transition of malonic acid at 360K is specially interesting (24). At ordinary temperatures, the unit cell of malonic acid contains two cyclic dimeric rings orthogonal to each other; above 360K, the two hydrogen-bonded rings become similar as evidenced from infrared and Raman spectra. Hydrogen bonds in the high-temperature phase are on the average weaker than those in the low-temperature phase. The phase transition occurs at a higher temperature (366K) in the fully deuterated acid, and the vibrational bands show a positive deuterium isotope effect. It appears that the transition is governed by librational and torsional modes of the hydrogen-bonded rings (around 90 and 50 cm^{-1}, respectively, below the transition temperature), which show a tendency to soften (Fig.6).

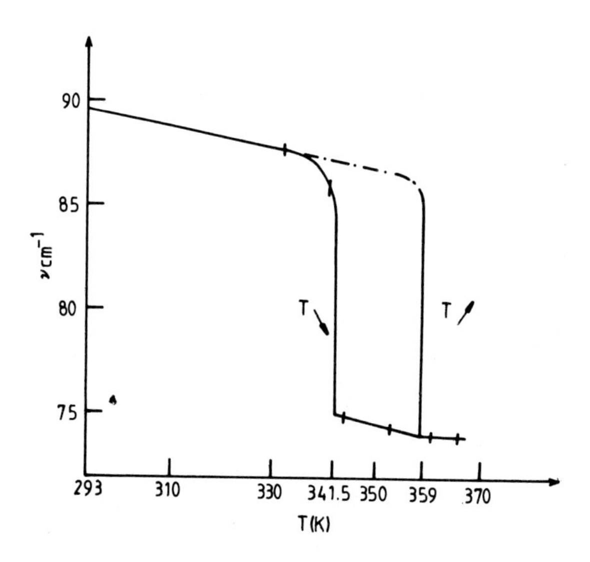

Fig. 6 Temperature dependence of Raman modes of malonic acid (from ref.24).

4. ORGANIC PLASTIC CRYSTALS

The characteristic thermodynamic feature of plastic crystals which led to the recognition of their existence by Timmermanns is that the entropy change in the crystal - plastic crystal transition is far greater than the entropy change accompanying the melting of the plastic crystal (Table 1). In addition to the low entropy of melting, plastic crystals usually melt at a relatively high temperature; for example, plastic crystalline neopentane melts at 257K while n-pentane, which does not form the plastic state melts at 132K. In some cases, plastic crystals directly sublime without melting, camphor and hexachloroethane being typical examples.

TABLE 1
Structure and thermodynamic properties of some plastic crystals[a]

Substance	a(Å)	T_t(K)	T_m(K)	ΔS_t(R)	ΔS_m(R)	E_a(kJ mol^{-1})
t-Butyl chloride	8.62	183	245	4.38	0.95	3.0
Neopentane	8.78	140	257	2.22	1.46	4.2
Cyclohexane	8.76	186	280	4.34	1.10	6.4
Hexamethyldisilane	8.47	222	288	5.28	1.25	6.3

[a] All the plastic phases listed in the table possess FCC structure: T_t, crystal-plastic crystal transition temperature; ΔS_t, entropy change at T_t; T_m, melting temperature; ΔS_m, entropy change at T_m; E_a, activation energy for molecular reorientation obtained from NMR spectroscopy.

Because of the orientational freedom, plastic crystals usually crystallize in cubic structures. It is significant that cubic structures are adopted even when the molecular symmetry is incompatible with the cubic symmetry. For example, t-butyl chloride in the plastic crystalline state has a fcc structure even though the isolated molecule has a three fold rotation axis which is incompatible with the cubic structure. Such apparent discrepancies between the lattice symmetry and molecular symmetry provide clear indications of the rotational disorder in the plastic crystalline state. Beside simple organic molecules, Π - donor - Π - acceptor complexes (Pyrene-Trinitrobenzene, Fluoranthene-Trinitrophenol) also show evidence for a pseudoplastic crystalline state intermediate between the crystalline and molten states (25). Molecular rotation in plastic crystals is rarely free; it appears that there are more than one minimum potential energy configurations which allow the molecules to tumble rapidly from one orientation to another, the different orientations being random in the plastic crystal.

As the name suggests, plastic crystals are generally soft, frequently flowing under their own weight. The pressure required to produce flow of a plastic crystal, as for instance to extrude through a small hole, is considerable less than that required to extrude a regular crystal of the same substance. The subject of plastic crystals has been reviewed fairly extensively and we shall briefly discuss the nature of orientational motion (5,6).

Existence of a high degree of orientational freedom is the most characteristic feature of the plastic crystalline state. We can visualize three types of rotational motions in crystals: free rotation, rotational diffusion and jump reorientation. Free rotation is possible when interactions are weak and this situation would not be applicable to plastic crystals. In classical rotational diffusion (proposed by Debye to explain dielectric relaxation in liquids), orientational motion of molecules is expected to follow a diffusion equation described by an Einstein type relation. This type of diffusion is not known to be applicable to plastic crystals. What would be more appropriate to consider in the case of plastic crystals is collision-interrupted molecular rotation.

The rotational diffusion model was generalized by Gordon (26) to include molecular reorientation in angular steps of fairly large size. According to this model, the molecules are supposed to freely rotate during the intervals between collisions. Two limiting cases have been discussed by Gordon, both of which involve angular diffusion steps (of arbitrary size). These are commonly known as the J-diffusion and the M-diffusion models. In the former, the direction as well as the magnitude of the angular momentum vector of the molecule follow Boltzman distribution due to collisions. In the M-diffusion model, the orientation of the angular momentum vector is randomized, but the magnitude is unchanged. In Gordon's model applied to linear molecules, the general diffusion equation follows classical mechanics at short times. At longer times, the rotational frequency is unchanged in successive steps in the M-diffusion model while it follows Boltzman distribution in the J-diffusion model.

In the model proposed by Hill (27) and Wyllie (28), molecules exhibit librations (perturbed angular oscillations), having been trapped in potential wells. Fluctuations in orientation of neighbouring molecules result in a change in the shape of the potential well and consequently give rise to angular motion of molecules with large amplitudes. It is possible that diffusion of the potential well on the surface of a sphere gives rise to reorientational motion of molecules. If the potential surface is related to the crystal space group, molecular reorientation will be associated with damped librations. Rotations of the crystal space group will give rise to allowed potential wells.

No single model can exactly describe molecular reorientation in plastic crystals. Models which include features of the different models described above have been considered. For example, diffusion motion interrupted by orientation jumps has been considered to be responsible for molecular reorientation. This model has been somewhat successful in the case of cyclohexane and neopentane (29, 30). What is not yet completely clear is whether the reorientational motion is cooperative. There appears to be some evidence for coupling between the reorientational motion and the motions of neighbouring molecules. It appears that there is scope for comparative experimental studies employing complementary techniques which are sensitive to autocorrelation and monomolecular correlation.

The nature of rotational motion responsible for orientational disorder in plastic crystals has been investigated by means of a variety of experimental techniques. There can be coupling between rotation and translation motion, the simplest form of the latter being self-diffusion. The diffusion constant as given by the Einstein relation depends on the magnitude of molecular velocity as well as the persistence velocity. Both radioactive tracer and NMR methods have been employed to measure the coefficients of self-diffusion in plastic crystals. In many cases, the two methods yield similar results since they refer to a similar process on the time scales of the velocity correlation function. The diffusion coefficients are anomalously high (compared with ordinary organic crystals) and the softest crystals exhibit the highest diffusion coefficients.

NMR spectroscopy provides spin-lattice (T_1) and spin-spin (T_2) relaxation times. Making appropriate assumptions with regard to the magnetic interactions responsible for the relaxation process, these relaxation times can be related to molecular motions. Since nuclear spin relaxation results from all processes which cause a fluctuation in the magnetic field at the nucleus, the correlation function will generally correspond to more than one kind of motion involving all possible interactions. If the relaxation process is essentially due to dipolar interactions, then NMR spectroscopy gives information on rotational diffusion. If nuclear quadrupole relaxation is dominant, then we obtain information on molecular reorientation. In Table 1, activation energies for molecular reorientation obtained from NMR spectroscopy are listed.

Thermal or low energy neutron scattering experiments have been most valuable in throwing light on molecular motion in plastic crystals. These experiments measure changes in the centre of mass of a molecule. Diffusion constants obtained from neutron experiments differ widely from those obtained from tracer experiments since neutron scattering is mainly determined by rotational diffusion. The inelastic part of the scattering provides information on the rotational motion while the elastic part is related to the translational motion. If the reorientational motion is diffusive, we can employ a diffusion equation similar to that mentioned earlier, the width of the quasi-elastic peak being related to rotational diffusion. One often employs different functions to see if rotational motion occurs through jumps. A good example of a neutron study of plastic crystals is that on _t_-butyl cyanide by Frost et al (31).

One of the most direct methods of examining reorientational motion of molecules is by far infrared absorption spectroscopy or dielectric absorption. In the absence of vibrational relaxation, the relaxation times obtained by IR and dielectric methods are equivalent. In both these techniques we obtain the correlation function, $< \mu (t_0) . \mu (t_0 + t) >$, for the motion about a specific molecular axis with respect to the principal rotation axes and the direction of vibration. Rayleigh scattering as well as Raman scattering are most useful in the study of molecular orientational processes. With the aid of isotropic and anisotropic Raman scattering measurements, orientational processes can be delineated from other mechanisms which cause line broadening. Analysis of infrared and Raman bandshapes provides information on the short-term and long-term behaviour of the correlation function.

Correlation times and activation energy parameters obtained from different techniques may or may not agree with one another. Comparison of these data enables one to check the applicability of the model employed and examine whether any particular basic molecular process is reflected by the measurement or whether the method of analysis employed is correct. In order to properly characterize rotational motion in plastic crystals it may indeed be necessary to compare correlation times obtained by several methods. Thus, values from NMR spectroscopy and Rayleigh scattering enable us to distinguish uncorrelated and correlated rotations. Molecular disorder is not reflected in NMR measurements; to this end, diffraction studies would be essential.

Computer simulation studies have thrown some light on the reorientational motion in plastic crystals. This aspect is briefly covered in the next section.

5. COMPUTER SIMULATION

Monte Carlo and molecular dynamics are two important techniques employed for the computer simulation of fluids and solids. In these techniques, properties of a finite system of particles interacting _via_ a known interparticle potential are evaluated. The principle of the Monte Carlo method is to perform a stochastic averaging of the properties by means of the Metroplis importance sampling technique. Most of the Monte Carlo calculations have been carried out in a cell of constant volume, Gibbs' petit-canonical or NVT ensemble. However, in recent years calculations in the isothermal isobaric ensemble in which the cell varies in volume during the simulation but is of fixed shape have appeared in the literature. In molecular dynamic calculations, Newton's equations of motion are solved numerically for a system of particles. As the total energy of the system is conserved, the ensemble generated in the simulation corresponds to the microcanonical or NVE ensemble. The main advantage of the Monte Carlo method is that the isothermal-isobaric calculations are closer to the experimental conditions of fixed temperature and pressure. Dynamical properties, on the other hand, are readily obtained from molecular dynamics.

The constant energy molecular dynamics and the canonical ensemble Monte Carlo techniques cannot be used to study phase transitions resulting from change in temperature or pressure, since the shape of the simulation cell is not permitted to change during the course of the simulation. Andersen (32) extended the molecular dynamics method to allow for the variation in the size of the cell. Parrinello and Rahman (33) further extended it to allow for the variation in shape and size of the cell. This has enabled studies of phase transitions by the molecular dynamics method (34). The Monte Carlo method in the canonical or in the isothermal-isobaric ensemble calculations does not permit variation of the shape of the simulation cell. The Monte Carlo method has been generalized by Yashonath and Rao (35,36) to enable study of phase transitions and related phenomena, by allowing for the variation in shape of the simulation cell. In molecular systems this has been done by including the orientational coordinates in terms of the Euler angles, making it possible to investigate polymorphic or structural phase transitions in solids by the Monte Carlo method. During the simulation of solid carbon tetrachloride, it has been found that the cell rotates in space. The rotation of the simulation cell can occur in variable-shape Monte Carlo as well as in variable-shape molecular dynamics studies.

The generalized molecular dynamics method wherein the shape of the cell can be continually varied through a transition has been employed to investigate phase transitions of ionic as well as molecular crystals especially those involving the orientationally disordered (plastically crystalline) state. For example, transitions from the crystalline to the elastically crystalline state transitions of CF_4 and bicyclo-(2.2.2) octane have been investigated by Klein and coworkers (37, 38). The generalized Monte Carlo method involving a variable-shape, variable-volume cell has been employed by Yashonath and Rao (36, 39) with appropriate pair potentials to investigate transitions from the crystalline to the platically crystalline states of CCl_4 and adamantane or tricyclo [3,3,1,1] decane. In

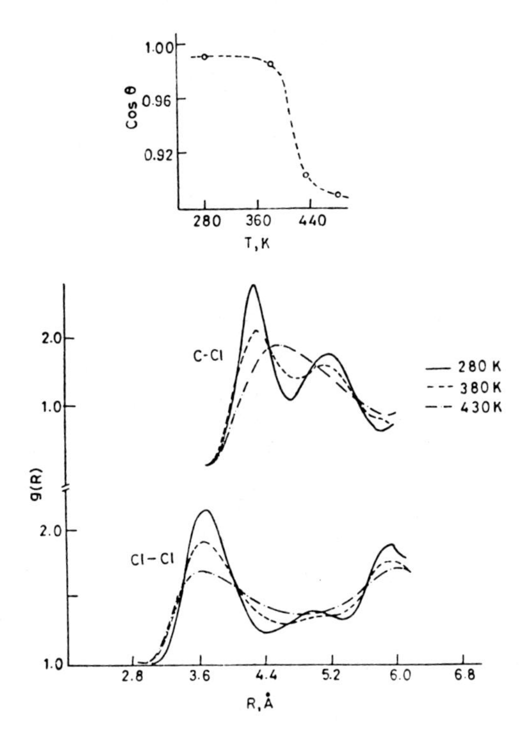

Fig. 7 Radial distribution functions of CCl_4 at 1 GPa. Inset shows a plot of Cos θ against temperature (from ref. 36).

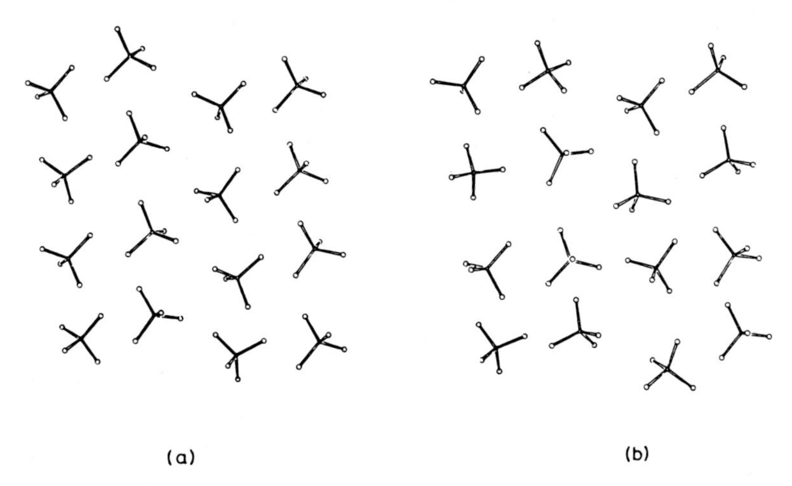

Fig. 8 Snapshot of the molecular arrangement looking down the b axis at (a) 380K and (b) 430K in CCl_4 (from ref. 36).

Fig. 7, the radial distribution functions of CCl_4 are shown at different temperatures. Above the transition temperature of ~410K, the peaks become broad as expected in the orientationally disordered phase. A snapshot of the molecular rearrangement (Fig. 8) clearly reveals the disordered nature of the high temperature phase of CCl_4. The cosine of the angle between the molecular C_3 axis and the crystallographic [101] direction

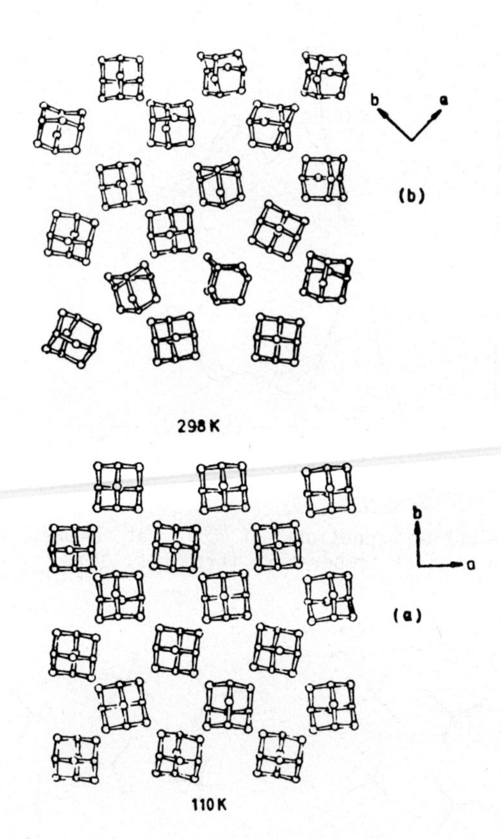

Fig. 9 Instantaneous arrangement in adamantane looking down the c-axis. Orientation disorder is seen at 298K (from ref. 39).

clearly shows evidence for the existence of the phase transition (Fig.7). A similar study on tricyclo [3,3,1,1] decane or adamantane has enabled us to establish the occurrence of a phase transition from the ordered, low temperature, tetragonal phase to a high temperature orientationally disordered, cubic phase. A snapshot of the molecular arrangement in adamantane is shown in Fig. 9 to demonstrate the changes occurring during the transformation. These two examples should suffice to demonstrate how the generalized Monte Carlo method can satisfactorily describe phase transitions in solids when appropriate pair potential parameters are employed. Preliminary studies have been carried out on the interesting phase transition of biphenyl which is accompanied by a change from the planar to the non-planar conformation of the molecule. A variety of other phase transitions can be examined by computer simulation techniques (e.g. liquid crystals). Organic solid state reactions, packing in organic structures and even crystal structures can be predicted by means of such simulation studies.

Monte Carlo studies on liquid and glassy isopentane have been carried out most successfully by employing transferable intermolecular potential functions (40), in order to understand the nature of the glass transition and the structural changes accompanying glass formation. It was of interest to examine not only the structural factors responsible for the transition but also the structural changes occurring, if any, when the glass is annealed. Thermodynamic properties, radial distribution functions, coordination number distribution etc. calculated for the liquid were in reasonable agreement with the known experimental data. By quenching the liquid, we have obtained the glass-transition temperature from the temperature variation of the intermolecular energy, orientational contribution to the intermolecular energy, volume and the heat of vaporization. Radial distribution

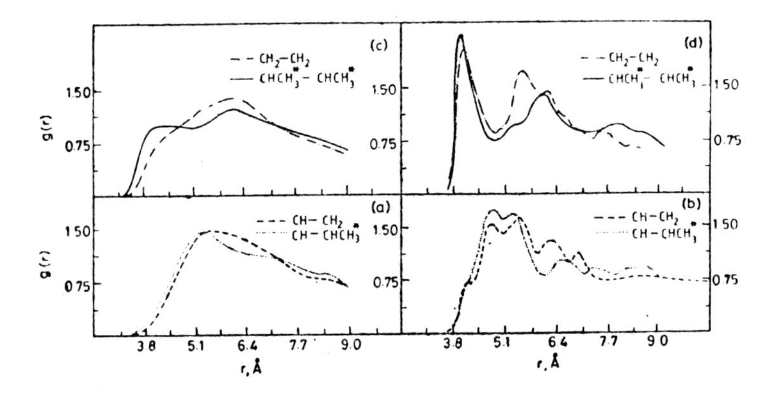

Fig. 10 Radial distribution functions between inaccessible CH and other groups in (a) liquid and (b) glassy states of isopentane at 301K and 30K respectively. Rdfs between peripheral groups in the liquid and glassy states are shown in (c) and (d) respectively (from ref. 40).

functions (Fig. 10) suggest that the structure of the glass is primarily influenced by geometrical factors. In Fig. 11, we show stereoplots of

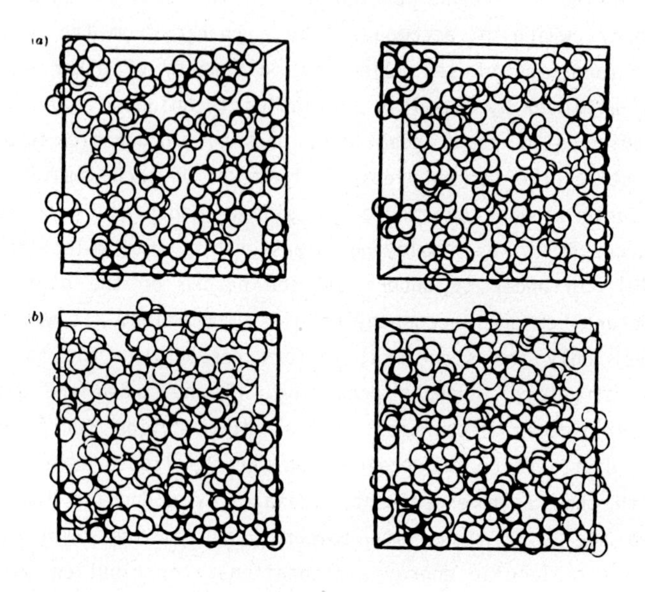

Fig. 11 Stereoplots of liquid (a) and (b) glassy states of isopentane at 301K and 30K respectively. Small voids can be seen in the glass as well (from ref. 40).

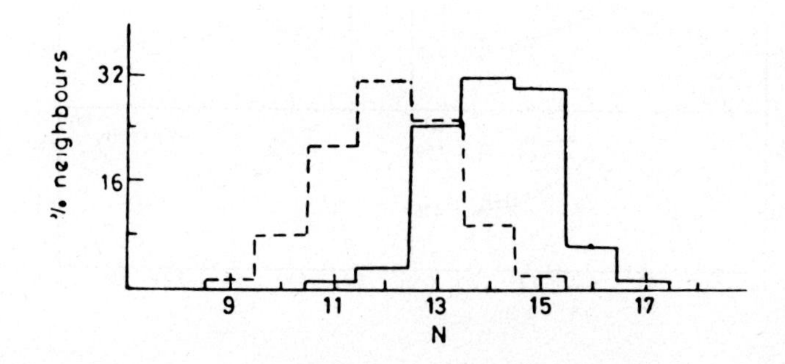

Fig. 12 Coordination number distribution for glassy isopentane at 30K (full curve) and the liquid at 301K (broken curve) (from ref. 40.).

the structure of the glass and the liquid. The histogram of the nearest neighbour distribution shows a shift towards higher coordination in the glassy state (Fig. 12). Interesting differences are found between the liquid and the glass in the dimerization and bonding energy distribution functions. There is a narrower distribution in the dihedral angle distribution function for rotation around the central C-C bond, after vitrification. A detailed analysis of the radial distribution functions (41) has shown that the reorientational contribution to the increase in intermolecular energy on vitrification is at least 50 per cent. Reorientational freezing plays a major role near the glass by a rearrangement involving rotation of the molecule.

ACKNOWLEDGEMENT

The author is thankful to the University Grants Commission and the Department of Science and Technology, Government of India, for support of research in this area.

REFERENCES

1. C.N.R. Rao and K.J. Rao, Phase Transitions in Solids, Mc-Graw Hill, New York, 1978.
2. C.N.R. Rao, Accts. Chem. Res., 17 (1984) 83.
3. K.G. Wilson and J. Kogut, Phys. Repts., 12C (1974) 77.
4. D.M. Hanson, J. Chem. Phys., 63 (1975) 5046.
5. J.N. Sherwood (Ed), The Plastically Crystalline State, John Wiley, New York, 1979.
6. C.N.R. Rao in "Topics in Molecular Interactions", Elsevier, Amsterdam, 1985.
7. C.N.R. Rao and J. Gopalakrishnan, New Directions in Solid State Chemistry, Cambridge University Press, 1986; also see C.N.R. Rao, K.J. Rao and J. Gopalakrishnan, Ann. Repts. Phys. Chem., Royal Society of Chemistry , 1986.
8. R. Parthasarathy, K.J. Rao and C.N.R. Rao, Chem. Soc. Rev., 12 (1983) 361.
9. H. Teranchi, T. Sakai and H. Chihara, J. Chem. Phys., 62 (1975) 3832.
10. M. Rey-Lafon and R. Lagnier, Mol. Cryst. Liq. Cryst., 32 (1976) 13.
11. J.M. Robertson and A.R. Ubbelohde, Proc. Roy. Soc., London, A167 (1938) 136.
12. P. Coppens and G.M.J. Schmidt, Acta Cryst., 18 (1965) 62, 654.
13. W. Jones, J.M. Thomas and J.O. Williams, Phil. Mag., 32 (1975) 1.
14. I.C. Paul and K.T. Go, J. Chem. Soc. B., (1969) 33.
15. G.M. Parkinson, J.M. Thomas, J.O. Williams, M.J. Goringe and L.W. Hobbs, J. Chem. Soc. Perkin 2 (1976) 836.
16. S. Ramdas and J.M. Thomas, J. Chem. Soc. Faraday 2, 72 (1976) 1251.
17. C.R. Theocharis, W. Jones and C.N.R. Rao, J.C.S. Chem. Commm., (1984) 1291.
18. S. Ganguly, J.R. Fernandes, D. Bahadur and C.N.R. Rao, J. Chem. Soc. Faraday 2, 75 (1979) 923.
19. F. Romain, P. Tougard, B. Pasquier, N. Le Calve, A. Novak, A. Peneau, L. Gnibe and J. Ramakrishna, Phase Trans., 3 (1983) 259.
20. C.N.R. Rao, S. Ganguly and H.R. Swamy, Croat. Chim. Acta., 55 (1982) 207.
21. C.N.R. Rao, S. Ganguly, H.R. Swamy and I.A. Oxton, J. Chem. Soc. Faraday 2, 77 (1981) 1825.
22. S. Ganguly, K.J. Rao and C.N.R. Rao, Spectrochim. Acta, 41A (1985) 307.
23. L.J. Bellamy and R.J. Pace, Spectrochim. Acta, 19 (1963) 435.
24. J. de Villepin, M.H. Limage, A. Novak, M. Le Postollec, H. Poulet, S. Ganguly and C.N.R. Rao, J. Raman Spectrosc., 15 (1984) 41.

25. H.R. Swamy, S. Ganguly and C.N.R. Rao, Spectrochim Acta., 39A (1983) 23.
26. R.G. Gordon, J. Chem. Phys., 44 (1966) 1830.
27. N.E. Hill, Proc. Phys. Soc. London, 82 (1963) 723.
28. G. Wyllie, J. Phys., C4 (1971) 564.
29. R.E. Lechner, Solid State Commun., 10 (1972) 1247.
30. L.A. De Graaf and J. Scieskinski, Physica, 48 (1970) 79.
31. J.C. Frost, A.J. Leadbetter and R.M. Richardson, Disc. Faraday Soc., 69 (1980) 32.
32. H.C. Andersen, J. Chem. Phys., 72 (1980) 2384.
33. M. Parrinello and A. Rahman, Phys. Rev. Letts., 45 (1980) 1196.
34. M.L. Klein, Ann. Rev. Phys. Chem., 36 (1985) 525.
35. S. Yashonath and C.N.R. Rao, Mol. Phys., 54 (1985) 245.
36. S. Yashonath and C.N.R. Rao, Chem. Phys. Lett., 119 (1985) 22.
37. S. Nose and M.L. Klein, J. Chem. Phys., 78 (1983) 6928.
38. E. Neusy, S. Nose and M.L. Klein, Mol. Phys., 52 (1984) 269.
39. S. Yashonath and C.N.R. Rao, J. Phys. Chem., 90 (1986) 2552.
40. S. Yashonath and C.N.R. Rao, Proc. Roy. Soc., London, A400 (1985) 61.
41. S. Yashonath and C.N.R. Rao, J. Phys. Chem., 90 (1986) 2581.

Chapter 11

MOLECULAR MOTIONS IN ORGANIC CRYSTALS: THE STRUCTURAL POINT OF VIEW

A.GAVEZZOTTI and M.SIMONETTA*

1. INTRODUCTION: THEORIES OF CRYSTAL STRUCTURE

Must a molecule have a shape? Most organic and practicing chemists, including crystallographers, may be inclined to answer "yes", but, of course, any scientist with a knowledge and understanding of quantum chemistry knows that the correct answer is "no". In fact, what one is confronted with in chemical problems is molecular stationary states, which do not require a definition of molecular structure, and which can in principle be probed themselves by experiment (ref. 1). Even if the most chemically attractive outcome of molecular quantum mechanics (the electron density) is called for, first principles warn that any wave function is defined as non-zero in all space, and no such idea as a boundary or wrapping surface is inherent to it.

After this principle statement, however, we will proceed to develop theories of crystal structure that rely almost entirely on the concept of molecular shape. Should we tackle the many-body problem of large molecules in condensed phases by the above sketched, extreme quantum mechanical attitude, we would meet immediate failure. Nevertheless, let it be clear that molecular structure (of which we shall offer a somewhat expanded definition) is not an intrinsic property; in other words (ref. 1), "among the new properties created by the many-body system are the size and shape of an individual atom or molecule". X-ray crystallography provides a unique opportunity to put one's finger on this truth; it is not by chance that the view of a molecule as a three-dimensional, solid object has developed at the same time and at the same pace as X-ray crystallography (ref. 2), to the point that ingenuously built, portable molecular models are nowadays commonplace on the desk of every chemist. The conceptual counterpart of these solid objects takes often the name of steric

*Deceased 6 january 1986.

factor.

Molecular shape consists ultimately of nuclear positions plus van der Waals radii. From this point of view, one can then proceed to discuss crystal statics and dynamics in terms of molecular and crystal structure. This idea, whose simplicity equals its powerfulness, was originally put forward by the Kitaigorodski school in the sixties (see ref. 3 for a recent summary). Crystal packing is then a problem in self-recognition of molecular shapes, while molecular motions and rearrangements, as well as reactivity, can be described mechanistically in terms of interlocking spheres, hard interatomic contacts, in what may be called an extreme solvent effect. Let us point out at once that, in this perspective, empty spaces play quite the same role as filled ones in determining the anisotropic properties of molecular solids.

The molecular shape model performs satisfactorily when phenomena that occur in the thermal excitation domain are considered. What has been said so far is less useful, for example, for crystals of charge-transfer complexes, or of conducting polymers. These substances are electronically delocalized, and their properties depend mainly on the behaviour of their mobile electrons. These problems are a sizable topic in themselves, and will not be dealt with in this chapter.

The mathematical tool which is commonly used to quantify the steric factor is that of atom-atom potentials. These can be derived by various calibration techniques, and other chapters in this book deal with such procedures. Atomic radii can be elaborated upon to account for directional effects (ref. 4), but the crucial point is that the basic description of crystal structure afforded by such methodologies survives small changes in their parameters. What simple methods hopefully bring about is understanding, and there is no parameter-dependent understanding.

2. THE WORKING MATERIALS

2.1 Databases

Organic crystal structure is encoded in a language which we do not yet understand fully. It is a basic principle of cryptology, however, that any secret code must yield if enough coded material is available for analysis. X-ray diffraction provides this material in abundance: with 73000 entries in the Cambridge Data Files, of which 51000 with full nuclear coordinates (as of october 1984), and an increase rate of about 5000 entries per year for the organocarbon database, the road is well paved. These data are often used to gain understanding of molecular structure (ref. 5), but they may be used to probe crystal structure as well (see for example ref. 6). The same can be said for studies of crystal dynamics and crystal reaction pathways, although the subset of dynamically characterized structures in the crystallographic data banks is scarcely populated - one reason being that crystallographers sometimes shy away from the intricacies of disordered structures when many neat and straightforward ones are on the waiting list.

Other clues may come from studies of polymorphism, which is sometimes a static picture of the result of dynamic processes in crystals. We will show some examples of a strict connection between molecular mobility and phase transitions in organic crystals.

2.2 Thermal studies and spectroscopy

Differential scanning calorimetry (DSC) provides a reliable way of detecting structural anomalies in crystals, and its use is becoming widespread. One global probe of crystal stability is of course the heat of sublimation, and a compilation (with a simple method for the evaluation of this quantity by a sum over group increments) has been given by Bondi (ref. 7).

Spectroscopic techniques - IR, UV, NMR, ESR and others, both in traditional form or specifically adapted to the solid state, are widely applied to study crystal dynamics, and the reader will find many examples throughout this chapter.

2.3 Previous resource papers and literature search

This chapter is intended to give an overview of dynamic effects that occur in organic crystals, with an emphasis on the computational techniques for their analysis. We have previously undertaken much the same task (ref. 8), but the speed at which the subject is developing makes a four-years old paper an obsolete one. Other resource papers, dealing with topics treated in this chapter, will be cited in the appropriate context. A literature search on dynamic effects in crystals has been conducted to cover the years 1980-1985, but it is well known that the yearly flow of chemical literature makes completeness a chimeric undertaking. We believe however that the examples we report are able to give a complete sampling of the many branches in this field.

2.4 Calculations

A number of calculations on dynamic effects, either taken from previous work or performed ad hoc to illustrate a given computational technique, will be presented throughout this review. Section 4 will provide the reader with the means to perform such calculations for himself. Today, ideas are often embedded in computer programs, and the expanding communicability among computers gives a good opportunity to export and circulate working models of such ideas.

3. MOLECULAR VIBRATIONS AND THERMAL PARAMETERS

3.1 Structure and vibration

Once the molecular structure cast of mind is accepted, we may proceed to analyze crystal dynamics in terms of bond vectors (available from atomic coordinates) and atomic vibrations; small oscillations around the points which represent nuclear positions are usually described, in the X-ray crystallographic language, by atomic thermal parameters. It should be remembered that the time scale of the diffraction experiment is of the order of 10^{-6} seconds, while the time scale of thermal molecular motions is 10^{-12} seconds, so that only mean-square vibration amplitudes are accessible. These are usually displayed in the form of libration ellipsoids. The beautiful pictures which are so commonplace in crystallographic papers do, at the same time, catch and deceive the eye, since each ellipsoid can only tell about the motion of a single atom, while the human observer is eager to integrate somehow over the whole collection of ellipsoids to find an often unexisting correlation.

Fortunately, there are mathematically objective method of finding such correlations. The mean-square amplitude of the atomic displacement in any direction R is given by $R^T UR$, where U is the atomic vibration tensor from the X-ray analysis. From here on, the problem of correlating motions is just a problem of relative phases of the displacements. In the popular rigid-body analysis procedure (refs. 9-10), the phasing is assumed to arise from annihilation of intramolecular motion, and the success of the ensuing least-squares treatment is considered a measure of molecular stiffness. An excellent introduction to the intriguing world of the interpretation of the dynamic behaviour of atoms and molecules in crystals has been given (ref. 11).

Another popular model for the interpretation of vibrational parameters is the socalled "rigid bond" test (ref. 12). It is stated that the difference in mean-square displacement for atoms X and Y along an X-Y bond direction is zero when the bond is "rigid". For an example of the chemical implications of such interpretations, we may mention the statistical analysis of difference vibrational parameters, ΔU, in the study of conformationally labile molecular systems, such as the following (ref. 13):

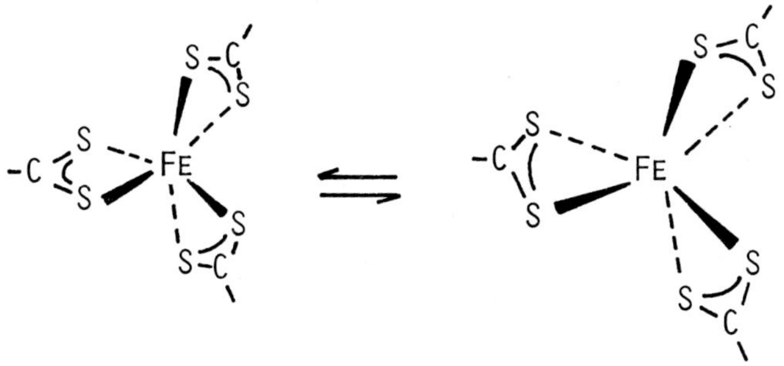

For this equilibrium, the value of ΔU(Fe-S) is strongly correlated to the Fe-S distance, over a range of about 30 different compounds, and is maximum for those compounds which show an intermediate geometry between the two extremes.

Along this way, the path is open to connections with molecular vibrational spectroscopy. A study (based on X-ray thermal parameters) of internal molecular motions in organic molecules (ref. 14) yielded torsional amplitudes and quadratic force constants for hundreds of librating groups in different crystal structures.

3.2 The effect of temperature

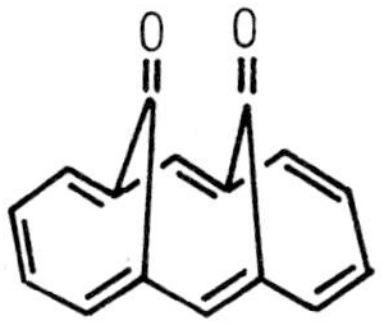

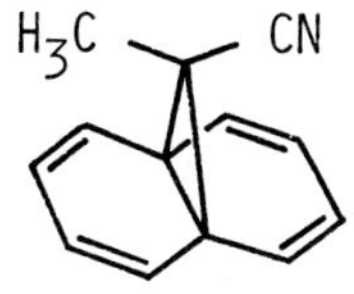

Of course, all the above is strictly a function of temperature. To quote just an overall example, isotropic temperature factors B for non-hydrogen atoms in compound 1 at 16 K are in the range 0.8-1.0 (ref. 15), compared to usual values of 3-4 at room temperature. An example of how temperature effects can influence molecular conformation along a soft bond stretch coordinate is provided by 2, in which the distance between bridgehead carbon atoms decreases from 1.712 Å at room temperature to 1.640 Å at 173 K (ref. 16).

Temperature affects also intermolecular forces; in the case of the inclusion compound of B-cyclodextrin with N-acetylphenylalanine methyl ester (ref. 17) crystal structure studies at four different temperatures between 110 and 297 K reveal a striking difference between the molecule in the cavity, which is loosely bound to the substrate by van der Waals forces, and dramatically increases its oscillations as the temperature increases, and the backbone of host molecules, which is much more rigid due to lattice forces, and is much less affected by temperature changes. All this is quite well understandable on simple inspection of the drawings of the atomic libration ellipsoids.

We may note at this point that if one wants to move from purely structural aspects to more physico-chemically meaningful studies, the analysis of temperature effects is a must. The spreading of reliable apparatus for T-dependent X-ray crystal structure determinations is most welcome in this respect.

3.3 Lattice dynamics and molecular dynamics

For the calculation of lattice vibrations, one can consider a periodic collection of molecules immersed in a potential energy field which has in principle both intra- and intermolecular components, and proceed to solve the equations of motion for the whole system, subject to crystal symmetry. The mathematical procedures for doing this are by now rather well established (see ref. 18 for an overview), and will not be described here. In most cases the intramolecular vibrations do not couple significantly with the intermolecular ones, and the computational procedure is greatly simplified. Lattice modes can be described with good accuracy in this way.

These calculations couple fruitfully with experimental IR and Raman crystal spectroscopy; this subject deserves in itself an extended treatment, which will not be attempted here. Incidentally, it may be pointed out that such subtle investigations of the vibrational properties of molecular crystals give hints also on large-amplitude molecular motions and phase transitions (topics which will be treated extensively in further sections, from a somewhat different point of view).

The problem of course arises of giving a quantitative form to the potentials, and the first choice for molecular crystals are the atom-atom potentials (see also section 4). A large number of parameters are available in the literature for building such potentials, all of which perform almost equally well in describing the fundamental properties of organic crystals (sublimation energies, equilibrium molecular positions), but, if one aims at quantitative agreement with experiment for lattice frequencies, the simple 6-exp functions are not always fit to the purpose. Dipole-dipole interactions have been added for succinic anhydride (ref. 19); quadrupole interactions were introduced for the study of naphthalene (ref. 20); distributed dipoles were used for azabenzene (ref. 21); multipole-multipole interactions plus a special term for the hydrogen bonds have been used in the lattice dynamics of acetylene (ref. 22). This last case illustrates a typical shortcoming of highly elaborated potentials: the good results for one system are obtained at the expense of loss of transferability, to the point that the same potential was not adequate for the study of a second phase of the same compound. Finally, the effects of non-central (three-body) terms on the interatomic potential have been explored (ref. 23), and were found to be negligible on the phonon spectrum of phenanthrene and chlorinated benzenes - although they were not negligible on lattice energies.

If intramolecular contributions to the potential are turned off, the molecule is classified as "rigid" , and intramolecular vibrations do not couple with intermolecular ones. A substantial amount of work has been carried out in our laboratory to take into account non-rigidity, and temperature factors for aromatic crystals have been succesfully calculated in this way (refs. 24-25). In the case of p-terphenyl, where libration about the interring bonds couples with lattice modes, the procedure was indispensable to obtain a good agreement with anisotropic thermal parameters from neutron diffraction (ref. 26). More detail on the computational procedures can be found in ref. 27.

Ultimately, one can proceed to consider a collection of independent molecules immersed in a potential field, and to solve the equations of motion for each molecular object; the way to the vibrational properties and to the thermodynamics of the whole system is then found through some time and space average of the single molecular motions. This procedure (broadly speaking called molecular dynamics) leads to a proper simulation of order-disorder transitions, phase transition in the solid, and melting, but is of course very time-consuming. For naphthalene, the computational problem was solved (ref. 28) by the use of an array processor, in which each of the 4096 processing units was in charge of a single molecule. This field is developing fast, and substantial progress is expected in the future.

4. COMPUTATIONAL MODELS

4.1 Extended molecular structure

Before we proceed to a survey of studies of large-amplitude molecular motions in crystals, it is appropriate to describe in more detail some applicable computational techniques. In the interpretation of such phenomena, as outlined in the introduction, we shall adopt a structure-oriented view of the laws that govern crystal statics and dynamics. Therefore, also the models and quantifiers we use to describe the condensed phases of organic compounds will be strongly biased by our shape-and-size prejudice.

We first introduce the concept of extended molecular structure which, starting from atomic coordinates as given by X-ray diffraction analysis, takes into account not only bond lengths and angles, but also molecular bulk and shape, through the use of atomic radii. The derivation of molecular volumes, V_M, and molecular surface, S_M, is then rather straightforward (refs. 29-30). The overall (Kitaigorodski) packing coefficient, C_K, is then:

$$C_K = Z\ V_M\ /\ V_C\ , \tag{1}$$

where Z is the number of molecules in a cell of volume V_C. A crude shape quantifier, the exposure ratio, is given by

$$X_R = S_M\ /\ V_M\ ; \tag{2}$$

for a sphere of radius R, X_R = R/3, and, by defining the overall molecular radius, R_M, as

$$R_M = (3V_M\ /\ (4\pi))^{1/3} \tag{3}$$

that is, the radius taken as if the molecule were a sphere of radius R_M, one can compare X_R with $R_M/3$ and obtain a crude estimate of molecular sphericity. Other combinations of the quantities V_M, S_M, R_M are being explored to find correlations with crystal packing or polymorphism and mesogenic ability.

4.2 Packing energy

The packing potential energy (PPE) of the crystal can be computed as

$$PPE = \sum_i \sum_j \left(\sum_k A\ \exp(-B\ R_{i,k_j}) - C\ R_{i,k_j}^{-6}\right) = \sum_i E_{ai} = \sum_j E_j \tag{4}$$

where i and k run over atoms in the molecule, j runs over surrounding molecules,

E_{ai} (the atomic relevance) is the amount of PPE due to atom i in the central molecule, and E_j is the energy for the interaction of the central molecule with the j-th surrounding one. The packing energy, PE, is equal to one half of PPE (to be compared with the sublimation energy). There are some subtleties if the molecular unit does not coincide with the asymmetric unit (see ref. 31). A nice correlation exists between S_M and PPE (ref. 30). Some more interesting results can be obtained by breaking down S_M and V_M into atomic contributions:

$$S_M = \sum_i S_{ai} \quad ; \qquad V_M = \sum_i V_{ai} \tag{5}$$

S_{ai} is found to correlate with E_{ai}, since the more exposed (higher-surface) atoms make more contributions to the PPE than unexposed ones. Along these lines, one can speculate on relationships between molecular and crystal structure (ref. 32), but we shall not discuss this point further, since it bears rather upon crystal statics than dynamics. One further quantity, that may be called atomic sensitivity, is defined as

$$F_{ai} = (dE_{ai}(R) \;/\; dR)_{R \sim 3\mathring{A}} \tag{6}$$

and can be computed by partitioning E_{ai} into contributions from spherical shells of radius R around atom i (ref. 33; see Fig.1). This quantity is qualitatively connected to the force exerted by the intermolecular field on atom i, although it lacks completely directional information.

4.3 PPE activation barriers for rotation

When a molecule or a molecular fragment rotates around some axis in the crystal, PPE variations along the displacement coordinate can be evaluated by eq.(4), where the position of the central molecule is changed as required, while the positions of the surrounding ones are unchanged. This gives a PPE profile that can interpret or predict molecular motion, or static and dynamic disorder. Alternatively, a cooperative motion can be simulated by allowing first-neighbour molecules to relax after the displacement of the central one. This usually lowers substantially the activation barrier.

The PPE profiles could be used, in principle, also for the interpretation of small-amplitude thermal motions, that is, the libration tensor components from rigid-body or other kinds of analysis of the thermal motion, but a warning should be issued. The method is valuable to gain information on the number of minima and on the barriers between them along large-amplitude molecular jumps,

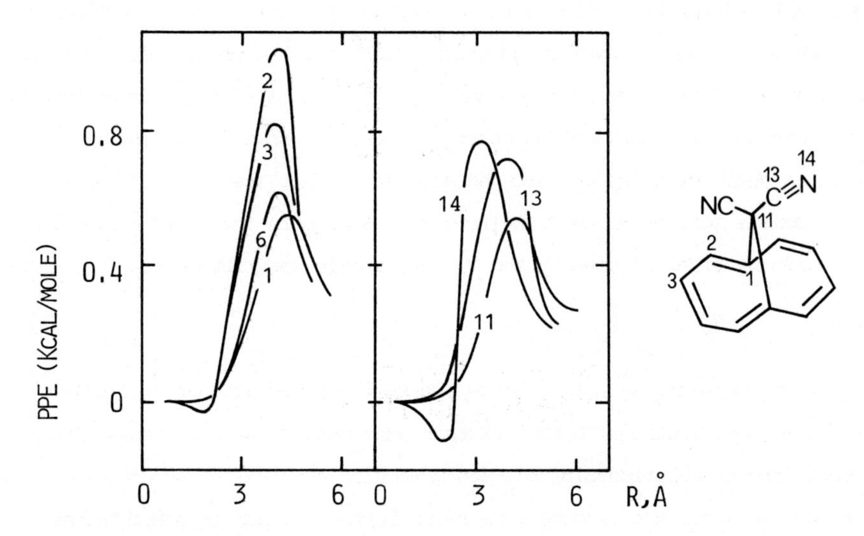

Fig. 1. Packing potential energy in shells of radius R around each atom in a molecule. The slope of the curve at R∿3 defines the F_{ai} (see text and ref. 33).

but it is hardly sensitive to the energetics of displacements by a few degrees (the typical librational amplitude of a molecule in an organic crystal). Nevertheless, it is sometimes true that large libration tensor components are a harbinger of possible large-amplitude molecular motion. Lattice-dynamical calculations as described in section 3.3 are much more advisable for the study of molecular libration in crystals.

4.4 Packing analysis by volume analysis

To say that reactivity in the solid state requires free space is at least a good working hypothesis. If this free space is provided by random crystal defects, our shape and size arguments cannot be applied in a systematic manner. There are however many examples in which the crystal matrix actually steers the reaction in a predictable and coherent manner, and an explanation of these effects in terms of shape and size of the molecule or of the cavity surrounding it is well worth trying. To this aim, one must investigate the cavity to a detail which cannot be provided by such overall quantities as C_K or PPE. The idea arises then of computing the packing density in volumes smaller than the crystal cell - eventually, in elementary volumes, V_i, throughout the crystal (an effective value for V_i being around 0.5 Å^3). This yields a packing density map from which holes and channels of free space can be spotted by visual inspection, and checked against the molecular motions expected for the solid-state reaction. The guideline in this procedure is simply that these motions (and, in turn, the reaction) will be easy if they bring atoms and fragments towards free space. A PPE profile for these motions can be computed as described in the foregoing sections, and what can be a subjective impression in the perusal of packing density maps can be given, in some cases, a more quantitative basis.

We may mention here that a fruitful use of this approach is in the field of inclusion compounds, since the size of the cavities in the host lattice can be computed and matched to the molecular volume of guests; the best location for these guests (often disordered) can be found by a combination of cavity shape analysis and PPE calculations, as illustrated in Fig. 2. Two possible orientations for the guest bromobenzene molecule were detected in this way (ref. 34).

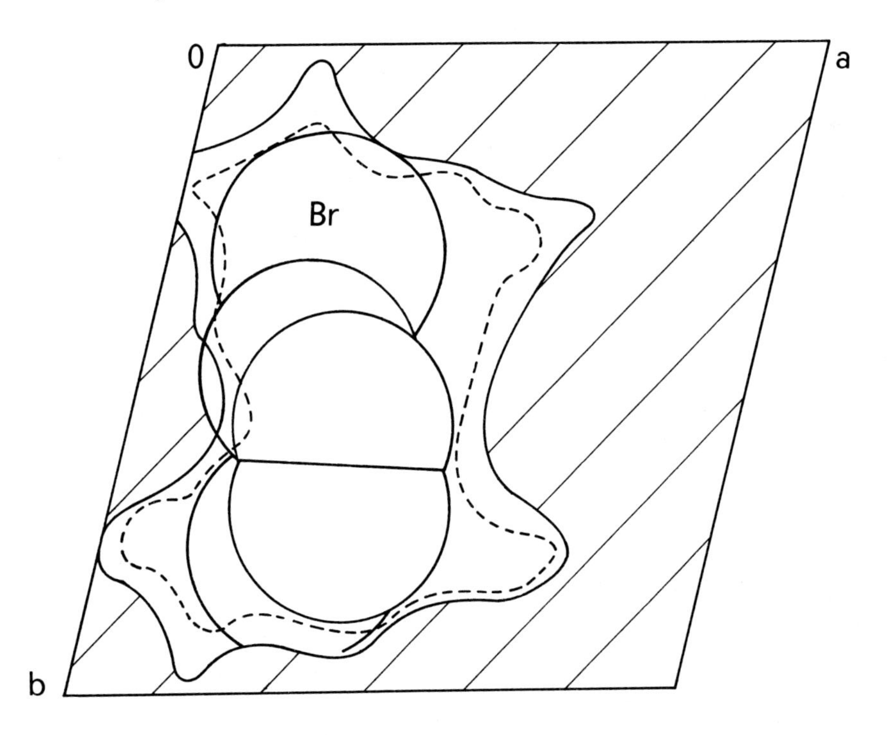

Fig. 2. Packing density map for the 1,1'-binaphthyl-2,2'-dicarboxylic acid - bromobenzene (1:1) clathrate. The dotted line encloses the cavity. The van der Waals spheres of part of the guest bromobenzene molecule are also shown. The orientation at 180° for bromobenzene (Br atom downwards in the picture) was found to be also possible by PPE calculations, as can be guessed by the inspection of the free space in the cavity. See ref. 34.

4.5 Feasibility of the calculations

The mathematical details of the above described techniques can be found in refs. 6,29,30, and will not be repeated here. Examples of the applications and a critical evaluation of the performance can be found in refs. 8,31,33, and throughout this chapter.

At least two pitfalls in the crystal force field we have described in this section must be mentioned. One is the lack of a suitable treatment for non-van der Waals potentials - those that are, broadly speaking, called "electrostatic" potentials. It is postulated that such long-range forces are less effective in relative energies for localized displacements. A more accurate discussion can be found in ref. 35. The second shortcoming is the lack of a model potential for the hydrogen bond, and this is a more serious problem, since the geometrical and energetic range spanned by these bonds in crystals is so wide (ref. 36) that it may defy attempts of a sound and general treatment for quite a while.

The computing techniques so far described are coded in OPEC (Organic Packing Energy Calculations), a computer program developed in Milano since 1973, and by now widely distributed among solid-state organic chemists and crystallographers (see also ref. 29). It has succesfully been transferred on any kind of computer, from IBM PC to CRAY. It requires neither large core memory nor relevant computing times, but rather some caution and, in many cases, much personal feeling for the problem, since it performs very seldom in standard applications, but lends itself to a choice among many options. Widespread use of the program is encouraged since only in the long run its scope and capabilities will be better assessed.

5. DISORDER

5.1 Thermal motion versus disorder

It is often very difficult, if not impossible, to distinguish between anisotropic thermal motion and disorder in crystals when anomalously large temperature factors appear. For triphenylbromomethane (ref. 37) a head-to-head arrangement of pairs of C-Br bonds is found along threefold axes. The Br...Br distance is very short, and thermal ellipsoids for Br atoms are elongated perpendicularly to the intermolecular contact (see Fig. 3a). It may well be that the ellipsoids are in fact the best description, compatible with the refinement procedure, of a random distribution of the Br atoms around the axes, the threefold symmetry

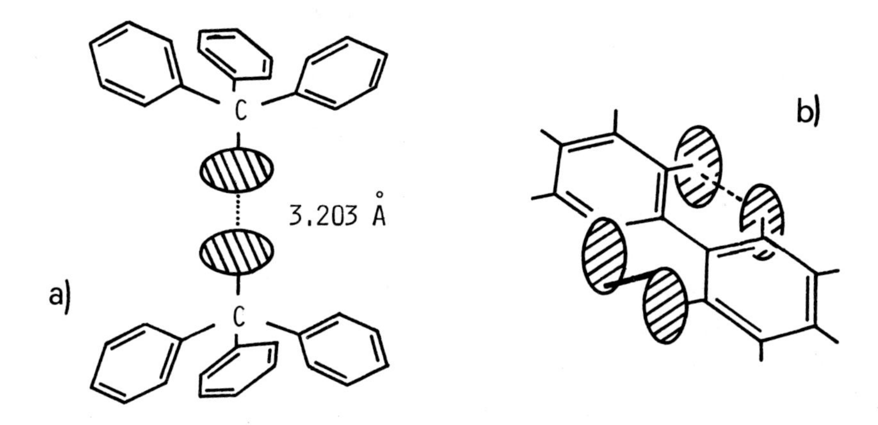

Fig. 3. A scheme (after ref. 37) of the arrangement of triphenylbromoethane (a) molecules in the crystal (Br atom ellipsoids denoted by dashed areas); and mean instantaneous distance (solid line) versus distance between mean atomic positions (broken line) in biphenyl, (b). After ref. 38.

being obeyed in an average sense. A similar effect is seen, this time intramolecularly, for biphenyl, whose planar structure in the solid state has been explained (ref. 38) by assuming that, due to thermal motion perpendicular to the vector joining the ortho hydrogen atoms (Fig. 3b) the mean instantaneous distance between them is larger than the distance between mean atomic positions. To quote just another example from the rich literature on this subject, in the complex between anthracene and TCNB the diffraction data could be interpreted either by a very anisotropic libration for anthracene, or, slightly better, by the superposition of two anthracene orientations 12° apart (ref. 39; see Fig. 4). Apparently, this is connected to a phase transition: the low-T phase of this compound consists of stacks in which the anthracene molecules are rotated ± 6° with respect to the main axis of TCNB.

5.2 Static versus dynamic disorder

The cases we have mentioned so far correspond to a potential energy diagram as in Fig. 5a. The choice of one or the other description is a matter of taste, and a good subject for academic disputation. The case is quite different when the potential energy diagram looks as in Fig. 5b. Here large displacements or jumps of the molecules are necessary, if the disorder is dynamic in nature. The height of the barrier can be measured from temperature-dependent NMR studies, or calculated by PPE methods, as previously described. The value of ΔE(dis)

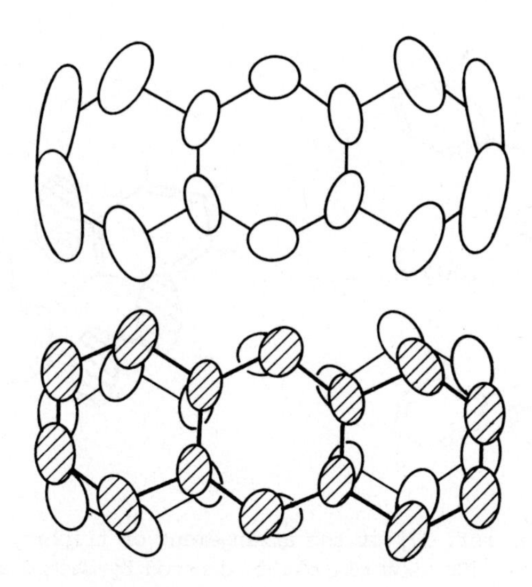

Fig. 4. Above: anisotropic model, below: disordered model for anthracene in its molecular complex with TCNB. Adapted from ref. 39.

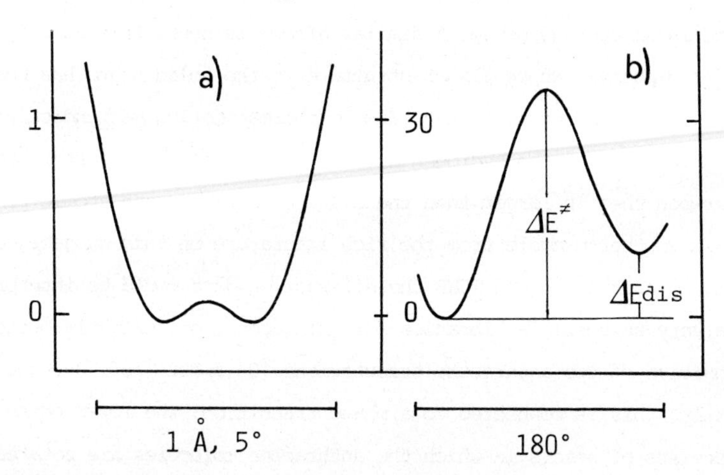

Fig. 5. a) Potential energy diagram for thermal vibration and positional disorder. b) Potential energy diagram for static versus dynamic disorder. Units are Kcal/mole. The abscissa is the molecular displacement.

determines the population of the secondary minimum, and can also be calculated as described above; the concept of molecular shape can be fruitfully advocated for a preliminary estimate of this quantity. In fact, the host must accommodate a guest (rotated) molecule, and it can do so only if shapes are compatible. To illustrate this point, we may mention the case of dibenzofuran (ref. 40), carbazole (ref. 41), and fluorene (ref. 42). Fig. 6 shows the molecular shapes and the corresponding PPE profiles, calculated as in section 4.3. For dibenzofuran,

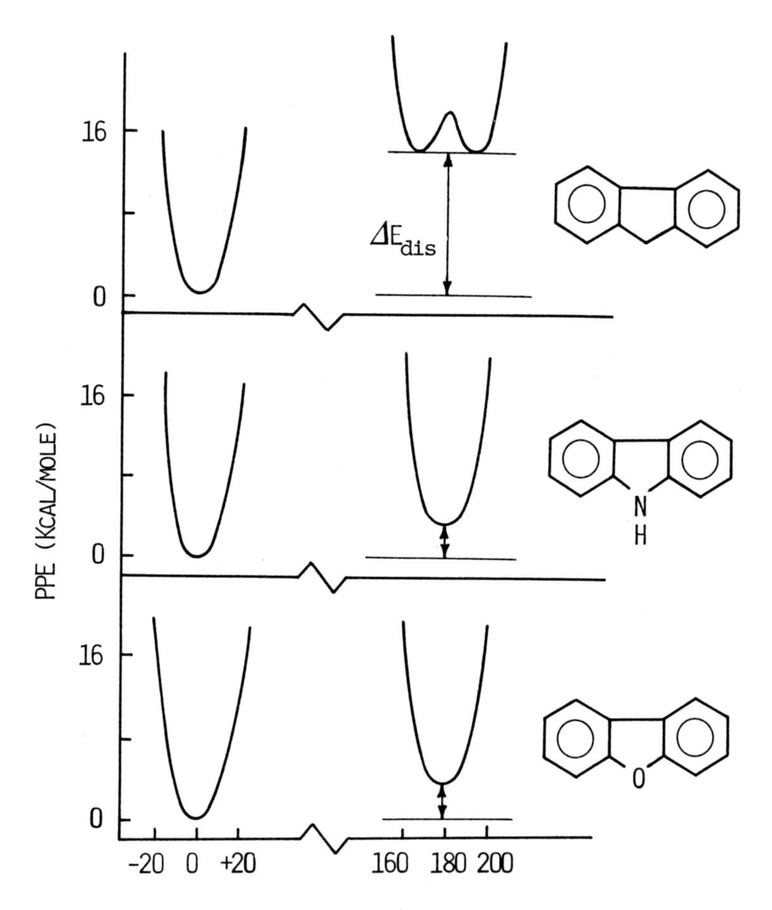

Fig. 6. PPE curves for in-plane rotation for fluorene, carbazole and dibenzofuran in the crystal. Carbazole and dibenzofuran can be disordered (small ΔEdis) The barrier for 180° rotation is very high, so the disorder must be static.

as the molecular shape inspection reveals, there is a chance for disorder, which is indeed seen in the crystal structure (ref. 40). No disorder was found for carbazole and fluorene, although the PPE minimum at 180° for the first compound looks promisingly low; in any case, there is no chance of dynamical disordering, so that crystals formed under equilibrium conditions may give a completely ordered structure. A small difference in shape, like that between fluorene and carbazole, can make a rather sizeable difference in the chance of disorder. For comparison, in naphthalene ΔEdis is zero by symmetry, and the height of the barrier indicates that disorder may be dynamic; NMR measurements show this motion in naphthalene crystals just below the melting point (ref. 43).

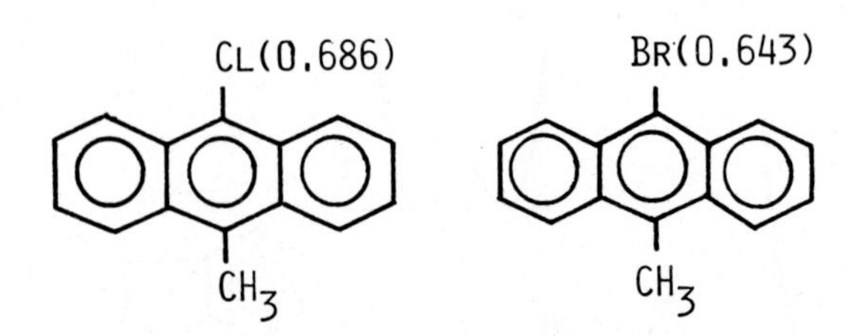

Fig. 7. Disordered 9-halo-10-methylanthracenes. Numbers in parentheses are the occupancy factors of the halogen atom (refs. 44, 45).

Fig. 7. shows another case of positional (static) disorder that could have been guessed from the molecular shapes and volumes (see refs. 44,45). A more spectacular case is shown in the scheme below:

A true mistaking of molecular structure has occurred: a disordered, less symmetrical molecule of 3 simulates the more symmetrical 4 by adopting 30 statistically distributed orientations in the crystal, in each of which the line connecting the methylene carbon atoms is parallel to a dodecahedral edge (ref. 46).

As provocatively stated by Marsh (ref. 47), it should be remembered that order and disorder in a crystal, as seen by X-ray diffraction studies, are in fact results of the solution and refinement of the structure itself,and,as such, depend on how these procedures are carried out in detail. This brings in the question of how subjective disorder in crystals can be. Ref. 47 reminds that often a centrosymmetric-non centrosymmetric dilemma conceals an order-disorder dilemma.

Although we cannot offer here experimental or computational evidence, it is very likely that disorder in the substituted anthracenes will be static, and disorder in 3-4 will be dynamic, since also the height of the barrier in Fig. 5b can be estimated from the molecular shape. For globular molecules, when rotation becomes widespread and only long-range positional order of molecular centers is preserved, one reaches the so-called plastic phases, or ODIC (Orientationally

DIsordered Crystal) phases. All this is of course to a large extent a function of temperature (it should be remembered that in the crystal also the height of the barriers in Fig. 5 is a function of temperature, through changes in cell dimensions). In the following sections, many examples of these and related phenomena will be described.

6. MOLECULAR ROTATIONS IN CRYSTALS

6.1 Results from wide-line NMR experiments

In contrast to solution experiments, where sharp and well resolved NMR bands are obtained, proton NMR spectra of solid samples show broad, featureless adsorptions. However, when molecular motion is present in the solid, these bands narrow in a way that is, to some extent, predictable for each type of motion; ref. 48 is an excellent introduction to these topics, and we shall not report further detail here. The important quantities in the NMR analysis of molecular motion in the solid state are the bandwidth, the second moment, and various types of relaxation parameters; for instance, the variation in second moment caused by a given type of molecular motion in a given crystal structure can be calculated and compared with experiment. From the relaxation behaviour of the nuclei, barriers for the motions can be obtained. Of course, the analysis must be carried out as a function of temperature. Fig. 8 shows the typical behaviour of linewidth and relaxation parameters when molecular motion is present in the crystal.

Many examples of such direct observation of molecular motion in crystals have been reported, and the following sections report on that. The interested reader will find its way through these examples to many other results in the vast literature on this topic. Two points should be mentioned:a) the observation of the motion is a direct one, but the inference on the type of motion is often an indirect one; b) the observed relaxation behaviour is averaged over the ensemble of all protons in the molecule, since the resolution is low; hence, only group motions can be detected. Nevertheless, variable-T ^{1}H NMR spectroscopy is a tool of unique importance for the study of molecular motions in solids.

(i) Rotation of methyl, t-butyl and similar groups

Numerous investigations by solid-state NMR techniques have been performed to study these rotations. One of the main results is that methyl rotation is ubiquitous in organic crystals, and is hindered by a barrier of about 3 kcal/mol,

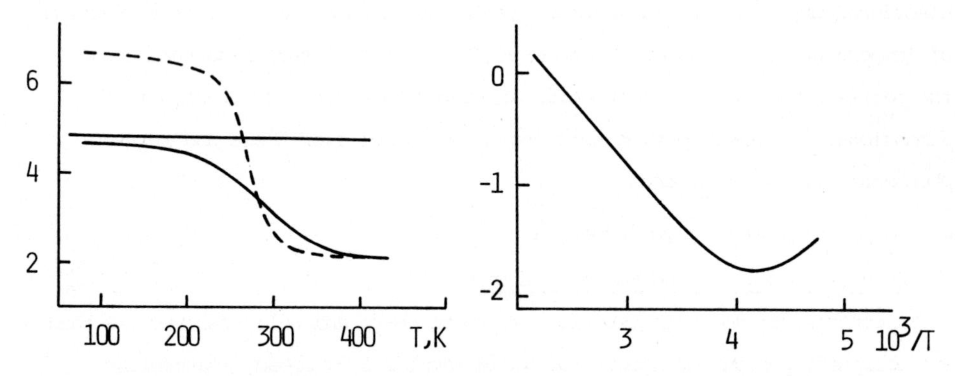

Fig. 8. Left: temperature dependence of the linewidth (dotted line, G) and second moment (full curve, G^2) for the proton resonance in pyrene-pyromellitic dianhydride complex: the horizontal line is the rigid lattice second moment. Adapted from C.A.Fyfe,J.Chem.Soc.Faraday II, 70(1974)1633-1641. Right: log of spin-lattice relaxation time (T_1) of coronene in the solid. Adapted from C.A. Fyfe, B.A.Dunell and J.Ripmeester, Can.J.Chem. 49(1971)3332-3335.

mainly of intramolecular origin. The shape and volume variation during this rotation is very small, so it is not surprising that the reorientation is quite insensitive to intermolecular forces. It is often very difficult to distinguish methyl from t-butyl rotation (see ref. 49, however). Table 1 reports some data for methyl and t-butyl reorientation barriers, as derived from variable-T NMR experiments.

A few comments to the data in Table 1 are appropriate. For dimethylsulphoxide, two barriers were detected, corresponding to two crystallographically inequivalent methyl groups. In the t-butylamine-tropolone adduct, the t-butyl and methyl barriers are very different, while in substituted t-butyl phenols a range between 1.6 and 9 kcal/mole comprises all mixtures of the two motions. In p-toluidine, the barrier was attributed to a mixture of methyl and amino group rotation (8.8 kcal/mole). The unusually high barrier to methyl rotation in 4-methylphenanthrene is due to the intramolecular effect of the protruding 5-hydrogen, while in the 4,5-dimethylderivative a gearlike motion is thought to be possible. In triethylammonium chlorides, there is a substantial increase in the barrier when the ethyl groups participate in the motion; the decrease of the barrier to reorientation of the whole cation with increasing size of the counterion is correlated with the increase in cell volume. Note how the barrier to whole cation reorientation is higher for the less symmetric triethylammonium cation than for the globular tetramethylammonium cation.

TABLE 1

Barriers (kcal/mole) to rotation of methyl (Me), tert-butyl (t-Bu) and similar groups in organic solids, from temperature-dependent NMR studies

Compound	Barrier, Me	Barrier, t-Bu	Others
MeSOMe [a]	4.15 - 5.0	-	-
t-BuNH_3^+ [b]	2.99	10.5	7.23 (NH_3^+)
t-BuNO_2 [c]	3.5	3.6	11.1 (diffusion)
$(CD_3)_3SO^+$ [d]	-	11.1 $(CD_3)_3S$	-
t-BuX, X=D [e]	-	1.20	-
X=Cl	-	1.68	-
X=OD	-	1.39	-
X=CH_3	-	0.78	-
$(Me)_3N^+CH_2CH_2OH$ [f]	3.5 - 5.6	7.4 - 11.1[g]	5.1 - 9.1 w.m.[h]
$(N(Me_4))^+$ [i]	2.8 - 3.8	-	4.5 - 5.0 w.m.
1,4-di-t-Bubenzene [j]	3.4 - 5.4 (+t-Bu)		-
1-Me-4-aminobenzene [k]	8.88 (+NH_2)	-	-
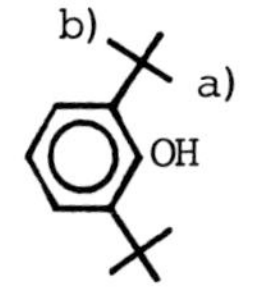 b) a) OH	a) 3.2-4.0[j] b) 8.0-9.0	-	-
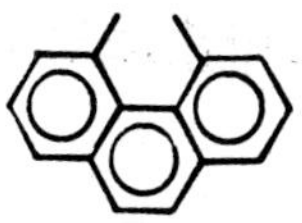	2.27 - 2.70[l]	-	-
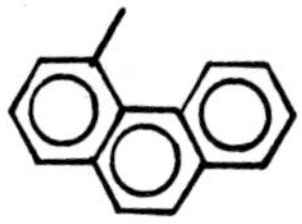	5.05[l]	-	-
$(CH_3CH_2)_3NH^+Cl^-$ [m]	<3.6	-	6.6-13.8 (Me+C_2H_5)
Br^-		-	5.7-11.7
I^-		-	4.4-10.1
Hexamethylethane[n]	3.2 (+t-Bu)		2.2 w.m.; 2.4[o]
Perchloroethylene [p]	-	-	12.4 w.m.
Perchloroethane	-	-	18.4 w.m.
CF_3CCl_3 [q]	3.8 (CF_3)	-	1.4 w.m.
$MeCCl_3$ [q]	4.3	-	2.7 w.m.

(continued)

TABLE 1 (Continued)

[a] J.A.Ripmeester,Can.J.Chem.59(1981)1671-1674.

[b] C.A.McDowell,P.Raghunathan and D.S.Williams,J.Mag.Res.24(1976)113-123; adduct with tropolone.

[c] T.Hasebe,N.Nakamura and H.Chihara,Bull.Chem.Soc.Japn.57(1984)179-183.

[d] L.J.Schwartz,E.Meirovitch.J.A.Ripmeester and J.H.Freed,J.Phys.Chem.87(1983) 4453-4461.

[e] For molecules trapped in hydrates; J.A.Ripmeester,Can.J.Chem.60(1982)1702-1705.

[f] C.A.McDowell,P.Raghunathan and D.S.Williams,J.Chem.Phys.66(1977)3240-3245.

[g] Trimethylammonium group.

[h] W.m. meaning whole molecule motion.

[i] L.K.E.Niemela and J.E.Heinila,Chem.Phys.Letters 82(1981)182-184.

[j] P.A.Beckmann,Chem.Phys.63(1981)359-375.

[k] R.Sircar,S.C.Mishra and R.C.Gupta,J.Mol.Struct.73(1981)209-214.

[l] K.Takegoshi,F.Imashiro,T.Terao and A.Saika,J.Chem.Phys.80(1984)1089-1094.

[m] Z.Pajak and J.Radomski,J.Mol.Struct.81(1982)283-288.

[n] A.R.Britcher and J.H.Strange,J.Chem.Phys.75(1981)2029-2040.

[o] Doped with adamantane.

[p] Y.N.Gachegov,A.D.Gordeev and G.B.Soifer,J.Mol.Struct.83(1982)109-112. ^{35}Cl NQR.

[q] T.Tsukamoto,N.Nakamura and H.Chihara,J.Mol.Struct.83(1982)277-280. ^{1}H, ^{19}F and ^{35}Cl NMR.

The last part of Table 1 shows results for substituted ethanes. Here, CX_3 and t-butyl group rotations mix with whole molecule reorientations in the plastic (ODIC) phases of these compounds. In hexamethylethane, the barrier for this last motion is sensitive to doping with adamantane, since this molecule is large enough to interfere with the highly cooperative molecular motions of the host (while, for the opposite reason, it is insensitive to doping with t-butyl-chloride, which is too small). For perchloroethane, the barrier was assigned to whole molecule reorientation since it is too small to be attributable to CCl_3 rotation. The same holds for perchloroethylene. Also in CF_3CCl_3 and in CH_3CCl_3 the intermolecular barrier to rotation in the plastic phase is substantially lower than the intramolecular barrier to rotation about C-C bonds.

Many of the substances mentioned in Table 1 form different crystal phases in the temperature range scanned for the NMR investigations. The reader is referred to the references for details on the thermal behaviour of these compounds.

(ii) Rotation of larger molecules

Table 2 reports some recent results on molecular rotations in solids. The reader may see refs. 8 and 48 for other interesting examples. As can be seen, a wide variety of molecular structures allow rotation in the solid state, and some barriers are surprisingly low. The question arises therefore, whether there is a simple structural criterion to predict this kind of mobility, and which rotation axes are most likely. A credible guideline can be proposed again on the basis of molecular shape; it has been stated that "those axes which cause the fewest atoms to move and to transcribe closed curves for the smallest radii will have the lowest barriers to rotation" (ref. 50). That is, motions are likely for which the process induces a minimum disturbance of the intermolecular niche that hosts the molecule in the crystal.

Of course, this guideline is of prime importance when doing PPE calculations for these motions (see refs. 6,8,33,50,51,52, where it has been applied more or less implicitly). Note that the statement involves only molecular parameters and dimensions, and does not involve the crystal field, so that the prediction of rotation in the solid state can be made using only the structure of the free molecule.

That some molecular rotations in crystals are determined only by molecular shape is demonstrated by the fact that the same motion may be present in different environments. For example, in-plane ring rotations are detected in small cyclic molecules (see Table 2) but also in metallocenes and π-complexes (see Table 3); PPE calculations were used to analyze these motions (ref. 53). It was also shown that the barrier to rotation for solvate benzene molecules is the same as in the pure benzene crystal (ref. 54).

(iii) In-plane rotation of naphthalenes

Some NMR data for this kind of molecular motion in solids is reported in Table 4. The barrier for naphthalene itself is rather high, and rotation starts just below the melting point; on the other hand, the 1,8-disubstituted compounds show a much smaller barrier. The tetramethylderivative again has a rather large barrier. Thus, the height of the barrier depends in a subtle (and not fully understandable) way from the nature of the substituents. More calculations and more experiments are needed to clarify this important point.

The calculated barriers (ref. 52) are in reasonable agreement with the

TABLE 2

Activation energy barriers (kcal/mole) for molecular rotations in organic crystals, from temperature-dependent NMR studies.

Compound	Barrier	Type of motion	Footnote
Benzene	4.2	In-plane	a
Furan	2.7	In-plane	a
SO_2 (cyclic)	5.5-12.4 12.0-21.0	Twofold axis Isotropic rotation and diffusion	b
Perfluorocyclobutane	2.7-6.7	Quasi-isotropic rotation	c
Cyclohexanol	3.4	Overall molecular rotation	d
1,3,5-trichloro-trifluorobenzene	4.3	Threefold jumps	e
p-polyphenyls	20.0-31.2	180° ring rotation	f
Ortho-and para-carborane	6.5-9.6	Isotropic reorientation	g

[a]Ref. 51.

[b]D.W.Kydon,A.R.Sharp,M.E.Hale and A.Watton,J.Chem.Phys.72(1980)6153-6157.

[c]^{19}F NMR; E.Szczesniak and J.R.Brookeman,Mol.Phys.48(1983)1221-1228.

[d]P.L.Kuhns and M.S.Conradi,J.Chem.Phys.80(1984)5851-5858.

[e]Y.Yoshioka,N.Nakamura and H.Chihara,J.Mol.Struct.111(1983)195-199; ^{19}F and ^{35}Cl NMR, and high-resolution spectra.

[f]B.Toudic,J.Gallier,P.Rivet and Y.Delugeard,Chem.Phys.99(1985)275-283.

[g]P.Beckmann and A.Leffler,J.Chem.Phys.72(1980)4600-4607.

observed ones (ref. 55); it can be said that the PPE method is doing rather well on this subject.One very puzzling feature of the data shown in Table 4 is that rotation for the 1,8-dimethyl derivative is calculated to be easy, but is not observed by spin-lattice relaxation techniques. If the NMR data are correctly interpreted, and if they refer to the same crystal phase that was considered in the calculations, rather than a metastable polymorph, the discrepancy remains without a plausible explanation.

Figures 9 and 10 show some calculated PPE profiles for the naphthalenes, as

TABLE 3

Barriers to ring rotation (kcal/mole) in metallocenes and complexes in the solid state.

Compound	Barrier	Technique	Footnote
Ferrocene T>164 K	1.1	IQNS[a]	b
T<164 K	2.0	NMR	b
gas phase[c]	0.9		
$Cr(CO)_3(\eta^5 C_4H_4S)$	2.6	IQNS	d
$Cr(CO)_3(\eta^6 C_6H_6)$	3.7	IQNS	d
$Mn(CO)_3(\eta^5 C_5H_5)$	4.0	IQNS	d

[a] Incoherent quasielastic neutron scattering.

[b] A.B.Gardner,J.Howard,T.C.Waddington,R.M.Richardson and J.Tomkinson,Chem.Phys. 57(1981)453-460, and references therein; the same was found for nickelocene.

[c] By electron diffraction.

[d] K.Chhor,J.F.Boucquet,G.Lucazeau and A.J.Dianoux,Chem.Phys.91(1984)471-477, and references therein.

TABLE 4

Observed (by NMR,spin-lattice relaxation) and calculated (by PPE) activation barriers for in-plane rotation of naphthalenes. Refs. 52,55, unless otherwise stated. Kcal/mole units.

Compound	Observed	Calculated
Naphthalene	21.7	27.9
1-chloro-8-methyl	4.5	-
1-bromo-8-methyl	4.7	-
1,4,5,8-tetramethyl	17.4	16.9
1,5-dimethyl	16	35[a], 40[b]
1,8-dimethyl	not observed	10.5
1,5-difluoro	13.3[c]	11[a,b]

[a] With cooperation of surrounding molecules.

[b] Ref.56.

[c] O.Lauer,D.Stehlik and K.H.Hausser,J.Mag.Res.6(1972)524-532.

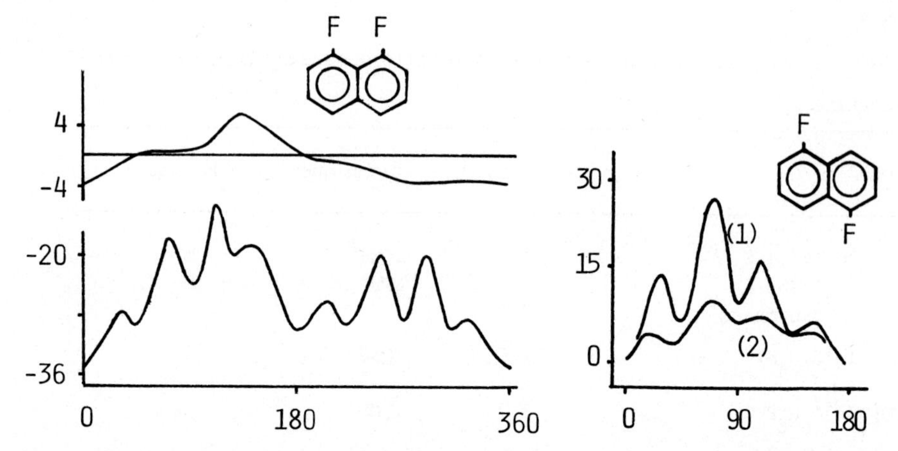

Fig.9. Left: PPE curves for in-plane rotation in 1,8-difluoronaphthalene. Total energy (below) and dipole-dipole energy (above). Right: total PPE for rotation in 1,5-difluoronaphthalene, without (curve 1) and with (curve 2) cooperation of surrounding molecules. After ref. 56.

well as the profile for ring rotation in biphenyl. The calculated barrier for this last compound is still somewhat larger that the observed one for polyphenyls (see Table 2). Some cooperative motion is not being accounted for in the calculations.

Molecular complexes formed by naphthalene and other aromatic molecules also show a large amount of in-plane rotational freedom in the high-temperature crystal phases (see refs. 57-58). This again confirms that rotational freedom is more related to molecular shape itself than to a particular intermolecular field.

6.2 High-resolution CP-MAS NMR spectra

By the skillful choice of experimental conditions, it has recently become possible to observe solid-state NMR spectra whose resolution is comparable to that of solution spectra. The so-called Cross-Polarization Magic Angle Spinning technique is described in refs. 59 and 60, with a choice of chemical applications. The main advantage of high-resolution is, of course, that NMR goes back to its full analytic potentiality; that is, it does discriminate nuclei accord-

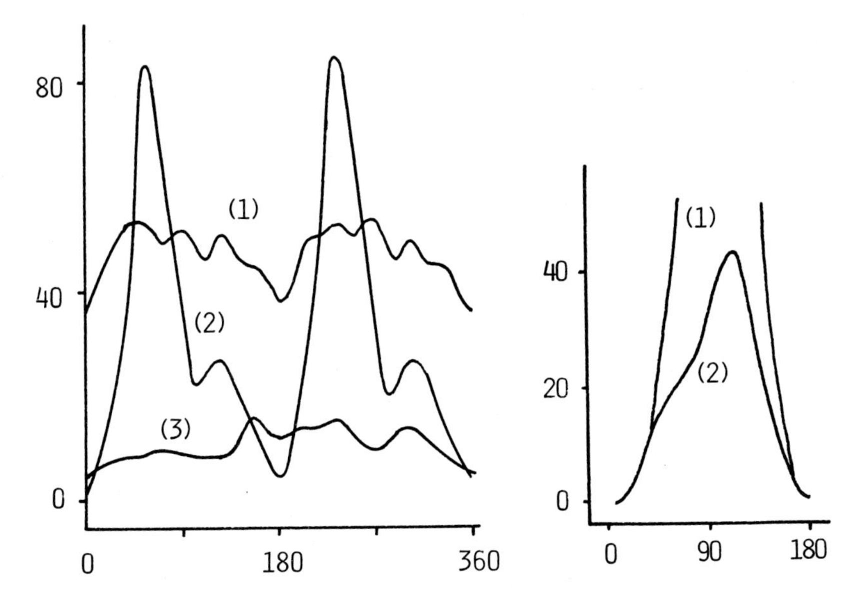

Fig. 10. Left: PPE curves for in-plane rotation of 1,4,5,8-tetramethylnaphthalene (1), 1,5-dimethylnaphthalene (2), and 1,8-dimethylnaphthalene (3). After ref. 52. Right: PPE curve for rotation around the C-C central bond in biphenyl, without (1) and with (2) cooperation of surrounding molecules. After ref. 56.

ing to their chemical shifts. An example of how this can be exploited in the study of molecular motions in the solid is shown in Fig. 11, and depends on the non-observation of different signals; the fluxional properties of these systems can thus be detected. Furthermore, at least in principle, relaxation data could be obtained for each atom in the molecule, and, if studied as a function of temperature, could open the way to the discrimination of types of motion at the level of single atoms or of molecular fragments.

CP-MAS experiments are carried out on dilute nuclei, typically on ^{13}C in natural abundance. The technique is relatively new, the first examples dating back to the late '70s; it involves rather extreme experimental conditions and critical apparatus, and is not, generally speaking, at a routine level. It looks like, however, it will develop into one of the main tools in the elucidation of molecular mechanisms for solid-state processes. In the following scheme:

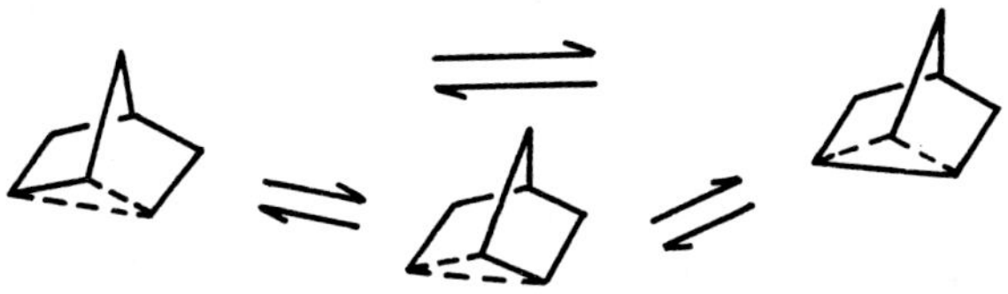

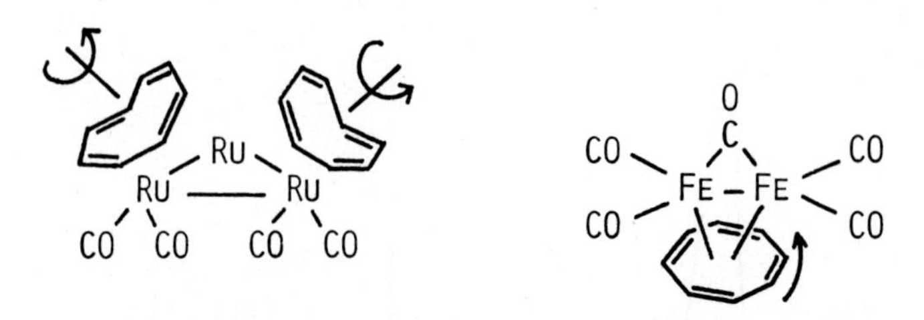

Fig. 11. The ^{13}C CP-MAS NMR spectra of these compounds show only one line for the CH carbons, indicating a fluxional (rotational) behaviour of the rings. The motion is not revealed by the X-ray analysis.

is shown a hydride shift studied by high-resolution NMR. The lineshape analysis as a function of temperature for each of the carbon atoms yields an activation energy equal to the solution one, but indicates a 3- to 4-fold decrease of the rates with respect to solution measurements.

Refs. 61 and 62 report on how the combined use of ^{13}C high-resolution spectra and spin-lattice relaxation methods (over the temperature range 4.2-363 K) allowed the determination of the number and type of the possible orientations of a methanol molecule in a clathrate with β-quinol, and of the barriers separating these orientations. Ref. 63 shows how high-resolution NMR can distinguish conformationally non-equivalent carbon atoms in cycloalkanes in the solid, and how the dynamics of the conformational exchange can be followed in detail. It also hints (see also references therein) at how the ^{13}C chemical shift can be correlated to packing effects.

6.3 Molecular motions in adamantanes

Adamantane and its derivatives are nice globular molecules, that form plastic phases at high temperatures. They have been studied extensively by a combination of NMR and IQNS (Incoherent Quasielastic Neutron Scattering) techniques; some results are shown in Table 5. With the exception of the iododerivative, all high-T phases are isomorphous, face-centered cubic structures, and the lattice parameter is in obvious correlation with the size of the substituent. The parent compound reorients about all fourfold axes in the crystal; the substituent introduces a unique axis, that defines two different motions, a rotation about this axis and an overall tumbling motion. Roughly speaking, what one can see in Table 5 is that rotation about the unique axis becomes easier the bulkier the substituent, while the opposite holds for reorientations of the C-substituent axis. This last motion is blocked (on the experiment timescale)

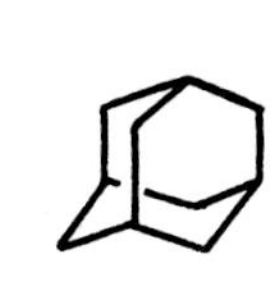

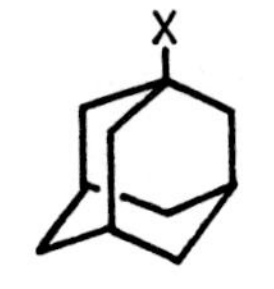

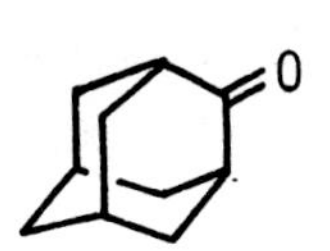

Adamantane 1-derivative 2-adamantanone

TABLE 5

Data on the structure (fcc, space group Fm3m, Z=4) and barriers to rotation (kcal/mole) for the high-temperature phase of adamantane derivatives.

Compound	Cell edge, Å	Barrier to jumps[a]	Barrier to tumbling
Adamantane [b]	9.45	2.8	2.8
2-adamantanone [c]	9.524	2.3	5.0
1-cyanoadamantane [d]	9.813	2.0	very high
1-fluoroadamantane [e]	9.54	3.1	5.9
1-chloroadamantane [f]	9.974	2.5	5.1
1-bromoadamantane [g]	10.1	-	7.6
1-iodoadamantane [g,h]	-	1.2	-

[a]About the unique C-substituent axis.

[b]J.P.Amoureux,M.Bee and J.Virlet,Mol.Phys.41(1980)313-324.

[c]M.Bee and J.P.Amoureux,Mol.Phys.47(1982)533-550.

[d]M.Bee,J.P.Amoureux and A.J.Dianoux,Mol.Phys.41(1980)325-339.

[e]M.Bee and J.P.Amoureux,Mol.Phys.50(1983)585-602.

[f]M.Bee and J.P.Amoureux,Mol.Phys.48(1983)63-79.

[g]J.Virlet,L.Quiroga,B.Boucher,J.P.Amoureux and M.Castelain,Mol.Phys.48(1983) 1289-1303.

[h]Orthorhombic, space group $Pmn2_1$.

for the cyanoderivative. These studies offer a chance of establishing a simple relationship between molecular structure and crystal structure and dynamics.

All motions are considerably slowed down, if not entirely stopped, in the low-T phases of these compounds. On the other hand, near the melting point also translational diffusion sets in. Detailed calculations on the rotational and translational motions for adamantane have been carried out (ref.64 and 65) by the PPE method, and their outcome compares favourably with the experimental results. Again, it is demonstrated that a skillful application of such calculations can yield much insight into the dynamic properties of molecular crystals.

7. PHASE TRANSITIONS

7.1 Polymorphism in organic crystals, and ODIC phases

While some crystal structures are stable over a very wide temperature range (for instance, compound 1 has only one crystal phase between 16 and 300 K), some other display a striking propensity to modification. To quote just one example, the phase diagram of solid thiophene was studied by calorimetry and X-ray crystallography (ref. 66), also as a function of pressure, but the matter is far from settled, since many phases have been detected, and it is not always easy to distinguish stable from metastable ones. As already mentioned in this chapter, a shape factor is at work: for example, globular molecules exhibit ODIC structures (see refs. 67 and 68 for ethane derivatives), and chain molecules like the n-alkanes form the so-called "rotator" phases, in which the chain motion is a complex admixture of rotation about the elongation axis and of kink formation. These phases have been extensively studied, by X-rays (refs. 69 and 70), and IR spectroscopy (refs. 71 and 72); a study of the phase behaviour of n-triacontane by a combination of X-ray, Raman, NMR, dielectric relaxation, and IQNS has revealed a step-like decrease in the degree of ordering, resulting from the successive onset of rotational jumps, translational jumps along the chain axis, and creation and diffusion of intrachain defects (ref. 73). A Monte Carlo calculation on the rotator phases of n-alkanes (ref. 74) has revealed the key role of domain formation, the average size of a domain being about 30 Å at 400 K. This remarkable structural flexibility makes rotator and plastic phases a very interesting, although challenging, target for both experimental and theoretical investigation.

X-ray crystallography is still a powerful tool for the study of disordered phases, but a number of intricacies are in the way. First, such phases often exist in exotic temperature ranges, and crystals must be grown in temperature-controlled vessels directly on the diffractometers; besides, the usual formalism for the representation of atomic scattering factors is usually inadequate to compute structure factors. Thus, more elaborate formalisms, that take into account the peculiar types of motion in plastic crystals, must be used. Refs.

75 and 76 are an example of the use of symmetry-adapted linear combinations of spherical harmonics, and of isotropic gaussian distribution functions for the molecular librations.

The question of what molecular structure generates what kind of phases is still a largely unanswered one. There have been attempts to correlate molecular shape to the ability to form ODIC phases in terms of the ratio between minimum distance between molecular centres, d_m, and maximum molecular diameter; D_m; $d_m/D_m > 0.81$ is claimed to be characteristic of ODIC precursor structures (ref. 77). While this is by no means a complete and stringent criterion, it points in a promising direction; more work along these lines is certainly a good investment.

7.2 Ways of looking at phase transitions

Some general features of phase transitions in solids, from a macroscopic kinetic and thermodynamic point of view, are summarized in ref. 78. Much more difficult is to ascertain the actual mechanism at a molecular level (these are perhaps better understood for inorganic than for organic crystals; see again ref. 78). In the simplest case, the overall structural motif remains unchanged, and the transition is just the onset of some molecular rotation, or a change in speed of this rotation. For example (ref. 79) t-butyl cyanide undergoes fast reorientation around the C-C-N axis in the solid in a threefold potential between 190 and 280 K, and, at 233 K, a phase transition occurs, which is thought to be associated with a slowdown of the rotational motion (plus some translational motion). Ref. 80 is a molecular dynamics study of this system.

The ideal way in which such phenomena should be studied is by a combination of calorimetry, X-ray diffraction and NMR. Tris(hydroxymethyl)aminomethane, 5, was studied in this way (ref. 81). Only one phase transition, at 407 K, was detected by DSC in the 5-450 K range. A single-crystal X-ray analysis was carried out at five temperatures for the low-T phase and at 423 K for the high-T phase. From the analysis of the molecular librations and of the crystal packing, a likely transition path was formulated as an increase of the distance between layers of H-bonded molecules, together with a reorientation of the molecules within the layers to form the disordered, high-T phase.

There are a number of lucky cases in which one can get as close as possible to the true molecular mechanism, and pinpoint the most likely molecular motions. Ultimately, the smoothest possible transition is one in which the crystalline edifice collapses to the new phase in an entirely correlated fashion; such a motion is the limit of zero frequency, at the transition temperature, of one special lattice mode, called soft mode. The eigenvector of this mode, as computed for example by lattice dynamics, gives the actual atomic displacements. One good - but by no means unique - example of this is malononitrile, 6, where the space group changes from $P2_1/n$ to $P\bar{1}$, and the eigenvector of the soft mode consists of opposite rotations of the two molecules related by a screw axis in the higher-symmetry structure (ref. 82).

We may view such processes as just special points along the thermal paths of structurally invariant (in an overall sense) systems. In other cases, it is necessary to postulate discontinuous structural changes at a zone boundary during the transition. In squaric acid, 7, a high-T tetragonal structure transforms into a low-T monoclinic one by molecular reorientation and proton ordering in the H-bond network, but no soft modes were detected by lattice-dynamics calculations (ref. 83). Drifting further away from cooperative phase transitions, we may mention the case of tetrafluorodiiodobenzene (ref. 84), where a clear geometrical relationship between the two cells and the molecular orientations in the two phases was found, but the molecular displacement required for the transition would be as much as 3.7 Å. Ultimately, the two phases may grow each for itself, and there may be no path between the two without complete crystal disruption.

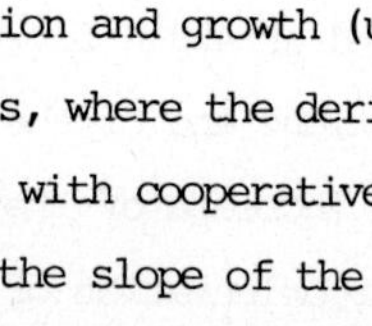

$C(CH_2OH)_3NH_2$ (5) $N \equiv C-CH_2-C \equiv N$ (6) squaric acid (7)

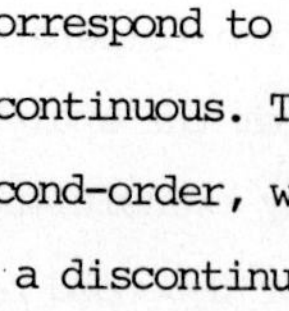

Nucleation and growth (unpacking and repacking) correspond to first-order transitions, where the derivatives of G(T,P) are discontinuous. Transformations associated with cooperative mechanisms are called second-order, where there is a change in the slope of the derivatives of G, but not a discontinuity. Let it be said at once that both the definitions and the distinction between the two are not unambiguous. Note however that the steric factor, which we have mentioned at the beginning of this chapter, requires in principle that first-order transitions start at packing discontinuities or lattice defects, since any large amplitude molecular motion requires free space in the surroundings of the molecule. The introduction to ref. 85 and references therein can be consulted for a discussion of modeling and theories of phase transitions.

A fruitful way of looking at phase transitions is in terms of lattice instabilities. An ordinary phase transition is then a temperature- or pressure-induced instability, and a solid solution of compounds A and B, which collapses from the structure of A to that of B at a critical composition, is subject to chemically induced instability. Raman studies of solid solutions of 1,4-dibromo- and 1,4-dichloronaphthalene reveal that some dynamic interaction topology is preserved through the structural transition, and the X-ray analysis reveals the geometrical foundations of this continuity (ref. 86). Studies in this direction allow a unique access to the influence of chemical structure on crystal structure. Structural stability against the change of a substituent on a fixed molecular core has been analyzed (ref. 87) in terms of available space at the substituent position versus substituent bulk. It may well be that this is a way to the definition of a substituent effect in crystal chemistry.

7.3 Calculations on phase transitions

The main purpose of the structural chemist is to correlate molecular structure to chemical behaviour; that is, answer the question: why is it that a molecule with a given structure transforms in a given way? The same concept can be used in crystal chemistry, and the following is one such example.

Biphenyl has received perhaps more attention than any other molecule in its solid-state behaviour; it provides an almost ideal model to probe intermolecular versus intramolecular forces, the first being manifest through packing effects at the phenyl rings, the second through a competition between interring conjugation and steric repulsion of ortho-hydrogens. A paradigmatic piece of theoretical work on this system has been carried out (ref. 38; incidentally, this reference also carries an almost complete account on the literature on solid biphenyl). At higher temperatures, the biphenyl molecule averages to a planar structure (see Fig. 3 and the ensuing discussion). Below 38 K, after a phase transition, the thermal motion is reduced, and a twisted (by 10°) conformation freezes out. The transition temperature is then the temperature at which thermal motion is no longer enough to carry the molecule through the barrier in a double-well potential. Ref. 38 makes this interpretation clear in structural terms and provides the computational means (including T-dependent intermolecular potentials) to make the explanation quantitative.

The atom-atom potential method was used to model the two known phases cf 2,4-hexadiyne (ref. 88), and to obtain thermodynamic properties and phonon dispersion curves for comparison with experimental values. The strong molecular anisotropy and uneven electron density lead to partial failure; the agreement between calculated and measured quantities was only qualitative. Thus, caution should be exerted when using this method for quantitative purposes on systems that differ substantially from those on which the atom-atom parameters were calibrated.

The ideal way in which the phase transition energetics should be studied is by a comprehensive calculation of the total energy along the molecular displacement that leads to the transition. This can be done only for second-order transitions, and even there, through idealized models. For example, an effort in this direction has been made for the conversion of trans-trans into trans-cis diacetamide:

$$CH_3-C(=O)-N(H)-C(=O)-CH_3 \rightleftharpoons CH_3-C(=O)-N(H)-C(=O)-CH_3$$

The two crystal structures were described in a common crystallographic system and atom-atom potentials were used to compute the energy for the isomerization in the crystal (ref. 89). Of course, some assumptions had to be made on the actual path for the transformation in the solid, but it appears that the calculated barriers are low enough to explain the ease with which diacetamide undergoes isomerization in the crystal.

Some phase transitions may occur under the influence of external forces. A computational scheme has been proposed (ref. 90) to take into account such forces (hydrostatic pressure, normal or shearing stresses, electric field) in the calculation of crystal energies, by addition of suitable terms to the van der Waals energy. Minimizing this global potential gives a structure distorted by the external force. The effect of such forces on the orientation and conformation of the molecules in the crystal can be computed, and an obvious extension would lead to the simulation of phase transitions induced by external forces of various kinds.

7.4 Adequacy of structural models

To what extent molecular shape and size models are just oversimplifications is demonstrated by chloropentamethylbenzene, where four phase transitions were found in the 3-300 K range, all involving an admixture of methyl and overall molecular rotation that partly defies analysis (ref. 91). For tetracene, the structural change in the transition (ref. 92) is just a small molecular reorientation, without appreciable change in cell setting, but the transition mode depends on thermal history of the sample and even on the sample mounting on its support. Thus, a simple structural change is not granted to give a simple phase transition, and vice versa. But the concept itself of phase transition seems to be a fuzzy one for some organic compounds. Ref. 93 carries a striking list of compounds that form glassy solids, being in a state of matter which may be very far from thermodynamic equilibrium, and hence very far from the structural description into which the thermodynamic requirements of equilibrium have been recast in this chapter. In fact, the organic solid-state chemist is faced with metastability more often than is thought. There is a sharp contrast between the anisotropy of condensed matter and the scarce selectivity that packing forces display among many different anisotropic arrangements.

8. THEORIES AND CALCULATIONS FOR SOLID-STATE REACTIVITY

8.1 Preliminary discussion

Organic solid-state reactivity has grown into an exploitable branch of chemistry in such fields as the manipulation of drugs, polymer synthesis and materials science. We will restrict ourselves here to the still rather few examples where a theoretical treatment of the reaction path, or at least of the precursor crystal and molecular structures, has been attempted.

From the theoretical point of view, the challenge is a formidable one, since so many degrees of freedom, side effects and boundary conditions must be taken into account. We may divide solid state reactions according to the effects caused by the first reacted molecule on the surrounding matrix. If these are highly disruptive, so that strong and long-range relaxation (actually, reconstruction) of the lattice is necessary, then most available computational methods fail, mostly due to the geometrical complexity of the problem. The same applies if reaction is initiated at a defect point, whose structure is in most cases unknown and unpredictable. If, on the other hand, the host lattice can accommodate the reacted guest without too much effort, then the process is amenable to theoretical simulation. Note that the difficulties arise mostly from the geometrical, rather than from the energetic, point of view; we have reasonably good potentials, but it is very difficult to model the reaction and reconstruction paths.

Looking at solid-state reactivity from the vantage point of the first reacted molecule is applying a sort of reverse topochemical principle. The computational procedure amounts to evaluating the mutual compatibility of a candidate product molecule and the undistorted (or moderately distorted) lattice. We will review in the following some examples of this.

8.2 Electronic excitations

Calculations on excitation, lattice relaxation and reactivity have been reported for 9-methoxyanthracene in host 9-cyanoanthracene (ref. 94). The 6-exp interatomic potential was augmented by a dispersive potential in the inverse sixth power of the distance between molecular centres, to model the effect of molecular electronic excitation. Lattice relaxation was included by allowing successive shells of molecules surrounding the excitation site to reach equilibrium positions, and it was found that two potential energy minima are possible, one leading to association of two host molecules, and the other to host-guest complex formation. It is remarkable however that simple PPE calculations on the statics of the mixed system gave the correct structural result, that is, head-to-head arrangement of substituents is favoured over head-to-tail in the precursor, unperturbed lattice.

8.3 Polymerizations

It has long been recognized and widely accepted that the primary requirement for solid-state polymerizations is a favourable arrangement of reacting groups in the crystal (Fig. 12a). A sometimes neglected, but necessary condition is however that the motion of side groups after the initial reaction step is not forbidden by lattice pressure. A careful analysis for the conditions of polymerization of bis-p-toluenesulphonate of 2,4-hexadiyne-1,6-diol, 8, a compound that forms almost continuous solid solutions with its polymer, has been conducted by neutron diffraction studies of the monomer units (ref. 95).

$$CH_3\text{-}C_6H_4\text{-}SO_2\text{-}O\text{-}CH_2\text{-}C{\equiv}C\text{-}C{\equiv}C\text{-}CH_2\text{-}O\text{-}SO_2\text{-}C_6H_4\text{-}CH_3 \qquad 8$$

The criterion for mobility was packing tightness, as gauged by the number of short intermolecular contacts at the side groups; this revealed that these groups are relatively mobile, and do not interfere with the polymerization-producing atomic motions at the triple bonds.

7-methoxycoumarin was found to dimerize in the solid state, even though the double bonds are unfavourably aligned in the crystal (see Fig. 12b). The PPE for the molecular rotation needed to bring the two bonds parallel was calculated (ref. 96); although the results are not fully conclusive, a relevant molecular mobility (8 kcal/mole for molecular displacements as high as 28° and 12° rotation on two axes) was revealed. Ref. 96 also carries a discussion of cases of solid-state polymerizations where the vectors of the reacting bonds are not parallel.

8.4 Steric compression control

Recently, McBride (ref. 97) and Scheffer and coworkers (ref. 98) have proposed mechanistic ideas on the feasibility of solid-state reactions that enlarge to some extent Cohen and Schmidt's topochemical principles. Steric compression is a force, acting at or near the reaction site, that impede or favours atomic motions along a reaction coordinate, not necessarily by action on the reacting atoms. For the decarboxylation of diacylperoxides (ref. 97), this force is provided by the detaching CO_2 molecule, that steers the newly formed radicals to

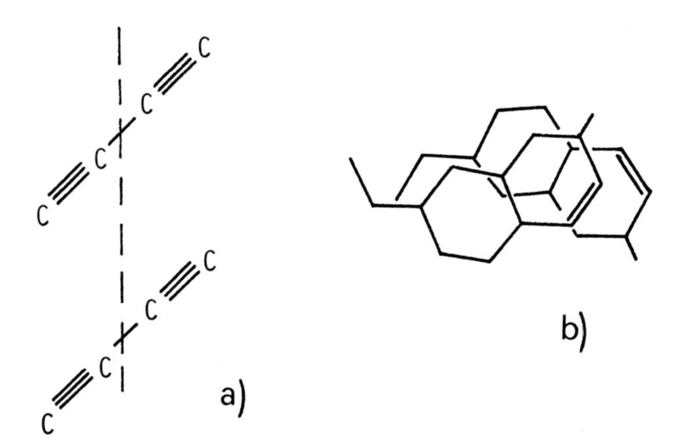

Fig. 12. a) Favourable arrangement of parallel reacting triple bonds for the polymerization of diynes. b) Unfavourable arrangement of non-parallel double bonds in 7-methoxycoumarin (after ref. 96). This compound is nevertheless reactive in the solid state.

rotations or translations, for some of which the PPE curves have been calculated (ref. 99, and Fig. 13). In the photochemical hydrogen migration reactions of α,β-unsaturated cyclohexenones (ref. 98), the resistence of the environment against pyramidalization at the attacked carbon atom is enough to forbid, in some cases, an otherwise fully topochemical reaction.

The experimental studies of solid-state reactivity are very demanding. We want to stress here that, whenever the geometrical modeling of the displacements is feasible, PPE calculations and cavity analysis can give straightforward answers to questions that may otherwise require some rather hectic experimentation.

8.5 Cavities

When probing the intermolecular response of the lattice to a reactive displacement, the concept of cavity is helpful. Methods for the evaluation of cavity size and shape in crystals were proposed independently, and almost simultaneously, in ref. 29 and by Ohashi and coworkers (see ref. 100 and references therein). These last authors were able to show that the feasibility of crystal racemization reactions of cobaloxime complexes was directly connected to the size of the cavity available for reorientation of side chains.

Of course, PPE calculations on mobility and the analysis of empty space complement each other, in the sense that motion towards cavities is energetically less demanding than motion in the direction of filled space. Both approaches have been succesfully applied to the study of molecular motions in the solid-state reactivity of organic peroxides (refs. 29,54,99). While these methods require a somewhat subjective blend of intuition and systematics, when the geometrical details of the motions are not obvious, they offer a unique possibility to rule out large numbers of impossible mechanisms and to quantify (to some extent, at least) the intermolecular energetics of the reaction. We may add that crystal forces other than steric ones (like electrostatic forces) are less sensitive to local displacements, and hence less important in steering molecular motions at reactive sites. This, of course, is not true when full opposite charges develop, or when hydrogen bond formation or cleavage is involved in the reaction.

8.6 Reactions in channels

As already mentioned (section 4), guest molecules in inclusion compounds can be located by PPE calculations (see also ref. 101) and cavity analysis. The

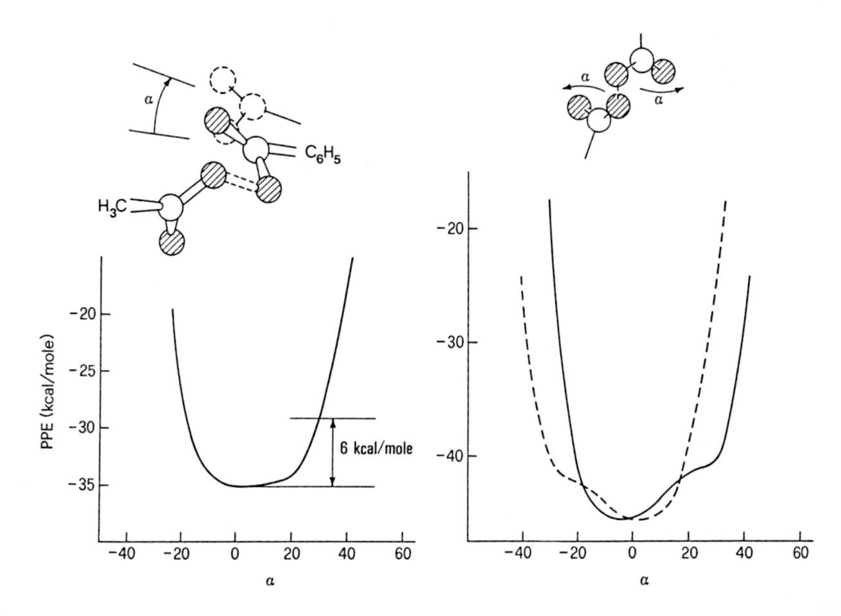

Fig. 13. PPE curves for in-plane rotation of C_6H_5COO groups resulting from peroxide bond breaking in acetylbenzoylperoxide (left) and dibenzoylperoxide (right). Rotation into cavities is reflected in the shoulders or secondary minima in the curves. After ref. 99.

rigid structure of the host channels can be exploited to perform chiral synthesis; the guest mobility in cages is usually high enough to allow reactions to occur. A beautiful example of this is the photoaddition of guest acethophenone to host deoxycholic acid (ref. 102), where X-ray and neutron crystallography and PPE calculations were used to follow a net 180° reorientation of the acetyl group in the cavity to perform a specific addition. NMR studies (ref. 103) had shown that molecular motions of different kinds are allowed for the guests in deoxycholic acid channels, from methyl rotation to ring inversion of cyclic compounds, up to some isotropic reorientations for specific lattice sites.

The free space requirement is obvious for heterogeneous reactions in which gas molecules must enter or escape from the reacting crystal. The photochemical oxidation of thioketones was examined for many different compounds, and the ease of reaction was correlated to the presence and the cross-section of crystal channels (ref. 104). For the solid-state hydrolisis of 2-methyl-4H-benzoxazin-4-one, 9, the presence of channels was found to be instrumental (ref. 105), since water molecules can diffuse through them by a sort of H-bond walking mechanism.

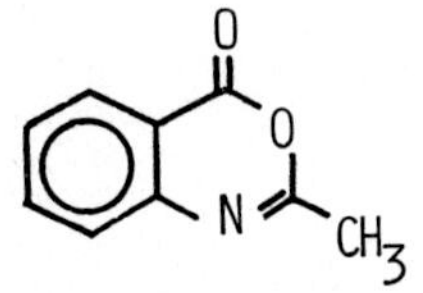

9

These are cases in which the packing analysis and cavity and channel analysis method (ref. 29) can be fruitfully applied.

9. CONCLUSION

The multiform aspects of the packing and dynamics of condensed organic matter have been reviewed, through a somewhat weighted sample of the available literature. X-ray structure analysis, when applicable on single crystals, still being the method of choice, many other experimental techniques are now in more or less widespread use, and new ones are being tested.

Intermolecular potentials are being built upon the basic atom-atom attraction-repulsion formula by addition of electrostatic terms, of electronic excitation terms, of external forces, and of temperature-dependent terms. From the zero-level packing potential energy calculations, the theoretician is now moving on to more sophisticated and comprehensive models of organic crystal structures, like lattice dynamics or molecular dynamics, this last one being the most promising for the future, when computational facilities will increase. Crystal symmetry is a great help in the calculations: solid-state reactivity, which requires the modeling of a permanent breakdown of this symmetry, is much more difficult to analyze in simple geometrical terms. On the whole, it can be said that crystal and molecular mechanics calculations are becoming more and more familiar also to experimentalists, and, as usual, the best results can be obtained by combined use of practice and theory.

The progress of a crystal structure from an ideally frozen lattice to the melt is, quite often, not discontinuous. As sketched in Fig. 14, a molecular ensemble evolves, in a sense, by addition of spatial disorder, through what may be called an order-disorder continuum. Some substances may touch all the steps, cycling through many crystal phases, others may choose but one route to the melt.The same substance may walk different paths in Fig. 14, depending on sometimes unpredictable circumstances, or local conditions. Thermodynamics and kinetics dictate, sometimes in a contradictory fashion, the terms of this beautiful and challenging complexity.

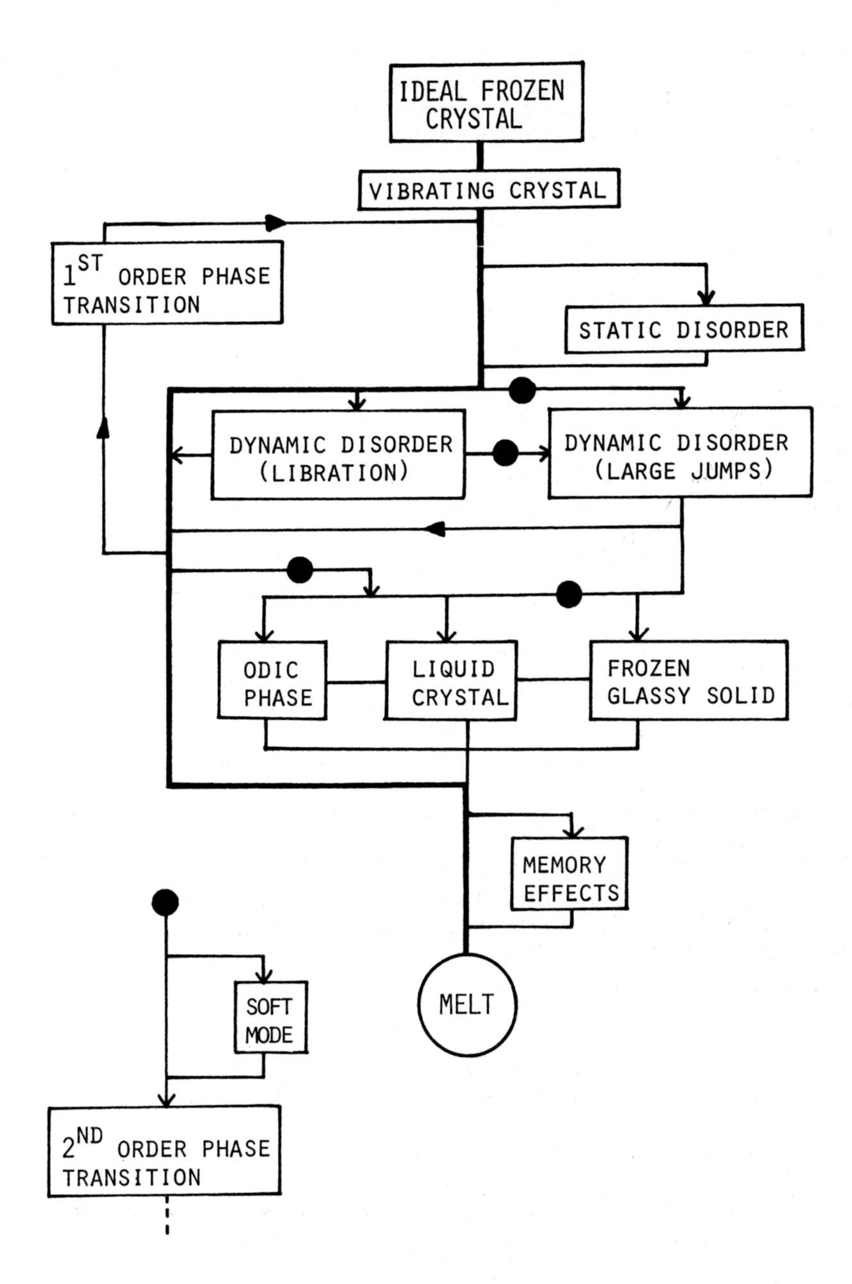

Fig. 14. The order-disorder continuum of organic condensed matter. The progress coordinate, downwards, is an admixture of (can be measured in units of) temperature, pressure, molecular shape, rms molecular displacements, external strains, order parameters of various kinds, and others. The heavy line is the path for a crystal that shows no phase anomalies. At any point in the diagram can solid-state reactions occur.

REFERENCES

1 R.G.Woolley, J.Am.Chem.Soc., 100(1978)1073-1078.
2 G.Natta and M.Farina, Stereochimica, EST Mondadori, Milano, 1968.
3 A.I.Kitaigorodski, Molecular Crystals and Molecules, Academic Press, New York, 1973.
4 S.C.Nyburg and C.Faerman, Acta Cryst., B41(1985)274-279.
5 F.H.Allen, O.Kennard and R.Taylor, Acc.Chem.Res., 16(1983)146-153.
6 A.Gavezzotti, Nouv.J.Chimie, 6(1982)443-450.
7 A.Bondi, J.Chem.Eng.Data, 8(1963)371-381.
8 A.Gavezzotti and M.Simonetta, Chem.Revs., 82(1982)1-13.
9 D.W.J.Cruickshank, Acta Cryst., 9(1956)754-756.
10 V.Schomaker and K.N.Trueblood, Acta Cryst., B24(1968)63-76.
11 H.-B.Bürgi and J.D.Dunitz, Acc.Chem.Res., 16(1983)153-161.
12 F.L.Hirshfeld, Acta Cryst., A32(1976)239-244.
13 K.Chandrasekhar and H.-B.Bürgi, Acta Cryst., B40(1984)387-397.
14 K.N.Trueblood and J.D.Dunitz, Acta Cryst., B39(1983)120-133.
15 R.Destro, A.Gavezzotti and M.Simonetta, in preparation.
16 R.Bianchi, T.Pilati and M.Simonetta, J.Am.Chem.Soc., 103(1981)6426-6431.
17 J.J.Stezowski, in: J.J.Stezowski (Ed.), Proceedings of the Symposium on Molecules in Motion, Lexington, Kentucky, may 1984; Trans.Am.Cryst.Assoc., 20(1984)73-82.
18 S.Califano, V.Schettino and N.Neto, Lattice Dynamics of Molecular Crystals, Springer-Verlag, New York, 1981.
19 D.Bougeard, R.Righini and S.Califano, Chem.Phys., 40(1979)19-23.
20 R.Righini, S.Califano and S.H.Walmsley, Chem.Phys., 50(1980)113-117.
21 Z.Gamba and H.Bonadeo, J.Chem.Phys., 75(1981)5059-5066.
22 M.Marchi and R.Righini, Chem.Phys., 94(1985)465-473.
23 I.A.Remizov, V.G.Podoprigora, A.N.Botvich, T.A.Kharitonova and V.F.Shabanov, Chem.Phys., 92(1985)163-168.
24 C.M.Gramaccioli, G.Filippini and M.Simonetta, Acta Cryst., A38(1982)350-356.
25 C.M.Gramaccioli and G.Filippini, Acta Cryst., A39(1983)784-791.
26 C.M.Gramaccioli and G.Filippini, Acta Cryst., A41(1985)361-365.
27 C.M.Gramaccioli and G.Filippini, Acta Cryst., A41(1985)356-361.
28 R.G.Della Valle and G.S.Pawley, Acta Cryst., A40(1984)297-305.
29 A.Gavezzotti, J.Am.Chem.Soc., 105(1983)5220-5225.
30 A.Gavezzotti, J.Am.Chem.Soc., 107(1985)962-967.
31 A.Gavezzotti, OPEC, Organic Packing Energy Calculations, Milano, 1985 (unpublished; available for distribution).
32 G.R.Desiraju and A.Gavezzotti, in preparation.
33 R.Bianchi, A.Gavezzotti and M.Simonetta, J.Mol.Struct., 135(1986)391-401.
34 M.Czugler and A.Gavezzotti, in preparation.
35 Z.Berkovitch-Yellin and L.Leiserowitz, J.Am.Chem.Soc., 104(1982)4052-4064.
36 R.Taylor, O.Kennard and W.Versichel, Acta Cryst., B40(1984)280-288.
37 A.Dunand and R.Gerdil, Acta Cryst., B40(1984)59-64.
38 W.R.Busing, Acta Cryst., A39(1983)340-347.
39 J.C.A.Boeyens and D.C.Levendis, J.Chem.Phys., 80(1984)2681-2688.
40 W.J.Reppart, J.C.Gallucci, A.P.Lundstedt and R.E.Gerkin, Acta Cryst., C40(1984)1572-1576.
41 R.E.Gerkin and W.J.Reppart, Acta Cryst., C42(1986)480-482.
42 R.E.Gerkin, A.P.Lundstedt and W.J.Reppart, Acta Cryst., C40(1984)1892-1894.

43 S.McGuigan, J.H.Strange and J.M.Chezeau, Mol.Phys., 49(1983)275-282.
44 R.D.G.Jones and T.R.Welberry, Acta Cryst., B37(1981)1125-1126.
45 R.D.G.Jones and T.R.Welberry, Acta Cryst., B36(1980)852-857.
46 O.Ermer, Angew.Chem.Int.Ed.Engl., 22(1983)251-252.
47 R.E.Marsh, Acta Cryst., B42(1986)193-198.
48 C.A.Fyfe and R.P.Veregin, in: J.J.Stezowski (Ed.), Proceedings of the Symposium on Molecules in Motion, Lexington, Kentucky, may 1984; Trans.Am.Cryst. Ass. , 20(1984)43-59.
49 J.A.Ripmeester and C.I.Ratcliffe, J.Chem.Phys., 82(1985)1053-1054.
50 C.A.Fyfe and D.Harold-Smith, J.Chem.Soc.Faraday II, (1975)967-984.
51 W.E.Sanford and R.K.Boyd, Can.J.Chem., 54(1976)2773-2782.
52 F.Imashiro, K.Takegoshi, A.Saika, Z.Taira and Y.Asahi, J.Am.Chem.Soc., 107(1985)2341-2346.
53 D.C.Levendis and J.C.A.Boeyens, J.Cryst.Spectr.Res., 15(1985)1-17.
54 A.Gavezzotti, Tetrahedron, in press.
55 F.Imashiro, K.Takegoshi, S.Okazawa, J.Furukawa, T.Terao and A.Saika, J.Chem.Phys., 78(1983)1104-1111.
56 A.Gavezzotti and M.Simonetta, Acta Cryst., A32(1976)997-1001.
57 J.A.Ripmeester, A.H.Reddoch and N.S.Dalal, J.Chem.Phys., 74(1981)1526-1533.
58 T.Inabe, Y.Matsunaga and M.Nanba, Bull.Chem.Soc.Japn., 54(1981)2557-2564.
59 C.S.Yannoni, Acc.Chem.Res., 15(1982)201-208.
60 J.R.Lyerla, C.S.Yannoni and C.A.Fyfe, Acc.Chem.Res., 15(1982)208-216.
61 S.Matsui, T.Terao and A.Saika, J.Chem.Phys., 77(1982)1788-1799.
62 J.A.Ripmeester, R.E.Hawkins and D.W.Davidson, J.Chem.Phys., 71(1979)1889-1898.
63 M.Möller, W.Gronski, H.-J.Cantow and H.Höcker, J.Am.Chem.Soc., 106(1984) 5093-5099.
64 C.A.Fyfe and D.Harold-Smith, Can.J.Chem., 54(1976)769-782.
65 C.A.Fyfe and D.Harold-Smith, Can.J.Chem., 54(1976)783-789.
66 D.André, P.Figuiere, R.Fourme,M.Ghelfenstein, D.Labarre and H.Szwarc, J.Phys.Chem.Solids, 45(1984)299-309.
67 J.P.Amoureux, M.Foulon, M.Muller and M.Bee, Acta Cryst., B42(1986)78-84.
68 P.Gerlach, D.Hohlwein, W.Prandl and F.W.Schulz, Acta Cryst., A37(1981) 904-908.
69 J.Doucet, I.Denicolò and A.Craievich, J.Chem.Phys., 75(1981) 1523-1529.
70 I.Denicolò, J.Doucet and A.F.Craievich, J.Chem.Phys., 78(1983)1465-1469.
71 G.Zerbi, R.Magni, M.Gussoni, K.H.Moritz, A.Bigotto and S.Dirlikov, J.Chem.Phys., 75(1981)3175-3194.
72 M.Maroncelli, S.P.Qi, H.L.Strauss and R.G.Snyder, J.Am.Chem.Soc., 104(1982) 6237-6247.
73 B.Ewen, G.R.Strobl and D.Richter, J.Chem.Soc.Faraday Disc., 69(1980)19-31.
74 T.Yamamoto, J.Chem.Phys., 82(1985)3790-3794.
75 D.Hohlwein, Acta Cryst., A37(1981)899-903.
76 W.Prandl, Acta Cryst., A37(1981)811-818.
77 M.Postel and J.G.Riess, J.Phys.Chem., 81(1977)2634-2637.
78 C.N.R.Rao, Acc.Chem.Res., 17(1984)83-89.
79 J.C.Frost, A.J.Leadbetter and M.R.Richardson, J.Chem.Soc.Faraday Disc., 69(1980)32-48.
80 M.Ferrario, I.R.McDonald and M.L.Klein, J.Chem.Phys., 83(1985)4726-4733.
81 D.Eilerman and R.Rudman, J.Chem.Phys., 72(1980)5656-5666.
82 M.T.Dove and A.I.M.Rae, J.Chem.Soc.Faraday Disc., 69(1980)98-106.

83 D.Bougeard, E.J.Samuelsen and D.Semmingsen, J.Chem.Phys., 82(1985)1454-1458.
84 S.L.Chaplot, G.J.McIntyre, A.Mierzejewski and G.S.Pawley, Acta Cryst., B37(1981)2210-2214.
85 L.J.Soltzberg, S.M.Cannon, Y.W.Ho, E.C.Armstrong and S.J.Bobrowski, J.Chem.Phys., 75(1981)859-864.
86 J.C.Bellows, E.D.Stevens and P.N.Prasad, Acta Cryst., B34(1978)3256-3261.
87 N.W.Thomas, S.Ramdas and J.M.Thomas, Proc.Roy.Soc.Lond., A400(1985)219-227.
88 M.Batley, S.Mraw, L.A.K.Staveley, A.H.Overs, M.C.Owen, R.K.Thomas and J.W.White, Mol.Phys., 45(1982)1015-1034.
89 F.Bayard, C.Decoret, J.Royer and Y.G.Smeyers, personal communication.
90 W.R.Busing and M.Matsui, Acta Cryst., A40(1984)532-538.
91 H.Gyoten, Y.Yoshimoto, T.Atake and H.Chihara, J.Chem.Phys., 77(1982) 5097-5107.
92 U.Sondermann, A.Kutoglu and H.Bässler, J.Phys.Chem., 89(1985)1735-1741.
93 H.Suga and S.Seki, J.Chem.Soc.Faraday Disc., 69(1980)221-240.
94 D.P.Craig, R.N.Lindsay and C.P.Mallett, Chem.Phys., 89(1984)187-197.
95 J.P.Aimé, J.Lefebvre, M.Bertault, M.Schott and J.O.Williams, J.Phys. 43(1982)307-322.
96 M.M.Bhadbhade, G.S.Murthy, K.Venkatesan and V.Ramamurthy, Chem.Phys.Letters, 109(1984)259-263.
97 J.M.McBride, Acc.Chem.Res., 16(1983)304-312.
98 S.Ariel, S.Askari, J.R.Scheffer, J.Trotter and L.Walsh, in: M.A.Fox (Ed.), Organic Phototransformations in Nonhomogeneous Media, ACS Symposium Series No. 278, pp. 243-256 (1985).
99 A.Bianchi and A.Gavezzotti, Chem.Phys.Letters, in press.
100 A.Uchida, Y.Ohashi, Y.Sasada, Y.Ohgo and S.Baba, Acta Cryst., B40(1984) 473-478.
101 S.Candeloro Desanctis, Acta Cryst., B39(1983)366-372.
102 C.P.Tang, H.C.Chang, R.Popovitz-Biro, F.Frolow, M.Lahav, L.Leiserowitz and R.K.McMullan, J.Am.Chem.Soc., 107(1985)4058-4070.
103 E.Meirovitch, J.Phys.Chem., 89(1985)2385-2393.
104 P.Arjunan, V.Ramamurthy and K.Venkatesan, J.Org.Chem., 49(1984)1765-1769.
105 J.Vicens, C.Decoret, J.Royer and M.C.Etter, Isr.J.Chem., 25(1985)306-311.

Chapter 12

INTERATOMIC POTENTIALS AND COMPUTER SIMULATIONS OF ORGANIC MOLECULES IN THE SOLID STATE

S. Ramdas and N.W. Thomas

1 INTRODUCTION

The past decade has witnessed significant advances in the modelling of organic molecules with digital computers. The manner in which the computer is used can vary according to the needs of its user: the quantum chemist is ever seeking faster and more powerful processors to perform accurate *ab initio* calculations, whereas chemists of other backgrounds have begun to appreciate the value of computers in enhancing their understanding of the chemical behaviour of matter.

Chemistry is, by nature, an empirical subject, and Solid-state Chemistry in particular can be treated by theories which are predominantly empirical. For example, simple ideas such as the distinctions between ionic, metallic and covalent bonding can be incorporated into empirical atomic radii, as shown by Pauling, Goldschmidt and Zachariasen. To the majority of chemists, quantum-mechanical modelling of solids is a sphere of activity reserved for only a few specialists; and the critic might argue that only a few relatively simple chemical systems can be described, at the expense of a very large investment in computing resources.

Such sentiments are worthy of further consideration, since many of the advances in Solid-state Chemistry have resulted from intuitive ideas about the nature of atoms and the bonds they form. Indeed there is ample experimental evidence for the nature of atoms and their bonding patterns in the thousands of crystal structures that have been determined by X-ray diffraction over the past few decades. The length of a bond, or the position of an atom in a unit cell is, to a large extent, incontrovertible. These quantities may be regarded as observables, the foundation upon which the rationalization and imagination of the scientist can be based.

An example of the contribution of the computer to progress in Solid-state Chemistry is given by the contemporary development of techniques in Molecular Graphics. Whereas some workers, rather disparagingly, regard the computer merely as a means of generating pictorial representations of crystal structures or electronic charge distributions, others are far more ambitious in their designs for the computer: progress in Artificial Intelligence (AI) should enable the computer to tackle problems with an approach comparable to that of a human being. Developments are being made in computer-aided drug-design (CADD), for example, where the computer itself is able to induce useful relationships between primary, secondary and tertiary structures in this complex field. In the AI approach, the computer reasons in terms of its "experience" (stored in a database), rather than by a long chain of deduction based upon a few fundamental assumptions about the nature of molecules in the solid state. The advent of 'affordable' array-processors and super-computers has also given an impetus to real-time modelling involving statistical mechanics (including Monte Carlo procedures), molecular dynamics (for the introduction of temperature, time, entropy and other probabilistic considerations), and molecular orbital calculations (ref. 1).

Our aim in this chapter, however, is to consider how the computer can be of value to the chemist in developing an understanding of a wide variety of organic systems. In particular, we shall review the development and application of semi-empirical atom-atom potentials in Organic Solid-state Chemistry. Given the ease of determining crystal structures by conventional diffraction techniques, one may wonder about the wisdom of using empirical and parametrized models of atomic interactions in structural predictions. In many respects, however, the knowledge of a crystal structure is just the beginning of a study, as questions concerning the intermolecular forces in the crystal remain. These forces have a direct bearing upon the conformations of the molecules as well as their modes of packing. A quantitative description of interatomic forces would promote our understanding of several phenomena, for example the role of substituents in 'crystal engineering', the mechanisms of solid-state chemical reactions, and changes in packing in the vicinity of molecular impurities and defects.

The application of non-bonded potentials need not be restricted to crystalline materials; indeed their use in conformational analysis extends to the modelling of polymers and proteins, as well as to the design of drugs and synthetic membranes. A relatively new application is the simulation of interactions between organic molecules and zeolite frameworks. Since the modelling techniques and choice of potentials are specific to these systems, a separate section will be devoted to this particular application.

2 THE DESCRIPTION OF NON-BONDED INTERACTIONS

2.1 The evolution of the semi-empirical non-bonded potential

The molecular packing in organic crystals can be analysed and understood at various levels of complexity. In general structural terms, we may argue, quite correctly, that the crystal structure of a particular molecule depends upon considerations of close-packing; i.e. the resulting crystal structure is characterised by the dovetailing of individual molecules to maximise the fraction of space occupied. The question then arises as to how we might test for such a condition being obeyed in a given crystal structure. A natural step would be to model atoms as solid spheres with characteristic non-bonded or van der Waals radii; a favourable crystal structure would then be one with the maximum number of atoms in adjacent molecules touching, but neither compressing one another nor lying too far apart. We could, quite legitimately, assess the likelihood of a trial structure existing in practice by classifying atomic separations as follows:

For two atoms, A and B, of non-bonded radii r_A and r_B respectively, separated by a distance r,

a short contact would be one with $r < r_A + r_B$;
a long contact would be one with $r > r_A + r_B$;
an ideal contact would be one with $r = r_A + r_B$.

A successful model-building approach based on these ideas has been developed by Smith and Arnott (ref. 2). Although the method can, in principle, be applied to all covalent systems, it is intended primarily for modelling of helical macromolecules, such as polysaccharides. X-ray diffraction evidence from polymeric materials is often sparse, owing to poor formation of crystallites. Consequently a conventional crystallographic analysis is precluded, and the diffraction data are analysed in conjunction with further assumptions about atomic 'contact distances'. The computational formalism (termed LALS, meaning "Linked Atom Least Squares") treats a trial structure as a sequence of linked atoms. Instead of describing the position of each atom by coordinates in the unit cell, a 'root atom' is specified, and the positions of all the other atoms are defined in terms of bond lengths, bond angles and dihedral atoms with respect to precursor atoms in the linked network. Stereochemical considerations are introduced by requiring that a given trial model has no over-short non-bonded interatomic distances. This corresponds to a "repulsion only" model, and a simple quadratic repulsion is used, following Williams (ref. 3):

$$C = \sum_i k_i (s_i - d_i)^2 \qquad (s_i < d_i) \qquad (1)$$

In this equation, C is the repulsion function, which is minimised in order to obtain the best trial model; s_i is an interatomic distance in the trial model which is shorter than the desired minimum, d_i, and k_i is a weight. Although it is only interatomic repulsion that is modelled explicitly, it transpires that the interesting volume of configuration space is at short intermolecular distances, corresponding to an implicit inclusion of interatomic attraction in the model. Short non-bonded interactions of an attractive nature, for example hydrogen bonds, can be incorporated in an *ad hoc* manner, by specifying inner and outer limits within which the distance of the interaction is allowed to lie. Convergence to a structural model that does not violate diffraction data seriously can be achieved by minimising differences between observed and calculated structure factors in the refinement.

The strength of this approach is that the simple parametrization allows a large region of configurational space to be monitored with a minimum of computational expense. Provided that agreement of the model with diffraction data (expressed as the crystallographic R-factor) is the main criterion for assessing the validity of a trial structure, the method is an invaluable tool for crystallographers working on polymeric systems.

A.I. Kitaigorodsky has pioneered the development of interatomic non-bonded potentials applicable to organic crystals. In 1961 he put forward a universal interaction potential of carbon and hydrogen atoms (ref. 4), given by

$$U_{AB} = 14.6 \{ -0.04 (r_o/r)^6 + 8.6\text{x}10^3 \exp(-13r/r_o) \} \quad (2)$$

In this expression A and B can be either C or H (corresponding to carbon and hydrogen atoms respectively); r_o(C...C) was taken to be 3.8Å, r_o(C...H) was taken to be 3.15Å, and r_o(H...H) was set at 2.6Å. The potential quoted is scaled so that its units are kJ mole^{-1}.

This is a one-parameter potential, the parameter r_o being the equilibrium separation between atoms in a close-packed model. The importance of this development is that the principle of close-packing has now been expressed in potentiometric terms. The potential models atoms as being slightly compressible, rather than as hard spheres; a repulsive force exists between the atoms when $r < r_o$, and an attractive force when $r > r_o$. The functional form of the potential is also significant in that the repulsive term varies as exp $(-k_1 r)$ and the attractive term as k_2/r^6, where k_1 and k_2 are constants. This reflects that the dominant contributions to the non-bonded energy come from short-range repulsive ('orbital overlap') interactions and weaker, longer-range dispersive attractive forces. We refer to this type of potential as a (6-exp) potential.

Considerable developments have taken place since these early days, facilitated by the enormous growth in computing power. The current formulation of the atom-atom potential model has also been developed by Kitaigorodsky (ref. 5), and its emphasis is on calculating a packing arrangement which corresponds to a minimum in the intermolecular interaction energy. This is essentially a stricter approach than arguments based on close packing.

In the energy minimisation process, the parameters of the crystal structure are the six unit cell constants, together with three translations and three Euler angles for each asymmetric molecular unit. Further intramolecular variables, such as bond and torsion angles, may be introduced when dealing with non-rigid molecules. Conversely systems of higher space-symmetry will be characterised by fewer independent variables.

It should be noted that the most favourable trial structure corresponds, in principle, to the structure of minimum free energy. In minimising the interaction (or potential) energy, however, contributions of temperature and entropy to the stabilisation of a structure are being ignored. Some authors justify this by remarking that the interaction energy is equivalent to the free energy at temperatures close to absolute zero. That the majority of organic crystals neither exhibit phase changes nor undergo significant geometrical distortions in the range from absolute zero to room temperature suggests that calculation of the potential energy alone is adequate. An alternative justification put forward by advocates of the semi-empirical potential method is that the derivation of potential parameters from observed room temperature crystal structures necessarily includes temperature and entropy effects (manifested by molecular vibrations) in the potential parameters themselves.

A number of techniques exist, however, for the explicit representation of vibrational energy in organic crystals. Two common methods are the quasi-harmonic approximation and the cell model (refs. 6,7,8). The former requires a calculation of the full set of frequencies of vibration from the second derivative of the potential energy with respect to the shift of molecules from equilibrium. This approach is generally adequate at low temperatures; but with the manifestation of anharmonicity at higher temperatures, the cell model is to be preferred for interpreting the behaviour of molecular crystals at room temperature. This model regards the crystal as made up of many cells, each containing one molecule, whose centres form the lattice. The motion of each molecule is assumed to be independent of its neighbours, although anharmonicity is admitted into the calculations. Pertsin and Kitaigorodsky have successfully used this method to evaluate parameters of the phase transition in adamantane (ref. 9). It is also possible to include experimental data (from Raman spectra or neutron-scattering, for example) in the modelling of vibrational energy, but this is a protracted process.

Having considered briefly the steps involved in evaluating the free energy of a crystal structure, let us direct our attention towards evaluating the intermolecular potential energy. The atom-atom method assumes that the total interaction between any two molecules can be represented by the sum of all the pairwise atom-atom interactions between the molecules. In the (6-exp-1) formalism, which is currently popular, the potential energy of interaction between two atoms, i and j, is given by:

$$u_{ij} = -A_{ij}/r_{ij}^6 + B_{ij}\exp(-C_{ij}r_{ij}) + q_iq_j/r_{ij} \quad . \tag{3}$$

The molar lattice energy is therefore given by

$$U_{molar} = \tfrac{1}{2}\sum_{i}^{N_1}\sum_{j}^{N_2} -A_{ij}/r_{ij}^6 + B_{ij}\exp(-C_{ij}r_{ij}) + q_iq_j/r_{ij} \quad . \tag{4}$$

The suffix i identifies an atom in the reference molecule, j identifies an atom on any of the molecules surrounding that reference molecule in the crystal, and r_{ij} is the distance between atoms i and j. N_1 is the total number of atoms in the reference molecule, and N_2 the total number of atoms in coordinating molecules that are to be summed. It is customary to apply a cut-off distance of several Angstroms, provided that the sum is convergent for this distance. Convergence acceleration techniques (ref. 10) permit shorter cut-off distances than would otherwise be tolerable, thereby reducing calculation times. The factor of $\frac{1}{2}$ enters the double summation in order not to count pairwise interactions twice; note that the constants A_{ij}, B_{ij} and C_{ij} conventionally have units of kJ $mole^{-1}$ $Å^{-6}$, kJ $mole^{-1}$ and $Å^{-1}$ respectively.

Not all authors interpret potentials of this kind in quite the same manner. Kitaigorodsky (ref. 6) regards the form of such potentials as being justified solely by the way in which their use leads to calculations in agreement with experimentally derived data. Other authors attempt to ascribe a precise physical meaning to each term: the inverse sixth term is considered to be the London dispersion attractive term (ref. 11), and the exponential term is interpreted as repulsion occurring because of the overlap of electronic orbitals. This attitude implies that dispersive interactions are described adequately by a single inverse sixth term, whereas the rigorous London theory works at large distances only. A rigorous approach would require the inclusion of components inversely proportional to the eighth, tenth etc. powers of the distance. Similarly a rigorous analysis of orbital overlap repulsion is impossible even for two helium atoms, and so the exponential term cannot be ascribed a precise physical meaning. The extra r_{ij}^{-1} term is identified by some authors as the 'Coulombic' term in the potential. Our viewpoint, however, is that this term merely adds another parameter to the potential, to improve its modelling of long-range interactions, which do, undoubtedly, have a strong Coulombic component. Quantum-mechanical considerations support the view that the intermolecular potential energy is made up these three components, when applied to non-polar molecules, i.e. dispersive, orbital-overlap and Coulombic forces.

Non-polar molecules containing extended π-type orbitals are liable to π-π^* charge-transfer attractive interactions as well (ref. 12). It seems that no attempt has been made, however, to include interactions of this kind explicitly within the atom-atom potential formalism. A further complication arises for polar molecules in that a rigorous treatment requires the inclusion of polarisation terms, which are not pairwise-additive. However, the simplicity of the (6-exp-1) pairwise-additive potential is a distinct computational advantage for application to chemical problems (ref. 13).

The most abundant source of experimental data for the derivation of the parameters A_{ij}, B_{ij} and C_{ij} in the potentials is the large number of crystal structures that have been determined by diffraction methods. The assumption that the correct non-bonded parameters will reproduce the observed crystal structures correctly when the lattice energy of the structure is minimised permits the derivation of a set of self-consistent parameters for a range of chemically related compounds. D.E. Williams has developed a powerful technique for the derivation of these non-bonded potential parameters (refs. 14,15). The parameters to represent C...C, C...H and H...H interactions by a (6-exp) potential were derived from the crystal structures of some nine aromatic hydrocarbons, along with some non-aromatic compounds as well. The C-parameters, C_{CC}, C_{HH} and C_{CH}, were not allowed to vary, and a geometric mean combining law was assumed for the A and B constants; i.e. $A_{CH} = (A_{CC}A_{HH})^{\frac{1}{2}}$ and $B_{CH} = (B_{CC}B_{HH})^{\frac{1}{2}}$. Some authors have used alternative combining laws, but the geometric mean law is needed for the convergence acceleration technique developed by Williams (ref. 10).

The (6-exp-1) parameters for hydrocarbon interactions can be derived similarly (ref. 16), as can the (6-exp-1) parameters for N...N interactions (ref. 17). In this comparatively recent work, Williams and Cox fitted the potential parameters to the crystal structure of α-nitrogen (N_2) and nine crystal structures of azahydrocarbon molecules which _do not_ exhibit hydrogen bonding. A treatment of non-bonded and hydrogen-bonded interactions _together_ in a fitting procedure is likely to yield poorly separated parameters. These authors are eager to maintain the elegant simplicity of the atom-atom method in defining an average N...N non-bonded potential, transferable to the majority of nitrogen-containing molecules in the organic solid state. This aim is largely vindicated by the applicability of the potential, although explicit consideration of nitrogen lone-pair site charges appears to be necessary in aromatic heterocyclic systems (ref. 18).

It is not possible to derive invariant values of the charge-parameters q_i and q_j for a given type of atom-pair in the (6-exp-1) potential. Since the atomic residual charges depend on the bonding environments of a given type of atom in the different compounds, one must estimate q_i for each individual atom in a molecule. A number of methods exist for doing this; Williams and Cox, for example, have modelled N...N interactions by invoking electrostatic potential-derived (PD) net atomic charges. The electrostatic potential was calculated from M.O. wavefunctions formed from an STO-3G basis set (ref. 17).

Some authors have cast doubt on the validity of representing the molecular electrostatic potential by a set of atom-centred residual charges. Hirshfeld and Mirsky (ref. 19) regard this as a reasonable approximation, whilst acknowledging that atomic multipoles can give a more faithful rendering of the actual electrostatic potential. Price (ref. 20) has argued in favour of a distributed multipole analysis (DMA) of the charge density, and has performed calculations of this kind on a series of aromatic hydrocarbons, using 3-21G and double-zeta basis sets. By comparison, Oie et al. (ref. 21) have commented upon the difficulties associated with deriving residual charges by a quantum mechanical method, and prefer to derive the charges from bond moments; these vary between different types of bond, but they are, nevertheless, broadly transferable. Interestingly, Berkovitch-Yellin and Leiserowitz (refs. 22,23) have modelled the electrostatic potential energy in amides and carboxylic acids in terms of multipole expansions with moments derived from experimental X-ray electron density maps. The reader is directed to an article by Moss and Feil (ref. 24) for an assessment of this approach.

It is not our intention to provide an up-to-date list of atom-atom potential parameters, as in an earlier review given by one of us (ref. 25). We suggest, instead, that the reader consults the comprehensive review of Timofeeva, Chernikova and Zorkii (ref. 26), in which the work of several independent research groups is brought together.

2.2 Treatment of non-rigid molecules: the modelling of intramolecular interactions

The atom-atom potential method implies that the pair interaction is central; i.e it is dependent only upon the distance between atom-centres. This assumption is adequate for intermolecular calculations, where, in simple physical terms, we are considering the interlocking of atoms with spherical surfaces at points of contact. In the treatment of intramolecular non-bonded interactions, however, the assumption of central interactions is a poor one. This does not present a serious problem in the modelling of rigid molecules; but in the case of flexible molecules an accurate representation of conformational energies is desirable. It will be appreciated that the conformational energies of a particular molecule are dictated, to a large extent, by stronger bonded interactions. There is also a tendency for non-bonded potentials to give calculated barriers to internal rotation which are too small (ref. 6).

A full semi-empirical treatment involves the addition of further terms into the expression for the total potential energy of the crystal:

$$U^{total} = U^{bond} + U^{ang} + U^{op} + U^{tors} + U^{nb} + U^{hb} \quad , \qquad (5)$$

where U^{total} is the total potential energy; U^{bond} is the bond deformation energy; U^{ang} is the angle deformation energy; U^{op} is the out-of-plane energy; U^{tors} is the torsional strain energy; U^{nb} is the non-bonded interaction energy, and U^{hb} the hydrogen-bond energy.

Oie et al. (ref. 21) have used these parameters to model over a hundred large molecular systems. U^{bond}, U^{ang} and U^{op} are represented by harmonic functions in their work:

$$U^{bond}_{ij} = f^{bond}_{ij} (d_{ij} - d^{o}_{ij})^2 \tag{6}$$

$$U^{ang}_{ijk} = f^{ang}_{ijk} (\theta_{ijk} - \theta^{o}_{ijk})^2 \tag{7}$$

$$U^{op}_{ijkl} = f^{op}_{ijkl} d^2_l \quad . \tag{8}$$

In equation (6), d^{o}_{ij} is the equilibrium bond distance between atoms i and j, d_{ij} is the observed bond length in the molecule, and f^{bond}_{ij} is the empirically determined force constant. Similarly in equation (7), θ^{o}_{ijk} is the equilibrium ijk bond angle and f^{ang}_{ijk} the bending force constant. The parameters in equation (8) have similar meanings: f^{op}_{ijkl} is the force constant for the out-of-plane bend of atom l from the plane defined by atoms i,j and k, to which atom l is bonded; d_l is the distance of atom l from this plane, and the expression is applicable naturally to sp^2-hybridised atoms.

The functional form of U^{tors} was taken to be:

$$U^{tors}_{ijkl} = U_{ijkl} \{ 1 + g_{ijkl} \cos(h_{ijkl}\phi_{jk}) \} + U'_{ijkl} \{ 1 + g'_{ijkl} \cos(h'_{ijkl}\phi_{jk}) \} \quad . \tag{9}$$

The sum is taken over all dihedral angles ϕ_{jk}, in which there are bonds between atoms i and j, j and k, and k and l. The six parameters to be chosen are U_{ijkl}, U'_{ijkl} (torsional barrier heights), g_{ijkl}, g'_{ijkl} (plus or minus sign), and finally h_{ijkl}, h'_{ijkl}, which govern the periodicity of the barrier. U^{nb} was taken to be a (6-exp-1) potential, and the hydrogen bonding energy was modelled by a (6-exp) potential.

Although the introduction of further parameters allows a more accurate fit to the data, it may be that improvements at a fundamental level are needed for a more realistic treatment. Since an improved fit is generally obtained at the expense of computational elegance, choosing the appropriate number of parameters for a given problem is always a matter for careful consideration.

The relative contributions of <u>intra</u> and <u>inter</u>-molecular forces in determining crystal structure have been investigated by Hagler and co-workers (refs. 27,28).

They have introduced the term 'conformational polymorphism' to describe the phenomenon of achieving greater intermolecular energy-stabilisation at the expense of intramolecular stabilisation, with associated changes in crystal structure.

2.3 The treatment of hydrogen bonds within the framework of atom-atom potentials

The occurrence of hydrogen bonding within organic molecular crystals is worthy of special consideration. The two special features of H-bonded interactions are their strength and also their marked angular dependence. The strength of a hydrogen bond can be modelled adequately by a deep potential of large curvature at the minimum. Kitaigorodsky has advocated the use of the Morse potential (ref. 6),

$$U^{hb} = D \{ 1 - \exp (n(r-r_o)) \}^2 , \qquad (10)$$

where D is the hydrogen bond dissociation energy, n is an empirical parameter, and r_o corresponds to the equilibrium separation of the hydrogen atom bonded to the donor atom and the appropriate acceptor atom. Oie et al. (ref. 21) have found the (6-exp) potential to be an adequate parametrisation of the hydrogen bond, whereas Berkovitch-Yellin and Leiserowitz (refs. 22,23) have modelled H-bonding motifs successfully with (6-exp-1) and (6-9-1) potentials.

It is not straightforward to include the angular dependence of H-bonding explicitly within the atom-atom framework. An implicit inclusion of this factor may be achieved, however, by imposing strict distance-constraints (by means of a steep curvature at the minimum), and also by including pairwise donor-acceptor interactions in an evaluation of the net hydrogen-bonding energy.

The strict geometrical constraints of hydrogen bonding have enabled Leiserowitz and Hagler to rationalise the allowed crystal structures of primary amides in terms of hydrogen-bonded ribbons, layers and stacks, without recourse to atom-atom potentials (ref. 29). It follows that the ease with which a particular compound may form hydrogen bonds is determined by the spatial distribution of functional groups in the molecule; this has been illustrated by Thomas and Desiraju (ref. 30), in the case of the six isomeric dichlorophenols.

Danziger (ref. 31) has used geometrical criteria alone to model the hydrogen-bonded interactions between proteins and ligands, which are of great importance to our understanding of drug activity. He has developed a computer programme which is capable of identifying possible sites for hydrogen bonding to a given protein. Approaches of this kind are currently being adopted throughout the field of Computer-Aided Drug Design (CADD).

3 CRYSTAL ENGINEERING OF ORGANIC MOLECULAR SOLIDS

Considerable interest has been shown in the idea of 'crystal engineering', where a particular type of organic compound can be designed to pack in a manner that has an associated, desired reactivity in the solid state. For example, it has been proposed that strong intermolecular non-bonded interactions such as C=O...Cl, C=O...Ph or Cl...Cl pair-interactions, can 'steer' the molecules to pack in a predictable manner (refs. 32,33,34,35).

It is now generally accepted that the solid-state mechanisms of organic chemical reactions are characterised by a minimum of atomic and molecular movement. A consequence of this is that, for a given reaction, the structures of the intermediate and product molecules are largely constrained by the molecular and crystal structures of the reactants. A major drawback, however, to the widespread adoption of organic synthesis in the solid state is that the crystal structure of a given reactant molecule often cannot be predicted with confidence.

The authors of this chapter have investigated the feasibility of using atom-atom potentials to predict the packing of several families of organic compounds, using the PCK6 computer programme (ref. 10). One of our aims was to predict the molecular conformations and crystal packing of the compound 2-benzyl-5-benzylidene-cyclopentanone (BBCP; see Fig. 1a) and its derivatives with methyl, chloro and bromo groups substituted at positions X and Y. The interest in this particular family of compounds stems from the observed behaviour of the parent compound, BBCP: this crystal reacts with ultra-violet light to form a dimer in a reaction that is both topochemical and topotactic (ref. 36). We were able to develop our modelling approach by reference to the fifteen BBCP derivatives of known crystal structure (ref. 37), and so a stringent test of the validity of the model was laid before us. The general strategy was to find a method which produces self-consistent results when applied to these fifteen compounds; but the approach was also to be of sufficient generality to be applied to other molecular systems without modification.

The straightforward use of atom-atom potentials did not produce any startling results: with starting models close to known crystal structures, the trial structures simply refined to the crystal structures evaluated from X-ray diffraction. The equilibrium crystal structure obtained in a given energy minimisation is dependent strongly upon the starting trial model, and refinement into a false minimum is a frequent occurrence. The results of such a procedure do not, essentially, provide a better understanding of the packing trends in a family of compounds, except that their observed crystal structures correspond to one of perhaps several energy minima.

For the purposes of crystal engineering, however, we should really like to know the roles of the substituents - Cl, Me and Br in this case - in determining the crystal structures of the derivatives. To this end, non-bonded potentials were used to monitor the relative contributions of the different types of atom-pair, C...C, C...H, H...H, C...X, H...X, X...X (X = Cl,Me or Br), to the total packing energy. This procedure has also been adopted by Bernstein and Hagler (ref. 27) and Bar and Bernstein (ref. 38). Not surprisingly, it was found that C...C and C...H interactions accounted for seventy to eighty per cent of the total lattice energy; the relative contributions of C...X, H...X and X...X interactions, however, were found to be extremely sensitive to the set of potential parameters adopted. One approach towards resolving this difficulty would be to attempt to derive a set of self-consistent potential parameters for the BBCP set of compounds, such that their use in an energy-minimisation would reproduce the observed crystal structures. It was decided not to follow this path, however, as the 'transferability' of any parameters derived in this manner to other families of compounds would be open to question.

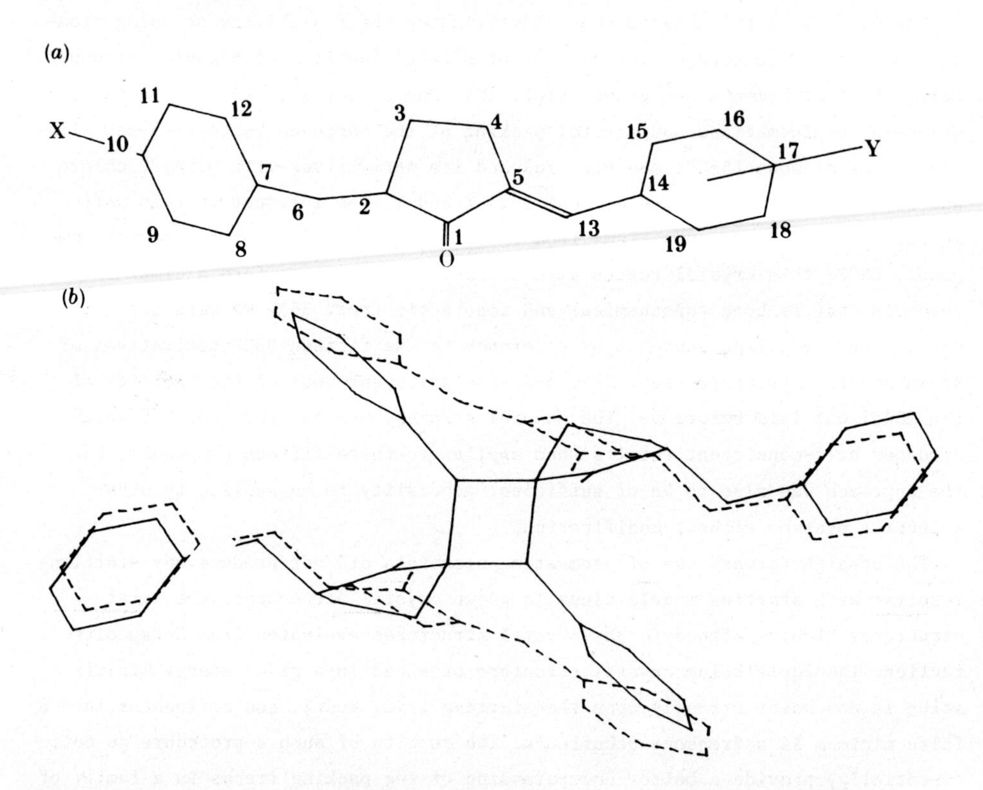

Fig. 1a The molecular structure and atomic numbering of BBCP and its derivatives. X denotes a substituent (Cl,Me,Br) in the para position of the benzyl phenyl ring, and Y a substituent (Cl,Me,Br) at ortho, meta or para positions of the benzylidene phenyl ring.
Fig. 1b Formation of a dimer molecule from two monomer molecules of BBCP.

Instead the authors decided to develop a framework of analysis in which the distances and directions of pairwise non-bonded interactions between a substituent and its coordinating atoms provide the basis of assessing the contribution of that substituent to stabilising the crystal structure. This approach yielded a good deal of physical insight into the role of the bromo group in stabilising the pCl B pBr BCP and B oBr BCP derivatives, for example (see Fig. 2). In both structures the bromo group interacts strongly with atoms in the benzyl phenyl ring; and in the former compound, these Br...C interactions are seen to cause a major conformational change in the molecular backbone (ref. 39). In other bromo-substituted derivatives, for example pBr BBCP and B mBr BCP, the bromo group was observed not to play a crucial role in their stabilisation. So we may anticipate the difficulties associated with deriving a C...Br non-bonded potential which is applicable to all the BBCP derivatives.

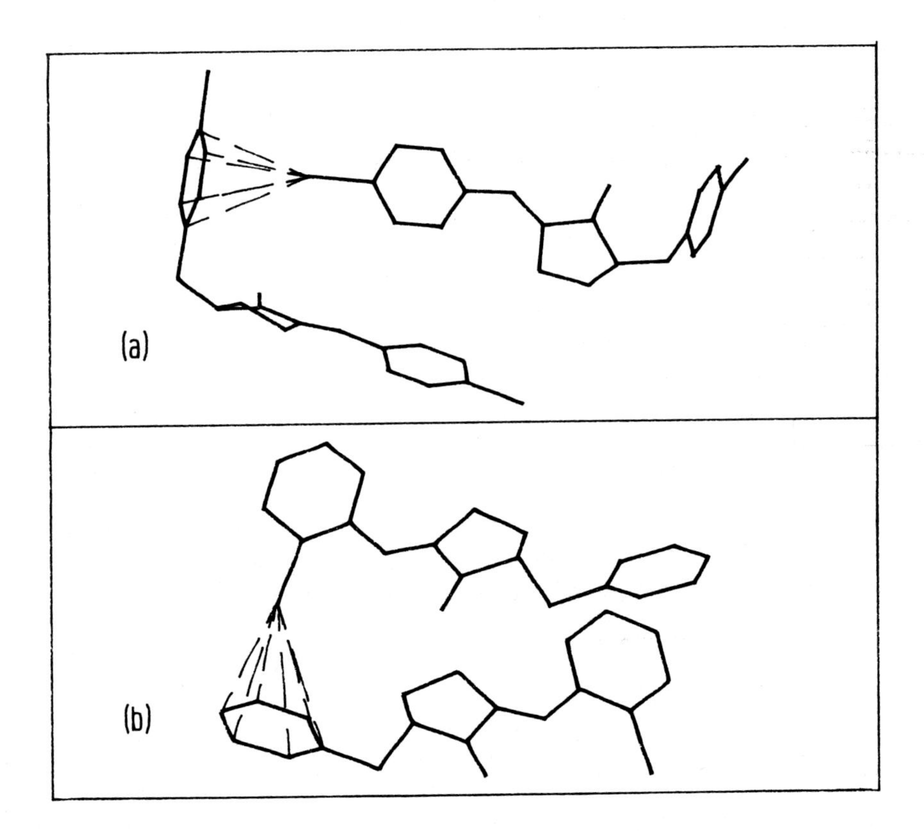

Fig. 2. Graphical representation of Br...C non-bonded interactions in (a) pCl B pBr BCP and (b) B oBr BCP as dashed vectors. Interactions between the bromo group and the carbon atoms of the benzyl phenyl ring are optimised in both compounds, giving rise to a conformational change in pCl B pBr BCP.

Another helpful comparison of the roles of different types of pairwise interaction in stabilising the derivatives can be made by plotting radial distributions of the distances of non-bonded interactions from atoms belonging to a reference molecule. The reader is directed to references 40 and 30 for further consideration of this technique in crystal engineering.

In order to achieve the aim of designing molecules to pack in a desired manner, however, we require far more than a post-rationalisation of why certain compounds pack in the way they do. The models we adopt must have some predictive power if they are to be of any use at all. That said, however, the framework of crystal engineering must be sufficiently flexible to accommodate new experimental observations, such as the occurrence of strong Br...C non-bonded interactions in some BBCP derivatives. The organic solid-state chemist wishes to answer questions of the kind: "What will happen to the crystal structure of compound X if a Y group is substituted for a hydrogen atom at position Z?" Before we attempt to answer such a question, however, it is necessary to consider how best to describe a particular crystal structure.

An unambiguous description of a crystal structure is given by specifying the space group and the atoms and their coordinates within the crystallographic asymmetric unit. If a comparison of several structures is required then the amount of data contained in such a description becomes too large to be meaningful: a more general description of crystal structures is needed. A direct and discriminating way of doing this is to compare their calculated energies of <u>molecular</u> interaction. For this purpose, it is adequate to use (6-exp) non-bonded potentials.

This calculation has been carried out for the fifteen BBCP derivatives of known crystal structure, using potential parameters quoted in reference 25. For each structure, the potential energies of interaction of a molecule (designated the 'central molecule') with its successively neighbouring molecules in the crystal are calculated. Typical results of such calculations are given in Table 1, for the compounds BBCP, *p*Cl BBCP and B *p*Cl BCP, with up to ten neighbouring molecules.

TABLE 1

Potential energies of interaction of the central molecule with successively neighbouring molecules a,b,c...j §

system ¶	a	b	c	d	e	f	g	h	i	j
BBCP	49.97	32.45	30.54	30.54	29.86	29.86	6.57	6.57	6.33	6.33
*p*Cl BBCP	50.76	35.19	33.71	33.71	31.31	31.31	15.42	4.50	4.50	2.91
B *p*Cl BCP	46.39	46.39	31.07	31.07	28.90	28.90	9.13	9.13	7.51	7.51

§ energies in kJ mol^{-1} ¶ meaning of a,b,c etc. defined in Fig. 4

From a graphical representation of this data on the energies of interaction (see Fig. 3), it is seen that BBCP and pCl BBCP have similar profiles, but that of B pCl BCP is quite different. Further, the magnitudes of the energies suggest that in all three of the above structures, the central molecule is coordinated most strongly by six molecules (a-f), with molecules g-j contributing much less to the total binding energy of the central molecule. Consequently, we may define a 'molecular cluster' as consisting of the central molecule and those six coordinating molecules that interact most strongly with it. The precise compositions of the molecular clusters are illustrated in Fig. 4.

One of the striking features of the BBCP family of compounds is that, of the fifteen known structures, six, including the parent compound, adopt a similar molecular cluster and associated intermolecular energy profile; these six compounds are the only photoreactive ones, owing to the proximity of C=C olefinic bonds within this cluster. An alternative molecular cluster is adopted by six of the other derivatives, with large C=C bond separations and consequent photostability. The remaining three derivatives are found to pack in their own, unique clusters (ref. 37).

By describing the crystal structures of the BBCP derivatives in terms of molecular clusters, we have adopted a conceptual framework of the appropriate complexity for the problem in hand. Our aim is to predict whether a given derivative will be photo-reactive or not; and given the correspondence of reactivity to type of molecular cluster, we must now establish the relationship between the chemical identity of a derivative and the molecular cluster it will adopt.

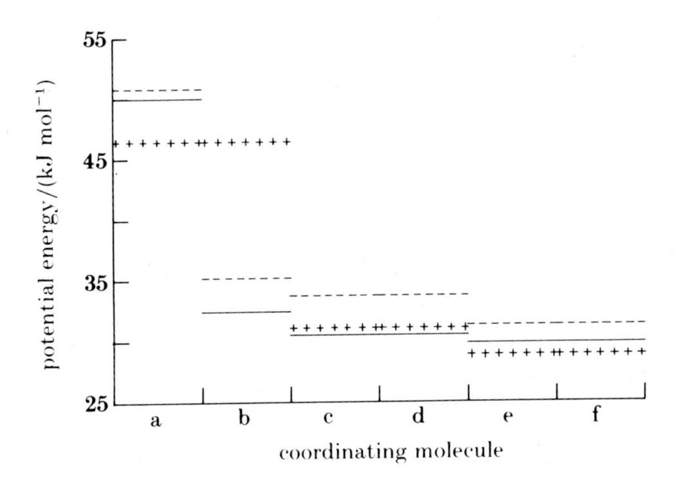

Fig. 3. Graphical representation of the data in Table 1. Letters a,b,c...f identify molecules in the cluster that are coordinated to the central molecule. ———, BBCP; -----, pCl BBCP; +++++, B pCl BCP.

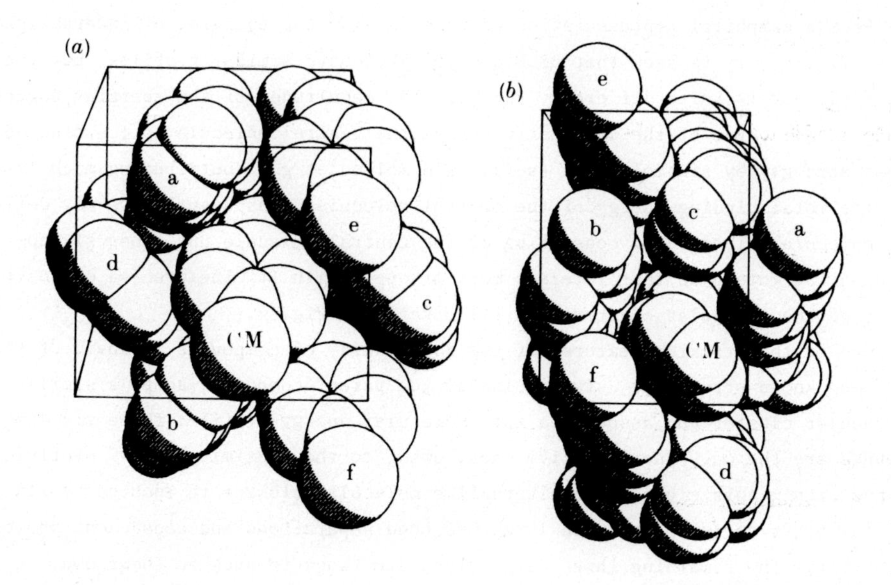

Fig.4a. The photoreactive molecular cluster adopted by BBCP, pCl BBCP and four other derivatives.
Fig.4b. The common molecular cluster adopted by six of the nine photo-stable derivatives of BBCP, including B pCl BCP.
The central molecules are labelled CM, and the coordinating molecules are labelled a,b,c...f, as in Fig. 3.

A simple approach is to assign a van der Waals radius to each type of atom (C: 1.70Å; H: 1.20Å; O: 1.40Å; Cl: 1.80Å; Br: 1.95Å), and calculate a local packing density for each hydrogen atom in the central BBCP molecule. Let us denote a hydrogen atom in the central molecule as atom i, and let atoms j represent non-bonded atoms belonging to other molecules <u>within the cluster</u>. The local packing density at atom i is then defined as:

$$n(i) = \sum_{j=1,6} 1/r_{ij}^2 \qquad (11)$$

In this equation, r_{ij} is the shortest distance from the centre of hydrogen atom i to the surface of the van der Waals sphere about atom j. The summation is performed over the six shortest non-bonded distances only, as longer distances correspond to atoms outside the first coordination shell of the hydrogen atom i. The simple functional form of the local packing density, n(i), was developed by equating the local packing density to the known bulk packing densities of two simple structures: f.c.c. and b.c.c. metals (ref. 41).

n(i) coefficients give an indication of the amount of 'empty' space surrounding a hydrogen atom in the crystal: the higher the value of n(i), the less likely it is that atom i can be substituted by a larger group such as Cl, Me or Br. Note that coordinating atoms j which belong to molecules outside the cluster are excluded from the summation; this is because the way in which these molecules coordinate the central molecule does not affect the basic type of molecular cluster that is adopted.

A consideration of all the known structures of BBCP derivatives enables us to define a critical local packing density; i.e. hydrogen atoms with densities above this value cannot be substituted by larger functional groups without the molecular cluster being disrupted. The critical local packing densities are found to lie in the range $0.934 < n(i) < 1.852$ for Cl and Me substitution, and $2.217 < n(i) < 2.490$ for Br substitution (ref. 37). The uncertainty in the magnitudes of the critical packing densities is a consequence of there being insufficient experimental data on which to base the model.

It is surprising that the larger bromo group has a higher critical value than the smaller chloro and methyl groups. Arguments based on volume alone, therefore, do not account entirely for the observed behaviour, and the propensity of the bromo group to coordinate itself closely with the carbon atoms of the benzyl phenyl group must be taken into account; this has already been discussed in connection with Fig. 2.

Application of this method to the BBCP system of compounds has enabled us to predict the types of molecular cluster that will be adopted by ten derivatives yet to be synthesized (ref. 37). Moreover, the predictive power of the method will be improved as more and more structures become known, since the critical packing densities are based solely on experimental data. There are also indications that this technique is applicable to other families of organic compounds, for example anthracenes and anthraquinones (ref. 42). By using the method it is possible to rationalise why the parent compound, anthracene, packs in a herringbone type of molecular cluster, whereas its 9,10-disubstituted derivatives form β-stacks.

It is clear from the foregoing discussion that sustained progress in crystal engineering depends upon the close interplay of experimental structural investigation and structural modelling techniques. It has been our intention to concentrate upon the latter in this section, and the reader is invited to consult the recent review by Desiraju (ref. 43) for an appraisal of current experimental work in this field.

4 APPLICATIONS OF SEMI-EMPIRICAL NON-BONDED POTENTIALS

4.1 Preamble

In general, the applications of atom-atom potentials can be divided into two broad categories: (i) phenomena involving a static lattice, in which the total configurational energy is minimised with respect to local variables in the computational model; and (ii) the various properties that depend upon the lattice dynamics of crystals. It is the latter type of application that presents a more severe test of the choice of potentials (ref. 44), and an excellent account of molecular dynamical simulations is given elsewhere in this book. It should be noted, however, that static lattice calculations, despite their simplicity, do reproduce the correct global energy minimum in the parameter space, even for surprisingly wide variations in the non-bonded potentials used (ref. 45).

Calculations involving the derivation of potential parameters and their subsequent verification through structure determination have been widespread in the literature. A typical example of work of this kind concerns the compound 9,10-diphenylanthracene. It has been shown (ref. 46) that the simultaneous comparison of the X-ray structure-refinement and the results from potential energy calculations actually hastens the solution of a crystal structure by direct methods. Such procedures are generally found to work well for rigid and partially rigid molecular systems (refs. 47,48,49). In this section, however, it is our intention to focus upon some of the novel applications which use the method of atom-atom potentials. Our choice of examples reflects both the variety of procedures and the range of chemical problems studied in the computer simulation of intermolecular interactions.

4.2 Enantiomeric intergrowths in hexahelicene

Pure specimens of hexahelicene grown from racemic solutions crystallise in a chiral space group (ref. 50). On the grounds of the phenomenon of spontaneous resolution, we should expect individual crystals grown in this way to be chiral. It has been discovered, however, that the expected 'chiral' crystals of hexahelicene are racemic, possessing enantiomeric excesses of the order of a few per cent (ref. 51). This puzzling behaviour, which is known to be exhibited by a few other materials (refs. 52,53), has remained a mystery. A possible solution to the problem was put forward with the suggestion that the individual racemic crystals are composed of intergrowths of alternating pure (+) and pure (-) layers (ref. 51); this would generate an essentially racemic 'composition'. Subsequent calculations confirmed this hypothesis (ref. 54), showing that enantiomeric intergrowths are, indeed, favoured on the (100) planes of the hexahelicene crystal structure (see Fig. 5). Other possible enantiomeric interfaces have higher excess energies; and it is not surprising that the plane for chiral turnover is identical to the plane of lowest surface energy in a pure crystal. It should be possible, through simulations of this kind, to predict whether molecules in a racemic solution will be resolved spontaneously, form enantiomeric intergrowths, or indeed crystallise into achiral space groups.

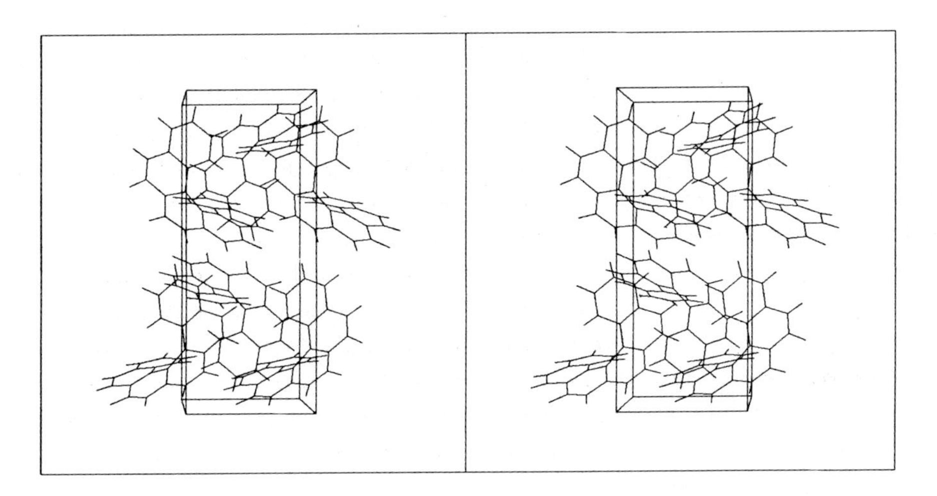

Fig. 5a. Stereo-diagram illustrating the packing of hexahelicene molecules within the unit cell, in the ac projection.

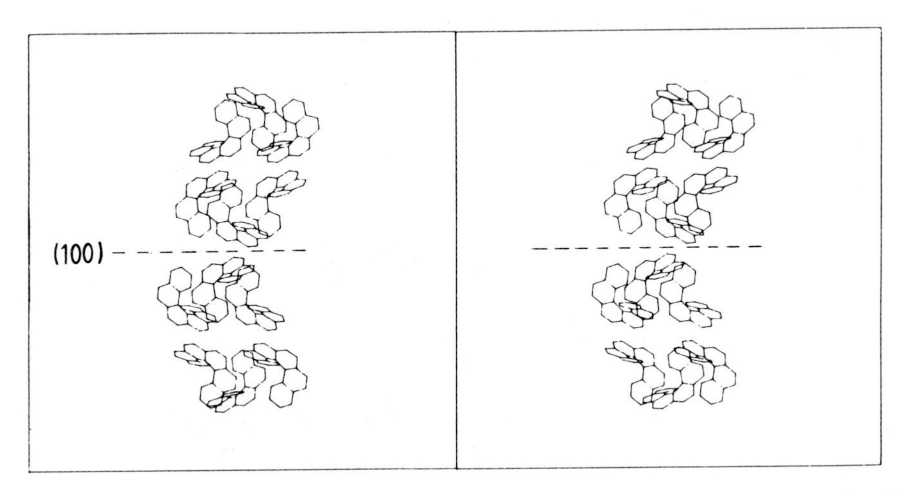

Fig. 5b. Stereo-diagram of the enantiomeric interface (100) viewed along [010]. The ease with which '+' and '-' molecules may be accommodated may be seen by comparison with Fig. 5a.

4.3 Impurity molecules in organic crystals

There have been a number of spectroscopic and crystallographic studies on doped organic molecular crystals (refs. 55,56,57); and it is known that the structures and photophysical properties of these mixed crystals are affected by the relative molecular geometries and electronic structures of the host and guest molecules. In particular, the nature of the deep traps formed by guest molecules and the shallow traps found in the host can be understood better if the exact geometry near the dopant molecule is known. Potential energy calculations on a test system of anthracene doped with tetracene have demonstrated the utility of the atom-atom method (ref. 47). The computer simulation consisted of a cluster of over forty host molecules, with the guest molecule located at the centre of the cluster; the guest molecule and a few of the neighbouring host molecules were allowed to 'relax' with respect to their rotational and translational parameters. The resulting orientation of the guest molecule and its perturbation of the host structure (ref. 58) were found to be in excellent agreement with the conclusions reached from ESR experiments (ref. 59). In order to gain more insight into the nature of the shallow traps formed in doped systems, it is necessary to correlate the size and shape of the guest with its influence on the host; it is anticipated that the atom-atom method will be invaluable in this context.

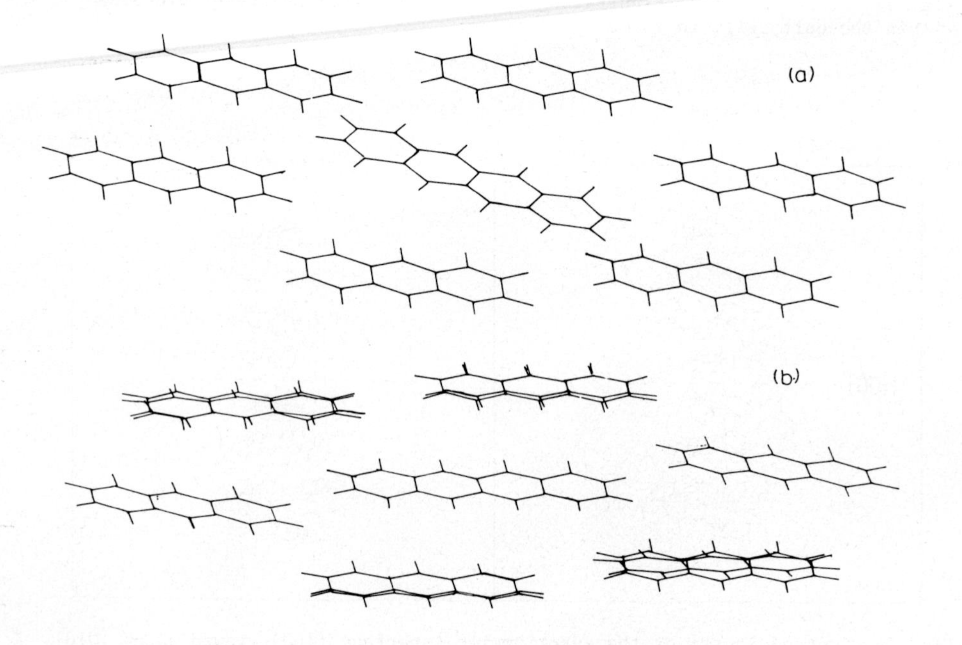

Fig. 6. The y^* projection of tetracene in anthracene. (a) the initial configuration; and (b) the final configuration, after minimisation of potential energy.

4.4 Channel and molecular complexes

There exist a number of organic compounds which crystallise with large cavities and channels within their structures. These are of interest, since small molecules, for example acetophenone, acetone and phenanthrene, can take up positions within the cavities and channels, to form complexes (ref. 60). Determination of the crystal structures of these complexes is rarely straightforward, however, since the small molecules exhibit both orientational and translational disorder (ref. 61). Some recent work has shown the positive role of computer simulations in this class of compounds.

Deoxycholic acid (DCA), for example, is known to form channel complexes with many molecules (ref. 62). One side of the molecule is hydrophobic, and the other, hydrophilic side displays a remarkable ability to associate in the solid state, by means of stable hydrogen-bonding schemes (see Fig. 7a). The compound is capable of crystallising in orthorhombic, tetragonal and hexagonal phases; in the orthorhombic phase, the interior surface of the channels is hydrophobic, whereas in the tetragonal and hexagonal phases the channel surfaces are hydrophilic. A study of a number of related orthorhomic DCA systems (ref. 63) has revealed that the size and shape of the channels (see Fig. 7b) depends upon the the mutual position along the y-axis and the separation along the x-axis of the two adjacent anti-parallel bilayers. Van der Waals energy calculations show that the observed host lattices do correspond to energy minima; they have also predicted further bilayer packing modes and identified some more occluded molecules that could go within the channels.

Another related study has been concerned with evaluating the energies of molecular complexation of a number of quinones and hydroquinones (ref. 64). The solid state energy of complexation is defined as the difference between the lattice energy of the complex and the stoichiometric sum of the lattice energies of the individual components, divided by the number of molecules that form the complex. It is to be expected that the calculated enthalpies of complexation will be less accurate, since they are obtained as small differences between large values of lattice energy. It is encouraging to note, however, that the calculated enthalpies have been of the right sign, and of the correct order of magnitude.

In the search for one-dimensional organic conductors (i.e. 'organic metals' like TTF-TCNQ, see Fig. 8), it is now recognised that there are two essential characteristics of such molecular systems: (i) the existence of segregated stacks of acceptor and donor molecules and (ii) the occurrence of a partial charge transfer, p, from the electron donor to the electron acceptor molecule. (In the TTF-TCNQ system, for example, p is equal to 0.59e). Once again it has been demonstrated that packing calculations are valuable tools in the prediction of structure and charge-transfer in TTF compounds.

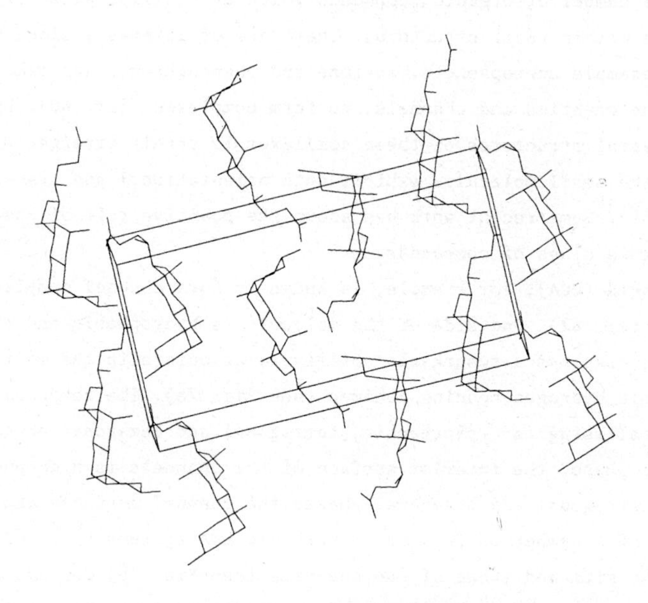

Fig. 7a. Crystal packing of DCA molecules, viewed along c

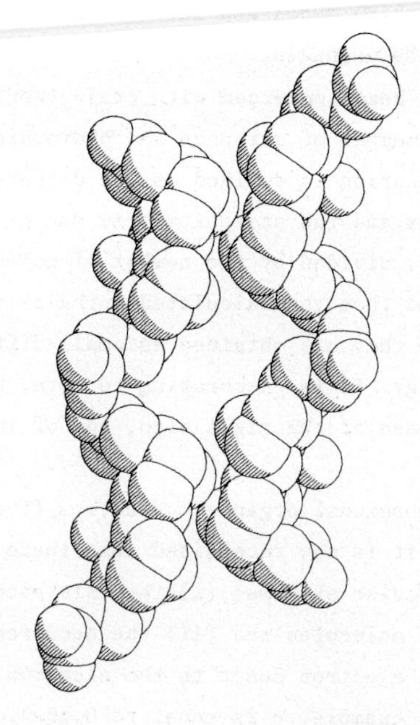

Fig. 7b. The size and shape of the channel running along c in DCA; this can accommodate small molecules like di-iodobenzenes and phenanthrene.

Govers and de Kruif (ref. 66) have used a simplifying approximation to evaluate the total electrostatic contribution from the stacking of TTF molecules. The approximation, which overcomes the poor convergence of the explicit summation of the $e_i e_j / r_{kij}$ terms, is as follows:

$$U_{elec} = U^{z'}_{elec} + (Ne/b) \{ \ln(N) - \ln(\tfrac{1}{2}z'+1) - 1 \} \quad . \tag{12}$$

In this equation, $e = \sum_i^n \sum_j^{n'} e_i e_j$; $U^{z'}_{elec}$ is the direct sum interaction $e_i e_j / r_{kij}$ between the atoms of the central molecule and those of z' neighbouring molecules. The quantity e is an intermolecular sum of the products of the charges on atoms i and j in a stack of N equidistant molecules at a distance b from one another. The aim of these calculations was to monitor the variation in intermolecular energy both with intermolecular separations within a stack, and also with changes in stacking pattern due to longitudinal and transverse slips. Calculations of this type (ref. 67) have been found to be more realistic than quantum mechanical studies of pairs of molecules (ref. 68), where computational restrictions prohibit the consideration of stacks or larger clusters of molecules.

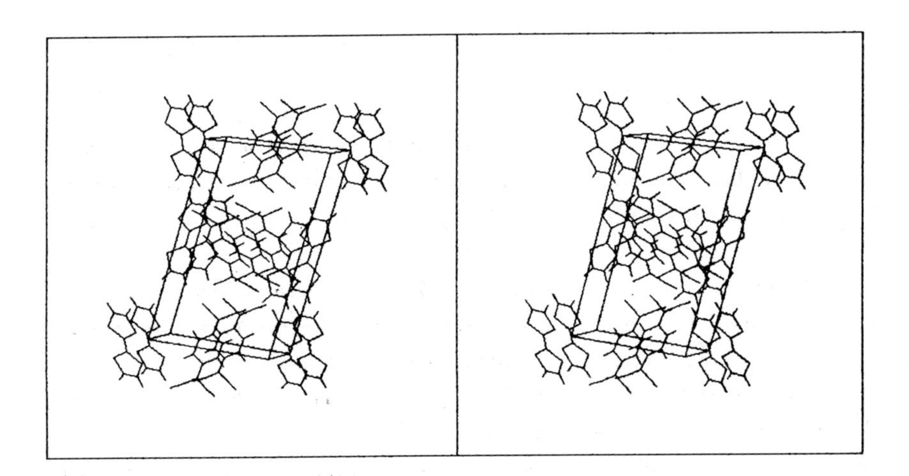

Fig. 8. Stereo-diagram showing the packing of TTF and TCNQ molecules in the solid state (see ref. 65).

4.5 Potential functions of siloxanes

Organo silicon compounds are worthy of special theoretical considerations (see ref. 69) both because of their unique physico-chemical properties, and also on account of the wide scope they offer for the derivation of new types of structures. These compounds can form polymers which are linear, cyclo-linear, ladder-like or spiro-cyclic; and the description of their structures could be essentially simplified if data were available on the potentials, and hence the structures of monomer fragments.

Siloxanes, unfortunately, are susceptible to large deformations of Si-O-Si angles and rotation about Si-O bonds (ref. 70). This necessitates a molecular mechanics type of simulation, using expressions of the form of equations (5) to (9) (see Section 2.2). A full description of the values of the empirical parameters is given in references 71 and 72; and some recent studies (ref. 70) have used the methodology to describe the conformational flexibility and cross-linking which are characteristic of these compounds.

4.6 Packing of polymer chains

In recent years we have seen the use of atom-atom potentials in the study and design of industrial polymers. Since perfectly crystalline polymers are difficult to produce, packing calculations provide an important additional tool in structural characterisation, in conjunction with X-ray and spectroscopic methods. Although polymers, as synthesized, may exhibit gross features characteristic of an amorphous material, it is recognised that short range order persists (certainly in stretched and oriented films); and in most cases, this leads to significant changes in macroscopic properties. Thus an understanding of this short range order is important in polymer design, and packing calculations are ideally suited to reveal this order.

The packing energy of a polymeric crystal can be regarded as a sum of two distinct contributions (see ref. 73): (i) that due to intra-chain conformations; and (ii) the inter-chain contribution arising from chain alignments. This separation of the two effects is valid because a polymer chain tends to adopt a conformation in the crystal that corresponds closely to one of its stable intra-chain conformations. Indeed each of the feasible intra-chain conformations may be used to construct trial lattice structures; minimisation of the inter-chain energy then provides a means of determining the most stable ordered configuration for a given chain conformation.

The conformations and packing modes of poly-acetylenes have attracted a number of theoretical studies (see refs. 74,75); and calculations have predicted correctly that all-trans chains pack with the lowest energy. The calculations also show that the trans-cisoid structure is more stable than cis-transoid (see Fig. 9). Calculations by Enkelmann et al. (see ref. 76) on trans-poly-acetylene have yielded a number of packing modes for the chains, clearly indicating the possibility of polymorphism in these materials.

Computational techniques have also been used to study the polymer-substrate interaction energies relevant to the epitaxial deposition of poly-ethylene on to graphite, sodium chloride and other surfaces (see ref. 77).

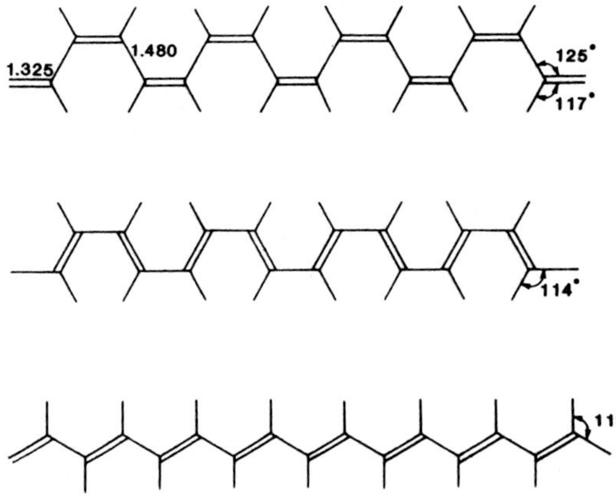

Fig. 9. The three types of chain-structure of polyacetylene

4.7 Interaction of organic molecules with zeolite frameworks

Zeolites are aluminosilicates with a three-dimensional framework of corner-sharing SiO_4 and AlO_4 tetrahedra. For each Al^{3+} ion in the framework, the condition of charge neutrality requires the existence of the appropriate number of non-framework ions like Na^+, Cs^+, Mg^{2+}, La^{3+} etc. These ions can be exchanged easily, and for catalysis, the NH_4^+ exchanged zeolites are calcined to yield the protonated form. In addition, zeolites contain plenty of water molecules within the structure. The variety of frameworks (of which over forty are known so far) give rise to channels and cages of varying shapes and sizes (see ref. 78), through which organic molecules and reaction products pass. The shape selectivity exhibited by the framework towards the incoming molecules, transition states and outgoing products (see ref. 79) governs, in most cases, the molecular sieving action and the catalytic reaction pathways. The behaviour of the zeolite catalyst can also be 'fine tuned' by varying the number of Al^{3+} ions in the framework, and consequently the concentration of non-framework ions as well.

The inclusion of aromatic molecules within zeolite frameworks is generally possible only at high Si/Al ratios; and it is this feature of the materials that permits the realistic simulation of interatomic interactions. Indeed, if a particular series of zeolites and organic adsorbates is being evaluated, it is appropriate to carry out calculations on an aluminium-free framework. This avoids a number of uncertainties in the description of the potentials and in the estimation of effective charges on the constituent atoms (see ref. 80). This approach allows us to consider the model zeolite essentially as a polymorph of silica (which also possesses a unique three-dimensional framework of SiO_4 tetrahedra).

There have been a number of studies for developing reliable interatomic potentials for silicates; and well established techniques in computer modelling (see ref. 81) have facilitated detailed predictions of structure, cohesive energies and defect energies. Some of the calculations have also shown how these properties change with temperature and pressure. Although some of the results have been unsatisfactory, particularly when two-body central force fields have been used (see ref. 82), accurate structures have been obtained for quartz-like systems. It has recently been shown, however, that greater reliability ensues when bond-bending terms are included in the potentials. Although the use of such potentials represents an accurate methodology, particularly in the modelling of framework distortions caused by organic molecules, the calculations become tediously long; and for many applications, the effort required may not be justified by the information sought. It should be noted that, if we restrict our queries to the location, optimum conformation and diffusion pathways of the adsorbents, then studies involving semi-empirical atom-atom potentials have been equally revealing. We shall now focus our attention on the description of these potentials.

In its simplest form, the potential energy of interaction between atoms of the zeolite framework and the organic molecules is represented by a (6-12-1) potential:

$$\phi_{ij} = -A_{ij}/r_{ij}^{6} + B_{ij}/r_{ij}^{12} + Kq_i q_j/r_{ij} \quad ; \qquad (13)$$

where atom i in the zeolite has a charge q_i and atom j in the adsorbate molecule has a charge q_j; r_{ij} is the distance between atoms i and j; and K is a unit conversion constant. The dispersion constant A_{ij} and the repulsion constant B_{ij} are obtained from the Muller-Kirkwood equations:

$$A_{ij} = 6mc^2 \frac{\alpha_i \alpha_j}{\left(\frac{\alpha_i}{\chi_i} + \frac{\alpha_j}{\chi_j}\right)} = \frac{3}{2} e^2 a_o^{\frac{1}{2}} \frac{\alpha_i \alpha_j}{\frac{\sqrt{\alpha_i}}{n_i} + \frac{\sqrt{\alpha_j}}{n_j}} \qquad (14)$$

In this equation, α is a polarizability, χ is a diamagnetic susceptibility, m is the electron rest mass, e is its charge, a_o is the radius of the first Bohr orbit, with n_i and n_j representing the numbers of electrons in the atoms. The repulsion constant B_{ij} is evaluated from the condition of ϕ being a minimum (in eqn. 13) at the equilibrium interatomic distance; this is considered to be the sum of the effective van der Waals radii of the interacting atoms. Thus

$$B_{ij} = \frac{r_o^6}{2} \left(A_{ij} + \frac{1}{3} \alpha_j q_i^2 r_o^2 \right) \tag{15}$$

It is generally expedient to treat the polarizability and the van der Waals radii as the parameters to be fitted; some of the published values for these parameters are given in Table 2 (see refs. 84,85,88,89).

TABLE 2

Some atom-atom potential parameters used in modelling interactions of zeolites with organic molecules

Atom or ion	$\frac{-\chi}{10^{-6}\ cm^3\ mol^{-1}}$	$\frac{\alpha}{10^{-24}\ cm^3}$	$\frac{r}{Å}$
$O^{-0.2}$	10.0	1.40	1.52
$O^{-0.15}$	9.9	1.25	1.52
O		0.85	1.52
N		0.97	1.55
H	2.0	0.43	1.35
C (sp^3)	7.4	0.96	1.80
C (sp^2)		1.34	1.80
C (sp)		1.42	1.80
Na^{+1}	5.6	0.30	0.98
K^{+1}	14.0	1.11	1.33

The charge is considered to be identical on all the oxygen atoms of the zeolite framework; and the numerical values of the mean charges on the oxygen ions and the non-framework ions are determined with due regard to the composition of the zeolite and the electroneutrality condition.

It will be noticed that potentials for the framework ions Si and Al are absent from the table. In the usual approximation, however, the atoms of the incoming organic molecules are not affected by the potentials of the Si atoms, since the cavities and channels are lined entirely by the van der Waals surface of the framework oxygens. This is clearly illustrated in Figs. 10 and 11, for a typical zeolite known as Theta-1.

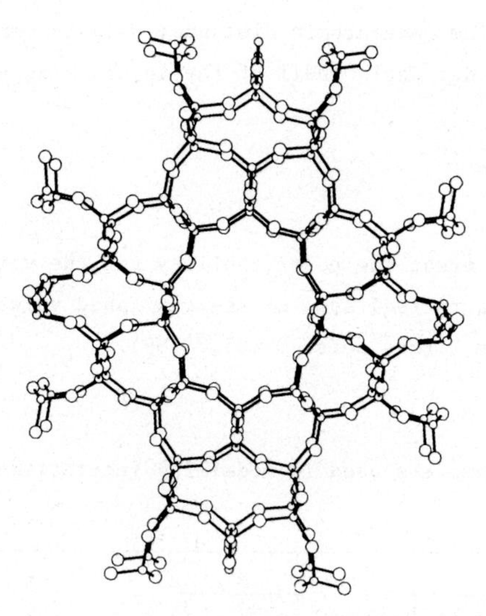

Fig. 10. The framework of zeolite theta-1. Note the three-dimensional network of SiO_4 units with corner-sharing oxygen atoms.

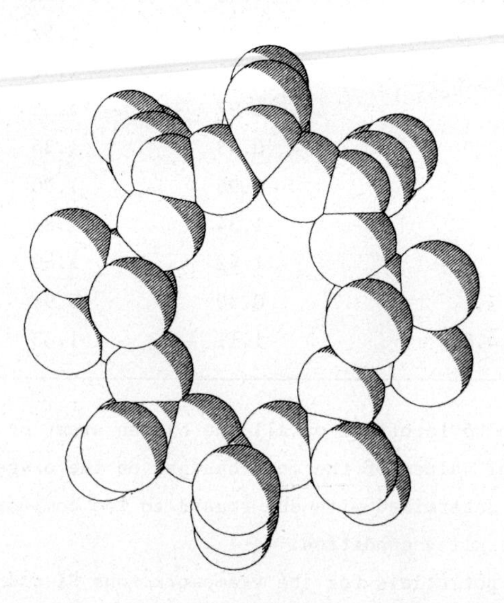

Fig. 11. The size and shape of the linear channel in zeolite theta-1, with oxygen ions lining the inner surface of the channel.

The following approximations are also made in the applications involving these potentials: (i) the zeolite framework stays rigid and intact in the presence of the organic molecules; (ii) the adsorbents keep rigid geometry corresponding to the minimum energy conformation; and (iii) the net atomic charges estimated from molecular orbital calculations are valid within the potential description given.

Despite the relative crudeness of the model, these calculations (see refs. 83,90) have yielded valuable insight on the positions of deepest minima, and also the most probable paths for the translation of the molecules. In addition, the thermodynamic characteristics of adsorption (Henry's constants, changes in internal energy, isoteric heats etc.) of several noble gases and hydrocarbons have been estimated on the basis of the calculated values of ϕ (see refs. 86,87). In this section, however, we shall highlight some of the recent applications which have a direct bearing upon the location of organic molecules within zeolite frameworks, and the complementarity of the simulations with diffraction techniques.

The framework of zeolite Y is shown in Fig. 12, together with the adsorption sites of methane determined by the calculations (see ref. 91); these are the sites at which the molecules have the most favourable interaction energy, and they will be occupied at low temperatures. Because of the inherent symmetry of the framework, there will be a number of such sites within a cage, not all of which will be occupied at a given time. Hence we should expect the experimental techniques to detect positional (as well as orientational) disorder in the system. Recent calculations involving the adsorption of benzene in zeolite Na-Y (see Fig. 13) have predicted two adsorption sites, one of which happens to be a local minimum. Neutron diffraction studies (see ref. 92) have indicated clearly that molecules are present on both these sites, one having a lower occupancy.

For the adsorption of molecules at ambient temperatures, it can no longer be assumed that molecules will be localized at the minimum energy sites. The change in internal energy of the system on adsorption is obtained by integrating over a large range of positions of the molecule in the zeolite framework. The Boltzmann factor accounts for the higher probability of occupancy of energetically favourable sites over others; and a rise in temperature increases the probability of occupancy of less favoured sites. We expect, therefore, a decrease in the energy of adsorption as the temperature is increased.

$$\Delta U^{ads} = \frac{\int_V \phi_{tot} \exp(-\phi_{tot}/RT)\, d\nu}{\int_V \exp(-\phi_{tot}/RT)\, d\nu} \tag{16}$$

The enthalpy of adsorption is given, therefore, by $\Delta H^{ads} = \Delta U^{ads} + RT$.

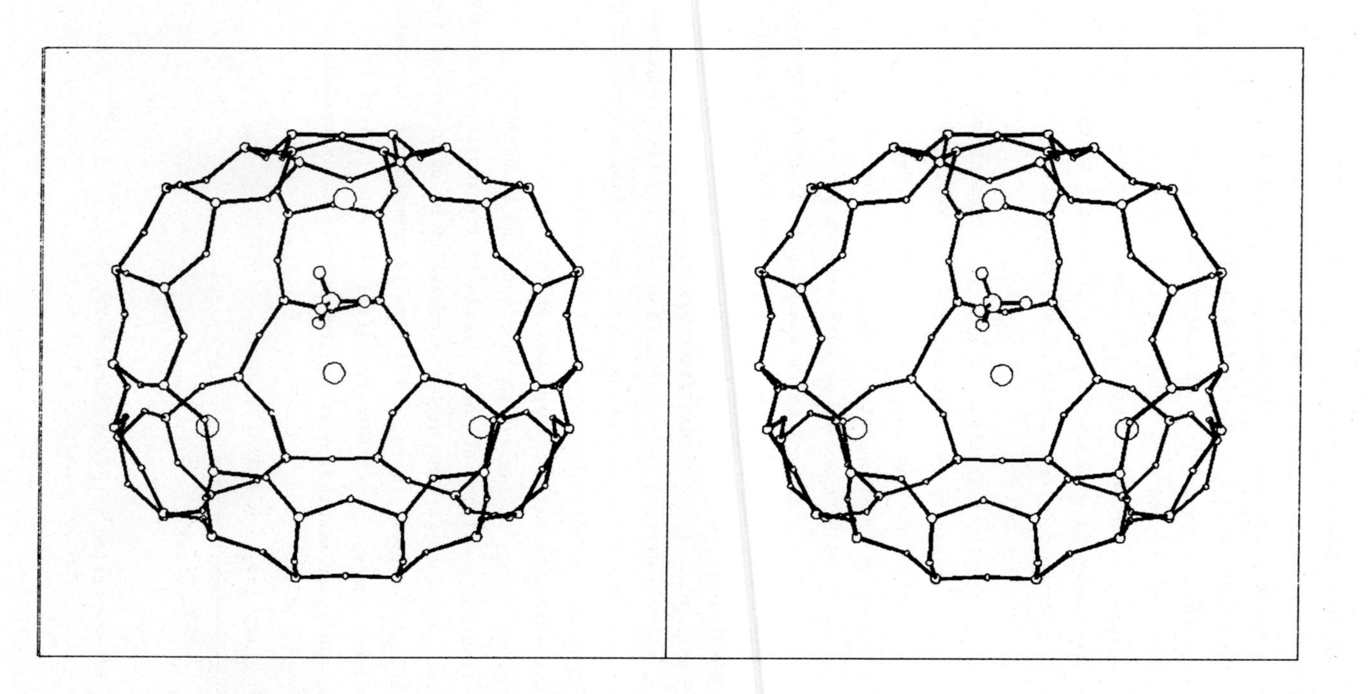

Fig. 12. Stereo-diagram showing the most probable location for a molecule of methane within zeolite Y containing La^{3+} as non-framework cations

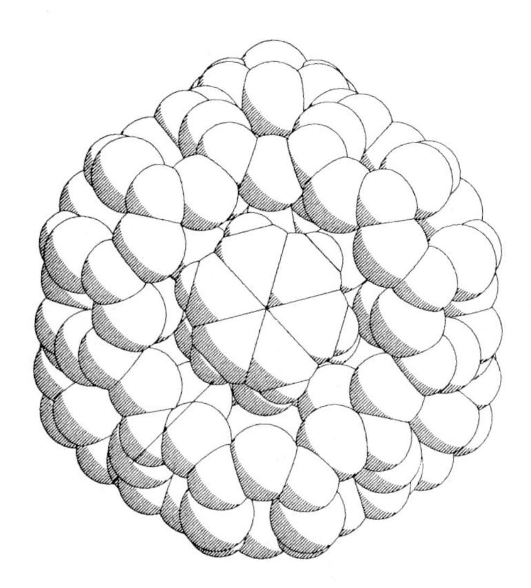

Fig. 13. Accommodation of a benzene molecule within the supercage of zeolite Y

The following table lists some of the results obtained on the adsorption of various hydrocarbons on to zeolite Na-Y.

TABLE 3

Experimental and calculated internal energies of adsorption on to zeolite Na-Y

Adsorbent	Calculated energy kJ mole^{-1}	Experimental energy kJ mole^{-1}
Methane	13.5	15.2
Ethane	21.5	23.3
n-propane	30.1	32.3
n-butane	35.2	37.4

These calculations have been carried out at a low loading of organic molecules, where it can be assumed that the molecules do not interact with one another (see ref. 91). In view of the established transferability of hydrocarbon potentials, realistic dynamical situations at high temperatures and pressures could yield valuable information about the influence of the channel and cage geometries on the actual process.

Good agreement has also been obtained between theoretical predictions and neutron diffraction studies in the determination of the adsorption sites of

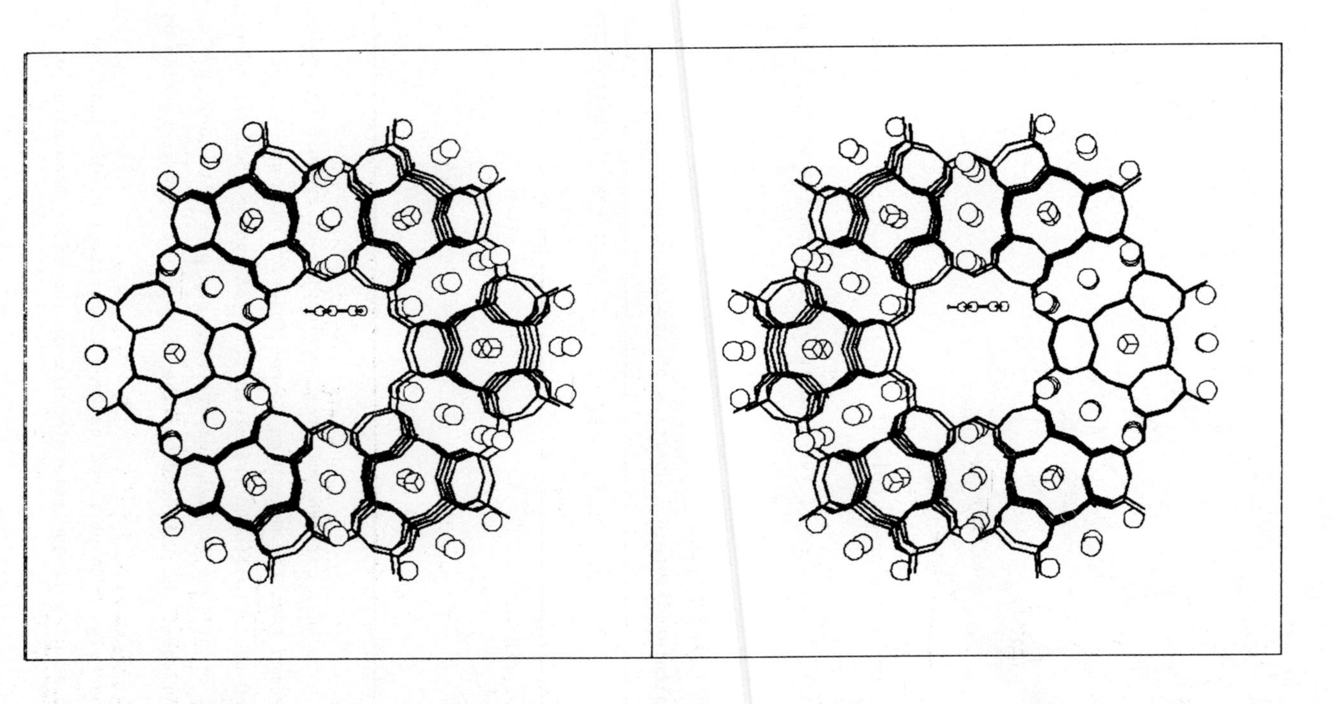

Fig. 14. Stereo-diagram showing the framework of zeolite L and the locations of all non-framework cation sites (of which only some are occupied by K^+ ions); the location and orientation of the pyridine molecule near one of the K^+ ions is confirmed both by calculations and also from neutron diffraction studies.

pyridine in K-gallo-zeolite at 4K. These calculations indicate clearly the existence of strong interactions between the K^+ ions and the quadrupolar molecule; they also show how benzene occupies a site different to that of pyridine in zeolite L (see Fig. 14).

Calculations have also been carried out for different adsorbates on silicalite (an aluminium-free zeolite ZSM5), in order to illustrate the orientations and pathways of the diffusing molecules. Fig. 15 shows the framework of silicalite, which is characterised by a straight channel along the view direction and a zig-zag channel parallel to the horizontal axis; a large cavity is formed at the intersection of two channels. (see ref. 80)

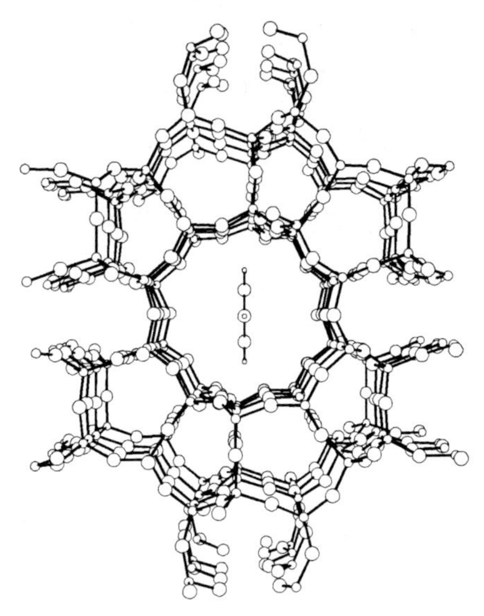

Fig. 15. A molecule of benzene in the straight channel of silicalite. For the energy calculations, the orientation of the molecule is varied such that the long molecular axis remains parallel to the channel axis b. It is the angle θ between the short molecular axis and the maximum diameter of the channel that is changed at regular intervals.

In Fig. 16 we show the energy contours (obtained with the molecular modelling package CHEMGRAF; see ref. 94) as the benzene molecule is moved along the straight channel. Comparison of this map with similar diagrams for the xylenes brings out the orientational restrictions imposed by the channels and their intersections.

The examples chosen indicate the utility of these simple atom-atom potentials in the description of these complex systems. It is clear, however, that there is scope for improvement of the potentials, as well as for the relaxation of the rigid framework approximation that has been described.

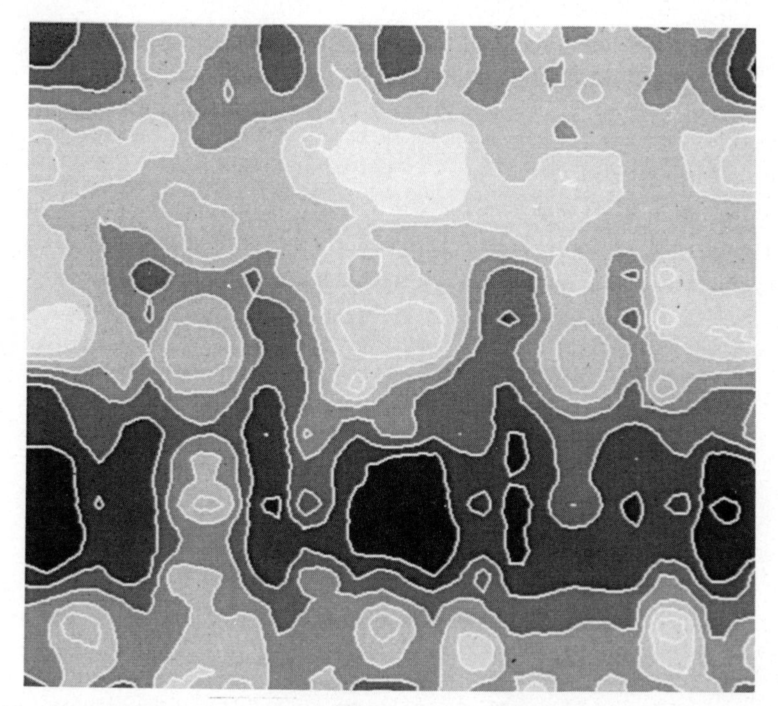

Fig. 16. Adsorption energy contours for all the possible orientations, θ, of the benzene molecule, as it moves along the channel. The contours are coded in grey levels, from black to white, to represent positive to negative values of adsorption energies.

5 CONCLUSION

We have considered several aspects of the semi-empirical atom-atom potential method, and how it can be used to enhance our understanding of molecular systems. It has been shown that the digital computer offers considerable scope for the modelling of the organic solid state and the interactions of organic molecules with zeolite frameworks, at various levels of sophistication. It is our belief that this approach will play an ever larger role in promoting our understanding of Chemistry in the solid state.

REFERENCES

1 D.N.J. White, J. Mol. Graphics, 3 (1985) 136
2 P.J.C. Smith and S. Arnott, Acta Cryst., A34 (1978) 3
3 D.E. Williams, Acta Cryst., A25 (1969) 464
4 A.I. Kitaigorodsky, Tetrahedron, 14 (1961) 230
5 A.I. Kitaigorodsky, Molecular Crystals and Molecules, Academic Press, N.Y., 1973
6 A.I. Kitaigorodsky, Chem. Soc. Reviews, 7 (1978) 133
7 G. Filippini, C.M. Gramaccioli and M. Simonetta, Chem. Phys. Letters, 35 (1975) 17
8 A.C. Holt, W.G. Hoover, S.G. Gray and D.R. Shortle, Physica, 49 (1970) 61
9 A.J. Pertsin and A.I. Kitaigorodsky, Mol. Phys., 32 (1976) 1781
10 D.E. Williams, Acta Cryst., A28 (1972) 629
11 F. London, Trans. Faraday Soc., 33 (1937) 8
12 A.K. Chandra and B.S. Sudhindra, Mol. Phys., 28 (1974) 695
13 J. Caillet and B. Claverie, Acta Cryst., A31 (1975) 448
14 D.E. Williams, J. Chem. Phys., 45 (1966) 3770
15 D.E. Williams, J. Chem. Phys., 47 (1967) 4680
16 T.L. Starr and D.E. Williams, Acta Cryst., A33 (1977) 77
17 D.E. Williams and S.R. Cox, Acta Cryst., B40 (1984) 404
18 D.E. Williams and R.R. Weller, J. Am. Chem. Soc., 105 (1983) 4143
19 F.L. Hirshfeld and K. Mirsky, Acta Cryst., A35 (1979) 366
20 S.L. Price, Chem. Phys. Letters, 114 (1985) 4
21 T. Oie., G.M. Maggiora, R.E. Christofferson and D.J. Duchamp, Int. Journ. Quant. Chem: Quantum Biology Symposium, 8 (1981) 1
22 Z. Berkovitch-Yellin and L. Leiserowitz, J. Am. Chem. Soc., 102 (1980) 7677
23 Z. Berkovitch-Yellin and L. Leiserowitz, J. Am. Chem. Soc., 104 (1982) 4052
24 G. Moss and D. Feil, Acta Cryst., A37 (1981) 414
25 S. Ramdas and J.M. Thomas, in M.W. Roberts and J.M. Thomas (Eds.), Chemical Physics of Solids and their Surfaces, The Chemical Society, London, 1978
26 T.V. Timofeeva, N. Yu. Chernikova and P.M. Zorkii, Russian Chemical Reviews, 6 (1980) 49
27 J. Bernstein and A.T. Hagler, Mol. Cryst. Liq. Cryst., 50 (1979) 223
28 P. Dauber and A.T. Hagler, Acc. Chem. Res., 13 (1980) 105
29 L. Leiserowitz and A.T. Hagler, Proc. R. Soc. Lond., A388 (1983) 133
30 N.W. Thomas and G.R. Desiraju, Chem. Phys. Letters, 110 (1984) 99
31 D.J. Danziger, J. Mol. Graphics, 3 (1985) 102
32 G.M.J. Schmidt, Pure Appl. Chem., 27 (1971) 647
33 M.D. Cohen and B.S. Green., Chemistry in Britain, 9 (1973) 490
34 J.M. Thomas, Phil. Trans. R. Soc. Lond., A277 (1974) 251
35 J.M. Thomas, S.E. Morsi and J.-P. Desvergne, Adv. Phys. Org. Chem., 15 (1977) 63
36 W. Jones, H. Nakanishi, C.R. Theocharis and J.M. Thomas, JCS Chem. Comm., (1980) 610
37 N.W. Thomas, S. Ramdas and J.M. Thomas, Proc. R. Soc. Lond., A400 (1985) 219
38 I. Bar and J. Bernstein, J. Phys. Chem., 86 (1982) 3223
39 W. Jones, S. Ramdas, C.R. Theocharis, J.M. Thomas and N.W. Thomas, J. Phys. Chem., 85 (1981) 2594
40 N.W. Thomas, S. Ramdas and J.M. Thomas, Mol. Cryst. Liq. Cryst., 93 (1983) 157
41 N.W. Thomas, Structure and Reactivity in Organic Crystals, Ph.D. Thesis, University of Cambridge, 1984
42 N.W. Thomas and J.M. Thomas, Mol. Cryst. Liq. Cryst., 134 (1986) 155
43 G.R. Desiraju, Proc. Indian natn. Sci. Acad., 52A (1) (1986), 379
44 C.M. Gramaccioli, G. Filippini, M. Simonetta, S. Ramdas, G.M. Parkinson and J.M. Thomas, JCS Faraday II, 76 (1980) 1336

45 G.N. Ramachandran, in E.R. Blout, F.A. Bovey, M. Goodman and N. Lotan (eds.) Peptides, Polypeptides and Proteins, John Wiley, N.Y., 1974
46 J.M. Adams and S. Ramdas, Acta Cryst., B35 (1979) 679
47 S. Ramdas and J.M. Thomas, JCS Faraday II, 72 (1976) 1251
48 S. Ramdas, J.M. Thomas and M.J. Goringe, JCS Faraday II, 73 (1977) 551
49 S. Ramdas, W. Jones, J.M. Thomas and J.-P. Desvergne, Chem. Phys. Letters, 57 (1978) 468
50 C. de Rango, G. Tsoucaris, J.P. Declercq, G. Germain and J.P. Putzeys, Cryst. Struct. Comm., 2 (1973) 189
51 B.S. Green and M. Knossow, Science, 214 (1981) 795
52 H. Wynberg, Acc. Chem. Res., 4 (1971) 65
53 R.H. Martin and M.J. Marchant, Tetrahedron, 30 (1974) 343
54 S. Ramdas, J.M. Thomas, M.E. Jordan and C.J. Eckhardt, J. Phys. Chem., 85 (1981) 2421
55 J.O. Williams and B.P. Clarke, JCS Faraday II, 73 (1977) 1371
56 D.P. Craig, L.A. Dissado and S.H. Walmsley, Chem. Phys. Lett., 46 (1977) 191
57 D.P. Craig, B.R. Markey and A.O. Griewank, Chem. Phys. Lett., 62 (1979) 223
58 S. Ramdas, Chem. Phys. Letters, 60 (1979) 320
59 H. Dorner, R. Hundhausen and D. Schmid, Chem. Phys. Letters, 53 (1978) 101
60 W.C. Herndon, J. Chem. Educ., 44 (1967) 724
61 S. Candeloro de Sanctis, E. Giglio, V. Pavel and C. Quagliata, Acta Cryst., B28 (1972) 3656
62 S. Candeloro de Sanctis, E. Giglio, F. Petri and C. Quagliata, Acta Cryst., B35 (1979) 226
63 S. Candeloro de Sanctis and E. Giglio, Acta Cryst., B35 (1979) 2650
64 H.G.M. de Wit, C.H.M. van der Klauw, J.L. Derissen, H.A.J. Govers and N.B. Chanh, Acta Cryst., A36 (1980) 490
65 Z.G. Soos, Ann. Rev. Phys. Chem., 25 (1974) 121
66 H.A.J. Govers and C.G. de Kruif, Acta Cryst., A36 (1980) 428
67 B.D. Silverman, J. Chem. Phys., 71 (1979) 3594
68 B.D. Silverman, J. Chem. Phys., 72 (1980) 5501
69 H. Oberhammer and J.E. Boggs, J. Am. Chem. Soc., 102 (1980) 7241
70 T.V. Timofeeva, I.L. Dubchak, V.G. Dashevsky, Yu. T. Struchkov, Polyhedron, 3 (1984) 1109
71 M.T. Tribble and N.L. Allinger, Tetrahedron, 28 (1972) 2174
72 N.L. Allinger, I.A. Hirch, M.A. Miller and I.J. Timinski, J. Am. Chem. Soc, 91 (1969) 337
73 S.K. Tripathy, A.J. Hopfinger and P.L. Taylor, J. Phys. Chem., 85 (1981) 1371
74 B.J. Orchard, S.K. Tripathy, A.J. Hopfinger and P.L. Taylor, J. Appl. Phys., 52 (1981) 5949
75 R.H. Baughman and S.L. Hsu, J. Polym. Sci. Polym. Lett., 17 (1979) 185
76 V. Enkelmann, M. Monkenbusch and G. Wegner, Polymer, 23 (1982) 1581
77 P.R. Baukema and A.J. Hopfinger, J. Polym. Sci. Polym. Phys., 20 (1982) 399
78 R.M. Barrer, Hydrothermal Chemistry of Zeolites, Academic Press, London, 1982
79 P.B. Weisz, Pure Appl. Chem., 52 (1980) 2091
80 S. Ramdas, J.M. Thomas, P.W. Betteridge, A.K. Cheetham and E.K. Davies, Ang. Chem., Intl. Edn., 23 (1984) 671
81 Computer Simulation of Solids, C.R.A. Catlow and W.C. Mackrodt (eds.), Lecture Notes in Physics, Vol. 166, Springer, 1982
82 C.S. Vemparti and P.W.M. Jacobs, Radiat. Eff., 73 (1983) 285
83 M.J. Sanders, M. Leslie and C.R.A. Catlow, JCS Chem. Comm., (1984) 1271
84 A.V. Kiselev and P.Q. Du, J. Chem. Soc. Faraday II, 77 (1981) 1
85 A.V. Kiselev and P.Q. Du, J. Chem. Soc. Faraday II, 77 (1981) 17
86 S. Furuyama and M. Nagato, J. Phys. Chem., 88 (1984) 1735
87 S. Furuyama, M. Miyazaki and H. Inoue, J. Phys. Chem., 88 (1984) 1741
88 A.G. Bezus, M. Kocirik and A.A. Lopatkin, Zeolites, 4 (1984) 346
89 A.V. Kiselev, A.A. Lopatkin and A.A. Shulga, Zeolites, 5 (1985) 261

90 P.A. Wright, J.M. Thomas, S. Ramdas and A.K. Cheetham, J. Chem. Soc. Chem. Comm., (1984) 1338
91 A.K. Nowak and A.K. Cheetham, Proceedings of the 7th International Conference on Zeolites, Tokyo, 1986
92 A.N. Fitch, H. Jobic and A. Renouprez, J. Chem. Soc. Chem. Comm., (1985) 284
93 P.A. Wright, J.M. Thomas, A.K. Cheetham and A.K. Nowak, Nature, 318 (1985) 611
94 CHEMGRAF, created by E.K. Davies, Chemical Crystallography Laboratory, University of Oxford; developed and distributed by Chemical Design Ltd., Oxford.

Chapter 13

CONFORMATIONAL POLYMORPHISM

JOEL BERNSTEIN

INTRODUCTION

"One of the most important tasks in structural chemistry today is...that of making critical assessments of the mass of already published material in the search for unifying ideas. To mention only two examples: we know almost nothing about the weak interactions that control preferred packing arrangements of molecules in crystals, and we are only beginning to understand the conformational complexities of organic molecules, although a vast amount of information on these and other untouched matters is stored away in the literature."[1]

It is just twenty years since Dunitz and Ibers thus challenged structural chemists to tap the rich sources of *chemical* information in the structural literature. That challenge has been met, but only to a limited extent. From the point of view of the organic solid state, the past two decades have witnessed the proliferation of high precision X-ray structure determinations of organic (and organometallic) substances[2]. The accumulation of a large base of structural data has revealed consistent trends in bonding and chemical connectivity, and has increased our confidence in the constancy of many of the features of molecular geometry which are crucial for our understanding of chemical and physical processes and phenomena. The collection of this information into a functional data file[3], with appropriate software for ready accessibility and analysis, has permitted statistical and geometric analyses of a scope which simply was not possible when Dunitz and Ibers wrote these challenging words.

The utilization of this structural information in a systematic way has provided a great deal of chemical information, from correlations in structure-activity relationships in biologically active materials[4] to mapping of reaction coordinates[5], from derivation of molecular force constants from anisotropic displacement parameters[6] to revision of the van der Waals atomic raaii for molecular crystal structures[7], characterization of the nature of hydrogen bonding[8], generation of model molecules or molecular fragments for the design of new drugs[9] and rationalization of protein structure stabilization via specific weak, long range interactions[10].

There has been significant progress in understanding the conformational complexities of organic molecules noted by Dunitz and Ibers, from both the theoretical and experimental points of view. Paralleling the increasing sophistication of computer hardware, semi-empirical and *ab initio* methods have been developed and refined to handle ever larger molecules so that the methods are no longer confined to the laboratories of the cognoscenti; they are now becoming standard tools of the modern chemist. A similar revolution has taken place with regard to the experimental tools for structural characterization, in particular by X-ray crystallographic methods.

A relatively small, but determined, effort has also been made to meet Dunitz and Ibers other challenge: understanding the weak interactions that control the preferred packing arrangements in crystals. The aim of this chapter is to present some of the manifestations of these weak interactions along with the problems and approaches to learning about them and how they might effect the conformations of molecules as we observe them in the solid state.

The proliferation of structural information is primarily due to the routine nature with which many (but not all!) crystal structure analyses may be carried out, which in turn has led to another important development. In the not very distant past, polymorphism - the existence of multiple crystal forms of a particular compound - was virtually ignored by most chemists, or at best considered a nuisance of little intrinsic scientific importance. Because of its industrial importance, examples of polymorphism and techniques for investigating and utilizing it come from those areas of chemical research where full characterization of a material is crucial in determining its ultimate use, e.g. in pharmaceutical[11], dyes[12], and explosives[13]. Various aspects of the subject have been treated in books[14] and a number of reviews[15]. The ubiquity of the phenomenon is still not generally recognized, although over twenty years ago McCrone suggested that virtually "every compound has differenct polymorphic forms...the number of forms for a given compound is proportional to the time and energy spent in research on that compound"[15].

The fact that polymorphism is a very widespread phenomenon was surprisingly not appreciated by many of those who concern themselves with the organic solid state, in spite of the fact that, for instance, the classic experiments of Cohen and Schmidt on cinnamic acids to demonstrate the topochemical principles were based on the existence of polymorphism[16]. Recently, however, there has been a growing awareness of the phenomenon, and in an increasing number of instances where polymorphism has been recognized, structural chemists have carried out the structure determination of the various crystalline forms of the material in question, rather than do one and ignore the rest. In many cases the results of such studies have led to some surprising observations; namely that the conformation of a molecule is not necessarily constant from one polymorph to another. This in turn has cast some doubts on the transferability of geometrical information, particularly conformational information, from the solid state where it is determined by X-ray methods to other phases for use in the interpretation of chemical, physical or biological behavior. The now recognized possibility of the variability of conformation from one crystal form to another (and by logical extension from one medium to another or even within a particular medium) does not necessarily invalidate our earlier assumptions about the general transferability of this geometrical information; it merely serves as a *caveat* in making and exploiting that assumption. Moreover, as we shall show below these are the very exceptions which provide us with ideal model systems for studying the interaction between molecules in organic crystals as well as the interplay between these interactions and the molecular geometry.

The first part of this chapter sets out the energetic framework in which we will operate. This is followed by a survey of a number of illustrative examples from the literature which demonstrate the interplay between crystal forces and molecular conformation. Some of the strategies and methods which may be employed to study the phenomenon are then described, concluding with some examples of the information which may be gleaned from studying conformational polymorphism as part of a systematic approach to the general question of the interplay between the geometry of a molecule and its environment.

1. CONFORMATIONAL POLYMORPHISM

1.1 Background

Mitscherlich is generally given credit for being the first to recognize the phenomenon of polymorphism, when he noted the difference in physical and chemical properties in different crystals of arsenates and phosphates[17]. Polymorphism (or more correctly allotropism) is reputed to have played an important role in history. Legend has it that the polymorphic transition of tin, from which the buttons of Napoleon's soldiers' uniforms were fashioned, was responsible for his defeat before the gates of Moscow in the winter of 1812, the low temperatures leading to the transformation from the metallic white form to the more dense gray form and a consequent crumbling of the buttons[18]. The widespread polymorphism of organic materials was recognized many years ago, and a number of very useful (but now unfortunately rather difficult to access) compendia of examples were published[19]. Many of these examples came from the synthetic organic literature at a time when chemists, probably by necessity, were generally much more aware and more observant of the the varieties of crystalline forms resulting from different crystallization procedures and conditions. As structural and analytical methods developed polymorphism itself became a subject of study, although there is still a great deal of truth in a remark made nearly fifty years ago by Buerger in a now classic paper, "...to most chemists it is still a strange and unusual phenomenon"[20].

Fortunately, to some extent the novelty of polymorphism is wearing off and an increasing number of chemists are now taking advantage of it to investigate directly, for instance, structure-property relationships such as magnetism[21], solid state spectral properties of dye aggregates[22] or electrical conductivity[23] in organic materials. The particular advantage of polymorphism in this regard is that the chemical identity of the material remains unchanged from one polymorph to another so that a direct correlation between activity and solid state structure may be made. Pseudopolymorphism is a term used to describe different crystal structures of a material in which molecules of solvation are also present. While not polymorphs in the strictest sense, they also provide structural situations (albeit with some added complications) which may be utilized to investigate the problems described here.

In a recent survey of the Cambridge File to extract information on polymorphic structures, over 1500 compounds were listed with the term 'form' in the title of the compound, indicating the presence of polymorphism, and of these the structures of at least two of these had been reported for 426 compounds. For molecular structures the natural question which arises in these known cases of polymorphic structures is whether the molecular structure differs from one crystal form to another, and if so whether the crystalline environment plays a role in bringing about and stabilizing these changes. To address this question it is important to examine the ranges of energy required to alter the various geometrical parameters within a molecule as well as bring about changes in the packing motif.

1.2 Molelcular Shape and Energetics

Molecular shape is defined by three different types of molecular parameters: bond lengths, bond angles and torsion angles. Variations in the molecular geometry of a molecule are then very simply defined as changes in these parameters: bond stretching or compression, bond bending or deformation and bond twisting or torsion. Typical force constants for bond stretching (in 10^5 dyne-cm^{-1}) range from 4.5 for the single bond in ethane to 15.7 for the triple bond in acetylene[24]. Assuming a Hooke's law dependence this translates to approximately 350 kcal-mol^{-1}-Å^{-2}, so that the distortion of a single bond of only 0.03Å (i.e. approximately three standard deviations of the average bond length determined by crystallographic methods) would 'cost' about 0.3 kcal-mol^{-1}. This factor rises approximately proportionally for double and triple bonds. Bond angle bending is less expensive. Mislow[25] has shown that for many carbon bond angles the empirical relationship for the potential energy (1) holds where θ is the bond angle. Thus an angular distortion of about

$$V_\theta \sim 0.01\ (\Delta\theta)^2\ \text{kcal-mol}^{-1}\text{deg}^{-2} \qquad (1)$$

10° involves the same amount of energy as the distortion of a single bond by about 0.05Å. Torsional changes involve the rotation about the bond axis; the barrier to rotation about the single bond in ethane is about 2.8 kcal-mole^{-1}, which is approximately the difference between the *trans* and the *gauche* conformations.The barrier to free rotations of methyl groups in dimethylacetylene is on the order of 0.5 kcal-mole^{-1}. Thus, as a rule of thumb bond stretching is roughly two orders of magnitude more expensive energetically than rotations about single bonds, with bond angle deformations falling in the intermediate range.

1.3 Intermolecular Interactions

On the intermolecular level in organic crystals one is dealing potentially with a variety of interactions, most of them quite weak compared to those involved in chemical bonding. A crystal structure corresponds to a free energy minimum which is *not necessarily the global minimum.* This will have important consequencecs later. Many structures have been found to exhibit disorder, which energetically must be accounted for by the inclusion of an additional entropy term in the determination of the free energy. As in any situation of this type the minimum in the potential energy represents a balance between the attractive and repulsive interactions mentioned below. The nomenclature of these interactions is quite variegated and the terms are not always clearly defined or distinguished from one another. Some in common usage include van der Waals interactions, London forces, dipole-dipole interactions (and higher terms), dispersion forces, steric repulsions, hydrogen bonds, charge-transfer interactions (also called donor-acceptor interactions), electrostatic interactions, etc.

There has been some convergence of thought about the use of these terms, at least among those who deal with 'crystal forces', especially from the computational point of view. Generally, the intermolecular interactions fall into three classes: 1) non-bonded, non-electrostatic (van der Waals, London, etc.); 2) electrostatic (coulombic); 3) hydrogen bonding. The lines of distinction between these general classes are not always particularly sharp so, for instance, hydrogen bonding has been treated by a combination of the first two general types of interactions[26]. Another important distinction is that the non-bonded and electrostatic interactions are generally treated as isotropic[27] while hydrogen bonds, by their very nature, are directional and anisotropic.

The non-bonded (van der Waals or London) forces are generally weak interactions between uncharged atoms or molecules. Separation of a molecule from its crystal environment requires overcoming all the attractive forces acting on it and is simply the sublimation energy of the crystal. Thus the magnitudes of these non-bonded interactions may be estimated from the sublimation energy of those organic crystals in which other interactions are essentially absent. For most of these substances the sublimation energy is roughly in the range 10-25 kcal-mol^{-1}. For a molecule surrounded by, say, 8-12 neighbors in a typical organic crystal, the interaction energy is then about 1-2 kcal-mol^{-1} per molecular neighbor; A sub-class of these interactions, which are generally more anisotropic in nature, is the charge-transfer (i.e. $\pi-\pi$ or $\sigma-\pi$) type, for which the energies involved rarely exceed 5 kcal-mol^{-1} [28]. The electrostatic interactions can vary over a much wider range, depending of course on the distance and on the degree of polarization of the molecule or parts thereof, which computationally is manifested in the assignment of 'partial atomics charges' to the various atoms. Intuitively, electrostatic interactions for, say, hydrocarbons might seem to be negligible; however, for many organic crystals they have been shown to comprise a significant part of the total lattice energy[29]. Hydrogen bond strengths are generally estimated to be in the range 1-10 kcal-mol^{-1}.

These intermolecular interactions all fall on the low end of the scale of energies required to bring about distortions of molecular geometry. This already suggests that if the crystal environment has any influence on the molecular geometric parameters then those most likely to be affected will be the torsion angles around the single bonds, rather than distortions in bond angles or bond lengths which require substantially larger energies to bring about significant changes. Hence, for the most part we will concentrate here on the changes in torsional parameters which may be due to the influence of the crystal environment[30].

Before dealing with conformational polymorphism *per se*, it is useful to define exactly what we mean by molecular conformation. Chemists have historically associated the idea of the conformation of a molecule with its shape with little or no regard for the energetics involved. The recent popularization of computational methods has modifed this view so that here, following the ideas of Dunitz[31], we will include energies into our considerations whenever possible, since they are a very important aspect of conformational polymorphism. In the most general way, conformations are those various shapes of molecules which arise by rotations about single bonds and correspond to potential energy minima. As Dunitz notes, a crystal structure yields information about the 'preferred conformations' (*sic* plural) and that any particular arrangement of atoms or conformation 'cannot be very far' energetically from the equilibrium structure of the molecule. Thus we will be concerned here with energies, in particular the differences among different conformations, as well as the geometric shapes of the molecules obtained from the X-ray experiment.

On this basis it is clear that a number of conformations of a molecule may be energetically equivalent, or nearly so. For example, Dobler recently surveyed 54 crystal structure analyses of the macrocyclic polyether 18-crown-6 and found that twelve different conformations are adopted[32]. When the observed conformations were compared with those calculated by force field calculations they were all found to lie within the energies of the best conformations. An important consequence of this and many similar observations is that different conformations may appear in different crystal structures, or for that matter in the same crystal structure when the number of molecules in the assymetric unit exceeds one. Certain crystal packing motifs are more favorable than others (e.g. planar molecules usually pack more efficiently than non-planar ones)[33] and this may lead to a predominance of one conformation in a crystal structure, while in solution a number of different conformations may be present, including, most likely, the one(s) in the crystal structure(s). It is important to emphasize here again that both the crystal structure and molecular conformation represent potential energy minima which are not necessarily unique, and at energies not far from the global minimum there may exist a number of possibile molecular geometries of very nearly the same energy for both the molecular conformation and the crystal structure. It is this proximity of molecular conformational energies which makes possible the existence of different molecular conformations in a single crystal structure or conformational polymorphism when they are in different crystal structures.

It remains now to justify the existence of different confromations in polymorphic organic crystals. The differences in lattice energy among different crystal forms of an organic compound can be expected to be in the range 1-2 kcal-mol^{-1} [14c,34], especially when van der Waals interactions dominate the structure. From the estimates of the magnitudes of intermolecular interactions this is clearly in the range of energy required to bring about changes in molecular torsional parameters about single bonds, but it is generally not sufficient to significantly perturb bond lengths and bond angles. Therefore, for those molecules which do possess torsional degrees of freedom, various polymorphs may exhibit significantly different molecular conformations. This, then, is the rationale which accounts for the fact that crystal forces may play a role in determining the conformation of a molecule and in particular for the phenomenon of *conformational polymorphism*[35,36]. The earliest use of this term in the literature appears to be by Corradini[35a]. In view of the fact that this paper is not generally available and indeed has been rarely cited, it appears useful here to review some of the terminology which was defined in the original Corradini paper and is relevant in the current context.

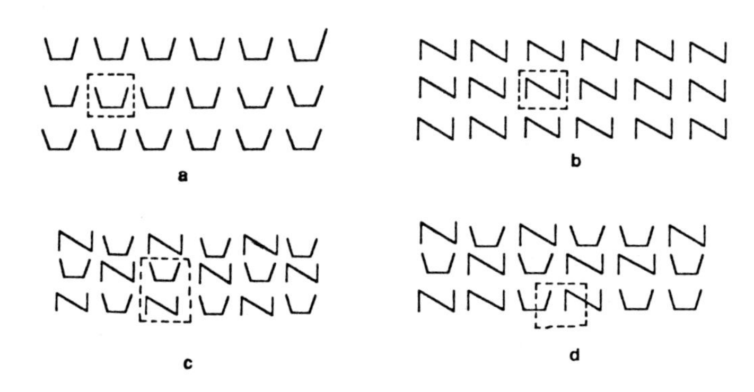

Figure 1. Schematic illustration of three possibilities for arrangement of molecules of the same chemical composition but different conformations in the crystal. For each case two conformations represent symbolically , say, cisoid and transoid dispositions around a single bond. The rectangle defined by a broken line represents one possible choice of the unit cell. (a) and (b), conformational polymorphs; (c) conformational isomorphism; (d) conformational synmorphism.

Conformational polymorphism is the existence of different conformers of the same molecule in different polymorphic modifications. Corradini specifies that the conformers are nearly isoenergetic and also includes as an example racemic and optically active crystals for the case of chiral molecules, but it is not necessary to include these qualifications with the general definition, which is illustrated graphically in Fig.1a,b.

Conformational isomorphism describes the existence of different conformers of a molecule in the same crystal structure, as a result of more than one molecule per crystallographic asymmetric unit. Thus each crystallographic site in the unit cell is always occupied by the same conformer and the ratio of conformers is defined by whole numbers as in Fig. 1c.

Conformational synmorphism describes the situation in which different conformers of a molecule are distributed randomly throughout the crystal lattice. Such a situation usually exists when two or more conformers have similar overall molecular shapes. Thus at any particular molecular site a number of conformations may be adopted, the relative population being determined by the relative intermolecular and intramolecular energies involved, as in Fig. 1d.

2. CRYSTAL ENVIRONMENT AND MOLECULAR BONDING AND GEOMETRY

As noted earlier, we are concerned with the general problem of the interplay between crystal forces and molecular geometry. We are now in a position to illustrate these interactions by appropriate examples. We present a number of cases in which the crystal environment clearly influences the molecular geometry. Most of these fall in the general category of conformational polymorphism, while others serve to demonstrate the existence of unexpected molecular geometries or even chemical bonding in crystal structures for a wide variety of molecules.The emphasis here is on small molecules where we can demonstrate differences due to changes in a small number of geometric parameters. Clearly, for larger molecules the effects can be cumulative, which of course can lead to significant changes in the overall shape of the molecule.

2.1 Evidençe for the Influence of Crystal Environment of Molecular Geometry and Bonding

The literature is rife with examples of environmental influences on molecular structure. Without in any way attempting to be exhaustive, we cite some here just to give an idea of the wide scope of problems encountered and treated.

The case of DL-α-amino-*n*-butyric acid **1** illustrates very nicely a number of the points already

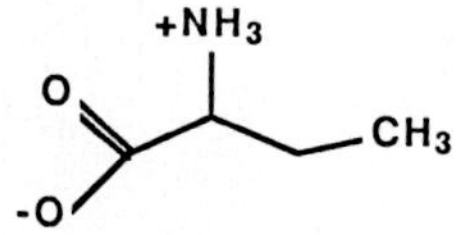

+NH3
O
CH3
-O

1

made. This material is known to crystallize in at least four polymorphic forms and IR spectra suggested the presence of conformational polymorphism[37]. Forms A (monoclinic $P2_1/a$) and B (tetragonal $P4_2/n$) are obtained by crystallization at room temperature by slow cooling from aqueous solutions and slow evaporation from aqueous ethanolic solutions respectively[38]. Form C (monoclinic I_2/a) is obtained by cooling below 201° K[39] and D (monoclinic I_2/a) by heating above 337°K[40]. The conformational parameter of interest is the torsion angle about the central C-C bond. The three staggered conformations (with respect to the nitrogen atom) in Figure 2 are exhibited to varying degrees in the four structures. In Form A there is an essentially equal distribution among the three conformations, while in B only the *trans* is found. Form C exhibits mostly *trans* with some *gauche* I and *gauche* II present. Form D on the other hand has a statistical distribution of *trans* and *gauche* II conformations.

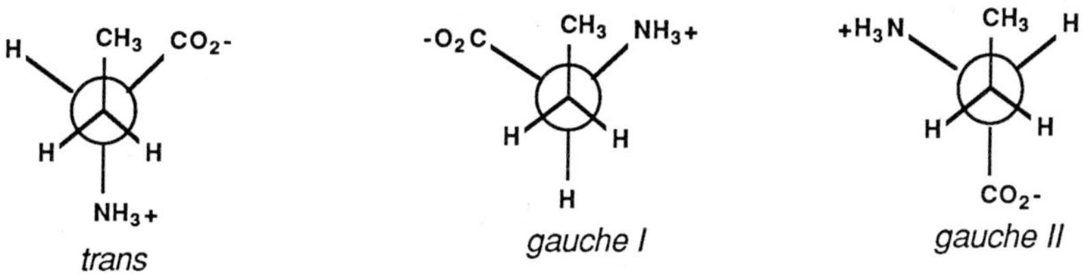

Figure 2. Conformers of DL-α-amino-*n*-butyric acid **1** observed in the various polymorphs.

In one of a series of very thorough studies involving various combinations of X-ray diffraction, neutron diffraction, and in some cases augmented by *ab initio* calculations, Jeffrey and his coworkers have investigated the geometric details of a number of small, but highly representative molecules in order to resolve discrepancies in the literature and provide precise structural data on these important molecules. These studies have often revealed effects of the crystal environment on the molecular structure. For instance, a recent neutron diffraction and *ab initio* study on deuteronitromethane **2** revealed[41] that while the molecule has approximate mirror symmetry the two N-O distances are not equal, which may be accounted for by a difference in the crystal environment. The oxygen which is involved in the longer N-O bond has four intermolecular O...D distances in the range 2.39-2.58 Å while the closest deuterium neighbor for the oxygen with the shorter bond distance is 2.69 Å. Although the question of whether these interactions should be called hydrogen bonds is still somewhat controversial[42,43], there is no question of the difference in environment and its influence on the molecular geometry.

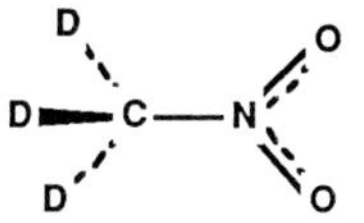

2 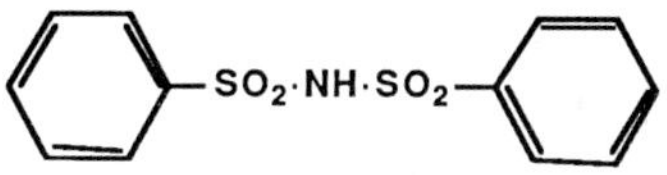 **3**

In a structural study of dibenzenesulfonamide **3** and its sodium salt Cotton and Stokely[44] found that the former has nearly C_2 (non-crystallographic) symmetry with the two phenyl groups on opposite sides of the S-N-S plane while the anion of the latter has approximate C_s symmetry with both phenyl groups on the same side of the reference plane of the central atom. The authors suggested that the *trans* configuration in the structure of the acid might be expected to minimize steric interference in the free acid, while the *cis* conformation in the anion allows a simultaneous approach of the sodium ion to the oxygens and electronegative nitrogen. More recently, Goldberg[45] studied the formation of inclusion structures from the 2,4-dichloro-5-carboxy derivative of **3** and found three different patterns of cocrystallization of the compound with a variety of guest molecules. In two of these, containing alternating layers of host and solvent, the host molecule adopts the approximate C_s symmetry while in the third type, a channel-like structure, approximate C_2 symmetry is observed. In the former, the energetically unfavorable intramolecular alignment of dipoles is compensated for by strong hydrogen bonding interactions between the dielectric layers of the solvent and hydrophilic functions of the host.

To complement the previous example in which the host molecule varies the conformation of the guest molecule in such inclusion compounds can be affected by the host. When 1,4-butanediol **4a** is the guest in the formation of inclusion complexes with **4b** and **4c**, it adopts two different conformations about the three carbon-carbon bonds, g^+ag^- in the former and *aaa* in the latter[46].

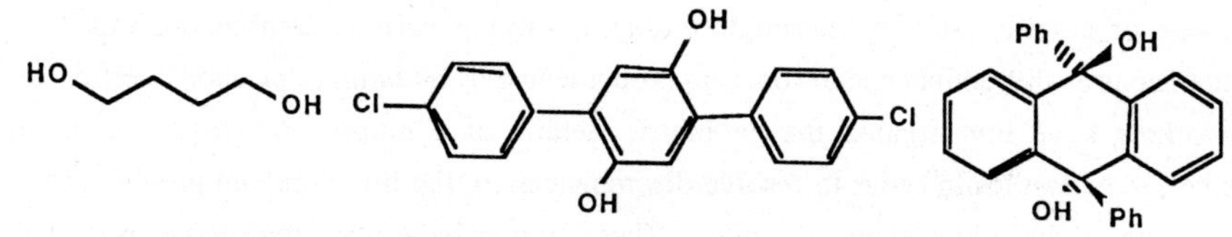

HO–(CH2)4–OH; Cl–C6H4–C6H2(OH)2–C6H4–Cl; Ph, OH / OH, Ph

4a **4b** **4c**

Interestingly, in both cases the guest lies on a crystallographic center of symmetry, which precludes the inclusion of chain-like diols with an odd number of carbon atoms.

Brock and coworkers have reported a fascinating case of the effect of crystal environment on the linkage isomerism in the palladium cation complex **5** for which they reported the structures of

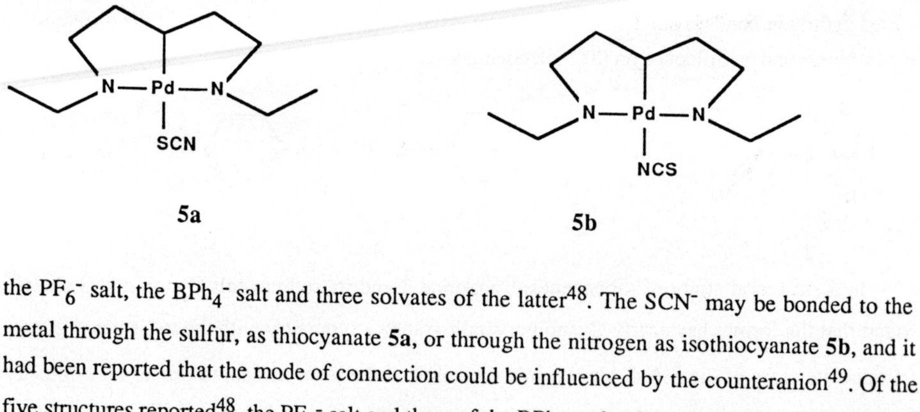

N—Pd—N, SCN; N—Pd—N, NCS

5a **5b**

the PF_6^- salt, the BPh_4^- salt and three solvates of the latter[48]. The SCN^- may be bonded to the metal through the sulfur, as thiocyanate **5a**, or through the nitrogen as isothiocyanate **5b**, and it had been reported that the mode of connection could be influenced by the counteranion[49]. Of the five structures reported[48], the PF_6^- salt and three of the BPh_4^- salts show the isothiocyanate isomer **5b**, whereas in the methanol solvate of the latter salt the thiocyanate isomer is present, due to the formation of hydrogen-bonded bridges with solvent molecules which are unique to this one structure of the five reported.

The pair of diastereomeric salts between ephedrine **6a** and R or S-mandelic acid **6b** shows a

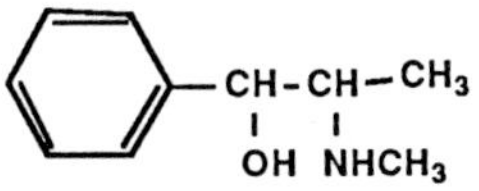

CH-CH-CH_3 / OH NHCH$_3$

CH-C(=O)OH / OH

6a 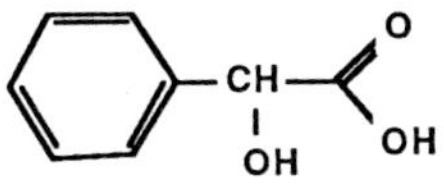 **6b**

combination of interesting phenomena[50]. The melting points differ by 66°C and the R mandelate is nearly 10% more dense than the S. The conformation of the ephedrine molecule, which has more degrees of conformational freedom, is very nearly the same in both salt complexes. On the other hand the mandelate anion, in addition to being of opposite handedness in the two structures, shows a difference in the the torsion angle about the exocyclic C-C bond of nearly 70°, so that the hydroxyl group is nearly *syn*-planar in the S complex and nearly perpendicular in the R complex. In both structures there is extensive hydrogen bonding as well as interactions between the phenyl rings. The main difference, however, arises from a close pairing of two mandelate ions in the R-mandelate which is not possible in the S-mandelate structure. Hence in this case it is clearly a difference in intermolelcular interactions which leads to the difference in molecular conformation.

2.2 Examples of Energetically Less Favored Conformations in Crystals

A fundamental question which often arises with regard to energetically less favored conformations is that of defining the energetically preferred conformation of the molecule, for purposes of comparison with the conformation found in the crystal. Many methods are now in routine use, including, on the spectroscopic side increasingly sophisitcated NMR techniques, IR and Raman in solution and microwave and electron diffraction in the gas phase. It is presumed that in solution or the gas phase at best that the minimum energy conformation is adopted and at worst that there is a mixture of conformations whose relative populations, of course, depend on their relative energies. From the theoretical point of view both semi-empirical and *ab initio* computational techniques have matured in the last decade to the point where they can provide rather reliable quantitative information on molecular energetics, especially when one is concerned with energetic differences in molecular conformations rather than in the absolute energy of any particular conformation. The combination of spectroscopic, crystallographic and computational techniques is properly becoming standard operating procedure for the structural characterization of molecules[51,52]. As a result both our knowledge of expected conformations and our confidence in predicting them are now quite well-founded. Consequently, for an increasing number of cases it is possible to state with some confidence that the molecular conformation observed in the crystal is not always that with the lowest energy.

For instance, in the crystal structure of the tetrapeptide Ac-Tyr-Pro-Asn-Gly-OH, the molecule adopts an approximately stretched conformation which is 5.5 kcal/mol^{-1} above the lowest energy folded conformation calculated via molecular mechanics[52]. The energy difference is just about that of an intermediate to strong intermolecular hydrogen bond, which could stabilize the otherwise unfavorable molecular conformation and indeed there are many hydrogen bonds in this structure. When the packing is dominated by weaker intermolecular forces, the difference in energy between the observed conformation and the calculated lowest energy one is naturally expected to be smaller. So, for (+)-butaclamol **7** in which the packing is dominated by van der Waals forces, the observed conformation is only 1.4 kcal-mol^{-1} higher in energy than the calculated minimal energy conformation[53].

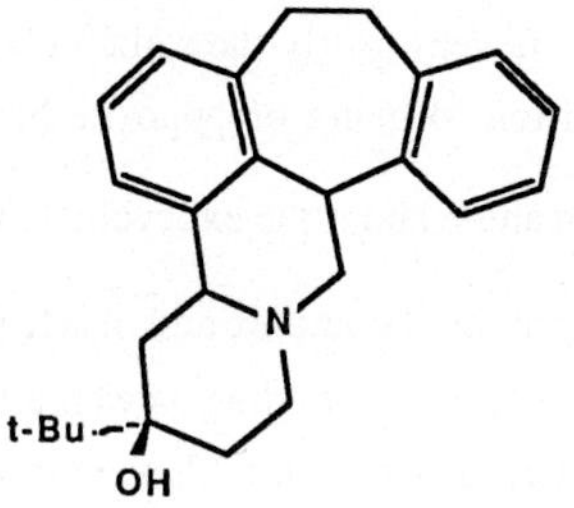

t-Bu
OH

7

A particularly dramatic example of how the crystal structure can stabilize an otherwise unfavorable example of the molecular conformation has been reported for **8**, in a combined X-ray and NMR study[54]. In the crystal structure, the molecule adopts two conformations **8a, 8b**, both

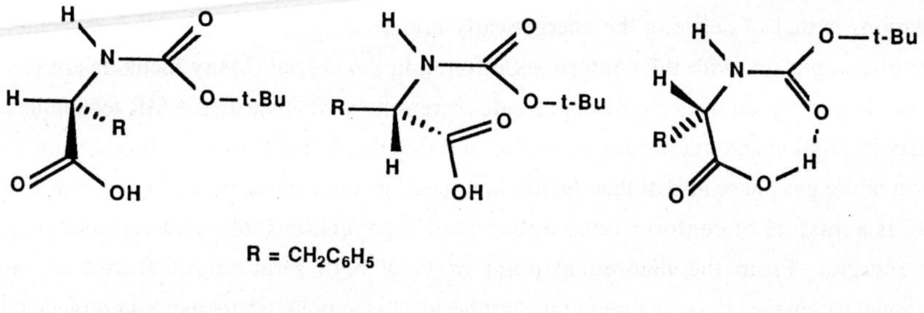

R = $CH_2C_6H_5$

8a **8b** **8c**

of which are E-conformers about the amidic bond. Upon dissolution and standing for a few hours at 50°C, the molecule undergoes almost a complete transformation to the Z-conformer **8c.** In the solid there are four intermolecular hydrogen bonds, while in solution there are two intermolecular and one intramolecular hydrogen bonds. While this competition between intramolecular and intermolecular interactions does become increasingly important with more polar molecules, or those that tend to form hydrogen bonds, the authors have shown with other examples that interactions other the competition between intra- and intermolecular hydrogen bonds play a role in this transformation[55].

The rotation of a phenyl group about an exocyclic bond has provided a large number of examples where higher energy conformations have been observed. Biphenyl **9** is of course the classic example, on which there has been a great deal of work, much of it summarized recently by Almenningen and coworkers[56]. The beauty of biphenyl and the reason for its choice as a model compound is that there is only one torsional parameter about an exocyclic bond, and the shape of the molecule changes dramatically with this parameter. Briefly, in the gas phase the torsion angle is 44.4°, while in the solid almost all experimental data are consistent with a model for a planar molecule which has large amplitude vibrations about the interring axis above 40°K, while below this temperature a non-planar molecule with torsion angle of ~10° is apparently frozen out. All the calculations[57] indicate that the non-planar conformation is energetically preferred by about 1.5 kcal-mol^{-1}, in agreement with almost all of the experimental results. The origin of the stabilization of the planar, or nearly planar conformation of the molecule in the solid state, the relationship between the mode of substitution, crystal packing and molecular conformation are still a subject of very active research[58].

The phenyl rings may be separated by a number of "spacers", thus increasing the number of torsional parameters. So, for instance, diphenylmethane **10**, with one spacer exhibits in the solid a helical structure that is calculated by a number of methods to be slightly higher in energy than a number of other possible geometries[59]. With two spacer atoms the number of torsional degrees of freedom may be still limited to two by inclusion of a double bond as in the three Schiff-base diazastilbenes **11**. Thus **11a** and **11b** are significantly more planar and consequently more energetic than **11c** in which there is an average rotation about the N-phenyl bond of ~50° and about the CH-phenyl bond of ~8° [60].

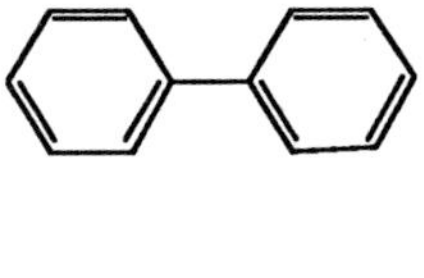

CH_2

9 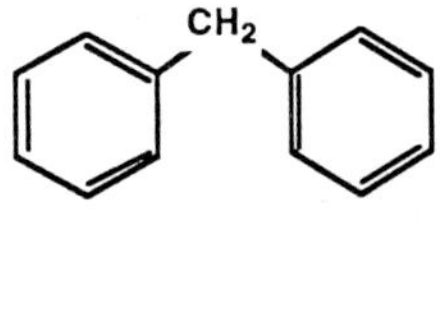 **10**

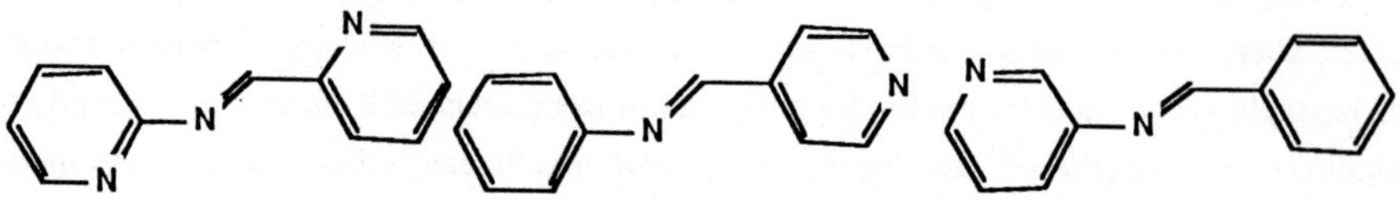

11a **11b** **11c**

A case with three spacer atoms between phenyl rings and four torsional degrees of freedom is represented by **12** which is dimorphic (monoclinic and orthorhombic)[61]. Stereoviews of the molecules in Figure 3 demonstrate the significant differences in conformation. This case is a good example of conformational polymorphism and provides a natural bridge to a more detailed discussion of that phenomenon in the next section.

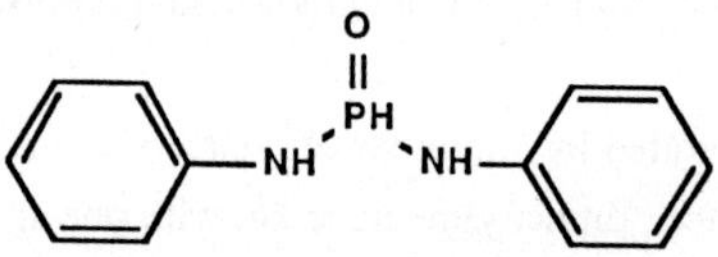

12

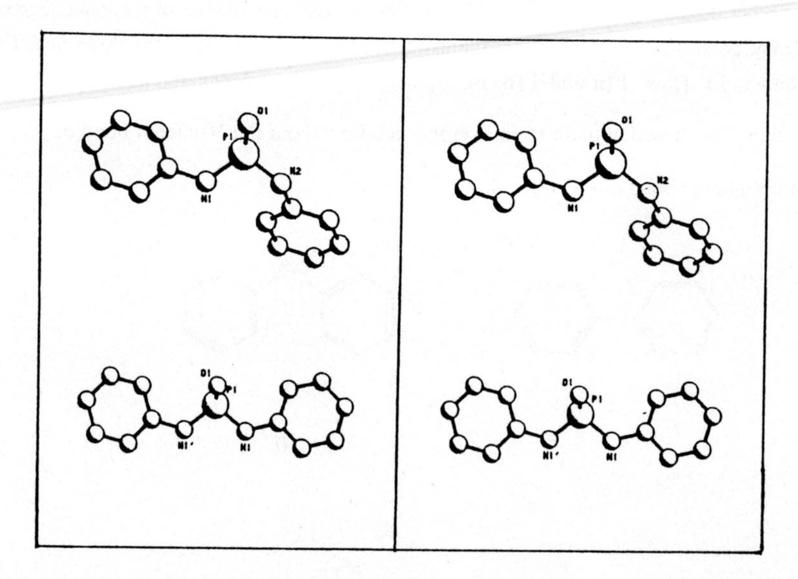

Figure 3. Stereoviews of the monoclinic (upper) and orthorhombic (lower) forms of **12**. In both cases the reference plane of view is N-P-N.

2.3 Conformational Polymorphism - Examples

In this section we cite a number of examples to demonstrate the rather wide variety of occurrences of conformational polymorphism. As the existence of the phenomenon has become increasingly recognized, an increasing number of investigators are carrying out the crystal structures of polymorphic materials in the search for variations in molecular geometry. Very often, the existence of conformational polymorphism can be predicted with a good deal of confidence from the results of physical methods other than full structure analysis. This may be important, for instance, when the existence of polymorphism has been established, but crystals suitable for single crystal studies on one or more of the crystal forms are not available. In addition to illustrating the variety of molecules which are already known to exhibit conformational polymorphism, this section will serve to illustrate many of the experimental and theoretical considerations treating such systems.

Polymorphism itself is readily and routinely detected by calorimetric methods[62], microscope hot stage methods[63], as well as, of course, by X-ray powder diffraction. Other physical methods such as IR, Raman and NQR and NMR spectroscopy and others are often employed, although not to the same extent as the former ones. Thus, for instance, on the basis of IR and Raman measurements conformational polymorphism has been predicted for N-propyl acetate[64], 1-bromopentane[65], 2-(4-morpholinothio)benzothiazole[66] and dibenzo-24-crown-8[67].

In a recent development related to this particular problem a number of groups[6,68,69] have been refining methods to extract information on internal molecular motions by careful analysis and evaluation of the anisotropic atomic vibration parameters which are obtained almost routinely in the course of a structure determination. In some cases these analyses have yielded different force constants for chemically related, but crystallographically unrelated groups on the same molecule. These results provide additional evidence for the influence of the environment on the molecule and potentially provide a rather sensitive tool for extracting details of the nature of these environmental influences, since in some cases it may be easier to extract information about differences in force constants than it will be to determine with confidence small differences in molecular geometry by conventional structural methods.

To cite a specific example it was possible to predict *a priori* the existence of conformational polymorphism in trimorphic iminodiacetic acid **13** on the basis of the IR spectra[70] the details of which were determined from the full crystal structures[71]. Of the four independent molecules in the three structures two are nearly identical; however, there are torsional differences of up to 30° about the C-N bonds. In two cases the hydroxyl oxygen is *trans* to the nitrogen, while in two cases it is eclipsed.

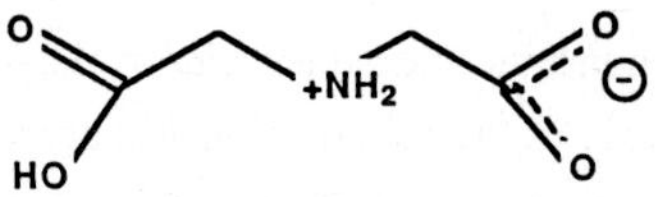

13

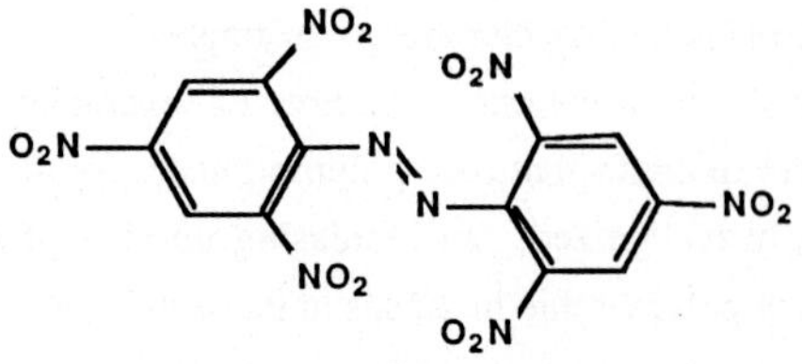

14

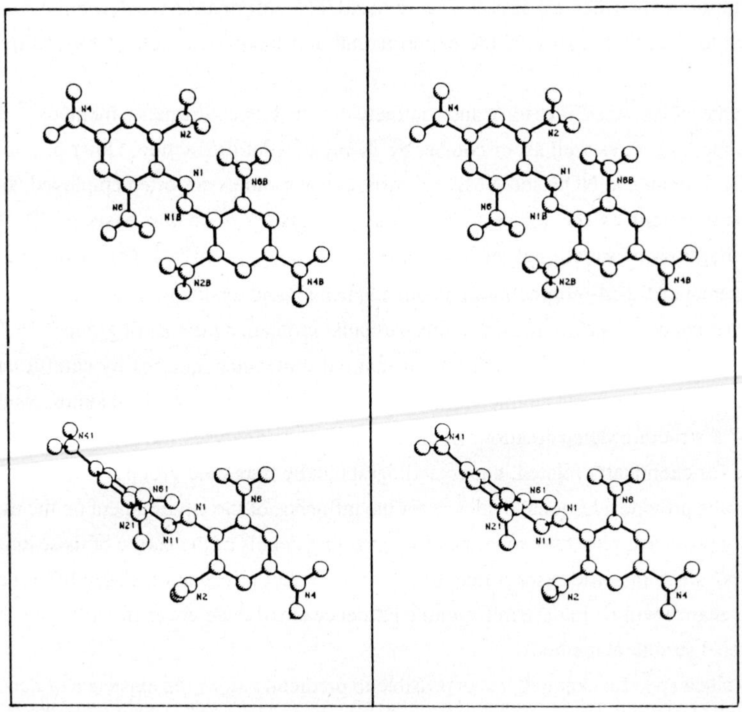

Figure 4. Stereoviews of the two forms of **14**, In both cases the molecule is plotted on the plane of the phenyl ring on the right hand side of the molecule.

When suitable crystals cannot be obtained for the full crystal structure analysis of all forms, crystallographic symmetry may also be useful in predicting the possibility of conformational polymorphism. Thus, in one of the forms of 2,2',4,4',6,6'-hexanitroazobenzene **14** the molecule lies on a crystallographic inversion center, requiring the two phenyl rings to be in parallel planes (Figure 4, upper), while in the second form the molecule lies on a general position. In this case it was possible to carry out the full structure determinations on the two forms[72] which verified that the lifting of the crystallographic restriction on the molecular symmetry in the second polymorph leads to a conformation in which there is an angle of 82° between the two rings (Figure 4, lower).

In the antiviral agent virazole **15**[73] and the biologically important nucleotide adenosine-5'-mono-phosphate **16**[74] there are significant differences in the torsion angles about exocyclic single bonds which can be easily seen in Figures 5 and 6. The latter illustrates a case in which the difference between computed and observed conformations provided a benchmark for the evaluation of the computational method. Platt and Robson[75] who studied this system concentrated on the most important torsion angle namely O-CH_2-CH-CH between the phosphate group and the sugar ring.They found that the *gg* conformation[76] observed in the monoclinic structure is 18.9 kcal-mol^{-1} higher than the minimum energy *tg* conformation, and suggested that this very large difference indicates a weakness of the computational method (orbital-centered force field) employed in the calculations. The large discrepancy may be partially overcome by including a dielectric term to account for the fact that a charged species is being studied. Such a correction reduces the difference to 4 kcal-mol^{-1} still in favor of the *tg* conformation. The torsion angle χ about the base-sugar bond differs by 46.8° and the sugar pucker is also different in the two forms, being C(2)'-*endo* in the orthorhombic form and C(3)' *endo* in the monoclinic form.

15 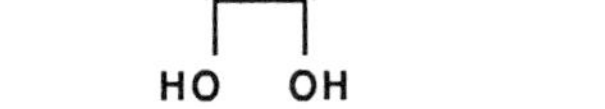 **16** 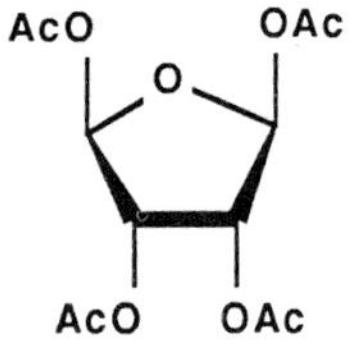 **17**

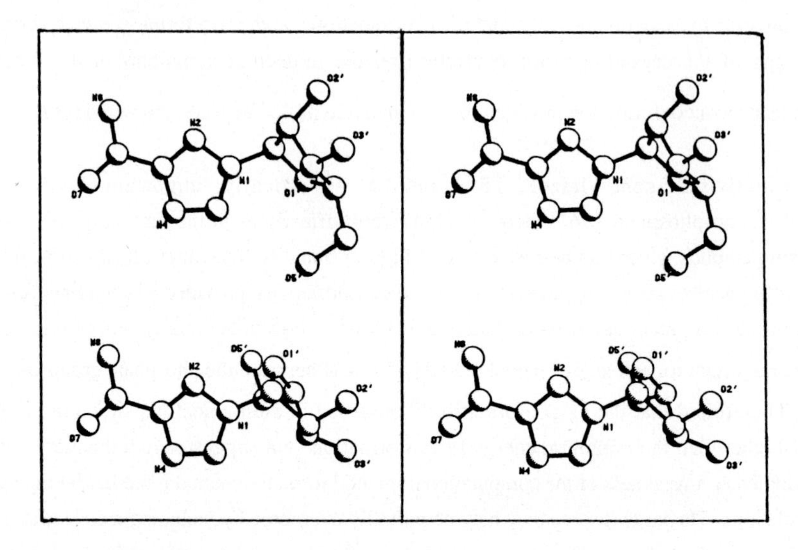

Figure 5. Stereoviews of the two forms of virazole **15**. In both cases the view is on the best plane of the triazole ring. For clarity, the carbons have been left unlabelled.

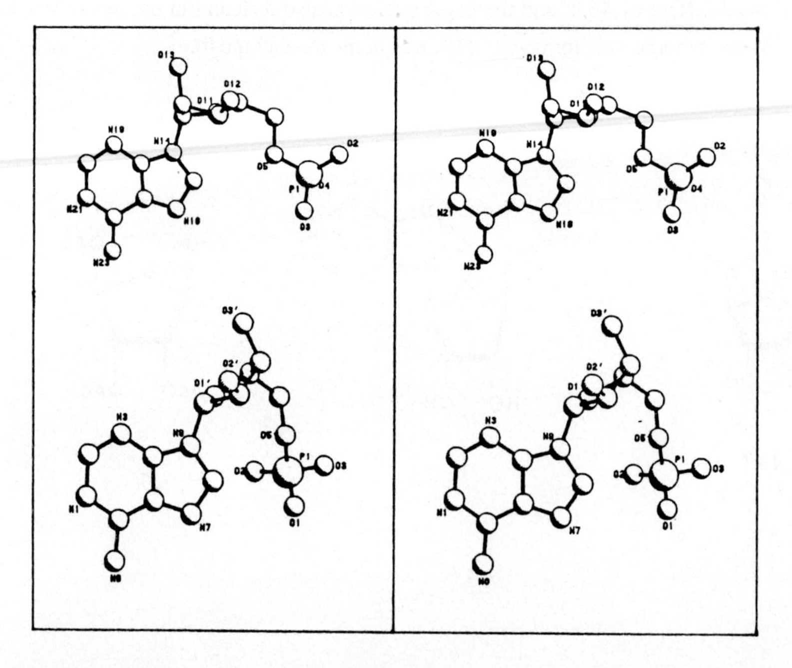

Figure 6. Stereoviews of the two forms of adenosine-5'-monophosphate **16**. Upper, monoclinic; lower, orthorhombic. In both cases the view is on the best plane of the six-membered ring of the base. For clarity carbons have been left unlabelled.

The conformational polymorphism and related crystal chemistry of dimorphic 1,2,3,5-tetra-*O*-acetyl-β–D-ribofuranose **17** are particularly interesting. For polymorphic materials, in general the density and melting point increase with stability[77]. Form A is the lower melting polymorph (329-331°K *vs* 358°K) but has a density per asymmetric unit which is higher than Form B by 1.5%[78]. Conformationally the puckering of the furanose rings is the same for both structures, but the torsional parameters for two of the four acetyl moieties differ between structures by values of 92.5° and 70° respectively. The conformations of the two molecules are shown in Figure 7. On the basis of calculations, the conformation in Form A is estimated to be ~3.9 kcal-mol^{-1} below that in Form B[78a].

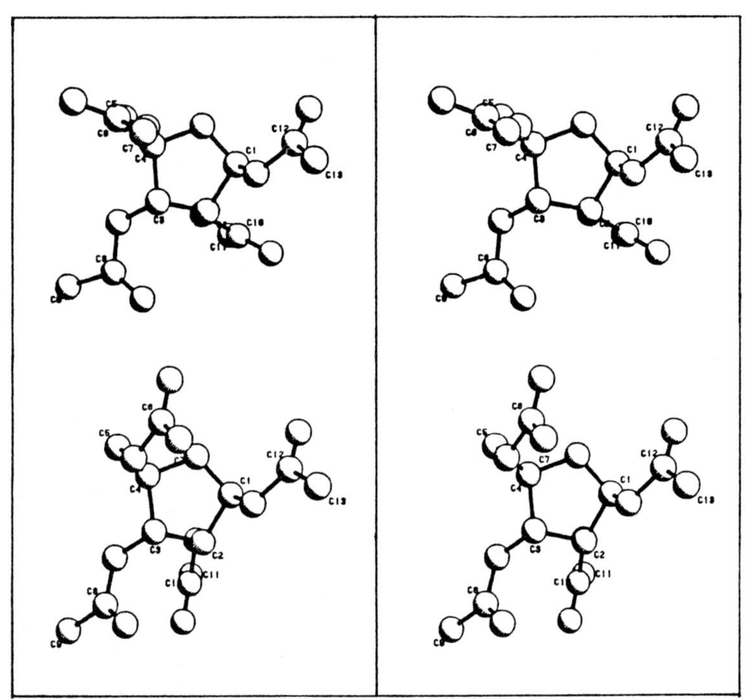

Figure 7. Stereoviews of the two forms of **17**. In both cases the view is on the best plane of C1-O-C4 of the furnose ring. Upper, monoclinic A form; lower, orthorhombic B form. For clarity only carbon atoms are labelled.

Another point of interest regarding polymorphism arose with respect to this molecule. Form A could not be maintained in a laboratory which had been contaminated with Form B. A number of similar cases were cited by Woodard and McCrone as examples of a general phenomenon of the 'disappearing crystal form'[79], but recent examples suggest that with proper care, seemingly 'lost', less stable polymorphic forms may in fact be 'recovered'[80-81].

Medium or large ring compounds often have a great deal of conformational flexibility. The geometric and energetic principles governing the variety of conformations possible have been given by Dunitz[82]. A very important point, demonstrated graphically in these papers, is that there is a coupling of bond angle deformations with torsion angles. For instance, for cyclobutane with D_{2d} symmetry (identical bond lengths and bond angles) the symmetry constraints are such that a change in bond angle of 1° (i.e. only slightly greater than the precision of most current routine crystal structure analyses) results in a total variation of torsion angle of 30° (±15°). Clearly the reverse is equally true so that even if bond angles do vary among conformational polymorphs, the variations may be below the precision of the crystallographic experiment.

One example of conformational polymorphism comes from the literature on energetically sensitive materials (explosives), which as noted above, is the source of many examples of the polymorphic behavior of solid organic materials. Octahydo-1,3,5,7-tetranitro-1,3,5,7-tetraazocine (HMX) **18** is known to crystallize in four polymorphs[83], of which the structure is known for

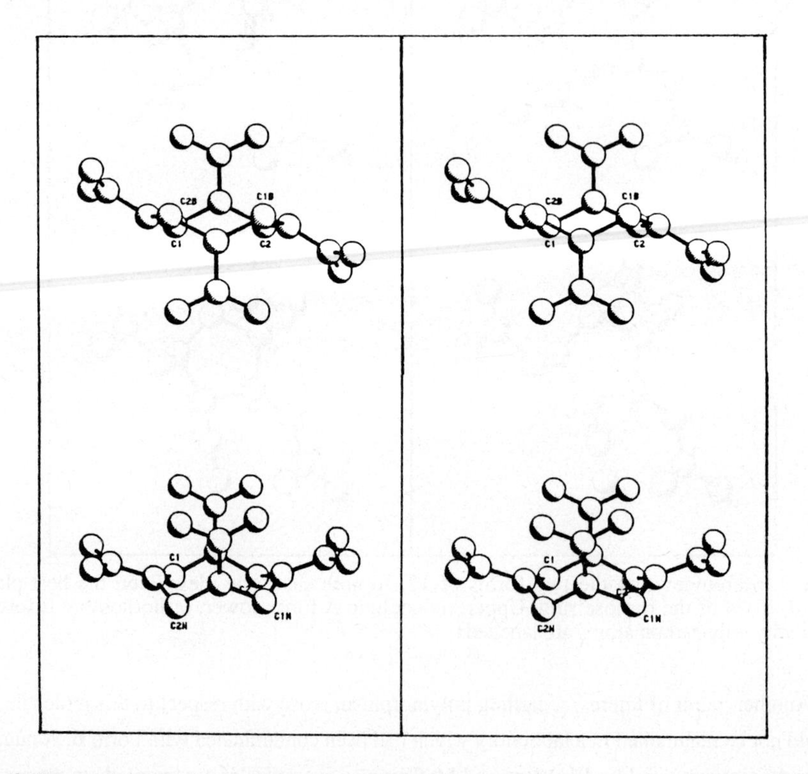

Figure 8. Stereoviews of two of the forms of HMX **18**. In both cases the view is on the plane C-N-C of the ring. Top, β-form; bottom, α-form. For clarity only carbon atoms have been labelled.

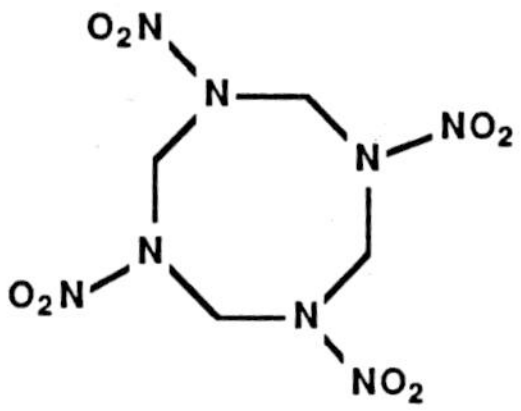

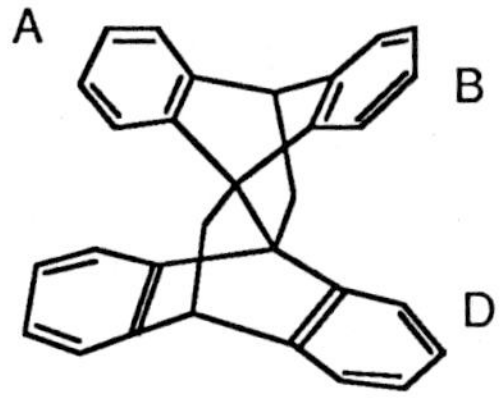

18 **19**

three[84]. In the β-form the ring conformation is chair (Fig. 8) which has the minimal intramolecular repulsions between NO_2 groups (four O...O distances of 3.10Å) and maximal C...N 1-3 interactions. The less stable α and δ polymorphs exhibit chair-chair conformations (Fig. 8), and all eight O...O distances are in the range 3.0-3.2Å. Apparently there is little bond strain involved in these changes since the IR spectra for all three polymorphs are similar. The intermolecular interactions have been analyzed on a semiquantitative basis using a coulombic model and do indicate that β is the most stable form, followed by α and γ. A more quantitative analysis of the conformation and lattice energetics would certainly be of interest here.

Even for cyclic systems which appear to be quite rigid there are examples of geometric differences between polymorphs. Both the stable triclinic[85] and the metastable orthorhombic[86] forms of lepidopterene **19** have been studied, the former with two half molecules per asymmetric unit and the latter with three full molecules per asymmetric unit. Here the density of the metastable orthorhombic phase (1.28 $g\text{-}cm^{-3}$) is slightly higher than that of the stable triclinic phase (1.26 $g\text{-}cm^{-3}$) which translates to a difference of 8$Å^3$ per molecule. As measured by the angle between the planes of rings A and B or that between C and D the molecules in the orthorhombic phase are more splayed than in the triclinic structure. The increase in density may be responsible for these differences in geometry, the latter in turn being responsible for differences in the excimer spectra polymorphs. Here the conformational polymorphism is utlized to probe the consequences of a molecular structural perturbation on the electronic properties, and for flexible molecules polymorphic systems such as this provide ideal candidates for this type of study[87].

Returning to smaller molecules, there are quite a few which exhibit conformational polymorphism and are accompanied by some unusual crystal chemistry. For instance, L-glutamic acid **20** is known to crystallize in two forms, α and β, the former having preferable properties for industrial applications[88]. Slow crystallization from solution yields the β form as does leaving the α form in solution for an extended period of time. It was found empirically that the industrially preferred α form may be obtained by addition of trace amounts of other amino acids or other compounds to the crystallization solution[89]. The study of such procedures in a systematic way has since been transformed into a very active field of research, and various polymorphs and/or crystal habits may be 'tailor made' by the appropriate addition of impurities which are *predesigned* on the basis of the crystal and molecular structures of the 'host' and the added impurity[90].

The conformations of the molecules found in the two crystal structures[91,92] are significantly different for all three of the torsion angles about the C-C bonds as shown in Figure 9. Hirayama and coworkers[92] have suggested a mechanism for the α→β transition. They have also shown that L-glutamic acid is a very conformationally flexible molecule, with at least six different conformations having been observed in various states of complexation.

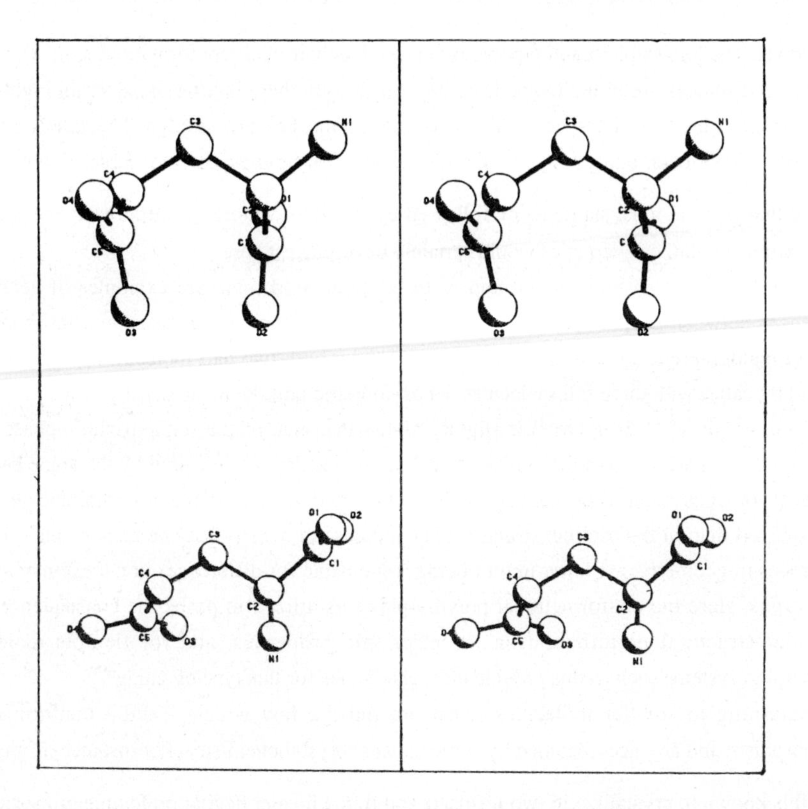

Figure 9. Stereoviews of the two forms of L-glutamic acid **20**. In both cases the view is on the plane of C2-C3-C4, which particularly highlights the conformational differences about C2-C3 and C3-C4. Upper, α form; lower, β form.

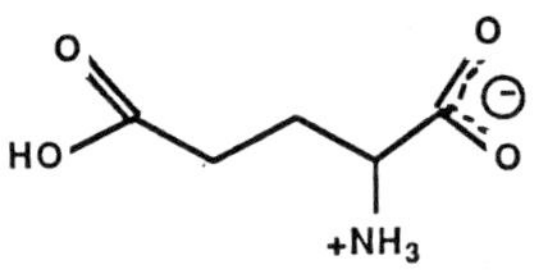

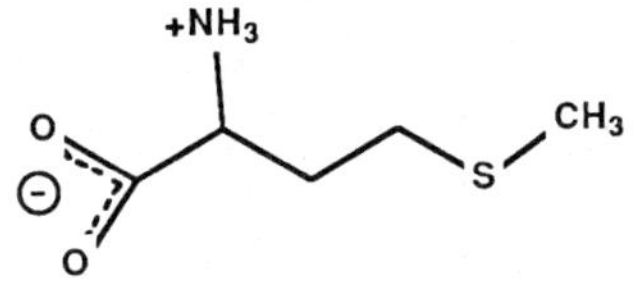

20 21

DL-methionine **21** is known to crystallize in two forms, for which the structures were originally determined from two-dimensional data[93]. A more recent refinement[94] employed three-dimensional data. The major difference between the conformations is the torsion angle about the biologically important C(4)-S bond, with values of 68.9° and 185.6° for the α and β forms respectively, such that the terminal methyl groups are *gauche* in the former and *trans* in the latter. This material apparently undergoes a topotactic thermal transition[95] from the β (*I*2/*a*) to the α ($P2_1/a$) form at 333°K, accompanied by a halving of the *c* axis. Both structures consist of similar layer units which are related by translation along the *c*-axis in the α form and by (a + b)/2 in the β form. In view of the general similarity between the two structures this is a case in which the reason for the large difference in this single torsional parameter may be more readily isolable than for more complex molecules and structures.

Another molecule of similar size is tetramethyl-β-oxoglutaric acid **22** which serves as a model compound for polydimethylketene with ketonic enhancement (Fig. 10)[96a]. In the triclinic form[96b] the torsion angles α and β are 66° and 205° respectively, while in the monoclinic form[96c] they are 32° and 277° a result which was predicted on the basis of conformational analysis.

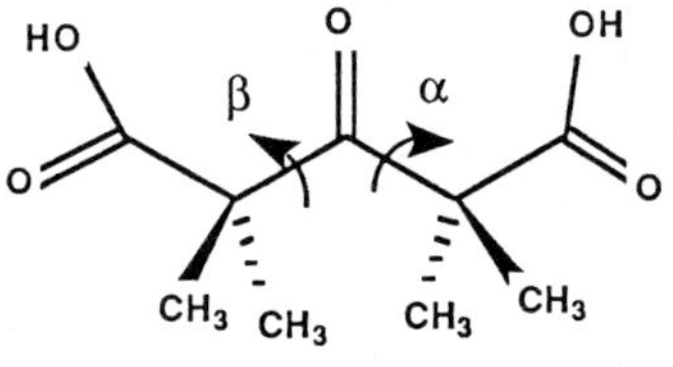

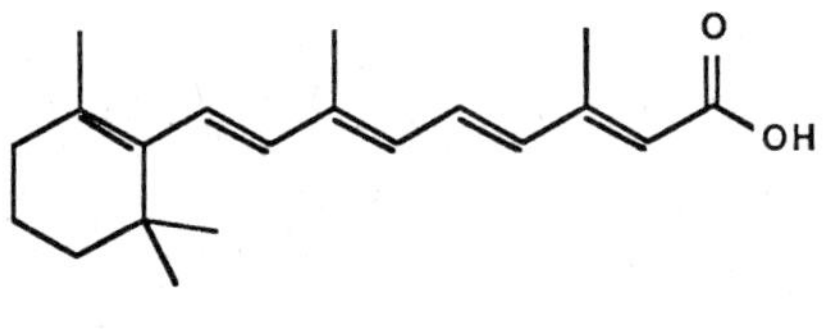

22 23

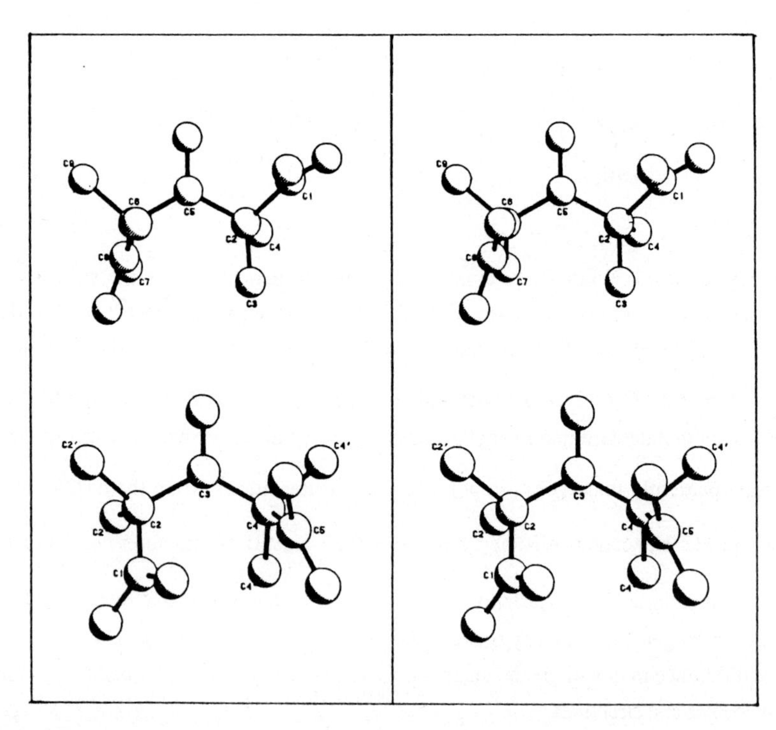

Figure 10. Stereoviews of the two forms of tetramethyl-β-oxoglutaric acid **22**. Upper, triclinic form; lower, monclinic form. In both cases the view is on the plane of the carbonyl carbon and its two carbon neighbors (numbered differently in the two structures).

A rather striking example has been reported for longer chain compounds, the triclinic and monoclinic modifications of vitamin A acid **23**, a member of the family of visual pigments. The cyclohexenyl ring has essentially the same conformation in both forms[97,98]. However, as seen in Figure 11 the triclinic form exhibits an s-*cis* conformation about the exocyclic bond and clearly a non-planar molecular conformation. The monoclinic form is s-*trans* about the same bond and the stereoview reveals a much more planar structure.

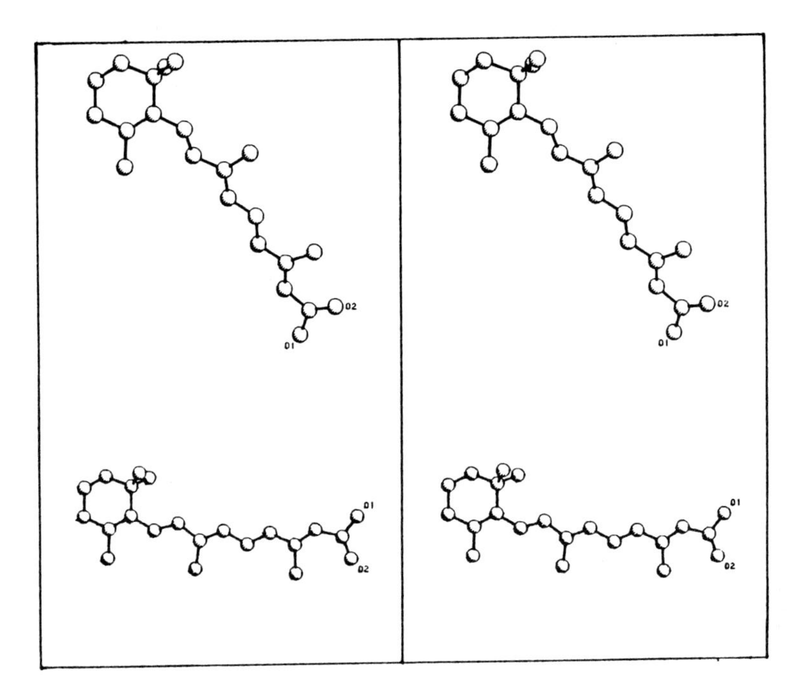

Figure 11. Stereoviews of the two forms of vitamin A acid **23**. Upper, triclinic form; lower, monoclinic form. In both cases the view is on plane of the three lower atoms of the cyclohexenyl ring.

In general the very extensively studied family of steroids[99] have shown a remarkable degree of conformational consistency[100]. Quite a few of them do exhibit polymorphism however[13], and as has been pointed out elsewhere[100], any variety observed in the conformations of these molecules can provide useful information on the range of conformations accessible and even on the mode of physiological action of a particular molecule or family of molecules. In view of the importance of these molecules, the description of a few examples of conformational polymorphism is certainly in order.

24 **25**

Cortisone acetate **24** is known to crystallize in at least three unsolvated forms[101] for which the structures of two, Form I[101] and Form II[102] have been reported. In both structures ring A is a

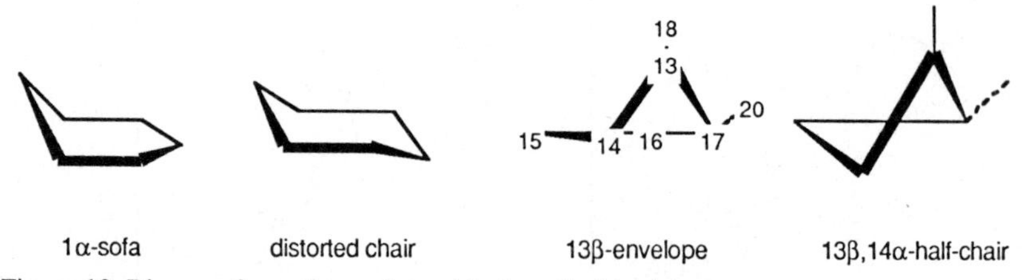

Figure 12. Ring conformations of steroids described in the text.

distorted 1α-sofa conformation while rings B and C are distorted chair . Differences arise in ring D, however; in Form I it exhibits a 13β-envelope and in Form II a 13β,14α-half chair conformation. The conformations of the molecules in both structures are compared in Figure 13. This figure deserves some comment since we believe that it is a particularly vivid method of comparing molecular conformations and should be employed with greater frequency. The atoms of the A,B, and C rings of the second structure were fitted to those of the first by a least squares procedure described by Nyburg[103]. The program (BMFIT) then produces coordinates for all remaining atoms in the framework of the first and these may be input into a stereo plotting program[104]. Thus it is possible to easily see the differences in the conformations of the side chains

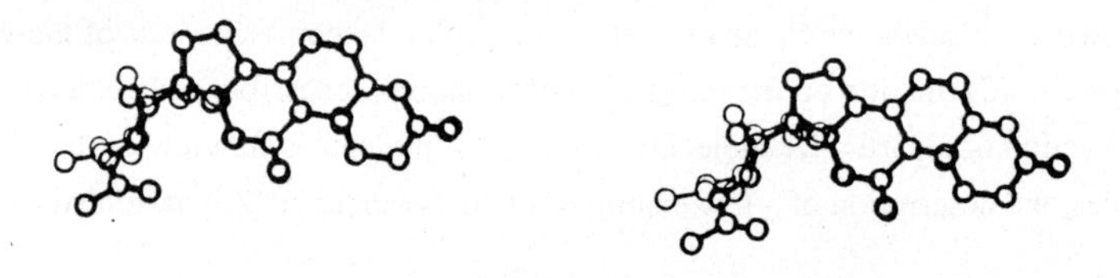

Figure 13. Stereoview of two superimposed molecules of cortisone acetate found in two different crystal structures. The atoms of the A,B, and C rings have been fitted to each other to emphasize the conformational differences in other regions of the molecule (see text).

of the two molecules. The major differences are about C(17)-C(20) and C(20-C(21) amount to 16° and 14° respectively, but the figure clearly represents the difference more dramatically than the numbers. Differences of a similar magnitude are observed in 6-bromo-testosterone acetate[105] and are manifested in different IR spectra.

Conformational polymorphism is manifested in other regions of the steroid framework as well. 19-nortestosterone **25** crystallizes in two monoclinic forms[106,107], both with two molecules in the asymmetric unit. In the structure by Bhadbhade and Venkatesan the A-ring is approximately 1α,2β-half chair for both molecules; in the form studied by Precigoux *et al* one of the two molecules is disordered, taking on either the 1α,2β-half chair or the 1β,2α-half chair conformation, while the A-ring in the second molecule adopts a conformation between 1α,2β-half chair and 1α-sofa.

The very rich structural chemistry of organometallic compounds also provides a number of interesting examples of conformational polymorphism. For instance, cyclic ligands have low barriers to rotation so that conformational changes might not be unexpected between polymorphs. Riley and Davis[108] have reported the structures of the dimorphic sandwich compound bis(2,6-dimethylpyridine)chromium (Figure 14), another case in which the molecular geometry is dictated by the crystallographic site symmetry. In the triclinic form the molecular site symmetry is C_i, requiring the 'trans' conformation, while in the orthorhombic form, the molecule lies on a 2-fold axis (site symmetry C_2), leading to the 'gauche' conformation.

A classic case has been presented by Foxman and coworkers[109] on the nickel complex **26**. This material is trimorphic, and in all three forms the expected square planar geometry for d^8 Ni(II) is found. The variety in bonding and packing is exhibited in an interesting way: Form II (for which only a single crystal has been prepared) and Form III have distinct packing arrangements, while Form I represents a combination of the previous two. Thus in a sense this trimorphic system is the structural representation of three points on the molecular energy-crystal energy hypersurface, the point represented by Form I lying on the 'packing coordinate`connecting the points represented by Forms II and III, although experimentally the three forms are not interconvertible.

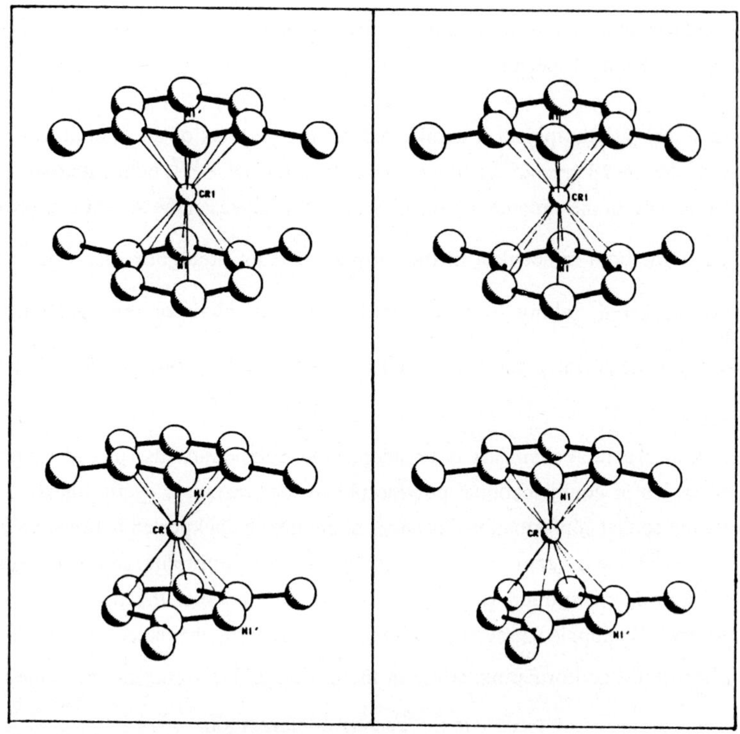

Figure 14. Stereoviews of the two conformers of bis (2,6-dimethylpyridine) chromium. Upper, triclinic; lower, orthorhombic. Carbon labels have been deleted for clarity.

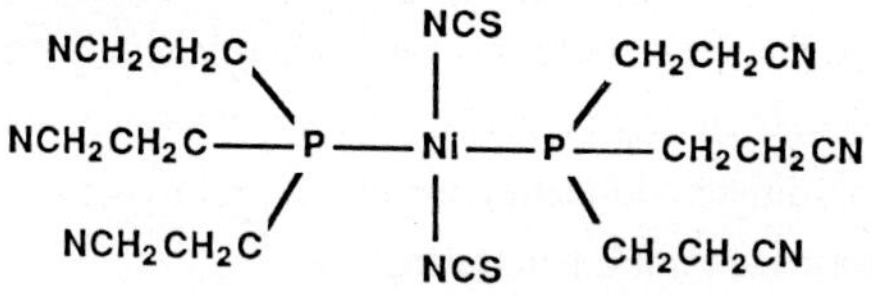

26 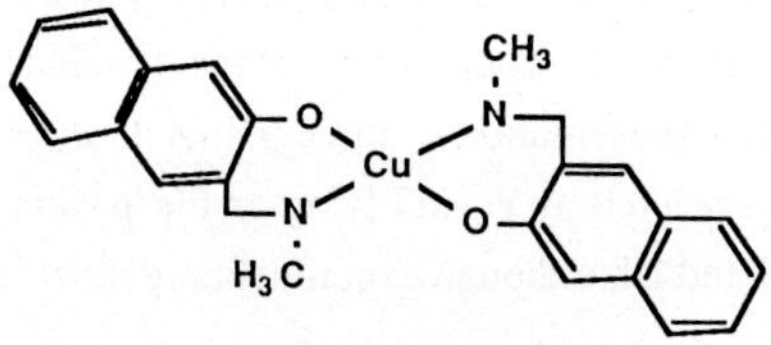 **27**

Form I, which exhibits disorder from which the characteristics of both Forms II and III may be extracted, has a 'pseudooctahedral' configuration, the additional two pseudo ligands coming from an intramolecular nitrile contact and an intermolecular sulfur contact. Form II has two of the former type of contacts to yield the pseudooctahedral coordination while Form III is pseudooctahedral by virtue of intramolecular Ni...S contacts. The four different conformations found in the three crystal structures (two molecules in the asymmetric unit in Form II) are shown in Fig.15.

The packing modes also differ significantly, a van der Waals solid being found in Form II, while infinite helical structures are present in Forms I and III. The helices are totally independent in Form I while they are cross-linked in Form III.

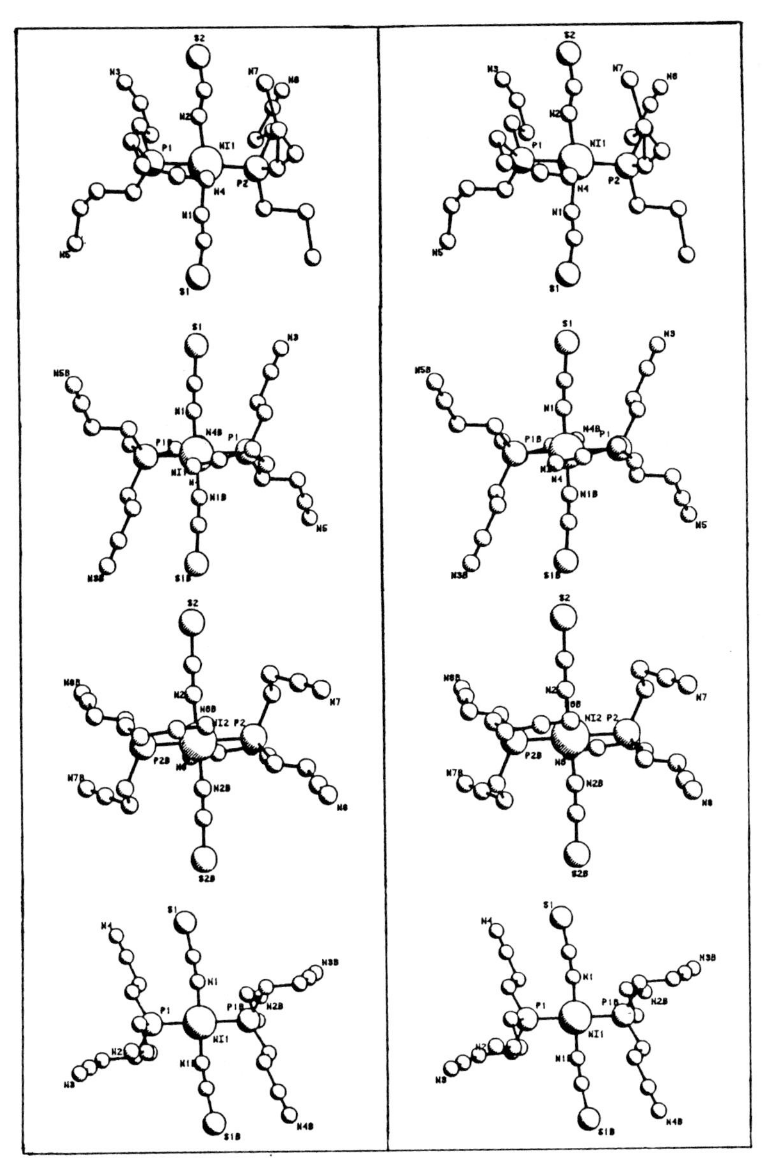

Figure 15. Stereoviews of the four different conformations of the nickel complex 26. In each case the view is on the best plane of the metal and the four atoms directly bonded to it. Carbon atom labels have been deleted for clarity.

In the first of these cases the coordination sphere of the metal is constant for the polymorphs, while in the second clearly there is a tendency to fill the available axial octahedral sites, although formally, at least there is no change in the coordination sphere among the nickel polymorphs. There are a number of examples where this is not the case, and while perhaps bending a bit the strict definition of conformational polymorphism, they do illustrate the balance between inter- and intramolecular forces. Thus, in some cases true polymorphism reveals differences in the coordination sphere, including the number of ligands involved. One of these is bis(N-methyl-2-hydroxy-1-naphthalidiminato) copper (II) **27,** known to crystallize in four forms[110,111], two of which are green and two of which are brown. In the brown γ form the Cu atom is 0.5Å from the plane of the naphthalene part of the ligand and there are dimers leading to a 5-coordinate (4+1 tetragonal pyramid) arrangement of ligands about the metal with an oxygen from the centrosymmetric dimer pair providing the fifth coordination site. In a second brown form (β)[112] the corresponding distance from the plane is 0.63Å and the metal is 6-coordinate (4+2, pseudooctahedral). One of the two green forms (δ)[113] exhibits a 4-coordinate square planar geometry with the Cu being only 0.09Å from the ligand plane, The authors prefer to attribute these variations not to crystal forces[110], but rather to kinetic effects in the crystallization process. At any rate, if there are differences in the molecular energies, something must act to stabilize the more highly energetic geometries, and our contention here is that the crystalline environment plays a significant role in this stabilization.

3. STRATEGIES AND METHODS FOR TREATING CONFORMATIONAL POLYMORPHS

Four strategies have been suggested for investigating the general problem of the influence of crystal forces on molecular conformation[34]:

1. Comparison of compounds in gaseous and crystalline states.
2. Comparison of the geometries of crystallographically independent molecules in the same crystal.
3. Analysis of the structure of a molecule whose symmetry in a crystal is lower than that of the free molelcule.
4. Comparison of molecules in different polymorphic modifications.

The advantages and disadvantages of these various approaches have been considered elsewhere[114], and will not be reviewed again here but to note that with the available computational and experimental techniques the strategy based on the utlization of conformational polymorphs appears to be the most viable one at this stage.This is very nicely demonstrated by a simple example (Fig.16) taken from a paper by Karle and Karle[115]. The figure shows five prolines, four natural and one synthetic, all exhibiting different conformations. The authors note

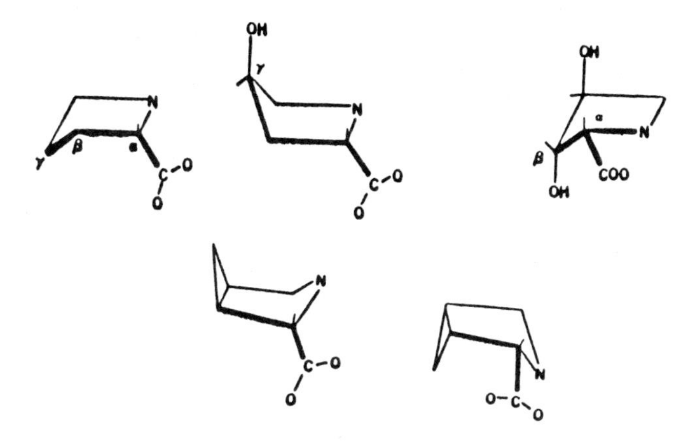

Figure 16. Examples of conformations observed in crystal structures of prolines (see ref. 115).

that even this figure does not represent the full gamet of possibilities for the conformations of a prolyl group in peptides and protein chains. It is clear that a wide variety of conformations are possible, and these vary not only with the nature of substitution, but also with the immediate environment of the proline group in any given chemical or structural situation. From the structural data on any combination of prolyl groups here, it would indeed be difficult to extract information about the environmental influences on the molecular conformation or separate it from the influence of the substituent. When utilizing conformational polymorphs to study this problem, the *only* variable is the crystal lattice and the desired information is, in principle, directly accessible, since the role of the substituent is eliminated by the very choice of a polymorphic system. In addition to the general extent of polymorphism noted above, utilization of the phenomenon has the distinct advantage over other approaches since the comparison of crystal structures and molecular conformations in different polymorphs of the same material is the most direct method for investigating the influence of crystal forces on molecular conformation.

In studying a system exhibiting conformational polymorphism, one is seeking the answers to two basic questions:

1. What are the differences in energy, if any, in the molecular conformations observed in the various crystal forms?

2. How does the energetic environment of the molecule vary from one crystal form to another?

To answer these questions, a typical study would proceed according to the following steps:

1. Determination of the existence of polymorphism in the system under study.

2. Determination of the existence of conformational polymorphism by the appropriate physical measurements.

3. Determination of the crystal structures to obtain the geometrical information - molecular geometries and packing motifs - of the various polymorphs.

4. Determination of differences in molecular energetics, by appropriate computational techniques.

5. Determination of differences in lattice energy and the energetic environment of the molecule by appropriate computational methods.

We proceed with an outline of each of these steps to indicate briefly the kinds of tools and techniques which can be applied.

The principle techniques for detecting polymorphism were given above. These are now routine analytical methods, and the application of a combination of them usually leaves little doubt as to the presence or absence of polymorphism. Nevertheless, the accumulation of evidence for conformational polymorphism in the absence of full crystal structure analyses is still fraught with some uncertainty. For many years IR spectroscopy has been the classic tool for determining the possibility of conformational differences in solids, and we have already cited some examples where conformational differences suspected on the basis of IR spectra were borne out by the full crystal structure analyses. Recent developments in NMR spectroscopy, particularly the crossed polarization magic angle spinning NMR technique indicate that this is a very powerful and sensitive probe of solid state molecular conformations and environments[116].

The identification and characterization of the polymorphs and conformational polymorphism is naturally followed by the determination of full crystal structure analyses, in order to obtain the geometric information - molelcular geometry and packing motif - required for the following steps. While the crystal structure analyses are generally routine, obtaining single crystals of suitable quality for all the polymorphs may be a serious problem, since many are often metastable. In fact, many of the polymorphic forms cited in the literature were discovered serendipiously, and it is to the credit of the investigators involved that they pursued the characterization of these newly found crystal forms. An increased awareness of the existence of polymorphism on the part of all those involved with handling solid materials would certainly contribute to the growing body of information on polymorphic materials.

On the presumption that the information regarding the existence of conformational polymorphism was correct, (i.e. that different conformations exist in the different crystal structures) the question then arises as to the magnitude of the differences in energy among the different molecular conformations. Fortunately, the techniques for determining these energies have evolved over the past two decades at a rate commensurate with those of crystal structure analysis and there are now many tools available for the detailed investigation of the molecular energetics. Many of the basic principles and models for the calculation of molecular energies are also applicable to the calculation of lattice energetics and similar progress has been made along the same lines in recent years. Worthy of note is the fact that many of the programs developed and refined during this same period have now become standard library programs which run on mainframe computers or superminis, are readily transferable, easily obtained, and hence may be utilized by almost any worker in the field. Some of these programs treat conformational parameters explicitly, while for others the multidimensional conformational energy surface is sampled by systematically varying the parameters in question and computing the energy involved.

The latter approach applies in general to molecular orbital calculations and may include automatic optimization procedures for producing the minimal energy conformation(s). These procedures naturally significantly increase the amount of computer time required for a particular calculation. Molecular orbital methods, from the simple Huckel approximation to ab initio methods are quite familiar and are well documented elsewhere[117].

The force field calculations which originated from vibrational spectroscopy have now matured into valence force field and molecular mechanics methods[118].Traditionally they have included specific terms for perturbations in molecular geometry which makes them advantageous for interpreting the energetics of molecular conformation in terms of particular molecular deformations. For full valence force field and molecular mechanics calculations the molecular energy is defined in terms of the same geometric parameters discussed at the beginning of this chapter, and now generally include the full range of torsional rotations, as well as non-bonded interactions of the Lennard-Jones (2) or Buckingham (3) forms where the lettered constants are empirical parameters for any pair of atom. Fairly efficient procedures have been developed to mini-

$$V(r_{ij}) = \frac{B}{r^{n}_{ij}} - \frac{A}{r^{6}_{ij}} \qquad n = 9, 12 \tag{2}$$

$$V(r_{ij}) = B' exp\,(-C' r_{ij}) - \frac{A'}{r^{6}_{ij}} \tag{3}$$

mize computing time[118e]. The choice of analytical forms for the force field, the determination of force field constants and the transferability of force fields and their constants among different chemical systems are still subjects of considerable interest and research. However, certain functions and parameters have proven useful in a variety of applications[118d,e,119] and valence force fields and molecular mechanics are now recognized tools in the investigation of molecular conformations.

In principle the extension of this approach from molecules to crystals is straightforward. The basic assumption is that the intermolecular interactions may be treated as the sum of atom...atom interactions which, again, are approximated by either the Lennard-Jones or Buckingham potentials[118e,120,121]. Williams[122] has demonstrated the need for inclusion of a coulombic electrostatic term in the atom...atom potential, even for hydrocarbon crystals, where such a contribution may approach 30% of the total energy. The Lennard-Jones and Buckingham potentials are then modified by adding a term which includes $q_i q_j / r_{ij}$, q_i, q_j being the charges on individual atoms i, j at a distance r_{ij}. These potential functions assume isotropicity of the atom...atom interactions; however, there is increasing evidence for their anisotropic nature[7,27,123].

In the actual calculation of lattice energies, either of the forms of the potential functions is employed to compute the lattice energy for a reference molecule using, in the function, empirically derived parameters which are characteristic of each of the atoms involved[124]. As long as there is some difference of opinion on the proper choice of potential function and the parameters used with them, one rather critical test is to employ both forms to test the dependence of the results on the functions and the parameters. In cases where the electrostatic interaction is included by assigning point charges to each atom, they may be estimated from bond dipole moments[125], quadrapole moments[126], or charge densities obtained from molecular orbital calculations at various levels of approximation[127].

It has been found that in general for van der Waals crystals an interaction radius of 12-20Å is sufficient for convergence[119d].The sublimation energy (i.e. the crystal 'binding energy' of a single molecule) is then minimized by altering the position and orientation of the reference molecule, as well as the unit cell parameters. If the starting model is a known crystal structure, usually only small perturbations in the structure result from the minimization procedure, and any crystallographic symmetry is maintained, even if not accounted for explicitly in the computational procedure. The resulting sublimation energy may then be compared to the experimental value to test the result and the reliability of the potential function as well as the parameters employed for each atom in the calculation. More detailed information on the nature of the intermolecular interactions may be obtained by partitioning the total lattice energy into non-bonded and electrostatic contributions. The fact that the calculation is based on the use of an atom...atom potential allows an additional partitioning into individual atomic contributions. As demonstrated below, this latter method of partitioning is particularly useful in the investigation of the relationship between crystal forces and molecular conformation.

It is important to note here some distinguishing characteristics of the computational approach to conformational polymorphism.With regard to the molecular energetics considerable computing effort is saved by noting that for the purposes of these investigation our interest lies really in those conformations which are observed in the crystal structures studied. Hence, rather than attempt to computationally sample many regions on the multidimensional potential energy surface, it is sufficient to compute the energies only for those few points actually studied. Moreover, our immediate interest in these studies is the differences in energies between various observed conformations rather than the total molecular energy, or even whether any particular observed conformation is at the global minimum. Clearly, in many cases the size of the molecule studied taxes the available software or computing power, especially for, say, *ab initio* calculations. However, the conformational energetics of large molecules may be obtained from molecular mechanics or *ab initio* calculations on model compounds representing the conformational parameters in question[136]. Alternatively, semiempirical molecular orbital methods often give good estimates of energy differences among conformations, even if other properties may not always be approximated very well.

For the lattice energies a good test of the calculation is a comparison with the sublimation energy, which may be measured fairly readily when sufficient quantities of material are available[128] or estimated from the sublimation energies of analogous model compounds and group contributions[129]. The differences in lattice energy, which are of more interest here may be determined calorimetrically[62] for experimental verification of the computed quantities.

4. AN EXAMPLE - The benzylideneaniline system

For a number of years we have been studying the system of diparasubstituted benzylideneanilines **28**, and our work on this system demonstrates the application of the strategy outlined above as well as the kind of information which may be extracted from these studies.

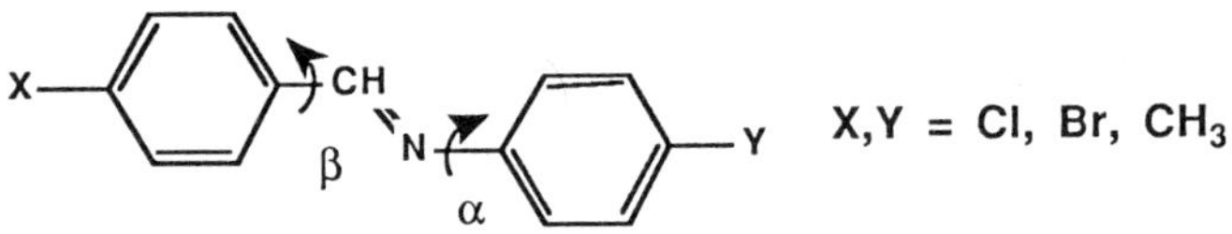

X, Y = Cl, Br, CH_3

28

The molecule has only two torsional degrees of freedom (α,β) about exocyclic bonds, which, as noted above, makes it an attractive candidate for conformational studies. The benzylideneanilines had aroused the interest of spectroscopists due to the marked difference between its solution UV absorption spectrum and that of its two isoelectronic analogues, *trans*-stilbene and *trans*-azobenzene[130]. There was general agreement that the latter two are essentially planar while the former is non-planar due to a repulsion between an *ortho* hydrogen on the anilino ring and that on the bridge carbon[131]. In the course of a survey of dichlorosubstituted aromatic compounds[132] it was noticed that the dichloro compound **28** X = Y = Cl crystallized in two different forms: nearly colorless triclinic needles and chunky pale yellow orthorhombic prisms.The difference in color suggested the possibility of different molecular conformations, an assumption which was subsequently borne out by both crystallographic and spectroscopic studies.[133-135].

Both forms exhibit disorder, he triclinic (planar: $\alpha = \beta = 0°$) form[133] about an inversion center and the orthorhombic (non planar: $\alpha = -\beta = 24.9°$) form[134] about a twofold axis. Hence the crystal structures reveal the existence of conformational polymorphism. The molecular energetics

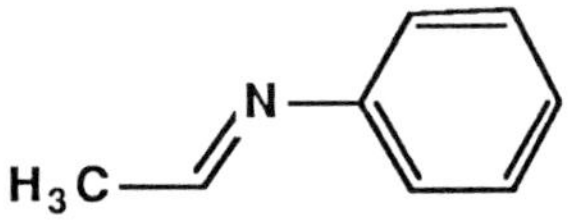

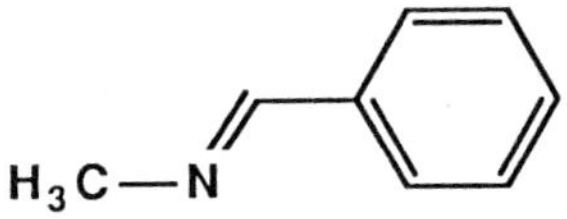

29 **30**

were computed in this instance using *ab initio* methods on model compounds **29** and **30** which have been shown to represent well the energetics of the full unsubstituted molecule **28** (X = Y = H)[136]. The calculations indicate that the minimum energy conformation is favored over the planar structure by ca 1.0-1.5 kcal-mol^{-1} and corresponds to a rotation about the N-phenyl bond (α) of *ca* 45° and virtually no rotation about the CH-phenyl bond (β). Note that this energy is also in the range of differences in energies between polymorphic crystal forms. Neither of the two conformations observed in this polymorphic system is the lowest energy conformation and that found in the orthorhombic structure is favored over the one found in the triclinic structure by about 0.5 kcal-mol^{-1}. The results from these molecular orbital studies are in accord with those from spectroscopic and the X-ray diffraction studies and in this case represent a significant improvement over the results obtained from semiempirical methods which in general have not been able to account properly for rotations about the exocyclic bonds[130].

Having established that there are differences in molecular energies corresponding to the differences in conformation, the next step is to determine the lattice energies of the various forms. In this case calculations were carried out employing three different potential functions, in part to test whether the results were indeed independent of the functions employed. Two Lennard-Jones (6-9 and 6-12) and a Buckingham potential were employed; in addition, the parameters for the chlorine potential were varied by ±10% to test the dependence of the results on this crucial atom[27].The absolute values of the total energies vary and, indeed, there is some variation in the magnitudes of the differences between crystal forms, but all trends are observed for all potential functions and all parameters. This has been our reassuring experience with virtually every system with which we have worked to date; however these potentials and parameters must be constantly tested in this manner to discover the limits of their applicability, and indeed cases of conformational polymorphism provide excellent testing grounds for these procedures and programs.

Returning to the case at hand, for all calculations the lattice energy calculations all favor the triclinic structure over the orthorhombic one, and by an energy which is compatible with the expected energy differences between polymorphs. Indeed, the triclinic form must have a lower lattice energy than the orthrhombic one in order to stabilize the more highly energetic planar conformation found in the former.

To this point we have dealt with the differences in total energies. Ultimately, we would like to be able to say something about the differences in specific intermolecular interactions which bring about changes in the molecular conformation on going from one polymorph to another. This is accomplished by partitioning the total lattice energy into its individual atomic contributions. Extracting this information is a relatively simple matter, since the total energy is computed from individual atomic contributions, and the programs may be simply modified to output this information. A careful analysis of these numbers, and a comparison among the various crystal forms of the individual contributions can reveal the differences in atomic contributions to the total lattice energy. Each of these numbers, of course, also contains contributions from many individual interactions between the atom in question and its neighbors, and, in principle, dissection of this information is also possible. Once again, the emphasis should be placed on trends and differences between polymorphs in individual atomic contributions or individual interactions rather than on the absolute values themselves. On moving from the overall lattice energy to individual contributions to individual interactions the numbers become increasing smaller and it is indeed questionable at the current stage of development whether use of the latter is really justified. However, the principle is still valid, and in the case of relatively strong anisotropic interactions such as hydrogen bonds, definitely warrants consideration.

Some of the findings from the specific case of the dichlorobenzylideneanilines deserve comment here, since they suggest further examination in other systems. We have found that the *relative* role of each atom's contribution to the overall energy is remarkably insensitive to the potential functions used. The order of the relative contributions of the partial energy to the total energy is the same for both crystal forms. This suggests that the environments of the atoms in the two crystal forms do not differ drastically in terms of energetics. As might be expected for a van der Waals solid no single atom makes an outstanding contribution to stabilizing the triclinic structure over the orthorhombic one; rather, the mode of stabilization is non-specific in that nearly all atoms make a small stabilizing contribution.This is true in spite of the fact that there are rather striking differences in the spatial arrangement of the molecules in the two structures[36].

Another aspect of crystal packing analysis arises with respect to these two structures, suggesting an additional application of these techniques to the investigation of organic crystal structures. It is quite common, in the analysis of crystal structures, to make note of interatomic distances (or 'contacts') which are shorter than the sum of the van der Waals radii, since these are presumably the important or dominant ones in determining the packing motif. In the triclinic structure there is a short Cl...Cl distance of 3.42Å, while the shortest in the orthrhombic structure is 3.79Å, which is slightly greater than the sum of the van der Waals radii. Hence, on a qualitative basis, at least, one might have expected that the contribution of chlorines in the triclinic form would be a major one in stabilizing that structure over the orthorhombic one. The analysis by partitioning suggests that this is not the case here; while the Cl is the largest contributor in both cases, the difference in the contribution is not significantly larger than for other atoms. The Cl...Cl interaction is a rather special one, as noted earlier[132], and is estimated to comprise about 3% of a Cl-Cl bond[137] or about 1.8 kcal-mol^{-1}. In view of the continuing interest in understanding its nature and utilizing it to 'engineer' crystals with desired structures and/or properties[138], the kind of analysis presented here can be quite useful to isolate and examine its nature and its ramifications.

The general success of this approach suggested additonal refinements to probe the extent of its applicability in a systematic way. These attempts have been based, for the most part, on the same benzylideneaniline system **28**, where the substituents in the *para* positions on the two rings have been manipulated to obtain the full set of three homodisubstituted and six heterodisubstituted analogues. This provides a very rich system of polymorphic and isomorphic structures for the investigation of the interplay between packing and conformation.

The dimethyl compound **28** (X = Y = CH_3) is trimorphic, also exhibiting conformational polymorphism[139-141]. In view of the similarity of the volumes of the chloro (19 $Å^3$) and methyl (24 $Å^3$) it might have been expected that there would be a correspondence between these polymorphs and those of the dichloro compound[142]. In fact, Desiraju and Sarma have recently shown that the so-called 'chloro-methyl exchange rule', based on these geometric considerations may break down for cases of nearly planar molecules or multiple chlorine substitution[143]. Two of the dimethyl structures (Forms I and III) are disordered with essentially planar molecular conformations[139,141], while in the third (Form II), the molecule adopts a conformation closely approximating that expected for the minimum energy noted above.

This system afforded the opportunity for a number of studies. Firstly, the computational methods may be tested on systems which exhibit disorder, and since the nature of the disorder is different in these two forms a further demand is put on the technique. The calculations[142] account for the relative stability of the two disordered forms compared to the non-disordered one, which is required to stabilize the more highly energetic planar conformation. This stabilization is shown to be due to the relatively favorable environment of the -CH=N- (bridge) atoms. The twofold and fourfold disorder in Forms I and III respectively introduce an entropy factor which contributes 0.41 and 0.83 kcal-mol^{-1} respectively to the total energy, and this no doubt contributes to the stabilization of these forms. The DSC data on the three crystal forms were consistent with the computational results.

Secondly, the lack of isomorphism between the dichloro and dimethyl system allowed us to test another aspect of the computational techniques. Since Form II of the latter exhibited the lowest energy molecular conformation and lacked disorder, it seemed perfectly natural to inquire as to whether the lattice energy calculations could account for the fact that we have failed to observe the dichloro compound in a similar structure. Computational substitution[144] of the dichloro compound into Form II of the dimethyl compound yielded a higher lattice energy for this proposed structure than for either of the two observed structures. Thus, the calculations can account for the 'observed nonexistence' of this conformational polymorph of the dichloro compound.

The 'chloro-methyl exchange rule' does hold, however, for the heterodisubstituted compounds **28** X = Cl, Y = CH_3; X = CH_3, Y =Cl which are isomorphous[145]. In fact, the two analogous methylbromo compounds are mutually isomorphous with these[146], suggesting that in some cases the exchange rule may be extended to bromine. None of these structures is isomorphous, however, with any of the dichloro or dimethyl polymorphs. In all of them the molecule is disordered about a center of symmetry and the molecule adopts a planar conformation. These systems provided even further testing grounds for the applicability and sensitivity of these computational methods[147] and they have proven to be quite versatile and reliable.

The remaining heterodisubstituted pair of molecules in this series **28** X =Cl, Y = Br and X = Br, Y = Cl present a fascinating case of isomorphism and the lack thereof, which may be another useful tool in efforts to design or 'engineer' crystals with particular molecular and/or packing properties. The two compounds are not mutually isomorphous. The former is isomorphous with the orthorhombic dichloro compound so that the molecule adopts the twisted conformation ($\alpha = -\beta = 24.9°$). The latter is isomorphous with the dibromo compound, in which there is disorder about a center of symmetry and the molecule is planar[148]. Hence, a systematic change in the mode of substitution can be employed to bring about a change both in the mode of packing and the molecular conformation[149].

5. CONCLUDING REMARKS

We have shown how the study of the polymorphic and isomorphic behavior of a systematically substituted family of compounds with a limited number of conformational parameters can yield a great deal of information of interest to chemists concerned with the organic solid state. The polymorphism provides the most direct means for examining differences in packing motifs, all other things being equal. In cases where they are not, the only additional variable is the molecular conformation, which consequently must be influenced by the crystal environment. A quantitative measure of this influence of the crystal environment may be obtained by computationally examining both the molecular and crystal energetics to sort out the specific interactions which bring about the observed changes from one crystal structure to the next. While the computational techniques have now reached a significant level of sophistication and new ones are being developed[150], their general applicability to a variety of problems, the transferability of potential forms and parameters from one problem to another, and their predictive power are problems and challenges which still remain. The application of these techniques to problems involving conformational polymorphism and related topics discussed here is a symbiotic one: the techniques are required for the investigation of the problems, and in their solution one can discover both their power and their limitations. The combination of chemical crystallography and computational chemistry is one that is proving to be of increasing value to organic solid state chemists and is sure to provide a great deal of information on, to repeat Dunitz and Ibers' words, "the weak interactions that control preferred packing arrangements of molecules in crystals, and...the conformational complexities of organic molecules..."

ACKNOWLEDGEMENT. I am particularly grateful to Hai Cohen for technical assistance, especially in the preparation of many of the figures and to Leah Shahal for a careful reading of the manuscript. This work was supported in part by a grant from the United States-Israel Binational Science Foundation (BSF).

REFERENCES

1. J. D. Dunitz and J.A. Ibers (Eds.), Perspectives in Structural Chemistry, Vol. I, J. Wiley & Sons, New York, 1967, p. vi.

2. There are now over 50,000 reported solved crystal structures, and with the essentially exponential increase in this number over the past few years, a number of 75,000 can be expected by 1990. F.H. Allen, Lecture Notes, International School of Crystallography XIth Course, "Static and Dynamic Implications of Precise Structural Data", Erice, Italy, June, 1985.

3. In particular, the Cambridge Structral Data File, see F.H. Allen, Acta Crystallogr. Sect. A, 36 (1980) C441; F.H. Allen, S. Bellard, M.D. Brice, B.A. Cartwright, A. Doubleday, H. Higgs, T. Hummelink, B.G. Hummelink-Peters, O. Kennard, W.D.S. Motherwell, J.R. Rodgers and D.G. Watson, Acta Crystallogr. Sect. B, 35 (1979) 2331; F.H. Allen, O. Kennard and R. Taylor, Acc. Chem. Res., 16 (1983) 146.

4. W.L. Duax, J.F. Griffin and D.C. Rohrer, in A.S. Horn and C.J. De Ranter (Eds.), X-ray Crystallography and Drug Action, Oxford University Press, Oxford, 1984, pp. 405.

5a. H.-B. Burgi, Angew. Chemie Int. Ed. (in English), 14 (1975) 460.
 b. J.D. Dunitz, X-ray Analysis and the Structure of Organic Molecules, Cornell University Press, Ithaca, 1979, pp. 301.
 c. H.-B. Burgi and J.D. Dunitz, Acc. Chem. Res., 16 (1983) 153.

6. K.N. Trueblood and J.D. Dunitz, Acta Crystallogr. Sect. B, 39 (1983) 120.

7.S.C. Nyburg and C.H. Faerman, Acta Crystallogr. Sect. B, 41 (1985) 274.

8a. R. Taylor and O. Kennard, Acct. Chem. Res., 17 (1984) 320.
 b. P. Murray-Rust and J.P. Glusker, J. Am. Chem. Soc., 106 (1984) 1018.

9. S.R. Wilson and J.C. Huffman, J. Org. Chem., 45 (1980) 560.

10. S.K Burley and G.A. Petsko, Science, 229 (1985) 23.

11a. J.K Haleblian and W.C. McCrone, J. Pharm. Sci., 58 (1969) 411.
 b. J.K Haleblian, J. Pharm. Sci., 64, (1975) 1269.
 c. J.A. Clements, Proc. Analyt. Div. Chem. Soc., (1976) 21.

12a. M.S. Walker, R.L. Miller, C.H. Griffiths and P.Goldstein, Molec. Cryst. Liq. Cryst., 16(1972) 203.
 b. C.H. Griffiths and A.R. Monahan, Molec. Cryst. Liq. Cryst., 33 (1976) 175.
 c. M.C. Etter, R.B. Kress, J. Bernstein and D.C. Cash, J. Am. Chem. Soc., 106 (1984) 6921.
 d. M. Tristani-Kendra, C.J. Eckhardt, E. Goldstein and J. Bernstein, Chem. Phys. Lett., 98 (1983) 57.
 e. D.L. Morel, E.L. Stogryn, A.K. Ghosh, T. Feng, P.E. Purwin, R.F. Shaw. C. Fushman, G.R. Bird and A.P. Piechowski, J. Phys. Chem., 88 (923) 1984.

13. For example, see R.J. Karpowicz, S.T. Sergio and T.B. Brill, I&EC Prod. Res. & Dev., 22, (1983) 363.

14 a. A.R. Varna and P. Krishna, Polymorphism and Polytypism is Crystals, J. Wiley and Sons, New York, 1966.
b. S.R. Byrn, Solid State Chemistry of Drugs, Academic Press, New York, 1983.
c. M. Kuhnert-Brandstatter, Thermomicroscopy in the Analysis of Pharmaceuticals, Pergamon Press, New York, 1971.

15. W.C. McCrone in D. Fox, M.M. Labes and A Weissberger, (Eds.), Physics and Chemistry of the Organic Solid State, Vol. I, Interscience, New York, 1963, p. 725.

16 a. M.D. Cohen and G.M.J. Schmidt, J. Chem. Soc. (1964) 1996.
b. G.M.J. Schmidt, J. Chem. Soc. (1964) 2014.

17. E. Mitscherlisch, Ann. Chim. Phys., 19 (1822) 350; *ibid.* 24 (1823) 264.

18. For a description of this transformation (in vivid medical terminology) see J.W. Mellor, A Comprehensive Treatise on Inorganic and Theoretical Chemistry, Vol. VII, Longmans, Green and Co., London, 1930, p. 300.

19 a. L. Deffet, Repertoire des Composes Organiques Polymorphes, Editions Desoer, Liege, 1942.
b. P. Groth, Chemische Kristallographie. 5. vols., Verlag- von Wilhem Engelmann, Leipzig, 1906.

20. M.J. Buerger and M.C. Bloom, Z. Krist., A96 (1937) 182-220.

21. S. Decurtins, C.B. Shoemaker and H.H. Wickmann, Acta Crystallogr. Sect. C, 39 (1983) 1218.

22. M. Tristani-Kendra, C.J. Eckhardt, J. Bernstein and E. Goldstein, Chem. Phys. Lett., 98 (1983) 57.

23. T.J. Kistenmacher, T.J. Emge, A.N. Bloch and D.O. Cowan, Acta Crystallogr., Sect B, 38 (1982) 1193; K. Bechgaard, T.J. Kistenmacher, A.N. Bloch and D.O. Cowan, Acta Crystallogr. Sect. B, 33 (1977) 417.

24. J.C.D. Brand and J.C. Speakman, Molecular Structure, The Physical Approach, E. Arnold, London, 1960, pp. 248.

25. K. Mislow, Introduction to Stereochemistry, W.A. Benjamin, New York, 1966, pp. 33.

26. A.T. Hagler, E. Huler and S. Lifson, J. Am. Chem. Soc. 96 (1974) 5319.

27. Note, however, that this also is an approximation which considerably simplifies the calculations using the model. Clearly, a precise treatment would have to take this anisotropy into account, and for small systems models have been proposed. [For instance, see T.L. Starr and D.E. Williams, J. Chem. Phys., 66 (1977) 2054]. The anisotropicity is evident from structural considerations for polarizable atoms such as chlorine [see J.A.R.P. Sarma and G. Desiraju, Chem. Phys. Lett., 117 (1985) 160; D.E. Williams and L.-Y. Hsu, Acta Crystallogr. Sect. A (1985) 296].

28. See, for instance, R. Foster, Organic Charge Transfer Complexes, Academic Press, London, 1969.

29 a. D.E. Williams, Acta Crystallogr. Sect. A, 30 (1974) 71.
b. S.R. Cox and D.E. Williams, J. Comp. Chem., 2 (1981) 304.

30. However, see M. Colapietro, A. Domenicano, G. Portalone, G. Schulz and I. Hargittai, J. Mol. Struct., 112 (1984) 141, where there is ample evidence for the influence of crystal forces on bond lengths and bond angles.

31. J.D. Dunitz, X-ray Analysis and the Structure of Organic Molecules, Cornell Univ. Press, Ithaca, N.Y., 1979, p. 312.

32. M. Dobler, Chimia, 38 (1984) 415.

33. A.I. Kitaigorodskii, Organic Chemical Crystallography, Consultants Bureau, New York, 1961.

34. A.I. Kitaigorodskii, Adv. Struct. Res. Diffr. Methods, 3 (1970) 173.

35a. P. Corradini, Chim. Ind. (Milan), 55 (1973) 122-129.
b. N.C. Panagiotoupoulis, G.A. Jeffrey, S.J. LaPlaca and W.C. Hamilton, Acta Crystallogr. Sect. B, 30 (1974) 1421.

36. J. Bernstein and A.T. Hagler, J. Am. Chem. Soc., 100 (1978) 673.

37. T. Ichikawa, Y. Iitaka and M. Tsuboi, Bull. Chem. Soc. Jpn. 41 (1968) 1028.

38. T. Ichikawa and Y. Iitaka, Acta Crystallogr. Sect. B, 24 (1968) 1488.

39. T. Akimoto and Y. Iitaka, Acta Crystallogr. Sect. B, 28 (1972) 3106.

40. K. Nakata, Y. Takaki and K. Sakurni, Acta Crystallogr. Sect. B, 36 (1980) 504.

41. G.A. Jeffrey, J.R. Ruble, L.M. Wingert, J.H. Yates and. R.K. McMullen, J. Am. Chem. Soc., 107 (1985) 6227.

42. J. Donohue, in A. Rich and N. Davidson (Eds.), Structural Chemistry and Molecular Biology, W.H. Freeman, San Francisco, 1968, pp. 443.

43 a. R. Taylor and O. Kennard, J. Am. Chem. Soc., 104 (1982) 5063.
b. S. Schroetter, D. Bougeard and B. Schrader, Spectroscopy Lett., 18 (1985) 153.

44. F.A. Cotton and P.F. Stokely, J. Am. Chem. Soc., 92 (1970) 294.

45. I. Goldberg, Personal communication.

46. F. Toda, K. Tanaka and. T.C.W. Mak, J. Incl. Phenom., 3 (1985) 225-233.

48. C.P. Brock , J. Huckaby and T.G. Attig, Acta Crystallogr. Sect. B, 40 (1984) 595.

49. J.L. Burmeister, H.J. Gysling and. J.C. Lim, J. Am. Chem. Soc., 91 (1969) 44.

50. R.O. Gould, S. Ramsey, P. Taylor and M.D. Walkinshaw, Ninth European Crystallography Meeting, Torino, Sept. 1985, Abstract #2-135.

51. See, for instance,
a. O.V. Dorofeeva, V.S. Mastrykov, N.L Allinger and A. Almennigen, J. Phys. Chem., 89 (1985) 252.
b. M.W. Barnett, R.D. Farrant, D.N. Kirk, J.D. Mershe, J.K.M. Sanders and W.L. Duax, J. Chem. Soc. (Perkin II), (1982) 105.

52. M. Cotrait, S. Geoffre, M. Hospital and G. Precigoux, Acta Crystallogr. Sect. B, 39 (1983) 754-760.

53. M. Froimowitz and S. Mattysse, Molec. Pharmacol., 24 (1983) 243-250; P.H. Byrd, F.T. Bruderlain and G. Humberm, Can. J. Chem., 54 (1976) 2715-2722.

54. H. Kessler, G. Zimmerman, HJ. Forster, J. Engel, G. Oepen and W.S. Sheldrick, Angew. Chem. Int. Ed. (In English), 20 (1981) 188.

55. H. Kessler, P. Kramer and G. Krack, Isr. J. Chem., 20 (1980) 188.

56. A. Almenningen, O. Bastiensen, L. Fernholdt, B.N. Cyvin, S.J. Cyvin and S. Samdal, J. Molec. Struct., 128, (1985) 59.

57. O. Bastiensen and S. Samdar, J. Molec. Struct., 128 (1985) 115.

58. See, for instance,
a. C.P. Brock and G.L. Morelan, submitted to J.Am.Chem. Soc.
b. W.R. Busing, Acta Crystallogr. Sect. A, 39 (1983) 340.
c. A. Horn, P. Klaeboe, E. Kloster-Jensen. B.N. Cyvin and S.J. Cyvin, Spectrochimica Acta, 41A (1985) 451.

59. J.C. Barnes, J.D. Paton, J.R. Damewood and K. Mislow, J. Org. Chem., 46 (1981) 4975.

60. M. Wiebcke and D. Mootz, Acta Crystallogr. Sect. B, 38 (1982) 2008.

61. M.L. Thomson, R.C. Haltiwanger, A. Tarassoli, D.E. Coons and A.D. Norman, Inorg. Chem. 21 (1982) 1287.

62. J.L. McNaughton and C.T. Mortimer, Int. Rev. Sci. Phys. Chem. Series 2, Vol. 10, Butterworths, London, p. 1

63 a. M. Kuhnert-Brandstatter, Thermomicroscopy in the Analysis of Pharmaceuticals, Pergamon Press, Oxford, England, 1974.
b. W.C. McCrone, Fusion Methods in Chemical Microscopy, Interscience, New York, 1957.
c. L. Kofler and A. Kofler, Thermo-Mikro-Methoden zur Kennzeichnung Organischer Stoffe und Stoffgemische, Wagner, Innsbruck, 1954.
d. D. G. Grabar, J.P. Hession and F.C. Rauch, Microscope 18 (1970) 241.
e. W.C. McCrone, Microscope 18 (1970) 257.

64. Y. Ogawa and M. Tasumi, Chem. Lett., (1979) 1411.

65. Y. Ogawa and M. Tasumi, Chem. Lett. (1978) 947.

66. J. Guzman and J. Largo-Cabrerizo, J. Heterocyclic Chem., 15 (1978) 1531.

67. P.E. Stott, C.W. McCausland and W.W. Parish, J. Heterocyclic Chem., 16 (1979) 453.

68. C.M. Gramaccioli, G. Filippini and M. Simonetta, Acta Crystallogr. Sect. A, 38 (1982) 350.

69. C.P. Brock, W.B. Schweizer and J.D. Dunitz, J. Am. Chem. Soc., 107 (1985) 6964.

70. Y. Tomita, T. Ando and K. Ueno, Bull. Chem. Soc. Jpn., 38 (1965) 138.

71 a. J. Bernstein, Acta Crystallogr. Sect. B, 35 (1979) 360.
b. C.-E. Boman, H. Herbertsson and A. Oskarsson, Acta Crystallogr. Sect. B, 30 (1974) 378.

72. E.J. Graeber and B. Morosin, Acta Crystallogr. Sect. B, 30 (1974) 310.

73. P. Prusiner and M. Sundaralingam, Acta Crystallogr. Sect. B, 32 (1976) 419.

74 a. S. Neidle, W. Kuhlbrandt and A. Achari, Acta Crystallogr. Sect. B., 32 (1976) 1850.
b. J. Kraut and L.H. Jensen, Acta Crystallogr., 16 (1963) 79.

75. E. Platt and B. Robson, J. Theor. Biol., 96 (1982) 381.

76. M. Sundaralingam, Jerusalem Symp. Quant. Chem. Quant. Biol., 5 (1973) 417.

77 a. A. Burger and R. Ramburger, Mikrochim. Acta [Wien], (1979)II 259.
b. A. Burger and R. Ramburger, Mikrochim. Acta [Wien], (1979)II 273.

78 a. M. Czugler, A. Kalman, J. Kovacs and I. Pinter, Acta Crystallogr. Sect. B, 37 (1984) 172.
b. V.J. James and J.D. Stevens, Cryst. Struct. Comm., 2 (1962) 609.
c. B.J. Poppleton, Acta Crystallogr. Sect. B, 32 (1976) 2702.

79. G.D. Woodard and W.C. McCrone, J. Appl. Cryst., 8 (1975) 342.

80. V.W. Jacewicz and J.C. Nayler, J. Appl. Cryst. 12 (1979) 396.

81. I. Bar and J. Bernstein, Acta Crystallogr. Sect. B 38 (1982) 121.

82 a. J.D. Dunitz in "Perspectives in Structural Chemistry", J.D. Dunitz and J.A. Ibers, eds. John Wiley and Sons, New York, 1968, Vol. II, pp 1-70.
b. J. D. Dunitz, J. Chem. Educ. 47 (1970) 488.
c. Ref. 31, Chapter 9.

83. T.B. Brill and C.O. Reese, J. Phys. Chem. 84 (1980) 1376.

84 a. C.S. Choi and H.P. Boutin, Acta Crystallogr. Sect. B., 26 (1970) 1235.
b. H.H. Cady, A.C. Larson and. D.T. Cromer, Acta Crystallogr. 16 (1963) 617.
c. R.E. Cobble and R.W.H. Small, Acta Crystallogr. Sect. B., 30 (1974) 1948.

85. J. Gaultier, C. Hauw and H. Bouas-Laurent, Acta Crystallogr. Sect. B 32 (1976) 1220.

86. H.-D. Becker, S.R. Hall, B.W. Skelton and A.H. White, Aust. J. Chem., 37 (1984) 1313.

87. See also, for instance, J. Bernstein, T.E. Anderson and C.J. Eckhardt, J. Am. Chem. Soc., 101 1979) 541.

88. S. Hirokawa, Acta Crystallogr. 8 (1955) 637.

89a. M. Hasegawa, N. Fukuda, H. Higuchi and I. Matsubara, Agric. Biol. Chem., 41 (1977) 49.
b. S. Hiramatsu, Nippon Nogii Kagaku Kaishi, 51 (1977) 27 (CA 87:68599d).
c. S. Hiramatsu, Nippon Nogii Kagaku Kaishi, 51 (1977) 39 (CA 87:68601d).

90. See, for instance, L. Addadi, Z. Berkovitch-Yellin, I. Weissbuch, J. van Mil, L. Shimon, M. Lahav and L. Leiserowitz, Angew. Chem. Int Ed. (in English) 24 (1985) 439, and references given therein.

91. M.S. Lehmann, T.F. Koetzle and W.C. Hamilton, J. Cryst. Mol. Struct., 2 (1972) 225.

92. N. Hirayama. K. Shirahata, Y. Ohashi and Y. Sasada, Bull. Chem. Soc. Jpn., 53 (1980) 30.

93. A. McL. Mathieson, Acta Crystallogr., 5 (1952) 332.

94. T. Taniguchi, Y. Takaki and S. Sakurai, Bull . Chem. Soc. Jpn., 53 (1980) 803.

95 a. J.Z. Gougoutas, Pure and Appl. Chem., 27 (1971) 305.
b. M.C. Etter, J. Am. Chem. Soc., 98 (1977) 5331.

96 a. P. Ganis, A. Panunzi and C. Pedone, Ric. Sci., 38 (1968) 801.
b. G. Avitabile, P. Ganis and E. Martuscelli, Acta Crystallogr. Sect.B 25 (1969) 2378.
c. G. Avitabile, P. Ganis and U. Lepore, Macromolelcules, 4 (1971) 239.

97. C.H. Stam and C.H. MacGillavry, Acta Crystallogr., 16 (1963) 62.

98. C.H. Stam, Acta Crystallogr. Sect. B, 28 (1972) 2396.

99 a. W.L. Duax and D.A. Norton, Atlas of Steroid Structure Vol. 1, Plenum Press, New York, 1975.
b. W.L. Duax J.F. Griffin and C.M. Weeks, Atlas of Steroid Structure Vol. 2, Plenum Press, New York, 1982.

100. W.L. Duax, M.D. Fronkowiak, J.F. Griffin and D.C. Rohrer in J. Jortner and B. Pullman (Eds.), Intramolecular Dynamics, D. Reidel, Dordrecht, 1982, pp. 502.

101. J.A. Knaters, A. de Koster, V.J. van Geerestein and L.A. van Dijck, Acta Crystallogr. Sect. C 41 (1985) 760.

102. J. P. DeClercq, G. Germain and M. Van Meerssche, Cryst. Struct. Comm., 1 (1972) 59.

103. S.C. Nyburg, Acta Crystallogr. Sect. B, 30 (1974) 251.

104. S.R. Wilson and J.C. Huffman, J. Org. Chem. 45 (1980) 560.

105. W.L. Duax, M. Numazawa, Y. Osawa, P.D. Strong and. C.M. Weeks, J. Org. Chem., 46 (1981) 2650.

106. G. Precigoux, B. Busetta, C. Courseille and M. Hospital, Acta Crystallogr. Sect. B, 31 (1975) 1527.

107. M.Y. Bhadbade and K. Venkatesan, Acta Crystallogr. Sect. C, 40 (1984) 1905.

108. P.E. Riley and R.E. Davis, Inorg. Chem., 15 (1976) 2735.

109. B.M. Foxman, P.L. Goldberg and H. Mazurek, Inorg. Chem., 20 (1981) 4368.

110. G.R. Clark, J.M. Waters, T.N. Waters and G.J. Williams, J. Inorg. Nuc. Chem., 39 (1977) 1971.

111. M. von Stackelber, Z. Anorg. Allg. Chem., 253 (1947) 136; E.C. Lingafelter,. G.L. Simmons, B. Morosin. C. Scheringer and C. Freiberg, Acta Crystallogr., 14 (1961) 1222.

112. G.R. Clark, J.M. Waters and T.N. Waters, J. Inorg. Nuc. Chem., 37 (1975) 2455.

113. D.W. Martin and T.N. Waters, J.C.S., (1973) 2440.

114. J. Bernstein in A.S. Horn and C.J. De Ranter, (Eds.), X-ray Crystallography and Drug Action, Oxford University Press, New York, 1964, pp. 23.

115. I.L. Karle and J. Karle, Int. J. Quant. Chem. Vol XII. Suppl. 1 (1977) 393.

116. For instance,
a. C.S. Yannoni, Acct. Chem. Res. 15 (1982) 201.
b. J. Schaefer and E.O. Stejskal in G.C. Levy (Ed.), Topics in in Carbon-13 Spectroscopy, Wiley-Interscience, New York, 1979.
c. R.E. Wasylishen and C.A. Fyfe, Ann. Repts. NMR Spect., 12 (1982) 1.
d. G.E. Maciel and M.J. Sullivan, ACS Monogr., 191 (1982) 319.
e. J. Schaefer and E.O. Stejskal, J. Am. Chem. Soc., 98 (1976) 1031.
f. E.R. Andrew, MTP Int. Rev. Sci. Phys. Chem. Sec. 2, 4 (1976) 173.
g. R.G. Griffin, Anal. Chem., 49 (1977) 951A.
h. M. Mehring, NMR - Basic Principles and Progress, Vol. 1, Springer, New York, 1976.
i. J.A. Ripmeester, Chem. Phys. Lett. 74 (1980) 536.
j. R.G. Bryant, V.P. Chacko and M.C. Etter, Inorg. Chem., 23 (1984) 3580.

117. For instance,
a. H.F. Schaeffer III (Ed.), Modern Theoretical Chemistry, Vol. 3, Methods of Electronic Structure Theory, Plenum, New York, 1977.
b. H.F. Schaeffer III (Ed.), Modern Theoretical Chemistry, Vol. 4, Applications of Electronic Structure Theory, Plenum, New York, 1977.
c. W.J. Hehre, Accts. Chem. Res. 9 (1976) 399.

118a. U.B. Berkert and N.L. Allinger, Molecular Mechanics, ACS Monograph No. 177, Washington, D.C., 1982.
b. M. Simonetta, Int. Revs. Phys. Chem., 4 (1984) 39.
c. N.L. Allinger, Adv. Phys. Org. Chem., 13 (1976) 2.
d. J.D. Dunitz and H.B. Burgi, Int. Rev. Sci. Phys. Chem. Ser. 2, 11 (1975) 81.
e. O. Ermer, Structure and Bonding 27 (1976) 163.
f. K. Mislow, D.A. Dougherty and W.D. Houshell, Bull. Soc. Chim. Belg., 87 (1978) 555.
g. D.B. Boyd and K.B. Lipkowitz, J. Chem. Ed., 59 (1982) 269.
h. A. Warshel in G. Segal (Ed.), Modern Theoretical Chemistry, Vol. 7, Plenum, New York, 1978, p. 133.

119a. A. Warshel and S. Lifson, J. Chem. Phys., 53 (1970) 582.
b. H.A. Scheraga, Chem. Rev., 71 (1971) 195.
c. A.T. Hagler, L. Leiserowitz and M. Tuval, J. Am. Chem. Soc., 98 (1976) 4600.
d. A.T. Hagler and S. Lifson, J. Am. Chem. Soc., 96 (1974) 5327

120. A.I. Kitaigorodsky, Molecular Crystals and Molecules, Academic Press, New York, 1973.

121a. D.E. Williams, Acta Crystallog. Sect. B, 29 (1972) 629.
b. V.M. Coiro, E. Giglio and C. Quagliata, Acta Crystallogr. Sect. B, 28 (1971) 3601.
c. G. Casalone, C. Marieni, A. Magnoli and M. Simonetta, Mol. Phys., 15 (1968) 339.

122. D.E. Williams, Acta Crystallogr. Sect. A, 30 ((1974) 71.

123a. G.L. Wheeler and S.R. Colson, J. Chem. Phys., 65 (1976) 1227.
b.S.C. Nyburg and C.H. Faerman, American Crystallographic Society, Annual Meeting, June, 1986, Abstract PB13.

124. For a general discussion of the atom...atom potential method see refs. 118b, 120. Other useful references:
a. A.I. Kitaigorodsky and K. Mirskaya, Mat. Res. Bull., 7 (1972) 1271.
b. K. Mirsky, Acta Crystallogr. Sect. A, 32 (1976) 199.
c. K. Mirsky and M.D. Cohen, J. Chem. Soc. Faraday II, 72 (1976) 2155, and references 1-7 therein.
d. K. Mirsky, Chem. Phys., 46 (1980) 445.
e. D.E. Williams, Acta Crystallogr. Sect. A, 28 (1972) 629.
f. D.E. Williams, Topics in Appl. Phys., 26 (1981) 3-40.
g. D.E. Williams and T.L. Starr, Comp. Chem., 1 (1977) 173.
h. S. Ramdas and J.M Thomas, Chem. Phys. Solids and Surfaces, 7 (1978) 31.
i. A.I. Kitaigorodsky, Chem. Soc. Revs. 7 (1978) 133.

125. A.T. Hagler, E. Huler and S. Lifson, J. Am. Chem. Soc., 96 (1974) 5319.

126. F.L. Hirshfeld and K. Mirsky, Acta Crystallogr. Sect.A, 35 (1979) 366.

127. S.R. Cox and D.E. Williams, J. Comput. Chem., 2 (1981) 304.

128. F. Daniels, J.W. Williams, P. Bender, R.A. Alberty, C.D. Cornwell and J.E. Harriman, Experimental Physical Chemistry, 7th ed., McGraw-Hill, New York, 1970, p. 53.

129. A. Bondi, J. Chem. Eng. Data, 8 (1963) 371.

130. T. Bally, E. Hasselbach, S. Lanyiova, F. Marschner and M. Rossi, Helv. Chem. Acta, 59 (1976) 486, and references cited therein.

131.H.-B. Burgi and J.D. Dunitz, Helv. Chim. Acta, 54 (1971) 1255.

132. G.M.J. Schmidt, Pure Appl. Chem., 27 (1971) 647.

133. J. Bernstein and G.M.J. Schmidt, J. Chem. Soc. Perkin II, (1972) 952.

134. J. Bernstein and I. Bar, J. Chem. Soc. Perkin II, (1976) 429.

135. J. Bernstein, T.E. Anderson and C.J. Eckhardt, J. Am. Chem. Soc., 101 (1979) 541.

136. J. Bernstein, M. Engel and A.T. Hagler, J. Chem. Phys., 75 (1981) 2346.

137. D.E. Williams and L.-Y Hsu, Acta Crystallogr.Sect. A, 41 (1985) 296.

138. J.A.R.P Sarma and G.R. Desiraju, Accts. Chem. Res., 19 (1986), In press.

139. J. Bernstein and I. Bar, Acta Crystallogr. Sect. B, 32 (1976) 1609.

140. I. Bar and J. Bernstein, Acta Crystallogr. Sect. B, 33 (1977) 1738.

141. I. Bar and J. Bernstein, Acta Crystallogr. Sect. B, 38 (1982) 121.

142. I. Bar and J. Bernstein, J. Phys. Chem., 86 (1982) 3223.

143. G.R. Desiraju and J.A.R.P. Sarma, Proc. Indian. Acad. Sci. (Chem. Sci.), 96 (1986) 599.

144. A.T. Hagler and J. Bernstein, J. Am. Chem. Soc., 100 (1978) 6349.

145. I. Bar and J. Bernstein, Acta Crystallogr. Sect. B, 39 (1983) 266.

146. I. Bar and J. Bernstein, unpublished results.

147. I. Bar and J. Bernstein, J. Phys. Chem., 88 (1984) 243.

148. J. Bernstein and I. Izak, J. Cryst. Mol. Struct., 5 (1975) 257.

149. I. Bar and J. Bernstein, Tetrahedr. Symp. in Print, 1987, In Press.

150. For instance, the introduction of the notion of molecular volume, molecular free surface and their calculation:
a. A. Gavezzotti, J. Am. Chem. Soc., 105 (1983) 5220.
b. A. Gavezzotti, J. Am. Chem. Soc., 107 (1985) 962.
c. R. Bianchi, A. Gavezzotti and M. Simonetta, J. Molec. Struct., 1986, In press.

Chapter 14

CRYSTAL ENGINEERING A 4 Å - SHORT AXIS STRUCTURE FOR PLANAR CHLORO AROMATICS

GAUTAM R. DESIRAJU

1. Introduction

The concept of crystal engineering has been a particularly valuable one in the field of organic solid state chemistry. The term was first used by Schmidt in his seminal 1971 review article (ref. 1) and as correctly predicted by him there, has grown from a collection of empirical rules to a systematic study of intermolecular forces (refs. 2-5). The importance of crystal engineering stems from the awareness that sustained development of organic solid state chemistry may not be expected unless it is somehow possible to predict crystal packings. The focus has therefore inevitably shifted from the mere obtaining of structures to the larger questions as to the factors and forces which are responsible for the adoption of these structures (ref. 6). Instead of asking 'why a particular structure?' one may as well ask 'why not another type of structure?' Unfortunately, a continuing obstacle to answering such questions is posed by the fact that these structures hinge on forces which are very weak.

Broadly speaking, a crystal structure of a molecular solid results from a play of dispersive and repulsive interactions and many structures may be rationalised, if not completely understood, in these terms (refs. 7,8). However, because of the shallow nature of the potential energy surface in the neighbourhood of the thermodynamically stable structure, alternative packings are possible and realised in practice as polymorphic modifications (ref. 9). Yet there is a strong impetus to research which attempts prediction of organic crystal structures since it enables a definition not only of new solid state reactions but also of new compounds with desired physical or electronic properties.

If only dispersive and repulsive forces were involved in stabilising a structure one should, in principle, be able to predict it since it would be strictly governed by the close-packing model of Kitaigorodskii (refs. 7,8). This situation is indeed realised for hydrocarbons; however, the ultimate choice of a stable packing for functionalised molecules seems to depend on the relative importance of other, directional forces such as dipole interactions or weak bonding. These effects all but render transferability of atom-atom non-bonded potentials impossible. Further, they may also vary drastically with molecular modifications that are apparently trivial, at least from the solution chemist's viewpoint. Realising such facts, the usual practice in crystal engineering has been to analyse the structures of a series of (what are felt to be!) closely related compounds so that a unifying structural principle may be made to emerge (refs. 10,11).

This has usually proved to be a happier task than outright prediction since it has been a common observation among crystallographers and structural chemists that certain intermolecular interactions in organic molecular solids are associated with only certain geometrical motifs. Another not unimportant factor which has facilitated such an approach to crystal engineering is the ready availability of a large amount of high quality crystallographic data in a computer retrievable format. The existence of the Cambridge Structural Database has meant that any structural conjecture may be tested and evaluated in statistical terms and that appropriate conclusions may be drawn with a high level of confidence (Ref.12).

This chapter probes the well-known steering ability of a chloro group when substituted on an aromatic moiety to direct the packing mode to one which is characterised by a crystallographic short axis of 4 Å (ref. 13). It has long been supposed that this so-called 'chloro-effect' is related to the short non-bonded Cl...Cl contacts which have been reported regularly in the literature (ref. 14). It will be shown here that these two structural phenomena are, in fact, closely linked and that the adoption of the 4 Å structure, by a large number of planar chloro aromatics may be understood if a weakly attractive and directional Cl...Cl interaction is assumed. These structures may be categorised on the basis of crystallographic criteria and it is hoped that the reader is drawn to a point where extrapolation, if only cautiously, to new chloro aromatics is possible.

2. The chloro effect - historical background.

The chloro effect was first observed by Schmidt and his co-workers Green and Leser who found that around forty chloro aromatic compounds, mostly planar and dichlorosubstituted, adopt the 4 Å or β-structure (refs. 1,13). The term β has been used by Schmidt for this packing mode since it is the one adopted by the β-modification of trans-cinnamic acid (ref. 15) and in this chapter we have used the terms 4 Å and β interchangeably.

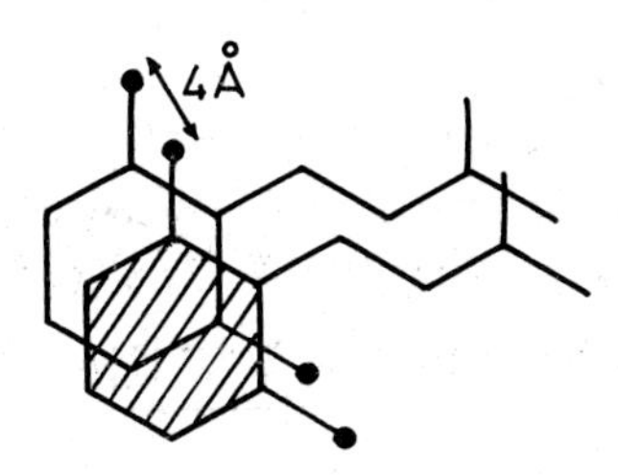

Fig. 2.1. The β-structure as typified by a schematic view of two stack-related molecules of 2,6-dichlorocinnamic acid.

While the reasons for this structural preference of chloro aromatic compounds were not very clear at that time, considerable progress was made in devising novel solid state reactions based on the chloro effect. An ingenious example is the eight-centre dimerisation of tetrachlorodibenzylidene acetone to yield a novel tricyclic cyclohexanedione (ref. 16). Another well-known reaction utilised the β-steering ability of the 2,6-dichlorophenyl group in a sequence which ultimately led to an enantiomeric excess of a chiral cyclobutane (ref. 17). All in all, the chloro effect seemed to be a reliable means of achieving a photoreactive β-packing where topochemical cycloaddition would result in a mirror symmetry cyclobutane. While much of the unit-cell parameter data collected by Schmidt and his co-workers remains unpublished, the 1984 version of the Cambridge Database reports about a hundred β-structures for planar chloro aromatic compounds. These compounds will be considered in some detail in this chapter.

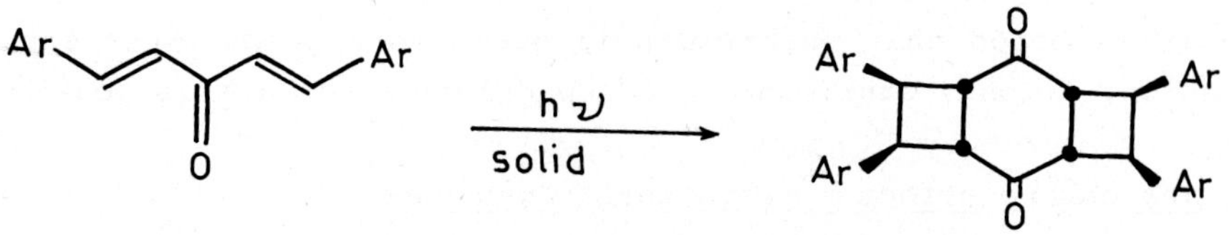

R = 3,4-dichlorophenyl

Fig. 2.2. Formation of a mirror-symmetry dimer from a β-structure dibenzylidene acetone (see ref. 16)

3. The role of Cl...Cl interactions

3.1 The nature of Cl...Cl contacts

Simultaneously, and for the most part independent of these developments in organic solid state chemistry, there has been much discussion, even controversy, on the nature of Cl...Cl non-bonded interactions in organic molecular crystals (refs. 14, 18). That the contacts are short seems not in doubt - the crystallographic literature is replete with references to 'anomalous', 'unusual' or 'exceptional' Cl...Cl contacts with distinct directional preferences (ref. 19). This categorisation as 'short' is valid whether Pauling's value for the chlorine van der Waals radius (1.80 Å) or more conservative values (1.75-1.80 Å) are used. It follows that the behaviour of chlorine cannot be modelled satisfactorily using isotropic atom-atom potentials for were such potentials to be used, these intermolecular distances would represent abnormally and unacceptably repulsive contacts. Explanations for such directional effects vary; it has been held that the Cl...Cl interaction may be likened to a weak bond (ca. 3% of a covalent bond (ref.18)) while alternatively, the van der Waals radius of chlorine has been assumed anisotropic (ref. 20). The approach of adjacent chlorine atoms in the crystal has also been likened to an electrophile-nucleophile interaction (ref. 21).

Conceptually and computationally, it is convenient to regard these short Cl...Cl contacts as manifestations of weak intermolecular bonding. In fact, there is much evidence that chlorine atoms in many molecular solids do not function as groups of a certain volume (close-packing) but rather through specific, anisotropic, electronic effects. In solid chlorine itself, the difference between the shortest Cl...Cl non-bonded contact (3.27 Å) and the next shortest contact (3.70 Å) is appreciable. These effects are enhanced in crystalline bromine and iodine both of which are isomorphous with chlorine. In solid iodine even the I-I stretch appears in the IR indicating weak intermolecular bonding. It has been pointed out that none of these three structures are of the simple van der Waals type because they are all layered with halogen atoms in planes parallel to (100) (ref. 18). The fact that solid fluorine does

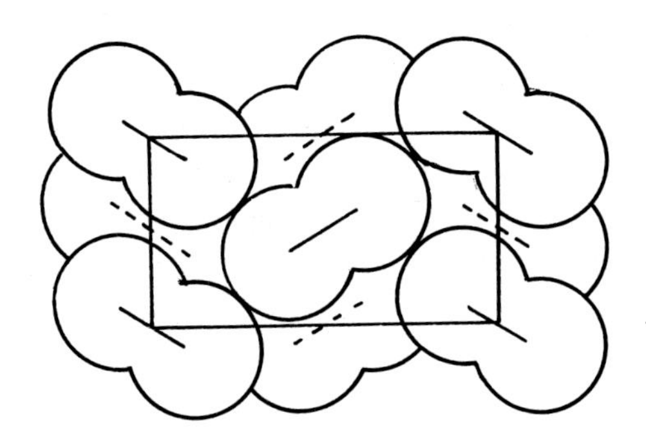

Fig. 3.1. The molecular packing in the crystal structure of chlorine (also bromine and iodine)

not adopt the layered structure of its congeners hints at d orbital participation in weak Cl...Cl bonding. Confirming this, the disposition of chlorine molecules in the crystal is such that the $2\pi_g^4$ HOMO makes a close approach to the $5\sigma_u^o$ LUMO resulting in a contraction of the corresponding intermolecular Cl...Cl distance. That Cl...Cl contacts must be attractive has also been inferred from a study of crystalline 1,4-dichlorobenzene where minimum thermal expansion coefficients were obtained in a direction which corresponds neither to the long molecular axis nor to the shortest intermolecular contacts but rather to a direction in which Cl...Cl contacts predominate (ref. 22).

3.2 Indirect evidence for Cl...Cl contacts being attractive; violations of the chloro-methyl exchange rule

It is almost axiomatic in organic crystal chemistry that molecules should be packed so that the projecting end of one molecule fits into the hollow space left by the adjacent one (refs. 7, 8). This close-packing principle of Kitaigorodskii rationalises the crystal structures of non-polar compounds and one of its consequences is the so-called chloro-methyl exchange rule. Since these two substituent groups have nearly the same volume (Cl, 20 Å^3; Me, 24 Å^3), a replacement of one by the other is not expected to change the crystal structure. Indeed the Cl-Me exchange is valid for a large number of (usually) non-polar, irregularly shaped aliphatic molecules where the major crystal stabilisation is expected from isotropic forces (refs. 23-26). However, the geometrical approach of this simple close-packing model breaks down to a greater or lesser extent when directional interactions are involved. For instance, the crystal structures of hexachlorobenzene and hexamethylbenzene, are quite different. In fact, the Cl-Me exchange is rarely if ever applicable to planar chloro aromatic compounds if all the Cl groups are replaced by Me groups (ref. 23). While hexachlorobenzene has a β-structure, hexamethylbenzene has a 'normal' close-packed structure. The significance of the failure of complete Cl-Me exchange in this compound is that it demonstrates that Cl...Cl interactions are attractive in nature and sufficiently directional in character to cause deviations from Kitaigorodskii close-packing.

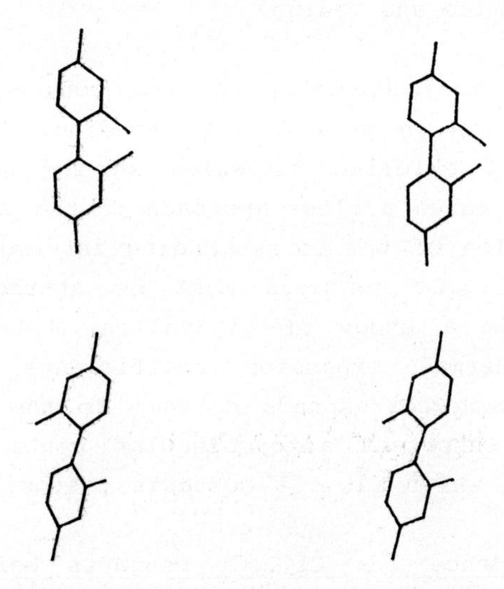

Fig. 3.2. Stereoviews of the molecules of 2,2'-dichlorobenzidine (top) and 2,2'-dimethylbenzidine (bottom) in the crystal. Note the close approach of Cl atoms in the top drawing.

Another very suggestive manifestation of such effects is shown by the pair of compounds 2,2'-dichloro and 2,2'-dimethylbenzidine. The crystal structures of these compounds are quite different as are the molecular conformations in the solid state. While the angle between the two rings in the dimethyl compound is 86°, the corresponding angle for the dichloro derivative is only 36° with the two intramolecular Cl atoms being only 3.36 Å apart. Further, one notes that for a large number of chloro-aromatic β-structures, substitution of a Cl group by the electronically similar though geometrically dissimilar Br group (volumes: Cl 20 $Å^3$, Br 26 $Å^3$) does not alter the crystal structure. Typical examples are: 1,4-dibromobenzene, hexabromobenzene, terephthaloyl bromide, 2-bromo-1,4-naphthoquinone and 2-bromobenzoic acid.

These anomalies and the unusually short intermolecular Cl...Cl distances in β-structures show up the limitations inherent in a purely geometrical approach and must be regarded as indicators for specific Cl...Cl intermolecular interactions that cannot be incorporated into the geometrical model. To summarise, the failure of Cl-Me exchange in chloro aromatic compounds is important confirmation of the attractive nature of the Cl...Cl interaction.

3.3 The β-mode as a distinct structure type: prevalence of short Cl...Cl contacts in chloro aromatic β-crystals

When a planar aromatic compound is translated in a crystal along a short axis, several broad classifications arise. Almost always, the length of the short axis is structure-type defining with the other axes being merely a function of molecular shape and size (refs. 27, 28). So while β-structures have short axes around 4 Å, non-β structures include the γ-mode (short axis between 4.7 and 5.2 Å), the classical herringbone mode (short axis betwen 5.2 and 6.3 Å) and various kinds of inversion modes (short axis between 7.0 and 8.0 Å). Though there seems to be some overlap between these non-β structures in the border regions of 5.2 Å and 7.0 Å, the β-structure is a sharply defined packing type since there is a pronounced paucity of planar compounds with short axes in the crystallographic 'no-man's land' of 4.2-4.7 Å. Even with the limited data available at that time,

Schmidt was aware of this 'empty region' (ref. 15). With the Cambridge Database, this awareness becomes a conviction. Of the approximately 600 chloroaromatic crystal structures considered (ref. 6), the tally of those with short axes less than 4.2 Å, between 4.2 and 4.7 Å and greater than 4.7 Å was around 100, 10 and 490 respectively.

This discontinuous distribution of short axis values indicates a sharp break in the mode of crystal packing in going from the β to the non-β structural types and simple geometry and energy calculations convey this essential difference in a semiquantitative fashion. In the β-structure, major stabilisation for a molecule is through its interaction with its two translational neighbours along the 4 Å stack axis while other neighbours interact only minimally. In the non-β structures, however, molecules are stabilised more uniformly and by a larger number of near neighbours. In effect, the 'molecular coordination' in a β-structure is two while that in the herringbone structure between six and ten (refs. 28-30).

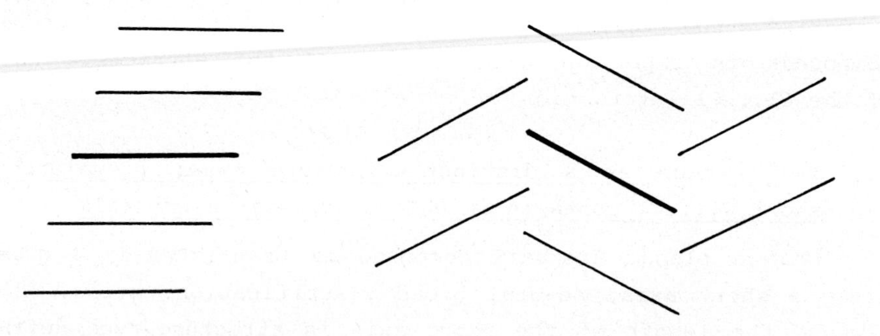

Fig. 3.3. Schematic views of the β (left) and herringbone (right) modes of packing for planar aromatics. The reference molecules are shaded. Notice the distinct molecular co-ordination in the two cases.

The β-structure for chloro-aromatic compounds is exemplified by the archetypes, hexachlorobenzene and triclinic 1,4-dichlorobenzene. The stereoview of the latter structure (see Figure 3.4) shows that the shortest Cl...Cl contacts are in the planes of the aromatic rings and that they are responsible for the formation of molecular sheets which are then stacked at van der Waals separation to optimise C---C contacts. This arrangement

Fig. 3.4. Stereoview of the crystal structure of the triclinic β-form of 1,4-dichlorobenzene. Short axis translated molecules are not shown for clarity.

is reflected in the large energy contributions from stack-translationally related molecules and much smaller contributions from all other near neighbours. In the case of hexachlorobenzene, Cl...Cl contacts of 3.72 Å result in the molecules being arranged along linear ribbons, rather than sheets. These ribbons are stacked, and successive stacks which are related by monoclinic two-fold screw axes, are held together by additional shorter Cl...Cl contacts of 3.51 Å as shown below in Figure 3.5. The key role of Cl...Cl contacts in stabilising this structure provides a convincing explanation for the lack of complete chloro-methyl exchange in this system.

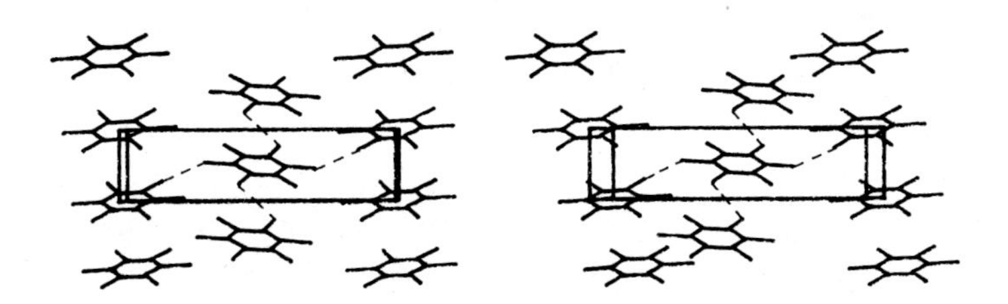

Fig. 3.5. Stereoview of the crystal structure of hexachlorobenzene. The short axis is vertical in the plane of the paper. Cl...Cl interactions are shown as dotted lines.

The nub of the issue in these β-structures is that while the major energy stabilization arises from the strong intra-stack C...C forces, the weak yet directional Cl...Cl forces which control the packing are interstack. The role of the Cl...Cl forces is thus to promote the adoption of two-dimensional motifs which may then stack to give the 4 Å short axis. In effect, these β-structures resemble solid chlorine in that they do not dovetail in three dimensions. In other words, close packing is achieved in two dimensions and the third dimension is filled out trivially by stacking. Although crystal growth need not in fact take place in this sequential manner, it is conceptually useful to consider two-dimensional close packing as a prerequisite to stacking and hence to β-structure adoption.

Interesting experimental corroboration for the directional preferences of Cl...Cl contacts and the lack of such preference for C...C contacts is obtained in the co-crystallisation of compounds such as 6-chloro-3,4-methylenedioxycinnamic acid and 2,4-dichlorocinnamic acid as a stoichiometric β-structure complex which is disordered along the 4 Å stack direction (ref. 31). The fact that the seemingly closely related 3,4-dichlorocinnamic acid gives solid solutions with the methylenedioxy compound only emphasises that the intermolecular Cl...Cl forces in such crystals must be rather specific directionally (ref. 32).

3.4 Classification of planar chloro aromatic β-structures

β-steering by the chloro substituent may be broadly ascribed to: (a) two-dimensional-motif stabilisation through Cl...Cl, and to a lesser extent, C-H...Cl interactions (b) C...C stacking of these motifs. We shall now consider, according to these criteria, the hundred or so β-structure chloro compounds with published crystal structures. These compounds were retrieved from the 1984 version of the Cambridge Database and are listed in Table 1. Almost all the molecules are aromatic and a considerable fraction are accurately planar. A few compounds containing methyl or methoxy substituents will obviously have some atoms outside of the mean plane, while a small number of biphenyls are significantly non-planar. These compounds are arranged according to broad chemical classes in the Table. Inspection of these classes reveals a wide range of benzenoid, heterocyclic and fused-ring compounds, side chains of different lengths, several types of functional groups (-OH, -CN, $-CO_2H$, $-NO_2$, quinones) and varying abilities for other structure-stabilising interactions (dipole-dipole, hydrogen bonding).

TABLE 1

Chloro substituted compounds adopting the 4 Å short axis-β-crystal structure as retrieved from the 1984 version of the Cambridge Structural Database.

Compound Name[a]	Short axis(Å) a,b,c	Space group	Shortest Cl...Cl distance[b] (Å)	Shortest (C-)H...Cl distance[c] (Å)	Packing mode[d]
Hexachlorobenzene	3.87 b	$P2_1/c$	3.51	-	lr
Pentachlorobenzene	3.86 b	$Pca2_1$	3.45	3.02	dcs
1,2,3,5-Tetra-chlorobenzene	3.85 a	$P2_1/c$	3.56	3.03	scs
1,3,5-Trichloro-benzene	3.91 c	$P2_12_12_1$	3.63	2.96	dcs
1,4-Dichlorobenzene	3.98 c 4.10 c	$P\bar{1}$ $P2_1/a$	3.42 3.73	3.13 2.98	ps scs
1,4-Dichloro-naphthalene	3.94 b	$P2_1/c$	3.62	3.00	lr
9,10-Dichloro anthracene	3.87 c	$P\bar{1}$	3.87	3.03	ps
2,3-Dichlorophenol	3.98 c	$P3_1$	3.58	-	Hydrogen bonded helices
2,4-Dichlorophenol	3.87 b	C2/c	3.51	-	
3,4-Dichlorophenol	3.86 c	$I4_1/a$	3.27	-	
3,4,5-Trichloro-guaiacol	4.08 b	$P2_1/c$	3.43	2.92	lr
1,4-Dimethoxytetra-chlorobenzene	4.08 a	$P2_1/c$	3.62	3.11	lr
2,5-Dichloroaniline	3.89 b	$P2_1/c$	3.36	-	scs
1,4-Diaminotetra-chlorobenzene	3.99 b	$P2_1/c$	3.57	-	scs
2,6-Dichloro-4-nitroaniline	3.73 a	$P2_1/c$	3.40	2.84	ps
2-Chloro-4-nitroaniline	3.87 c	$Pna2_1$	3.87	-	dcs
3,5-Dichloro-anthralinic acid	3.80 c	$P2_1/a$	3.53	-	ps
2-Chlorobenzoic acid	3.90 b	C2/c	3.90	-	scs
3-Chlorobenzoic acid	3.85 a	$P2_1/c$	3.85	2.92	ps
4-Chlorobenzoic acid	3.86 c	$P\bar{1}$	3.44	3.09	ps
2,4,6-Trichloro-cyanobenzene	3.97 a	$P2_1/c$	3.86	-	ps

(continued)

4-Chloro-benzonitrile	4.11	c	$P2_1/a$	-	-	ps
Terephthaloyl chloride	3.90	a	$P2_1/c$	3.72	3.10	ps
α-Chloroaceto-phenone	4.13	c	$P2_12_12_1$	-	-	dcs
1-Bromoacetyl-4-chlorobenzene	4.17	c	$P2_12_12_1$	-	-	dcs
5-Chlorosalicyl-aldoxime	3.90	b	$P2_1/c$	3.42	-	lr
4-Chlorocinnamic acid	3.89	b	$P2_1/c$	3.79	-	scs
6-Chloro-3,4-methylenedioxy-cinnamic acid	3.88	a	$P\bar{1}$	3.55	3.25	ps
2-Chlorobenzylidene-malonodinitrile	3.97	a	$P2_1/c$	3.97	2.99	-
2-Chlorobenzoyl-acetylene	3.97	a	$P2_1/c$	3.97	2.92	ps
1-(2,6-Dichloro-phenyl)-4-phenyl-butadiene	4.03	b	$P2_12_12_1$	-	3.09	dcs
1,6-Di-p-chloro-phenyl-3,4-dimethyl-hexatriene	4.18	b	C2/c	3.37	-	scs
N-(2,4-Dichloro-benzylidene)-aniline	3.93	b	$P2_12_12_1$	3.79	2.82	dcs
N-(p-Chlorobenzyl-idene)-p-chloro-aniline	3.93	b	$P\bar{1}$	3.42	3.13	ps
trans-2,2'-Dichloro-azobenzene	3.95	b	$P2_1/a$	3.95	3.00	scs
3,3'-Dichloro-benzidine	3.85	b	$P2_1/c$	3.35	-	scs
2'-Chlorobiphenyl-4-carboxylic acid	3.94	a	$P2_1/c$	3.94	2.89	scs
2-Chlorobiphenyl-4-carboxylic acid	3.90	a	$P\bar{1}$	3.90	3.09	scs
2'-Chloro-4-acetylbiphenyl	4.00	a	$P2_1/c$	4.00	2.96	scs
2,5-Dichloro-3,5-diamino-p-benzoquinone	3.79	a	$P2_1/n$	3.58	-	scs
2-Chloro-5-methyl-p-benzo-quinone-4-oxime	3.85	a	$P2_1/c$	3.81	-	scs

5-(2'-Chloroethoxy)-o-quinone-2-oxime	4.12 a	$P2_1/c$	3.55	2.75	ps
2,3-Dichloronaphthoquinone	3.83 c	$Pb2_1a$	3.83	2.93	dcs
2-Chloro-3-hydroxy-1,4-naphthoquinone	3.92 b	Pc	3.92	-	lr
2-Chloro-3-amino-1,4-naphthoquinone	3.93 b	Pc	3.93	-	lr
1,5-Dichloroanthraquinone	3.84 c	$P2_1/a$	3.84	-	scs
1,8-Dichloroanthraquinone	3.81 b	$Pca2_1$	3.81	-	dcs
2,3-Dichloroanthraquinone	3.81 c	$P2_12_12_1$	3.81	2.79	dcs
1-Chloroanthraquinone	3.99 b	$P2_1$	3.70	-	
7,10-Dichloroanthraquinone-oxadiazole	3.81 c	$P2_1/a$	3.81	-	
N-(p-Tolyl)-tetrachlorophthalimide	3.97 b	$P2_1/c$	3.47	-	
5-Chloro-2,1-benzisothiazole	3.93 a	$P2_12_12_1$	3.93	3.16	dcs
6,6'-Dichloroindigo	3.75 b	$P\bar{1}$	3.39	-	ps
6-Chloro-2-hydroxypyridine	3.89 a	$P2_1/c$	3.87	2.87	ps
6-Chloroquinoline	3.86 b	$Pca2_1$	3.86	-	dcs
5-Chloro-7-iodo-8-quinolinol	4.14 b	P2/a	3.91	-	scs
7-Chloro-8-methylcarbostyril	4.18 b	$P2_1/c$	3.67	3.30	lr
4-Amino-2,6-dichloropyrimidine	3.79 c	$P2_1/a$	3.47	-	
2-Amino-4,6-dichloropyrimidine	3.84 b	$P2_1/a$	3.43	-	Hydrogen bonded helices
2,5-Dichloro-3-methoxypyrazine	4.06 a	$P2_1/c$	3.60	3.05	ps
Hexachloroquinoxaline	3.92 b	$P2_1/c$	3.50	-	lr
1,4,6,9-Tetrachlorophenazine	3.95 a	$P2_1/n$	3.39	-	scs
2,3,7,8-Tetrachlorophenazine	3.85 a	$P2_1/c$	3.63	-	lr
Octachlorodibenzo-p-dioxin					

(continued)

Octachloro-dibenzo-p-dioxin	3.82 b	$P2_1/c$	3.49	-	scs
2,3,7,8-Tetra-chlorodibenzo-p-dioxin	3.78 a	$P\bar{1}$	3.58	2.88	scs
2,7-Dichlorodi-benzo-p-dioxin	3.88 a	$P\bar{1}$	3.40	2.96	ps
2-Chlorothio-xanthone	3.87 a	$P\bar{1}$	3.65	3.11	ps
bis-(m-Chloro-benzoyl)-methane	3.85 b	$Pca2_1$	3.63	2.79	dcs
p-Chlorobenzoic anhydride	3.89 a	$P2/c$	3.38	-	ps
bis-(3-Chloro-phenyl)-dithio-oxalate	4.03 a	$P2_1/c$	3.49	3.34	ps
p,p'-Dichloro-dithiooxanilide	4.08 b	$P2_1/n$	-	2.95	
2,2'-Dichlorodi-benzoylperoxide	3.84 c	$P2_12_12$	3.84	-	dcs
1-(2,5,-Dichloro-phenylazo)-2-hydroxy-3-naph-thoic acid-(4-chloro-2,5-dimethoxy)-anilide	3.96 b	$P\bar{1}$	3.64	-	ps
1,2,3,4-Tetra-chlorobenzo sesquifulvalene	3.83 c	$P2_1/a$	3.39	-	ps
Hexachlorofulvene	3.84 b	$P2_1/c$	3.45	-	lr
2-Chlorotropone	4.01 c	$Pna2_1$	3.59	2.80	dcs
Tetrachloro butadiene	3.87 a	$P2_1/c$	3.63	-	
Dimethyl-2,5-dichloro-muconate	3.97 a	$P2_1/n$	3.83	3.13	ps

[a] Other β-chloro-aromatics not given here include those isomorphous with hexachlorobenzene (11), with no co-ordinates supplied (10), or with disordered chloro groups (2).

[b] Cl...Cl distances longer than 4.0 Å are not given.

[c] C-H...Cl distances not quoted if hydrogen atom positions are not available in the Database.

[d] ps : planar sheet; lr : linear ribbon;
scs : singly corrugated sheet; dcs : doubly corrugated sheet.

In spite of this chemical variety, all these compounds share a common crystallographic feature, and the values of the stack axis have been tabulated. Remarkably, most of these structures have Cl...Cl and/or C-H...Cl contacts which may be considered as 'short' to some extent or other. The range of values for the shortest Cl...Cl contacts in this group of structures is from 3.27 to around 4.10 Å. Significantly, in those structures where the shortest Cl...Cl contacts are at the higher end of this range, additional stabilisation seems to be achieved through C-H...Cl 'hydrogen bonds' and these distances are also listed.

A study of the packing arrangements of the compounds in Table 1 shows that the manner in which Cl...Cl, C-H...Cl and C...C interactions are manifested in structures like hexachlorobenzene and 1,4-dichlorobenzene is fairly general and that these steering interactions are applicable to almost all the listed structures. This combination of intrastack C...C and interstack Cl...Cl and C-H...Cl forces is compatible with geometrical motifs which conform to triclinic and monoclinic symmetry and a very large majority of the structures in Table 1 belong to one of these two crystal systems. Three such motifs have been identified and are referred to as planar sheet, singly corrugated sheet and linear ribbon. A fourth motif which is termed doubly corrugated sheet is a consequence of orthorhombic symmetry. These four motifs are defined assuming a high degree of molecular planarity.

3.5 The planar sheet motif

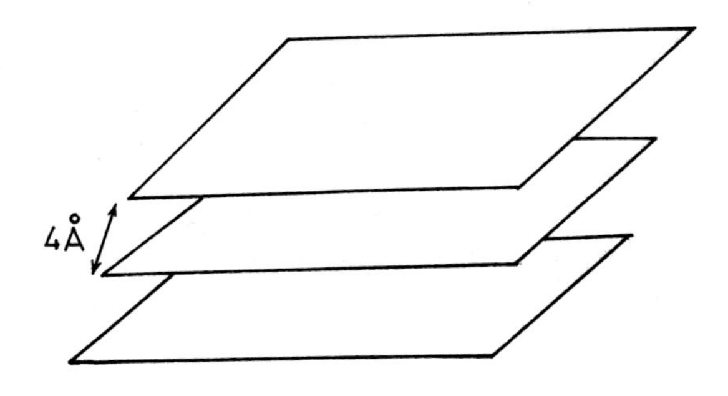

Fig. 3.6 The planar sheet motif for 4 Å-structures.

Conceptually, the simplest manner of optimising Cl...Cl interactions is in the formation of a planar molecular sheet with these sheets being stacked along the 4 Å axis to generate the crystal structure. Both triclinic and monoclinic structures show this planar sheet motif. Although a monoclinic β-structure may display other motifs too, the planar sheet is the only possibility for a triclinic crystal excluding those cases where non-parallel symmetry independent half-molecules lie on two distinct inversion centres in the space group $P\bar{1}$. If the sheet contains translation or translation and inversion related molecules, crystallisation is possible in the space groups P1 or $P\bar{1}$ respectively. Though there are no known examples in P1, Table 1 has large number of $P\bar{1}$ structures. This number includes 1,4-dichlorobenzene and 4-chlorobenzoic acid.

A planar sheet which contains screw (glide) related molecules is obtained in a monoclinic space group such as $P2_1/c$ if the short axis is non-unique and the molecular plane is parallel to or almost parallel to the unique axis. Table 1 shows that this situation is a most common one with a typical example being monoclinic 1,4-dichlorobenzene.

3.6 The linear ribbon and singly corrugated sheet motifs

If the 4 Å axis coincides with the monoclinic direction, the linear ribbon and singly corrugated-sheet motifs are the result. Figure 3.7 shows that the position of the screw axis (glide plane) relative to the molecular plane, determines largely which of these two motifs is adopted. In either case, the motif is stacked in the 4 Å-direction to complete the structure and since glide and screw operations are involved within the monoclinic framework, there is a preponderance of the space group $P2_1/c$. These two types account for many chloro aromatic β-structures. In the linear ribbon structure, Cl...Cl interactions may not only be ribbon forming but also occasionally, as in hexachlorobenzene, be ribbon linking. Such ribbons appear 'crossed' since the molecular planes must necessarily make large angles with the 4 Å unique axis. In the singly corrugated sheet motif, the sheets are stabilised by Cl...Cl interactions and in some cases like 2,5-dichloroaniline by weak hydrogen bonding (ref. 30).

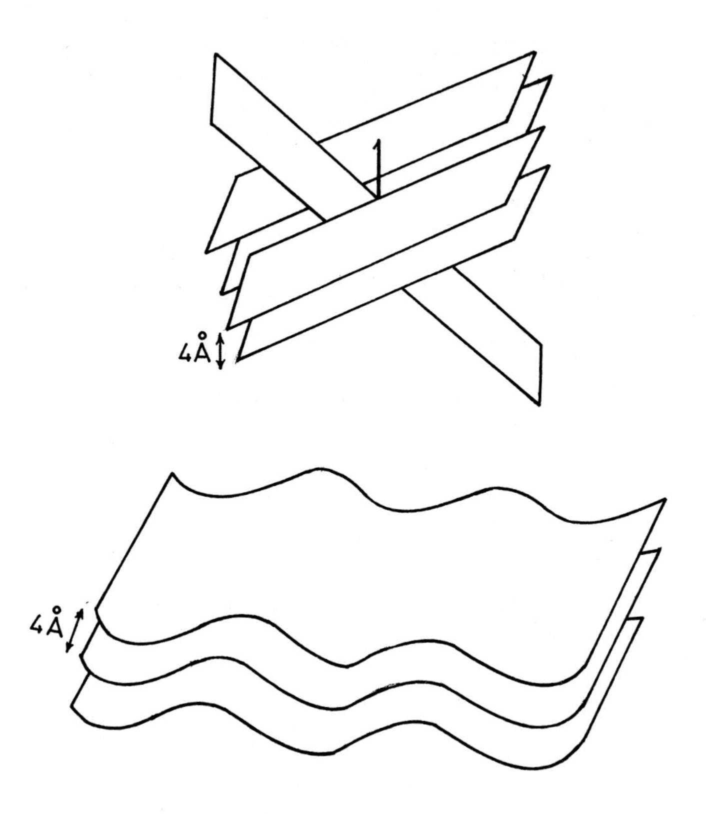

Fig. 3.7. The linear ribbon and singly corrugated sheet motifs for 4 A-structures.

A motif which approximates the singly corrugated sheet is obtained in monoclinic space groups with the short axis non-unique if the b-axis makes significant angles with the molecular mean plane. A type of corrugation arises in these structures as a distortion from the planar arrangement which might have resulted had the mean molecular plane been inclined only at shallow angles to the b-axis. Table 1 contains a few such compounds. Similarly, the listed biphenyls which might have been expected to crystallise in planar sheets on the basis of the criteria outlined above, display a somewhat corrugated motif on account of the ca. 50° dihedral angle between the phenyl rings. However, the non-planar nature of these molecules precludes an unambiguous classification and indeed this is the case for any non-planar β-chloro aromatic compound.

3.7 The doubly corrugated sheet motif

Fig. 3.8. The doubly corrugated sheet motif for 4 Å structures.

A doubly corrugated sheet may be obtained if mutually perpendicular screw axis are present and if the plane of the molecule is inclined to all the crystal axes. Such a possibility is illustrated by the fifteen orthorhombic structures in Table 1. Stacking of doubly corrugated sheets in these structures is similar to the planar sheet stacking described earlier. We note that the orthorhombic space groups in Table 1 ($P2_12_12_1$, $Pna2_1$, $Pca2_1$) are compatible with such an arrangement.

3.8 Orientation of Cl...Cl contacts

It is well-known (ref. 19) that Cl...Cl interactions in organic structures tend to assume one of two preferred geometries, either with both C-Cl...Cl angles equal and approximately collinear (160° ± 10°) or with one angle approximately perpendicular (near 80°) and a second nearly collinear (175°). In the light of the present structural classification of planar chloroaromatics, the first of these two geometries (type A) corresponds to the triclinic planar sheet situation while the second geometry (type B) is an essentially monoclinic motif, observed in the linear ribbon and singly corrugated sheet structures. While the type B contacts may be visualised as a dipolar electrophile-nucleophile interaction (ref. 21), the type A contacts are symmetrical and therefore cannot be accommodated by such a model. Again, there does not seem to be any marked tendency for type A contacts to be say, longer than type B contacts. Actually, some type A contacts are rather short as in triclinic

1,4-dichlorobenzene or 4-chlorobenzoic acid. Obviously a definitive theoretical picture of these contacts is still to emerge.

4. Crystal engineering of planar chloro aromatics

Since it has been possible to correlate a large number of chloro β-structures on the basis of directional in-plane Cl...Cl and C-H...Cl interactions which stabilise sheet and ribbon arrangements, one may reverse the argument and identify molecular features which favour the adoption of the β-structure for chloro-aromatic compounds. The following criteria, have been formulated after careful consideration of a large amount of structural data for several β and non-β crystals:

4.1 Molecular planarity

Although lack of molecular planarity need not necessarily prevent the stabilisation of sheets or ribbons, non-planar molecules cannot be stacked very effectively at 4 Å separation. Therefore all atoms, including hydrogen, which do not lie close to the molecular mean plane will hinder, to some extent, the adoption of the β-structure. Hardly any compound in Table 1 contains aliphatic residues and other groups which would prevent molecular planarity. It may also be mentioned, in this connection, that for the biphenyl compounds in Table 1, the angles between the ring planes are generally smaller than might be expected and might perhaps indicate a tendency towards a more planar conformation and through it to a Cl...Cl stabilised planar sheet β-structure. In the context of molecular planarity, the β-crystal structure of the significantly non-planar 2,4-dichlorobenzylidene aniline is also noteworthy. Here, doubly corrugated sheets are stabilised with very short C-H...Cl and short Cl...Cl contacts and 4 Å stacking of these sheets is possible since protrusions and hollows in these sheets mesh effectively.

4.2 Number of chlorine atoms

Other factors being the same, more Cl substituents will result in more Cl...Cl interactions and a greater likelihood of the β-structure. Thus, while chlorobenzene has a non-β structure, and 1,4-dichlorobenzene is trimorphic with one non-β and two β-forms, only the β-structure is obtained for 1,3,5-trichloro-, 1,2,4,5-tetrachloro-, 1,3,4,5-tetrachloro-, pentachloro- and hexachlorobenzene. Again, while 4-chloroaniline does not have the β-structure, 2,5-dichloro-, 2,4,6-trichloro- and 2,3,4,5,6-pentachloroaniline have this structure.

Cl Cl Cl Cl Cl Cl Cl Cl Cl Cl Cl Cl

Fig. 4.1. Structural formulae of hexachlorofulvene and hexachlorobenzene.

An interesting case of structural mimicry (see ref. 25 for a definition this term) is exhibited by hexachlorofulvene, C_6Cl_6, which is isomorphous with hexachlorobenzene. Inspection of the atomic positions in these two molecules reveals that the chloro groups in the fulvene form a nearly hexagonal net which approximates the accurately hexagonal net of chloro groups in hexachlorobenzene. The close geometrical similarity between these molecules with the same atom types occupying nearly the same relative positions enable Cl...Cl and C...C interactions to stabilise the two crystal structures in an almost identical manner.

This trend for a β-structure if more chlorine atoms are present is not valid for perchloro-fused ring hydrocarbons like octachloronaphthalene, decachlorophenanthrene, and compounds like perchlorofulvalene. In these cases, the large number of chloro substituents results in the planarity criterion not being satisfied.

4.3 Absence of stronger intermolecular interactions

Since Cl...Cl interactions are relatively weak, they are unable to steer crystal structures effectively when more dominant forces are present. In line with this, most of the molecules in Table 1 are fairly non-polar. However, there are important exceptions; several chloro substituted acids and phenols may adopt the β-structure if the weak Cl...Cl interactions are possible within the framework of O-H...O hydrogen bonding. For many other chlorophenols, chloroquinones and chlorobenzonitriles, the β-structure is not adopted. Another general observation concerns the importance of dipole-dipole interactions in crystal structures of chloro-aromatics with permanent dipole moments. The manifestation of such interactions is, in general, incompatible with 4 Å stacking.

Compounds where Cl...Cl interactions are pitted against weak electrostatic forces are good examples for the study of the relative importance of various intermolecular forces. Since several insights may be obtained by examining the crystal packings of these more polar derivatives, they will be discussed briefly.

4.3.1 Chlorophenols

Normally one would not expect a chlorophenol to adopt a 4 Å structure. This is because, the optimisation of O-H...O bonds, which is the main structural requirement, often precludes the manifestation of Cl...Cl interactions. Hydrogen bonding around a screw axis of length ca. 4.8 Å is the classical motif in the crystal structures of phenols and naphthols and is seen, for example, in 4-chlorophenol, 2,5-, 2,6- and 3,5-dichlorophenol. Some of these structures have been discussed in detail by Perrin et. al. in another chapter of this book. Within the framework of such hydrogen bonding, molecules must obtain as much stabilisation as possible from Cl...Cl and C...H interactions. This tendency is naturally reinforced when two -OH groups are present as in tetrachlorohydroquinone (ref. 30). Here, each end of the the molecule is secured by O-H...O bonds about a screw axis of 4.8 Å. However, when one of these hydroxyl groups is replaced by a chloro substituent, that is in the compound pentachlorophenol, dimorphism ensues. The increased number of chloro groups is paralleled by a greater importance of Cl...Cl interactions and therefore by a greater tendency to adopt the β-structure. So while the low temperature ordered form (short axis 4.9 Å) resembles tetrachlorohydroquinone and highlights the importance of hydrogen bonding, the high temperature polymorph which has six-fold rotational disorder involving

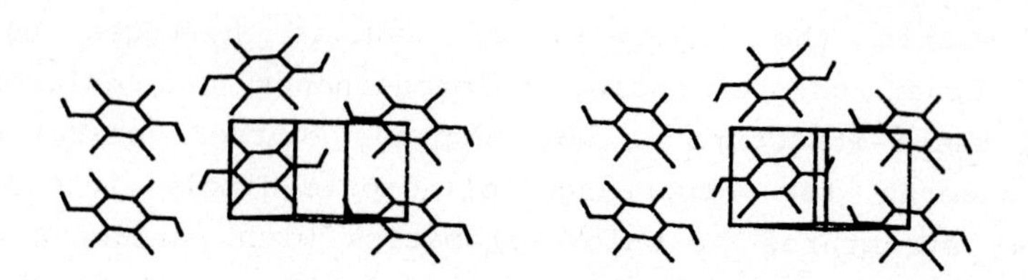

Fig. 4.2. Stereoview of the non-β crystal structure of tetrachlorohydroquinone (1,4-dihydroxy-2,3,4,5-tetrachlorobenzene). The hydrogen bonding direction is vertical in the plane of the paper. Notice that the molecules are anchored at the phenolic ends by a distinctive hydrogen bonding arrangement.

the ring substituents emphasises the Cl...Cl and C...C forces since it is a β-structure isomorphous with hexachlorobenzene. The disorder in the high temperature form shows that hydrogen bonding is unimportant there. Interestingly, pentachlorophenol forms solid solutions with hexachlorobenzene, isostructural with the latter; infrared spectroscopy of these mixed crystals shows both 'free' and hydrogen bonded hydroxyl groups, indicating short range ordering of the phenol molecules (ref. 33).

That these tendencies for β and non-β structure adoption are evenly matched in pentachlorophenol is confirmed by the fact that only β-modifications are seen for pentachloroaniline and pentachlorothiophenol in both of which hydrogen bonding must be weaker than in the corresponding phenol. Even 1,4-diamino-2,3,5,6-tetrachlorobenzene adopts the 4 Å-packing characterised by Cl...Cl and C...C interactions (ref. 30).

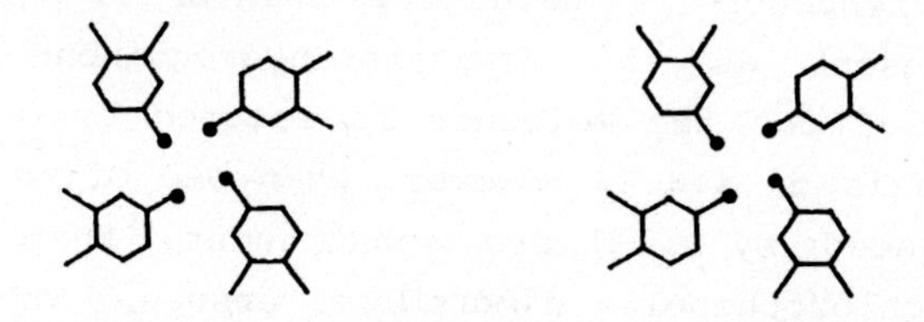

Fig. 4.3. Hydrogen bonding around a 4-fold screw axis in the crystal structure of 3,4-dichlorophenol. Oxygen atoms are indicated. This spiral arrangement results in a cylindrical motif along the screw axis direction.

The structures of 2,3-, 2,4- and 3,4-dichlorophenol are more curious (ref. 34). While the three other isomeric compounds adopt the 'normal' 4.8 Å short axis, these three are hydrogen bonded about a 4 Å screw axis and therefore have the β-structure. A significant observation here is that while the 2,5-, 2,6- and 3,5 compounds employ 2_1 axes, the 2,3-(3_1), 2,4-(pseudo 3_1) and 3,4-(4_1) adopt higher fold axes or their pseudosymmetric variants. These molecules are able to crystallise in these higher symmetries since their substitution pattern does not result in bad contacts from neighbouring hydrogen bonded molecules as would be the case if the 2,4-, 2,6- or 3,5- isomers were to adopt these 3_1 and 4_1 axes. Within the 2,3-, 2,4- and 3,4- compounds, it is expected *a priori* that the 3,4- compound should have the most gradual climb up the screw axis since it is free of ortho chloro groups. Thus it packs with a 4_1 axis rather than a 3_1 axis. These higher order axes permit stabilisation of helical stacks by strong C...C interactions thereby forming a β-structure. The shortest Cl...Cl contacts in these structures are again *inter* rather than *intrastack*; the chloro groups perform the function of pulling stacks together *via* such contacts. These molecules may be considered to crystallise in the form of hydrogen bonded helices having a hydrophilic (O-H...O) core and a hydrophobic (Cl) exterior. The weak Cl...Cl interactions are possible within the framework of a distinctive, if somewhat unusual, hydrogen bonding pattern.

4.3.2 Chloro aromatic acids

The crystal structures of the monochlorobenzoic acids and 4-chlorocinnamic acid are of some interest because of the occurrence of both carboxyl O-H...O hydrogen bonding as well as Cl...Cl contacts (ref. 35). It may be noted that although the hydrogen bonds are much stronger than the Cl...Cl interactions, their formation does not preclude the setting up of the weaker halogen contacts, since hydrogen bonding leads to the formation of planar centrosymmetric dimers. These dimers become, in effect, the building blocks of the structure and they may be linked through Cl...Cl interactions to give planar sheet (4-chlorobenzoic acid) or corrugated sheet (4-chlorocinnamic acid) motifs. Stacking of these motifs leads to the β-structure as described previously. While both 2- and 3-chlorobenzoic acids also adopt the β-structure, the shortest Cl...Cl contacts in these cases are greater than the van der Waals distance and it is possible that 'β-steering' is because of other directional interstack forces especially of the C-H...O type. In fact, an aromatic acid may adopt the β-structure even without chloro substituents, if such C-H...O interactions are significant. This is illustrated by compounds such as 3,4-methylenedioxycinnamic acid and its chloro analogue (refs. 10, 32).

4.3.3 Dipole stabilised inversion structures

A few rather simple planar chloro aromatic compounds do not adopt the β-packing in apparent violation of the generalisations presented in this chapter. For instance, 1,2,3-trichlorobenzene is the only polychlorobenzene out of the seven with solved crystal structures not to adopt the β-mode. Yet, the isomeric 1,3,5-derivative behaves normally. Another seeming anomaly is provided by the series of polychlorodibenzo-p-dioxins. While 2,7-dichloro, 2,3,7,8-tetrachloro and 1,2,3,4,6,7,8,9-octachlorodioxin have the β-structure, the 2,8-dichloro and the 1,2,3,7,8,9-hexachloro derivatives do not.

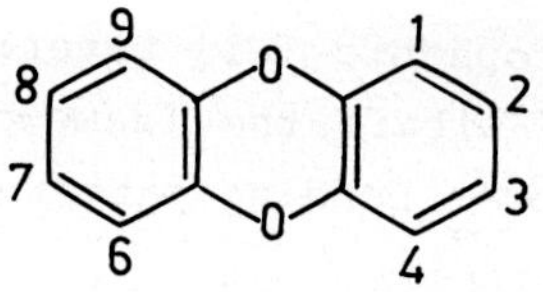

Fig. 4.4. Atom-numbering scheme for the polychlorodioxins.

These observations do not tally with the earlier conclusion that there is always a greater likelihood of a β-structure within a family of compounds if the number of chloro substituents increases. A consideration of the crystal structure of 1,2,3,7,8,9-hexachlorodibenzo-p-dioxin shows that while it is sheet based with sheets being built-up with Cl...Cl interactions as in the β-compounds, adjacent sheets which are stacked at van der Waals separation are related by inversion and not translation. Both translated and inverted arrangements probably afford the same extent of C...C stabilisation but the inversion motif is distinctly preferred here since the molecular dipoles on stack adjacent molecules would then be antiparallel (refs. 6, 36).

A similar situation prevails when the 1,2,3- and 1,3,5-trichlorobenzenes are compared. Corrugated sheets, held together by C-H...Cl and Cl...Cl contacts, are formed in both cases. But the sheets are inverted in 1,2,3-trichlorobenzene such that molecules may be considered to pack as van der Waals stacked diads within each of which the molecular dipole moments compensate. On the other hand, the molecular sheets in 1,3,5-trichlorobenzene are stacked translationally to obtain a β-structure since an individual molecule has a zero dipole moment. The crystallographic consequence of sheet inversion is therefore an axis of *ca*. 8.0 Å.

It is clear therefore that chloro groups in planar aromatic molecules need not always lead to 4 Å structures if the molecules have large dipole moments. Rather, the primary effect of the chloro groups is to form sheets or ribbons through Cl...Cl and C-H...Cl contacts. The 4 Å axis only results if it is energetically favourable to stack these two-dimensional motifs translationally. Sheet formation becomes undoubtedly easier if the number of chloro groups increases (as in the hexachloro dioxin above) but whether a translated or inverted stack sequence is adopted seems to depend more on the arrangement rather than the number of chloro groups in the molecule.

In general, the presence of a molecular dipole moment seems to favour the adoption of inversion structures. Thus while 2,3-dichloro-1,4-naphthoquinone is dimorphic with a and an inversion form and while 1,5-dichloro-9,10-anthraquinone has only a β-form, 2,3-dichloro-9,10-anthraquinone adopts an inversion structure exclusively.

4.3.4 Chloro-cyano aromatics

Another chloroaromatic sheet-based structure which is not stack translated along a 4 Å axis because of the dipole-dipole repulsion which would then result is 1,2-dicyano-3,4,5,6-tetrachlorobenzene. In fact, hardly any cyanochlorobenzene adopts the β-structure unless the number of chloro groups is large and/or that of the cyano groups small (2,4,6-trichlorobenzonitrile, 4-chlorobenzonitrile). In most of the other cyanochlorobenzenes with known crystal structures, the dominant interactions are the -CN...Cl contacts which are optimised rather than the weaker Cl..Cl interactions.

5. Conclusions

Planar chloro-aromatic compounds have a pronounced tendency to adopt the 4Å short axis β-structure. Such structures are stabilised by Cl...Cl, C...C and (where appropriate) by C-H...Cl and C-H...O interactions. The most important of these is the Cl...Cl interaction which is attractive and anisotropic. While observed Cl...Cl contacts in such β-structures are in the range 3.2 to 4.1 Å they almost always lie in or close to the molecular mean planes and are involved in stabilising molecular sheet or ribbon motifs which may be stacked at van der Waals separation to generate the entire structure.

It is interesting that a substantial number of crystal structures may be rationalised with the small number of assumptions summarised above. In the absence of a complete theoretical model for organic crystals, the steering group approach to crystal engineering outlined in this chapter offers many advantages. Indeed, the urge to categorise and classify is almost an instinctive one in chemistry since it may eventually lead to a definition of more basic and unifying concepts. While the chloro substituent is not even accorded the status of a full-fledged functional group in solution organic chemistry (as a brief inspection of Beilstein will reveal), we have seen here that it may definitely be considered as one in solid state chemistry. A steering group is to a collection of molecules what a functional group is to a molecule in isolation. Such analogies will possibly be extended and while one already recognises topochemistry, polymorphism and non-centrosymmetry as the solid state counterparts of reactivity, isomerism and chirality, it may become possible to redefine concepts such as homology for molecular crystals in the near future.

6. Acknowledgements

Financial support from the Department of Science and Technology, Government of India, in the form of Project 23(1)/81-STP-II is gratefully acknowledged. The author wishes to thank J.A.R.P. Sarma and T.S.R. Krishna for their assistance.

References

1. Schmidt, G.M.J. Pure Appl. Chem. 1971, **27**, 647.

2. Thomas, J.M. Philos. Trans. R. Soc. London, Ser. A. 1974, **277**, 251.

3. Thomas, J.M.; Morsi, S.E.; Desvergne, J.P. Adv. Phy. Org. Chem. 1977, **15**, 63.

4. Desiraju, G.R. Endeavour 1984, **8**, 201.

5. Desiraju, G.R. Proc. Indian natn. Sci. Acad. 1986, **52A**, 379.

6. Sarma, J.A.R.P.; Desiraju, G.R. Acc.Chem.Res. 1986, **19**, 222.

7. Kitaigorodskii, A.I. Acta Crystallogr. 1965, **18**, 585.

8. Kitaigorodskii, A.I. 'Molecular Crystals and Molecules', Academic Press, New York, 1973.

9. Bernstein, J. in 'X-Ray Crystallography and Drug Action' (Eds. A.S. Horn and C.J. De Ranter), Clarendon Press, Oxford, 1984, 23.

10. Desiraju, G.R.; Kamala, R.; Kumari, B.H.; Sarma, J.A.R.P. J. Chem. Soc., Perkin Trans.II 1984, **181**.

11. Thomas, N.W.; Ramdas, S.; Thomas, J.M. Proc. R. Soc. London. Ser. A. 1986, **400**, 219.

12. Allen, F.H.; Kennard, O.; Taylor, R. Acc. Chem. Res. 1983, **16**, 146.

13. Green, B.S.; Schmidt, G.M.J. Israel Chemical Society Annual Meeting Abstracts, 1971, 190.

14. Mirsky, K.; Cohen, M.D. Chem. Phys. 1978, **28**, 193.

15. Schmidt, G.M.J. J. Chem. Soc. 1964, 2014.

16. Green, B.S.; Schmidt, G.M.J. Tetrahedron Lett. 1970, 4249.

17. Elgavi, A.; Green, B.S.; Schmidt, G.M.J. J. Am. Chem. Soc. 1973, **95**, 2058.

18. Williams, D.E.; Hsu L.-Y. Acta Crystallogr. 1985, **A41**, 296.

19. Sundaralingam, M.; Sakurai, T.; Jeffrey, G.A. Acta Crystallogr. 1963, **16**, 354.

20. Nyburg, S.C.; Faerman, C.H. Acta Crystallogr. 1984, **A40**, c-113.

21. Ramasubbu, N.; Parthasarathy, R. Acta Crystallogr. 1984, **A40**, c-101.

22. Wheeler, L.; Colson, S.D. J. Chem. Phys. 1976, **65**, 1227.

23. Desiraju, G.R.; Sarma, J.A.R.P. Proc. Indian Acad. Sci. (Chem. Sci.) 1986, **96**, 599.

24. Jones, W.; Ramdas, S.; Theocharis, C.R.; Thomas, J.M.; Thomas, N.W. J. Phys. Chem. 1981, **85**, 2594.

25. Jones, W.; Theocharis, C.R.; Thomas, J.M.; Desiraju, G.R. J. Chem. Soc. Chem. Commun. 1983, 1443.

26. Theocharis, C.R.; Desiraju, G.R.; Jones, W. J. Am. Chem. Soc. 1984, **106**, 3606.

27. Nalini, V.; Desiraju, G.R. J. Chem. Soc. Chem. Commun. 1986, 1030.

28. Gavezzotti, A.; Desiraju, G.R. Submitted for publication.

29. Desiraju, G.R. Proc. Indian Acad. Sci. (Chem. Sci.) 1984, **93**, 407.

30. Sarma, J.A.R.P.; Desiraju, G.R. Chem. Phys. Lett. 1985, **117**, 160.

31. Sarma, J.A.R.P.; Desiraju, G.R. J. Chem. Soc. Chem. Commun. 1984, 145.

32. Sarma, J.A.R.P.; Desiraju, G.R. J. Am. Chem. Soc. 1986, **108**, 2791.

33. Masood, M.A.; Desiraju, G.R. Chem. Phys. Lett. 1986, **130**, 199.

34. Thomas, N.W.; Desiraju, G.R. Chem. Phys. Lett. 1984, **110**, 99.

35. Patil, A.A.; Curtin, D.Y.; Paul, I.C. Isr. J. Chem. 1985, **25**, 320.

36. Desiraju, G.R.; Sarma, J.A.R.P.; Krishna, T.S.R. Chem. Phys. Lett. 1986, **131**, 124.

SUBJECT INDEX